THE WELLAND CANALS
AND THEIR COMMUNITIES

JOHN N. JACKSON

The Welland Canals and Their Communities: Engineering, Industrial, and Urban Transformation

UNIVERSITY OF TORONTO PRESS
Toronto Buffalo London

386.47
J13w

 © University of Toronto Press Incorporated 1997
Toronto Buffalo London
Printed in Canada

ISBN 0-8020-0933-6

Printed on acid-free paper

Canadian Cataloguing in Publication Data

Jackson, John N., 1925–
   The Welland Canals and their communities :
   engineering, industrial, and urban transformation

   Includes bibliographical references and index.
   ISBN 0-8020-0933-6

   1. Welland Canal (Ont.) – History.   2. Welland Canal
   Region (Ont.) – Economic conditions.   3. Cities and
   towns – Ontario – Welland Canal Region – Growth – History.
   4. Urbanization – Ontario – Welland Canal Region – History.
   I. Title.

   HE401.W4J33 1997   386'.47'0971338   C96-931875-8

This book has been published with the help of a grant from the Humanities
and Social Sciences Federation of Canada, using funds provided by the Social
Sciences and Humanities Research Council of Canada.

University of Toronto Press acknowledges the financial assistance to its publishing program of the Canada Council and the Ontario Arts Council.

To Kathleen; Andrew, Karen, Caitlin, Emma; Paul; Susan, Bob, David

Contents

Maps and Tables

Maps

Tables

Preface

This book focuses on transportation – land-use interaction at the detailed local and regional levels of interaction. Its genre is urban historical geography. The core themes are the sequence of four canals that have crossed the Niagara Peninsula of southern Ontario, and the evolution and character of their related canal communities (see map 1, page 4).

The First Canal opened in 1829 between Lake Ontario and the Niagara River. This waterway was extended in 1833 to Lake Erie, and a new route now referred to as the Second Canal was completed by 1845. A Third Canal was added in 1887, and the Fourth Canal, then referred to as the Welland Ship Canal, which opened in 1932–3, became an integral part of the St Lawrence Seaway in 1959.

These waterways were first managed and controlled by the private Welland Canal Company of 1825. After the Act of Union in 1840, they came under the control of the Board of Works for the Province of Canada, and then, after Confederation in 1867, of the federal Department of Railways and Canals. They have been controlled by the St Lawrence Seaway Authority since 1959.

Many canal features have changed considerably over the decades. Different routes have been followed. The channel has been straightened, widened, and reduced in length. Sections have been abandoned for navigation, then reused for other purposes. The locks have become larger, deeper, wider, and fewer in number, and their structures have changed from wood to stone, then reinforced concrete. The vessels using the canal have progressed from sailing vessels to steam-powered propellers and large modern bulk carriers.

Cargoes passing through the canal have increased greatly in volume, as have the types of goods transported and the proportion of through to locally generated traffic. They have varied in character from the barrel economy of the pioneer period through to modern bulk commodities with their emphasis on wheat, iron ore, and coal. Vessels have been loaded and offloaded along the canal, or they now pass through the canal without stopping. Either way the flow of goods and traffic has placed important service demands on the canalside communities. The canal created ports at its entry points together with harbours and wharves along its length, with each now being different in size, location, and character from those of earlier years.

Crossing points of the canal have changed from wooden swing-bridges to lift-bridges, then tunnels, as the size of vessels and level of engineering technology have advanced. These crossings became focal points in the road, rail, and highway systems of transportation, and canalside settlements grew and were shaped at these points of convergence in the network of land communications.

Water supply for the canal was obtained first from the Grand River and then from Lake Erie as head ponds for the canal. As this flow was diverted from the canal at its locks and wherever the canal was at a higher level than the adjacent land, raceways became key elements in the canal scene. They attracted mills, then other forms of industrial development, and were later used for the generation of hydroelectric power. At a later period again, the canals became noted for their recreational–open space attributes and respected for their heritage contributions to Canada's urban-industrial history as part of Niagara's tourist industry.

As portrayed in one of the many studies of the canal by consultants providing guidance towards the achievement of a modern lake-to-lake parkway along the canal banks, 'the role the old canal played was similar to that of a main street upon which ships, rather than horses, wagons or later cars, moved through the town. The presence of the canal, with its ships and their crews, contributed to the town form and atmosphere and provided a variety and source of pleasure which otherwise would have been lacking. As the towns grew around and with the canal, it and its community were woven into the town fabric and became an integral, and dynamic part of it' (Parks Canada 1980).

This close degree of interrelationship, the interlocking of marine circumstances with those prevailing on the land, has taken on a variety of different forms through time. As with all North American cities, the communities that emerged along the Welland Canal have been shaped through a series of formative periods, which Yeates and Garner (1980) have depicted as follows:

- Up to about 1830, during which the location and spread of cities was influenced primarily by wagon-sail technology

- From 1830 to 1870, in which the innovations of the iron rail, canals, and steamboat greatly influenced the spread of cities
- From 1870 to 1920, when the steel rail, steam-driven ocean-going vessels, and industrialization, later promoted by hydroelectric power and massive foreign immigration, combined to produce the basic urban system that exists today
- From 1920 to 1970, when the automobile reshaped North American cities, and the innovation of air travel reinforced the interrelationships of the urban system that had developed earlier
- Post-1970, when the continental limited-access high-speed highway system, the prevailing use of the motor vehicle, air travel, and global transition to international economies were added to the established pattern of cities

Combining these major changes in urban expression with the canal sequence that has been identified, this narrative approaches the evolution of the Welland Canals and their associated communities through four distinct periods. Each inherits then amends the considerable environment that had been shaped during the preceding epoch.

After an introduction that reflects on the Welland Canal and its urban-industrial consequences at the world level of understanding, part 1 introduces the First Canal, which was constructed with wooden locks along an unintended route but initiated a new line of settlement across the Niagara Peninsula. Mills processed the products of local agriculture, including those from the upstream areas of the Welland and the Grand Rivers that were then being cleared and settled. Products were exported via the canal, often in containers produced locally and in vessels constructed in shipyards along the canal banks. The sturdier Second Canal followed along a comparable route, and the canal settlements continued to expand. The foundations of the modern urban-industrial system as a corridor of development along the canal were being established in this pioneer period, a story that is carried through to the mid-1850s.

Part 2 continues the story of the Second Canal, adds the Third Canal, and extends the discussion up to the outbreak of the First World War in 1914. This is the railway period of development, during which the steam vessel takes over from the sailing ship and most but not all of the canal communities advance through the addition of large-scale industry. Important growth factors included hydroelectricity and, after Confederation in 1867, protective tariff barriers that allowed the Niagara Peninsula to attract American manufacturing companies. The canal, important for water supply, the inflow of raw materials, and the shipment outward of goods, was now increasingly of national rather than regional importance.

Part 3 involves transition from the Third to the even more grandiose Fourth Canal, much larger than its predecessors and designed to follow a substantially different route. Major external events included two world wars and the intervening Great Depression, but overall the canal communities expanded through these disasters. Technologically, the period from 1914 to the late 1950s lay between the predominant railway and the predominant automobile periods of urban development, with the latter gradually taking command. The extended use of hydroelectricity and the streetcar also helped to spur urban expansion and the gradual merging of the former independent canal communities into one urban agglomeration.

Part 4 opens with the canal becoming part of the St Lawrence Seaway in 1959. As traffic tonnages steadily increased, proposals to enlarge the system were announced. A bypass was constructed around Welland and, when a bypass around St Catharines was proposed, the land was purchased, but this project was shelved because tonnage passing through the canal reached its maximum by 1979, then declined. In the meantime, as the historic canal communities became islands in the spreading urban matrix, the potentials of the old and new canals as historic, amenity, and tourist attractions came to the fore as a basis for action.

An epilogue draws together the various threads of argument that have been presented in parts 1 to 4, and interprets the canal as an evolving landscape of change.

In pursuing this double-barrelled theme of canal-urban interaction, I owe my first set of thanks to the many who through time have written about the canal or its associated settlements. The bibliography lists the many works that have been consulted.

For the canal itself, there are the records, publications, and archives of the Welland Canal Company, the Department of Railways and Canals, and the St Lawrence Seaway Authority. Particularly helpful have been the successive vice-presidents, from Malcolm Campbell to Camille Trépanier, of the Seaway's Western (now Niagara Region) office, and Seaway officers, particularly Gay Helmsley in Ottawa and Harley Smith in St Catharines.

The records in the local libraries, museums, and in regional and municipal offices along the canal, the advice of their well-informed staff, and the many encouraging suggestions from members of the local historical societies have been received with pleasure. Of particular note are the Special Collections, Map Library, and Reference Collections at Brock University.

To those who have read sections of the manuscript, my especial and most appreciative thanks are due. These stalwarts include Sheila M. Wilson, my co-author of *St. Catharines: Canada's Canal City* and past president of the Canadian Canals Society; Dorette Carter, Curator/Director of the Welland Historical Museum; Tim Elsley, Special Projects Officer, St Lawrence Seaway Authority – Niagara Region; and Joyce and Ron MacRae, as friends and neighbours.

Another stalwart is John Burtniak, Special Collections Librarian at Brock University and past president of the Canadian Canals Society, for his great understanding about the canal and its communities, and his ever-willing readiness to assist. Another memorable student of the canal is Colin Duquemin, who through his work at the St John's Outdoor Studies Centre has fascinated young and old with his enthusiastic accounts and his treasure trove of restored maps. Arden Phair, St Catharines Museum, has assisted greatly.

There can be no more dedicated person to the canal than Louis Cahill, who with persuasion and tact has pressed considerably for the recognition of its founder, William Hamilton Merritt, and for public understanding of the waterway as a national heritage resource.

Many others deserve a personal reference for their advice, comments, and information: Roland Barnsley, who introduced me to early documentation about the potential of a lake-to-lake parkway and the need for landscaping along the modern and former canal systems; Robert Ferguson at the University of Toronto Press, who has encouraged with helpful comments and steered the material comfortably through the publishing process (I also acknowledge the helpful and encouraging advice received from two unknown academic assessors, whose comments now flavour parts of the book); Hugh J. Gayler, Associate Professor of Geography, Brock University, for his understanding of the settlement patterns in the Niagara Peninsula; Brian and Lorraine Leyden, for their many enthusiastic contributions, helpful comments, and access to their newspaper files; and Bob Steele, of Proctor and Redfern, for technical data; Roberta M. Styran and Robert R. Taylor, who, currently working on a canal bibliography for the Champlain Society, have introduced me to several important facts and references; Craig Swayze, who helped with respect to the Canadian Henley Rowing Regatta; and Fran G. Wilks, who has made some diffident comments on a few small but nevertheless important points of information. At Brock University, Colleen Catling has provided advice on how to use my Macintosh Plus computer, and Jenny Gurski and Joyce Samuels have typed and retyped successive manuscripts.

Funding the research has been undertaken primarily on a personal basis, but the project also owes a considerable debt of gratitude to Henry Burgoyne, chairman of the St Catharines *Standard*, for allowing the residue of funding from *St. Catharines: Canada's Canal City* to be used for this publication. Finally, this book has been published with the help of a grant from the Humanities and Social Sciences Federation of Canada, using funds provided by the Social Sciences and Research Council of Canada.

To all who have assisted in any way, my very sincere thanks. I apologize and accept responsibility for the mistakes and errors of fact and interpretation that may remain.

This private, family-commissioned statue (1928) was located close to the First Canal and the Welland Canal Office of William Hamilton Merritt, the St Catharines merchant and mill owner who promoted the canal when water was needed to serve his mill on Twelve Mile Creek.

A dam across the Grand River created the canal community of Dunnville.

The centre of Welland grew around the timber aqueduct where the Feeder Canal crossed the Welland River. This became the aqueduct of the First Canal, to be superseded by the stone aqueducts of the Second and then the Third Canal. In this turn-of-the-century scene, the Second Canal aqueduct can be glimpsed behind that of its successor. Note also the mill and the commerical frontage facing the canal, the lock connection with the Welland River, and its nearby toll house.

This cross-section of the canal in St Catharines (1871) includes tow path, canal traffic, the canal bank with mill races and mills, and the backcloth of St Paul Street with commercial activities along the crest of the bank.

The Welland Mills in Thorold, built in 1846 and then the largest of their kind in Canada, producing from 200 to 300 barrels of flour a day.

The Muir shipyard, here at Port Dalhousie, was transferred to Port Weller when the Fourth Canal opened, and has survived there as the only Canadian yard on the Great Lakes.

The Hopkins swing-bridge at Port Colborne, named after the facing retail building, crossed the Second and Third Canals at the head of the harbour (c. 1900).

The canal lock-keepers were housed in buildings constructed close to the locks, as here in Merritton, where the canal crossed the Niagara Escarpment.

St Paul's Anglican Church at Port Robinson was built in 1841 during the construction period of the Second Canal. A part of its debt was paid off by Black soldiers stationed in the village to keep peace between rival gangs of canal workers.

Port Colborne harbour, c. 1900.

The departure of the *Garden City* passenger side-wheeler from Port Dalhousie to Toronto, c. 1910. Note the connecting links with the N.S.&T. electric streetcar system and the coal-handling facilities on the far bank in association with the Welland Railway.

Marine distances and trading linkages from the canal to ports around the Great Lakes and abroad are indicated on this sign at Lock 3.

The last surviving lock of the Third Canal at Port Dalhousie, c. 1969.

The Fourth Canal, under construction in 1914, curtailed the eastern growth of Thorold.

Water was diverted from the Third Canal at Allanburg to reservoirs at DeCew, where penstocks carried the flow across the Niagara Escarpment to the power station below.

The most notable feats of engineering on the modern canal are Locks 4, 5, and 6 of the Fourth Canal. Here, before the official opening, is the passage in 1930 of the first vessel through these Flight Locks.

After the Welland Canal By-Pass was constructed, a syphon carried the Welland River under the new channel at Port Robinson. Here a Russian freighter crosses the new syphon.

The towers and spans of the vertical lift-bridges that crossed the Fourth Canal were a commanding visual feature, as here on Main Street, Welland, c. 1930s.

The Queen Elizabeth Way crossed the Fourth Canal by the high-level Garden City Skyway, which replaced the Rolling Lift Bridge on the main highway between the United States and urbanized southern Ontario.

The Townline Tunnel at Welland, a combined highway-railway tunnel to carry land traffic under the canal.

The Algoma Steel plant in Port Colborne, one of many large-scale industries attracted to the Third and Fourth Canals.

Neptune's Staircase, where the locks of the Second Canal crossed the Niagara Escarpment at Merritton. When abandoned for through navigation, its water was used to supply industrial development that included the pulp and paper industry along the canal from Thorold to St Catharines.

Previously McKinnon Dash and Metal Works, then McKinnon Industries, and now General Motors, this plant in St Catharines has a long history of close association with the canal, starting on St Paul Street next to the Hydraulic Raceway and ultimately locating on the east side of the Fourth Canal.

THE WELLAND CANALS
AND THEIR COMMUNITIES

Introduction:

The Welland Canal within the World Experience of Canals

The Welland Canals (map 1, on page 4) express many of the forms and types of canals that have been developed since antiquity to improve upon the navigation abilities of natural rivers (Hadfield 1986). Canals conceived for the purposes of trade have changed the landscape through which they pass by their introduction of novel features such as the channel, its water supply, and the deposition of excavated materials. They introduce new crossing points and modify the earlier patterns of settlement by providing an impetus for development in the communities that follow the canal. Canals offer a fertile location for the promotion of industry, urban advance, and the regional process of development. They are urban catalysts, and the reverse operates when a canal is changed or removed from the landscape. Canals and their settlements are creatures of mutual interaction, one with the other.

From Rivers to Canal Improvement Schemes

Long before canals were invented, natural rivers played an important role in transportation the world over. In Canada this applied to many routes. The birch-bark canoe and the kayak carried the native peoples on exceptionally long journeys, and these routes were later often followed by the French and then the British fur traders.

The continent was crossed using its rivers, and people, their goods, and their possessions were conveyed across the intervening watersheds by portages of varying length and severity. One such arduous portage introduces the story of the Welland Canal, for the tempestuous middle length of the Niagara River, with its rapids, The Falls, and the gorge, was impassable to any type of vessel. The portage around these obstacles was in turn

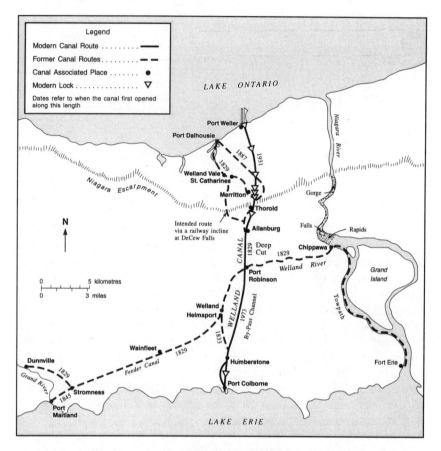

Map 1. *The Welland Canals and their communities.* The First Canal, completed in 1829 between Port Dalhousie on Lake Ontario and Chippawa on the Niagara River, was extended from Port Robinson to Port Colborne in 1833. It was displaced along approximately the same route by the Second Canal in 1845. The Third Canal in 1887 followed a new route between Port Dalhousie and Thorold, then broadly the same route as the Second Canal south to Port Colborne. The Fourth Canal, opened officially in 1933, extended from a new harbour at Port Weller on Lake Ontario along a substantially new alignment to Port Colborne. This route was changed in 1973 when the section of the Fourth Canal through Welland was abandoned and a bypass constructed around that city.

bypassed by the First Welland Canal of 1829 to provide a direct passage between Lake Ontario and Lake Erie. Amendments to this route have continued this function subsequently.

When a towpath was added to the natural watercourse, men or animals might pull vessels upstream against the current. This action applied to the Upper Niagara River, where a 'Breeze of Oxen' was used to reach Lake Erie. When the First Welland Canal reached Chippawa in 1829, vessels were towed up to the higher level of Lake Erie along a towpath now followed by the Niagara River Parkway.

Weirs might be added to a river. Of great antiquity, these held back the water to create a pond from which water might be diverted through a mill-race to serve a mill or mills. To avoid the hindrance caused to navigation by this dam, a movable section might be incorporated in its structure so that, when opened, boats could pass downstream with the flow of water. More difficult was the need to winch or otherwise haul vessels upstream against this flow of water. Even so, flash locks with vertically rising gates existed in many parts of Europe from the eleventh century on.

The resilient Chinese added slipways around a weir. A boat could now be hauled around the barrier from one operating level to another, but as this process might damage the keel of the boat and also posed a security risk where an immobilized boat might be attacked and its cargo pilfered, an improvement was made to add a chamber with gates at each end of the passage through the weir. These pound locks with cumbersome, vertically rising gates were used on the Grand Canal in China from about AD 1100. They followed in Europe from about the fourteenth century. The invention of the mitre gate that has been used on each of the four Welland Canals is attributed to Leonardo da Vinci (1452–1519), the Italian painter, architect, and engineer. Its V-shape pointed upstream to resist the pressure of water, and this weight pushing against the gates kept them firmly closed.

On the Welland, gates of greater size and weight have had to be designed as the volume of water to be impounded and the size of vessels have each steadily increased. The locks have served to cross the abrupt height difference over a very short distance across the steep slopes of the Niagara Escarpment. The Welland Canal is credited in particular for its 'Neptune's Staircase' of locks on the First and Second Canals, for the tier of locks on the Third Canal, and for the double series of flight locks and the longest lock in the world at Port Colborne on the modern canal.

A river might also be straightened for use by water transport through building a cut, with or without a lock, across a bend in that river. This advance in canal technology can be viewed at Welland Vale, St Catharines, where a meander of Twelve Mile Creek used for the First Welland Canal became a cut across the inside of the curve for the straightened route of the Second Canal (see map 2, page 31).

From cuts it was but a short step to bypass a difficult length of river, a feat first achieved in England on the Exeter Canal during the 1560s, then applied on the Manchester Ship Canal to bypass the rivers Mersey and Irwell between Manchester and Liverpool. In Canada, the St Lawrence Canals upstream from Montreal provide instructive examples for this type of canal-river improvement scheme where rapids impeded navigation at the Cascades, the Cedars, and most severely at Coteau du Lac upstream from Montreal. Using Cornish miners from England to undertake the rock cutting, four short military canals were constructed in the early 1780s by the government to a depth of 2.5 feet (0.8 m) for small boats (bateaux) carrying 30 to 40 barrels of flour. The cut at Coteau du Lac, 900 feet (274.3 m) long by 7 feet (2.1 m) wide, had three locks, and the four cuts were together about 1700 feet (518.2 m) in length. The Lachine Canal to the west of Montreal provides another example, this time bypassing the rapids and replacing a portage some 8 miles (13 km) in length.

Upstream on the St Lawrence system, the Welland Canal was also initiated as a bypass. The natural route for the flow of water out from Lake Erie via the Niagara River was quite impassable by any vessel. Navigable only along the upper and the lower rivers, the central length was bypassed first by a portage on the American side. Transferred to the Canadian side after American independence, the New Portage between Chippawa and Queenston in turn lost out to the First Canal of 1829 in a location some 12 miles (19.3 km) further inland from the river.

More difficult again than a bypass in the history of canal achievement was the connection of two river routes and their catchment areas over intervening higher ground. In France the Canal de Briare provided a link between the Loire and the Seine and helped to provide Paris with food. Its successful completion in 1642, with lock staircases and a long feeder to supply the summit with water, led to the reformulation of ideas for a canal link between the Mediterranean and the Bay of Biscay. When the Languedoc (now Midi) Canal opened over the high ground between the Garonne and the Aude in 1681, it featured three large aqueducts, a tunnel, a new port, curved locks to reduce earth pressure, and a high-level summit water supply.

Elsewhere in Europe, the Rhine became connected with both the Rhone and the Danube. England was crisscrossed by watershed canals: between the Trent and the Mersey; across the Pennines between Leeds and Liverpool; between the Thames and the Bristol Channel; and between the Forth and the Clyde in Central Scotland. In the United States, the Erie Canal that opened in 1825 between the Hudson River and Lake Erie provides a prime example for this type of canal.

Canadian counterparts include the Trent-Severn Waterway between Trenton on Lake Ontario and Port Severn on Georgian Bay. Its length of 240 miles (387 km) had 43 locks, and a marine railway. The earlier Rideau Canal that opened in 1832 had 45 locks and extended for some 124 miles (200 km) between Kingston on Lake Ontario and Bytown (now Ottawa) on the Ottawa River.

The Welland Canal had to offset the height difference of 327 feet (99.7 m) that existed between Lake Ontario and Lake Erie, and therefore had to overcome an elevation greater than for all other world canals. The distinctive profile of each successive canal has included a gentle change in levels and several locks along its northern length across the Ontario Plain, then a closely spaced cluster of locks to cross the precipitous brow of the Niagara Escarpment, and, depending upon which canal, a length with either no or few locks along its southern length between the Niagara Escarpment and Lake Erie.

The Erie Plain south from the escarpment to Lake Erie is flat, except for some morainic ridges close to the escarpment and a few east–west river valleys. The canal cut into and through these ridges, giving rise to a section between Port Robinson and Allanburg known as the Deep Cut – the only length of canal that has remained in the same position throughout the full history of the four different canals with their changing routes. The broadest river, the Welland, was crossed by an aqueduct at Welland on the first three canals, and then as the canal was deepened a syphon culvert took the river under the Fourth Canal.

The Welland Canal had also to avoid the highest land in the peninsula, the Fonthill Kame that lay to the west of the line selected for the First Canal and its successors. This physical feature explains why the shortest route for a canal across the Niagara Peninsula between Lake Erie and Lake Ontario has never been followed.

A final canal classification includes those that cross a watershed to connect two seas. The Kiel (1895), Gota (1832), Corinth (1893), Suez (1869), and Panama Canals (1914) fall suitably within this category, as does the earlier Welland Canal (1829), which connected two of the Great Lakes, depicted as Inland Seas because of their vast size, sailing distances, and depths. Further, the Welland Canal also became part of the world's international navigation system when in 1959 the St Lawrence Seaway opened as a deep-draught waterway extending 2340 miles (3700 km) inland from Montreal to Thunder Bay at the head of Lake Superior. Along this major international route of trade, the rise of 327 feet (99.7 m) along the Welland may be compared with 243 feet (74.1 m) downstream between the St Lawrence and Lake Ontario, and 23 feet (7.0 m) to pass between Lake Huron and Lake Superior.

The British Industrial Canals

The immediate precedent for the Welland Canal was the American Erie Canal that opened in 1825 across the watershed of northern New York State. American finance, engineers, contractors, and labourers that had worked on the Erie were also actively involved in the later Canadian achievement (Styran and Taylor 1992). Further, through the roots of Upper Canada in the British Empire, the canal network at home was well known to the administrators, the decision-makers, and many of the pioneer settlers through their earlier experiences in the United States or Europe. Canals, indeed, as part of general knowledge were also much referred to in contemporary documents and newspaper accounts. They were an intrinsic aspect of the cultural and technological transfer from Western Europe to North America as the new continent was colonized, and it is not surprising that many of the more evocative features on the First Welland Canal, such as its flights of locks, a proposed marine railway to offset the Niagara Escarpment, an intended tunnel at the Deep Cut, and an aqueduct across the Welland River, replicate design characteristics to be found on the British canal system.

Indeed, Thomas Telford, the British civil engineer, was consulted by William Hamilton Merritt, the founder of the Welland Canal enterprise, on one of his many visits to England to raise funds for the canal. British investors, including Telford, contributed to the canal enterprise, and Merritt when in England recorded visits in his journal to the Duke of Bridgewater's canal at Worsley and Runcorn. Francis Hall, who reported on a survey in 1824 for the Welland Canal, had worked with Telford.

The particular significance to Upper Canada of such links was that the industrial revolution in Great Britain had been nurtured by canals, spurred by the development of the steam engine with its voracious appetite for coal, and inspired by the Duke of Bridgewater, who had been fascinated by the potential of the Canal du Midi on a Grand Tour of Europe (Hadfield 1968). The duke owned coal mines at Worsley near Manchester. He financed a canal that was driven into the mines, where side cuts and different levels were excavated. The canal was taken over the River Irwell by an aqueduct, and wharves, docks, and warehouses were constructed at the Manchester end. To avoid the steep pull up the bank by loaded horse-drawn wagons, the canal was extended into a tunnel where a crane driven by a waterwheel brought the coal up to street level.

These works that started in 1759 owed much to James Brindley, the duke's canal engineer, and they initiated the rapid building of a canal network across the country. Over 4000 miles (6430 km) had been constructed by 1840, and canal acts and urban expansion became synonymous with each other. There were twenty-four canal acts of parliament between 1751 and 1790, 215 between 1791 and 1810, and 108 between 1811 and 1830, when the First Welland Canal was introduced to the then British colonial scene. As canals were added to the existing communities, every British town of importance before 1841 except Luton was either on the sea or an inland waterway, and their population had almost tripled over the preceding four decades.

The British canals fed into the ports, and the inland manufacturing towns and former quiet market towns received a new commercial or industrial base. They served industry, and factories of all types lined their banks. The canals were built to carry heavy, bulk commodities, especially coal, but their loads also included many other raw materials and industrial products: stone for road-making; building materials such as bricks, timber, and slate; fertilizer to spread on the land; mined products such as salt and iron; and clay for use in the Potteries. Bales of cotton and wool were brought from the ports to the textile towns, and finished goods sent to inland destinations and to the ports for distribution overseas. Wheat, corn, flour, and animals were transported to the urban markets, and machinery, beer, and earthenware reached their destinations by canal.

The canal companies built walled-in yards along their routes for the import and export of products. Some became trans-shipment points between canals of different width, or between the canal and the river systems of navigation. All were centres of local trade, fed by horse-drawn wagons for the distribution of goods received from the canal and the collection of goods destined for the canal. As cargoes were costed per ton-mile of travel, toll houses followed the canals, as did stables to house the tow animals and inns at the basins or where a road crossed the canal.

Boats to ply the canal were built by independent operators, either separately or in conjunction with the canal companies. Their size was determined by the minimum dimensions through the locks of the canal on which they worked, resulting in the famous narrow boats on the canal system centred on Birmingham, a growing manufacturing city located on an inland plateau and approached through its surrounding hills by tunnels and flights of locks. Some boats were designed to carry passengers, others carried the mail, and some were market boats that took people and their produce, including animals, to the market in the morning and returned at night after all transactions had been completed. If a demand existed, then a canal and boats were constructed to meet the expected circumstances, a truism that applied to the Welland Canal, which also gave birth to many shipyards during the nineteenth century.

The engineers who designed the canals were usually itinerant, moving from one project to another upon completion, and the contracts for construction along the line of the canal were let out to many small contractors. The canal company then employed such permanent staff as a clerk, a treasurer, an engineer to maintain and improve the system, and an agent or manager to promote its trade, together with their office and field staffs. Along the line of the canal were toll collectors, lock keepers, the wharfingers who managed the yards and basins, and the labour gangs required for canal maintenance. Boats had to be constructed then manned, and the tow animals had to be stabled, fed, and guided along the canal. These activities each meant employment and new incomes in the communities that arose along the line of each canal.

Every British canal changed the landscape through which it passed as it wound gently cross-country along the contours. Cuttings and embankments were rare, but the canals climbed hills by locks and by the occasional inclined plane, and they tunnelled through the more severe physical objects. As the canals existed before accurate maps, the surveyors had to undertake their own levelling, prepare alternative route possibilities, and assess the costs that were expected along each line of canal. Locks were inserted into each route to get from one level to another, streams were tapped or diverted to provide a supply of water, and reservoirs were constructed to conserve the supply of water and to meet the voracious demand at the locks for a regular flow. These features provided the blueprints that were to be faithfully copied when the Welland Canal was constructed across Ontario's Niagara Peninsula.

Canal Expansion in Europe and North America

As the industrial revolution spread into western Europe, the mania for canals expanded slowly throughout the eighteenth century, then more rapidly during the first half of the nineteenth century. 'Slowly a network of interconnected waterways was created, spreading outwards from the ports to the industrial areas, the capitals, and the interiors of France, Belgium, the Netherlands, and Germany, over which the barges passed with little regard for frontiers. A small system based on the Po was created in Italy; the Danube was given some artificial tributaries, and further east, canals were built in Russia, and rivers made passable' (Hadfield 1968).

In the United States the urge to build canals came later. Thirty American waterway companies had been incorporated by 1792, but by 1816 only about 100 miles (160 km) had been constructed. In the east several canals fed into tidewater. The most important, the Delaware & Hudson, which opened in 1828, carried coal from north-eastern Pennsylvania to the Hudson River. In 1817 steamers made their appearance on the Mississippi; as the falls at Louisville provided a serious obstacle, they were circumvented by a canal in 1823.

Inland, the Atlantic states wished to sell their manufactured goods to the expanding Midwest and import raw materials and food products from these localities. However, the Appalachian mountain barrier stood in the way of this dream, and the horse-drawn movement of goods was slow and costly until the Erie Canal was completed in 1825 between Buffalo at the eastern end of Lake Erie and the Hudson River (Whitford 1906). A narrow ribbon 363 miles (584 km) long, 4 feet (1.2 m) deep, and 40 feet (12.2 m) wide at the surface now connected the Upper Great Lakes with the Atlantic. As in England, the Erie Canal ushered in a period of abundant urban growth as new towns proliferated between Buffalo and Albany along the line of the canal. This manifestation was so great that the Erie Canal received the epithet 'Mother of Cities.'

The Welland Canal indeed had many successful precedents in western Europe and North America on which to draw for its achievement. During its period of conception from 1818 to 1829, canals symbolized urban and industrial advance. They were the features of stature by which nations progressed from agriculture and commerce into the expanding world of wealth, industrialization, external trade, and large cities. They received strong public support and blessings from the legislatures and the business groups of the day for that reason.

The Canal Era in Decline

But industrial advance through canals was not always to continue. The experience of steady progress was restricted as railways arrived and took over the movement of many commodities that had previously been handled by the canals. The British canals retained their traffic for about another twenty years, but they did not receive their proportionate share of the increased trade. They lost out to more versatile competition, and generally either declined or were abandoned when income failed to keep up with the costs of upkeep, repair, and maintenance. In North America the great Erie Canal, even though enlarged and improved in an effort to compete with railways, also gradually succumbed before this competition.

A marked exception to the more general trend of canal decline proved to be the Welland Canal. Here, the Welland Railway worked in conjunction with the Second Canal to transfer grain from Port Colborne to Port Dalhousie. It acted as the canal's saviour, and cargo volumes passing through the sequence of improved waterways continued to expand throughout the railway era until 1979, when the highest levels were recorded. Only then did the possibility of an uncertain business future start to emerge.

More generally than this distinctive Canadian experience, in Britain years, then decades, of decline set in. Many British canals were purchased then devoured by the railway companies, which deprived them of their route advantages and attracted their trade. But the canal companies were also in part to blame. They developed in a piecemeal fashion, the lock dimensions on the different canals varied considerably, and long-distance haulage frequently added the expense of trans-shipment from one system to another. By contrast, the railways, whether in Britain, Canada, or the United States, soon standardized their gauges and operated an integrated system across many routes.

Canal boats were slow moving and they could not operate during winter ice. Railway routes, more flexible in location, could generally be more easily and swiftly constructed, and they provided a faster, more regular, and often cheaper system than the canals. Canals became nostalgic memories and many were abandoned as a new era of rail supremacy took over most transportation needs.

Even so, some British canals survived through modernization. For example, the Grand Union combined with other companies to control all the main routes and their more important branches between London and Birmingham, and the Trent Canal was used to bring large craft carrying 600 tones (544 tonnes) up the Trent River to Nottingham. The Second World War led to some canal expansion but also to ruthless pruning elsewhere. In 1948 most of the remaining British canals were nationalized, passing into the control of the British Transport Commission. More decline and closures followed.

The saviour of canals was neither industry nor the transportation of goods – their historical function – but the use of the surviving waterways for pleasure, recreation, and leisure (Burton 1989; Squires 1979). A new canal era also dawned as restoration became an enthusiastic pastime for local volunteers. People took canal holidays. Old canal boats were converted for family use as houseboats, and new boats to serve the recreational market were constructed. Canal towpaths became footpaths for walkers, and the banks attracted hopeful anglers. Abandoned lengths of canal were restored, sections containing water were cleaned and improved, and suitable canal buildings were retained, reused, or converted to some beneficial use. Canal societies were formed to actively promote renewal of the canal scene. Revamping the canal heritage became an enduring theme in conjunction with urban renewal and redevelopment in the adjacent communities (Squires 1979).

A comparable process of canal rejuvenation took place in the United States, and in Canada many of the old canals received national support as Heritage Canals. In Ontario, this treatment was applied with enthusiasm to the Rideau Canal and the Trent-Severn Waterway. In Quebec the industrial Lachine Canal was rejuvenated because of tourism and in recognition of the canal's historical importance as a major contributor towards Canada's development.

On the Welland, the calibre of the remnant engineering works and the historical relevance of the associated canal communities and their older buildings has yet to be fully recognized. Some channels, locks, and other engineering structures have survived, but they are generally in a derelict state and mostly unrecognized for the possibilities of reclamation and reuse that exist. In other instances the channel and its locks have been filled and used for low-grade purposes such as outdoor storage or vehicle parking that provide no credit to the former presence of a major national waterway. One justification for this book is to provide the evidence for a more appreciative approach in the future.

The Canal Promotes Development

The man-made landscape of canal channels – with their locks, weirs, bridges, and ponds, and the ships and cargoes carried on the canal – may be viewed in urban terms for the settlements that they created and formed. As with the previous British canals and the Erie Canal, the Welland proved to be a major element in the creation of a new patterns of settlement along the line of channel (map 1).

Along the main line of the canal, only the rural service centres of St

Catharines and the hamlet of Humberstone preceded the waterway. A pattern of townships, with their gridiron framework of survey roads, and earlier Indian trails that had become pioneer routes crossed the land. The prior economic base of activity was agriculture as the forest cover was steadily cleared to support an expanding pattern of rural settlement, and grist and saw mills driven by natural streams had emerged. The life line of commerce via Lake Ontario and the Upper St Lawrence River was frozen over the winter months, and in the summer months it lay over treacherous rapids to the port of Montreal.

The canal itself began with an idea and the political and business will to carry the project through to fruition (Aitken 1954). Support had to be obtained and funds raised. William Hamilton Merritt, a St Catharines mill-owner and storekeeper, had these entrepreneurial abilities, and the first canal office was located in his home above the canal.

When the Welland Canal Company constructed its own office, this building was located close to Merritt's residence in an area that became the business centre of St Catharines. Here banks, stores, churches, and other service activities expanded. They owed their initiation to St Catharines as the centre for a rich agricultural hinterland, to the trade that flowed along the canal, to the availability of water power, and to this village becoming the primary centre of canal administration, a function retained to the present with the Western (now Niagara Region) offices of the St Lawrence Seaway Authority.

The process of change through the Welland Canal was advanced remarkably and irretrievably by the arrival of a low-wage, foreign-born labour force to clear the land, excavate the channel, and construct the canal. The area with its dispersed farming population had no surplus labour on which to draw, and besides, during the warm summer months when farming was the main preoccupation, the labour demand by the canal was at its highest.

The canal labourers had to be housed and fed, which initiated shack communities along the line of the canal and provided farmers and storekeepers with new opportunities for the sale of food and wares. The new workers from an alien land also strengthened the Roman Catholic church in a primarily Protestant locality. St Catharines as a see of Catholic authority owes this dimension of its character to these workers, as do several Catholic churches that located in communities along the canal.

Physically, the canal, usually a cut sliced into the landscape, might also be placed on an embankment, and it was carried over the Welland River on an aqueduct. A towpath followed each of the earlier canals until the advent of steam-powered vessels, and then bollards to permit their mooring and service roads along the canal became part of the public scene. In a later era, these service roads along the canal have been viewed for their potential to create a scenic parkway along the length of the canal, but this aim has not been fulfilled.

The landscape was also altered when the north–south channel cut across east–west flowing streams and impounded their upper waters. The spoil excavated from the channel and where canal structures were built was used to provide banks along the sides of the canal and to infill low land. Elsewhere it created new hills, and the shoreline was extended into both lakes.

The vicinity of the canal obtained a new and regular flow of water. The water supply was controlled by engineering mechanisms such as locks, which raised or lowered the boats on the canal as they passed from one level to another, and by sluices and waste weirs with valves or gates that regulated the flow of water. These mechanisms changed the route along the canal into a series of steps or reaches with different levels of water between each lock.

The water supply, the essential prerequisite for the movement of vessels (with its depth at the locks being the essential controlling factor over the size of the vessels that might use the canal), may also be used for other purposes. These have included the drainage of land, a source of supply for domestic, municipal, and industrial purposes, and the recreational use of the water bodies that have been created.

Mill races or flumes might also divert water from the canal to drive the wheels of industry. At a later period, water taken from the canal has also been used to generate hydroelectric power at the locks and at DeCew Falls. The canal as a water body with its abundant supplies of water thereby became a major contributor to urban and industrial development.

The canal as a trading artery involved the movement of vessels through its system and, in addition, the traffic flow generated activities that located along the canal and its feeder arteries. Canal traffic might use only a section of the canal for the conveyance of goods. When fees and tolls were levied for the passage of vessels through locks, the lengths of canal outside the area with locks enjoyed a preferential advantage for urban growth and industrial development, while those areas inside the lock system suffered a corresponding disadvantage.

The cargoes carried in vessels using the canal provided a further incentive for enterprises to locate along the canal, and industries have taken advantage of a canal situation to import some or all of their raw-material requirements. In addition, after manufacturing or processing, the product may be exported to its destination by water. The canal as a cheap mode of transportation in particular attracted the bulk movement of low-value goods for long-distance transport. It also provided a competitive element that could reduce the cost of moving goods by the rail or road modes of transport.

As an adverse factor, the direction and the extent of movement by canal vessels are more circumscribed than by other methods of transportation. The areas served are linear along the extent of the canal and its connecting systems of lake, river, rail, and ocean transportation. A wider dispersal of goods is feasible through either the highway or the railway mode, each with greater route flexibility and wider arcs of movement, but also usually involving higher costs.

Canals also suffer from the seasonal operating disadvantage that winter ice might close the channel, which means that canal and industrial equipment lies unused over a period of some four months, a period of enforced inactivity that requires the stockpiling of imports and exports by the industries that use the canal for transportation. Seasonal closure may also involve winter limitations on the amounts of water extracted from the canal for power and water-supply purposes.

The Welland Canal suffers from its cold weather and mid-continental locale, but more serious is the fact that its outlet heads north-east to the even greater winter severity of the frozen St Lawrence River. From the mid-1850s on, competitive railways offered the advantages of year-round transportation to Canadian and American Atlantic ports and then a shorter crossing of the Atlantic to European markets, but these opportunities lay beyond Canadian control and might be subject to external trading embargoes.

The vessels that used the Welland waterway provided a further resource. They had to be constructed, manned, serviced, and repaired, and then replaced either because of old age or as new technological circumstances came into play. Suitable locations along the canal for these purposes were embayments out of the line of shipping traffic and points where canal communities provided functions for the vessels that used the canal. The ports with lake or river access were especially favoured in this respect, and here the marine services and the manufacturing activities that became an important aspect of the local economy extended back into many subcontractors and the subcomponents that were required for each operational activity.

Canals, like rivers, as a water body present a physical barrier to land communications. They separate the banks through which they pass, and in a few instances, as formerly in Welland and now at St Catharines, this barrier impact has been reflected in the use of the canal as a municipal boundary. This barrier effect is offset where bridges carry land communications across the canal. In a later era tunnels, pipes, conduits, and mains might extend urban services under the canal, and overhead cables may carry services from one bank to another. In these circumstances the canal remains a barrier to land communications for much of its length, but the bridges, tunnels, and other points of crossing become focal points for the development of communities at the limited number of crossing points over or under the canal.

When settlement patterns and especially ports and harbours are under consideration, an important factor associated with the development of river valleys is the first bridging point inland from the sea, and then how that position has changed through time (Bird 1971). On the Welland Canal the equivalent situation arose at the two extremities of the canal, Port Dalhousie and Port Colborne. At Port Dalhousie, the mouth of Twelve Mile Creek was at first too wide to be crossed by a bridge, and the first upstream crossings were at Welland Vale some three miles (5 km) to the north near St Catharines and at St Catharines. With no road crossing at Port Dalhousie until the period of the Second Canal, this deficiency hindered growth at The Port, while that in St Catharines was encouraged.

Port Colborne did not have such restrictions placed on its early growth. The canal was crossed first at its entrance, and then at the head of the harbour, permitting growth around both sides of the harbour, whereas at Port Dalhousie the two sides remained separate until the railway introduced industry to the east side.

Dunnville provides a third example of this bridge phenomenon. Located on the Grand River and the port at the terminus of the Feeder Canal, much of its growth can be attributed to this place being the first inland bridging point on the broad Grand River inland from Lake Erie.

Canalside settlements as central points along the Welland Canal system grew to provide the service and other needs of passing vessels and, as the bridges focused land communications on these points, the services that arose also catered to the needs of the residents living in the rural hinterland areas of each canal community. Growth thereby tended not to be linear for extensive lengths along the canal banks, but rather along and across the canal at selected points of the system, with the bridges acting as pivotal points in this process of urban achievement and agricultural areas being left in the areas in between.

This potential for service growth was heightened considerably when the nodal site was combined with a situation where water power was also available from a lock or locks. Fortuitously, the dual circumstances that combine both a lock and a bridge often existed where the canal was at its narrowest. Communities developed nodally, with service and industrial activities around the bridges and the locks that provided the core of most canal settlements, then expanded linearly, with industrial development where open sites existed along the canal and its raceways. These circumstances advantageous for urban growth are selective and discriminating. They occur at certain locations along the canal, rather than being continuous along its length.

To these features of settlement growth must be added the ports and the harbours where ships take on or discharge their cargoes. On the Welland these wharfing facilities are located at each end of the canal, and where the canal crosses or is close to another waterway, as at Welland and Port Robinson. These areas are reflected in extended basins of water. On the Welland, these basins are man-made behind a sand bar and upstream from a weir across a former river mouth at Port Dalhousie, excavated at an entry point to the canal route at Port Colborne, and make use of a natural depression at Port Robinson. Docks and wharves may be added both at the ports and elsewhere along the canal, and breakwaters to protect a harbour from wind and waves may extend the canal into the adjacent lake, an engineering decision made to ease the flow of vessels into the harbour, but with repercussions on Lake Erie and Lake Ontario that affect the flow of lake currents and lead to either sedimentation or erosion along the lakeshore.

Economically, the sheltered harbour, whether natural or artificial, may accommodate either the transfer of cargo from one type of vessel to another or a transfer between the canal and a land mode of transport. Harbours are the important break-in-bulk points of the canal system where, as stated in a late-nineteenth-century study on the economics of transportation with reference to the location of towns and cities, 'population and wealth tend to collect ... The location of the greater number of commercial towns the world over, and from the earliest times to the present day, has been fixed at the point of juncture between land and water movements' (Cooley 1894). Two types of break are identified by Cooley: a mechanical break that requires only the physical transfer and temporary storage of goods, and a commercial break where the interruption causes a change in the ownership of the transported goods.

On the Welland a triple advantage existed where port activities supplemented by the availability of water power combined with one or more bridge crossings of the canal system. Such centres provided commercial, industrial, and service advantages that encouraged the foundation and then the growth and expansion of urban commitments. The canal was a major instrument of urban growth, a factor expressed both locally within that community and on a relative basis within the wider regional dimension. The reverse side of this equation is that a town nurtured by any of these three formidable factors in its heritage might also decline or stagnate if one or other of the initiating advantages was removed or changed. Nothing is fixed and static in the history of either canals, cities, or transportation.

The Changing Canal Scene

A potent force of change in the canal environments has been that of technological advance, including the transition from sailing vessels that had to be towed through the canal by teams of horses or oxen to the independence of steam-operated vessels that required coal for their motive power, and then to the diesel-operated vessel powered by oil. Repair and construction yards have had to adopt to different types and sizes of vessels on the canal, and through vessels have steadily relied less upon the canal for their servicing needs. The hulls of vessels have changed from wood to iron then steel, and, as they became larger and carried more cargo, the channel has had to be widened and deepened and the locks increased in size and capacity to accommodate this steady increase in the size of vessels. As these canal elements have changed, so too there has also been an inevitable series of related changes in the characteristics of the communities alongside the canal.

Changes in trade on or through a canal may have nothing to do with local circumstances along the canal itself. It is an external impact beyond the control of either the local community or even the canal authorities, as when the flow of wheat from the Prairies or iron ore from Quebec is changed drastically. A customary cargo might decline or grow through a changed demand. It may be displaced by some new product or commodity, or a supply from a cheaper alternative source. The weight of tolls, grants, subsidies, tariffs, and other subventions to operate the canal, as well as the ease by which the various commodities can cross national and international borders, are also subject to change and competition from other modes of transportation. This feature has been of particular importance to the Welland because, since inception, the canal has been viewed as an essential part of the national waterway route for exports and imports between the Upper Great Lakes and Europe.

When changes to the canal have been required, the channel and its new locks and bridges have necessarily been relocated in new positions in order to retain the flow across or through the old structure until the new is available. Then arises the critical question of what happens to the abandoned features. Are they retained and reused, removed, or allowed to decay to oblivion? And how does the community that has developed close to the canal in conjunction with one or more of these features as a major part of its urban fabric respond to these changes when channels, bridges, or locks are moved to a new location? Abandonment and reuse of the canal becomes as important as the former commercial activity when the urban side of the canal coin is under examination.

Communities may have lost one or more of their bridge crossing points, their water-power potential, and their canalside activities through widening the canal or changing its route. The through movement of vessels, their servicing needs, and their potentials for trade and business activity through their cargoes and servicing needs may have been transferred elsewhere. As a consequence, some canal communities such as Allanburg and Port Robinson have declined to being shadows of their former existence because of this regressive experience; Wainfleet, Dunnville, and Chippawa no longer lie on the waterway that provided them with so much sustenance.

Canal communities such as Port Dalhousie, inner St Catharines, Thorold, and Merritton have weathered the changed circumstances and even expanded as the formidable factor of geographical inertia takes over as a potent city-forming element. Activities have been retained after the original reasons for their introduction have been changed, modified, or disappeared. Once an industry or a town is established, it may regenerate continuously through new associations and traditions (Monkhouse 1965). The earlier circumstances have been superseded by the cumulative vitality of later developments. In exchange and as compensation, new and critical advantages of location have accumulated, including the investment of capital in site and building works or the establishment of ancillary services and dependent activities. A momentum of environmental change has been introduced, in which new developments and achievements have replaced, added to, and strengthened the former existence and character of a given place. The essential factor is that the assets of infrastructure introduced by the canal may over time have become more important than the canal, and in time may overshadow the earlier impetus initiated by the canal.

The new features added to the canal endowment have included the railway, hydroelectricity, the streetcar, and the automobile as each successively arrived on the scene to change the pre-existing landscape. There may be the advent of new industry, perhaps using one or more of the industrial buildings, the site, or the water privileges of an earlier enterprise, or using the canal to import raw materials or its flow of water for industrial purposes. There may be cultural change through the arrival of a new immigrant group, its skills, and social attitudes, events that may be expressed in the infrastructure through the provision of homes, churches, schools, and the services provided for the urban population and those residing in the rural hinterland.

A continuous process of change is introduced within each canal community through such events. Without this continuing replacement activity within cities, the Welland Canals Corridor of urban development would not have emerged from the small, widely spaced, mid-nineteenth-century communities that were formed and grew because of the canal.

At a later period, the canal communities with their buildings, having a character that represents decades of Canadian urban, industrial, and social history, have slowly been recognized as important heritage resources, but with less alacrity than in other parts of Canada or along the American and British canals. At the national level, William Hamilton Merritt as the founder of the Welland Canal is now depicted as the Father of Canadian transportation and commemorated for his initiatives. Historical plaques provide an introduction to certain canal features, and displays in museums and libraries serve as reminders that earlier canal circumstances have matured into themes of some importance.

Along the modern canal, the flow of vessels, the engineering works, and in particular the locks have taken on a new element of visitor and tourist appeal as the automobile, the shorter working week, and a higher standard of living have encouraged family travel. Viewing platforms have been provided at Locks 3 and 8 at St Catharines and Port Colborne on the modern canal, and a bulletin board at the former site describes the cargoes, the nationality, and the places of origin of the vessels expected to pass through the canal that day. The earlier canals and their settlements are also viewed increasingly for their heritage and recreational appeals, and considerable argument is now being entertained for a linking system of parkway and heritage routes along the modern and the old canals to present their credentials for public esteem and enjoyment.

The Welland Canal is not an isolated phenomenon. Its impact extends outwards to influence the urban development of Southern Ontario, and its repercussions spread across the international boundary to make important contributions within the North American economy around the Great Lakes. These more extensive aspects are only marginal to this text with its focus on the regional impact of the canal for the Niagara Peninsula.

Within this focus, the task in the ensuing analysis is to interweave the successive phases of canal development with the associated expansion and changing character of the canal communities. To the urban geographer, it is always important to understand and express the dynamic forces of settlement evolution as a continuum, yet with changing ingredients through time. It is essential to examine the foundations of the settlements under study, to portray their raison d'être, and to understand the early expression of each community in the terms of its form and its functions as the foundation from which all subsequent developments have emerged. The present can only be understood in terms of the past, and it is through the many public, business, and private decisions taken over the decades that a city or town takes on and attains its present form, quality, and status.

At the regional level, as all places do not grow at the same pace, it is essential to understand and explain how and why differences occur between adjacent communities that are apparently subject to the same forces and pressures, in this instance the Welland Canal as the catalyst for urban growth and industrial change within the common framework of a regional, provincial, and national society, yet also for the considerable changes that exist along its route.

As Griffith Taylor, one of Canada's first urban geographers to focus on the urban environment, wrote in 1943, 'in a genetic approach to urban geography, we are chiefly interested in the town as an evolving organism. We should know something as to the original site and nucleus, as to the reason for its founding, as to its growth with respect to its nucleus.' This approach, to assess the processes and stages in the evolution and development of towns and to appreciate the background impact of topographic and political controls, is essential to all forms of urban understanding and for the professions of urban and regional planning that seek to guide our communities into the future.

If the present can be explained only in terms of the past, it also becomes imperative to draw a double distinction. On the one hand, there are the original forces of prospective advantage that nurtured the Welland Canal, its industries, and towns into existence in the first place. And then there are the later cumulative advantages of site and situation that have continued to promote and foster that settlement. The First Canal, its marine features, and its commercial and industrial successes are past history, but the communities that live on to this day remain as living entities of the canal expression in the landscape.

PART 1

The Development and Impact on
Settlement of the First and Second
Canals to the Early 1850s

1

The First Canal: From Inception to Completion in 1829

The English proverb 'Many a slip between cup and lip' may be applied to the achievement of the First Welland Canal that opened in 1829. The alternative proposition is that 'if anything can go wrong, it will,' for the canal as achieved bore little relation to the route that had been planned and intended. The history of building the First Canal shows that many Fortuitous Acts of Fate (FAOFs) intervened to change the course of the route and the form of construction. With hindsight, these FAOFs also influenced all the subsequent urban, industrial, and community events along the canal system.

The critical need for a canal, how it was achieved, the alternative routes that were considered, and the enforced changes to expectations during its works program provide the content for this chapter. Also introduced are the abilities, the perseverance, and the dominant role in the canal's achievement of William Hamilton Merritt, the main instigator and constant proponent of the canal (Aitken 1954).

The Justification for a Canal

As the population of Canada by the early nineteenth century was confined to the banks of its rivers and the shores of its lakes, and as the lakes and small sections of the rivers were used for navigation, long under consideration as population and trade each increased was the need to avoid the lengthy and circuitous passage by water around the long, projecting peninsula of southern Ontario when travelling from the St Lawrence River to the heart of the continent. Various alternative routes existed: from Lake Simcoe to Lake Ontario and the Holland-Humber, the Holland-Rouge, and the Trent River routes, or via Lake Nipissing and the Ottawa River.

These routes were shorter and inland from the boundary that had been established in 1783 with the United States, but residents of the Niagara Peninsula were able to persuade the government that theirs was the preferred project. The distance was longer, but the potentials of water power and industrial development were each considerably greater.

The arguments for a canal across the Niagara Peninsula to connect its two bounding lakes with a navigable route were powerful and persuasive. Physically, the natural outlet of water flow from Lake Erie through the Niagara River was quite impassable. No vessel then or today could negotiate The Falls, with its upstream rapids, the boiling cauldron at the Horseshoe Falls, and the swirling waters through the gorge. Here was the greatest blockage of all on the St Lawrence–Great Lakes system that fed inland to the heart of the North American continent from the Atlantic Ocean and the St Lawrence River at Montreal (Heisler 1973; Creighton 1956).

A portage to bypass the Niagara River had been constructed on the American side at some unknown date by the native Indian peoples, and was later used intensively by explorers, merchants, the military, and by the French and later the British fur traders. Here was a major venue of conflict between rival British and French interests for control over water access to the continental interior of North America.

This portage was transferred to the Canadian side after the American Revolution, but transportation across this New Portage was costly, slow, and inconvenient (Seibel 1990). It required haulage over land in wagons or on Indian backs between Chippawa and Queenston, and trans-shipment at these two settlements, at first known appropriately as the Upper and Lower Landings. A further trans-shipment from wagon or river vessel to a sailing vessel might also be required at the lakeside terminals, Niagara (now Niagara-on-the-Lake) and Fort Erie.

As described in 1795: 'At the lower landing, Queenston, the vessels discharge their cargoes and take on furs brought from 300 to 1,500 miles back. I have seen four vessels of 60 to 100 tons burden unloading at once and sixty wagons loaded in a day for the upper landing at Chippawa Creek. Each wagon was drawn by two yoke of oxen and carried from 2,000 to 3,000 pounds ... Their goods they transfer to a batteau which carries them to Fort Erie' (Crysler 1943).

These onerous conditions with loading and unloading at each end required some form of remedial action. A canal across the Niagara Peninsula had been advocated in 1699 by Sebastien Vauban, the French military engineer (Taylor 1950). Based on a study of maps sent to him by the governor of New France, he wrote that 'Niagara Falls ... is tremendously high, but there is nothing there which cannot be corrected by men, and a canal 8 or 10 leagues [about 24 to 30 miles, 38.6 to 48.3 km] long with locks will remove the difficulties and will be able to communicate Lake Frontenac [Ontario] with Lake Erie for ships of from 60 to 200 tons [54.4 to 181.4 tonnes].' A second French engineer, De La Mothe, expressed a comparable idea in 1710.

Nearly a century later Robert Hamilton of Queenston proposed the construction of a 'Tract Road for dragging boats from the Welland Creek [now Welland River] to Fort Erie.' In 1799 a petition signed by Robert Hamilton and Thomas Clark of Queenston, and George Forsyth of Niagara, prayed 'for the authority of Parliament ... to construct a canal with locks to facilitate the passage of boats' between the rapids at Fort Erie and Queenston. This, later that year, became a 'Bill To Improve and Amend the Communications Between the Lakes Erie and Ontario, by Land And Water' for the levying of tolls on the merchandise, peltries, and stores passing through the canal in order to pay for its construction, but this possibility foundered on the rock of petitions from nearby townships that the scheme was 'monopolous and oppressive' (Cruikshank 1932). The Assembly dropped the question in 1800, and the project was not then taken up until it was promoted by William Hamilton Merritt in 1818.

The political scene was that, at the end of the American Revolution, the Niagara Peninsula and the east bank of the Niagara River, previously a continuous part of inland British North America, had been divided into two distinct national units. The new nation of the United States lay on the east bank of the river and the British province of Quebec (later Upper Canada, now Ontario) occupied its western bank. The region upstream from The Falls was not connected with the St Lawrence for trade by water, yet the area by the 1820s was being steadily populated and its trade inbound and outbound was increasing. (Background statements of the pioneer circumstances include Craig 1963; Glazebrook 1968; Guillet 1933; Howison 1821; Johnson 1975; Koene 1984; Stelter and Artibise 1977; and Wood 1975.)

To encourage population growth, trade, and economic activity, improvements to the lake and river systems of navigation were imperative. The Niagara Portage had served its day, and any bypass had to be on Canadian soil oriented towards Montreal and the St Lawrence if the Niagara area was not to become subservient to the American commercial empire centred on New York City.

The British lands of the Niagara Peninsula had been invaded across the Niagara River by American forces during the War of 1812. Though ultimately repelled, and with peace reinstated along the boundary, this invasion nevertheless clearly demonstrated that the Canadian side of the Niagara River was indefensible should renewed aggression occur. This statement of fact has enormous implications of direct concern for this narrative, for it helps to explain the inland location of the Welland Canal rather than an alignment around The Falls, as had been suggested and as was deemed to be feasible on the basis of land surveys. The selection of an inland route for the Welland Canal did, however, also involve important considerations of delicate political and military criteria with respect to the intentions of the American neighbour across the river (Strachan 1820).

The most suitable outlet on Lake Ontario for a canal was certainly not Twelve Mile Creek, as followed by the Welland Canal, but rather the Lower Niagara River. Despite the currents in the river and shoals offshore, the mouth and entrance to the river at Niagara provided the largest and deepest harbour of all on Lake Ontario. Even so, it was also indefensible in time of war, as both the river and the town lay within firing range of American guns at Fort Niagara on the opposite bank. After the War of 1812, a canal route to this vulnerable location could not be accepted by the British military and political leaders. The alternative of an inland location was necessary to satisfy the military arguments against the use of the Niagara River.

When Gourlay in 1822 presented his diagram of 'Practicable Water Communications between Lakes Ontario and Erie,' he argued that 'sooner or later, Lakes Erie and Ontario will be connected by a canal ... The plan should be such as to admit schooners of 100 tons [90.7 tonnes] burden and steamboats of 500 [453.6 tonnes]; – vessels sufficient to carry on the whole traffic, without unloading, from Quebec to the remotest shores of Lakes Michigan and Superior.' The various routes laid out included a line to the Niagara River at Queenston. 'If eternal peace shall reign ... the mouth of the Niagara River is, and probably ever will be, the best harbour on the south side of Lake Ontario. [However,] should any apprehension of war continue,' inland routes including from the Welland River to Twelve Mile Creek past St Catharines were suggested.

Apart from the use of the existing water system and its prospects for improvement, the transportation scene was that the Niagara Peninsula had few main roads, and those that existed were in a deplorable state. The long-distance carriage of passengers and freight had to be primarily by water, and the St Lawrence–Great Lakes water artery was the lifeline of Upper Canada. Its ocean outlet at Montreal was British, and its trade was to

England, mostly Liverpool. Montreal's hinterland extended across the Niagara Peninsula to the continental interior of North America, and as the whole of this vast extent of land was tributary to Montreal, British merchants such as Hamilton operated in the frontier centres along the Niagara River at Niagara, Queenston, Chippawa, and Fort Erie.

Suddenly, in 1825, this slender lifeline of British commerce inland from Montreal was threatened catastrophically by the opening of the Erie Canal. Construction of this waterway, though mooted since the turn of the century, did not commence until after the War of 1812. Its successful completion was necessary in American eyes to ensure that New York City, not Montreal, would capture all trade to and from the expanding West. If Lake Ontario was to be a British lake controlled from Montreal, then Lake Erie and the Upper Lakes had to be under American trading control by the merchants of New York.

The Canadian response was inevitable. The prospects of a competitive American canal rekindled pioneer enthusiasm for the prospects of a canal across the Niagara Peninsula. The story of the Welland Canal Company thereby becomes, 'in essence, the story of how Upper Canada met the challenge of the Erie Canal ... While the Erie Canal unified the western and eastern sections of New York State, it threatened to split the western and eastern sections of Upper Canada ... This was already the part of the province most American in character, and commercial dependence on an American transportation route was certain to have political consequences' (Aitken 1954).

As Aitken further remarks, 'The Erie Canal, finally completed in 1825, succeeded brilliantly in achieving what military force had failed to achieve: the destruction of Montreal's commercial empire around the lower lakes. No longer was western New York tributary to Montreal ... No longer ... [was] merchandise from Montreal sold on our borders for fifteen per cent below the New York prices. No longer did the steadily expanding exports of the Midwest have to find their way over the Niagara Portage, across Lake Ontario, and the rapids of the St. Lawrence. Montreal's commercial supremacy, based as it had been on a monopoly of the only natural water route (except the Mississippi) ... was at an end.'

Along with the physical necessity to bypass The Falls and the military imperative not to use the mouth of the Niagara River, there was also an immediate competitor and rival for trade that helped to spur the achievement of the Welland waterway. Upper Canada as British territory had necessarily to meet the economic and political challenges presented by the Erie Canal and American commercial assertions.

The Role of William Hamilton Merritt

Almost accidentally and unwittingly, an opportunity arose to counter-balance the nearby American presence and their trading pressures (FAOF no. 1), an event that closely involved a St Catharines mill-owner and businessman, Captain William Hamilton Merritt, who provided the necessary initiative and leadership qualities to raise funds and successfully guide the fledgling canal undertaking to fruition. (See National Archives of Canada, Merritt Papers; Aitken 1952, 1954; Carter 1923; Cruikshank 1925, 1932; Keefer 1920; McDougall 1923; Meaney 1980; Merritt 1875; St Catharines Historical Museum 1984; and Talman 1976.) As stated by a ship's captain, 'the canal was cut in consequence of the exertions of one Mr. Hamilton Merritt, a blushing, phrasing, plausible chief, that can gar folk believe that he can make stock-shafts bear plums' (Abbott 1844).

The times were propitious for this venture. When, in 1817 and 1818, committees in the townships of the Niagara District responded to questions posed by Robert Gourlay, Humberstone suggested that a canal could be cut from Lake Erie to Lyons Creek to connect with the Welland River and then the Niagara River. Similar concepts were identified by Fort Erie, Crowland, and Wainfleet. Fort Erie commented favourably on a canal projected at the entrance to the Niagara River to avoid its rapids. Grantham, under the chairmanship of William Hamilton Merritt, referred to the idea of diverting water from the Welland River into Twelve Mile Creek. Louth mentioned the possibility of a link to Lake Erie using Twenty Mile Creek. Grimsby considered that the upper waters of Forty Mile Creek might be linked to the Welland River and to the Grand River at Port Maitland. Gourlay mapped these possibilities, which were later published in his Statistical Account of 1822. Earlier in 1818, a lengthy and circuitous military canal had been proposed by James G. Chewett from the Grand River via Twenty Mile Creek to Burlington Bay.

Reality began to play a hand when Merritt's mill on Twelve Mile Creek at present-day Welland Vale (map 2) had insufficient water to grind the summer harvest of wheat. As the backbone of his business was now in severe jeopardy, in the now memorable words of his son's biography, 'A scarcity of water for his mill supply occurred, and was always uncertain in warm weather; so he believed a remedy for this could be found by having a communication with the Chippawa River' (Merritt 1875). (The Chippawa River or Creek is an earlier name and one still used to describe the Welland River.)

Merritt's mill had at first taken advantage of the perennial flow from

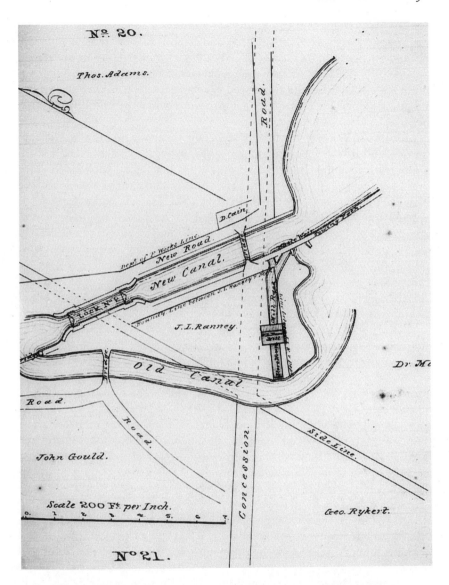

Map 2. *The site of Mr Merritt's mill on Twelve Mile Creek.* The First Canal
followed a river meander under the steep bluff of the valley side, whereas the Second
Canal crossed this meander loop. Water power diverted from the canal at a higher
level was carried to Merritt's former mill along a mill race, the height of the fall being
suggested by the need for a lock some 9 feet (2.7 m) deep on the Second Canal.

sand deposits south of the Niagara Escarpment via the eastern arms of Beaver Dams (now Beaverdams) Creek and from the Short Hills area in the
vicinity of Fonthill. But as settlement advanced and the forest cover was
denuded by clearance or agriculture, the water regime changed. The run-
off of water from the land increased, less water was held in the soil, and the
streams diminished in flow after a hot summer when their power was most
needed to grind the wheat harvest.

Merritt's solution to the lack of sufficient water was to draw water from
the Welland River through a mill race. This provision, described as 'a communication with the Chippawa,' brings in his substantial knowledge about
the lay of the land, its topography, and the drainage pattern. Essentially the
Niagara Escarpment intervened between Lake Ontario and Lake Erie. It
crossed the Niagara Peninsula from east to west and, through erosion
across its brow, was the physical cause of the impassable gorge and falls on
the Niagara River. South of the escarpment, drainage such as Beaver Dams
Creek, the Welland River, Lyons Creek, and the upper waters of Twenty
Mile Creek was in broad shallow valleys. As had been indicated in the presentations to Gourlay, the linkage of these east-flowing creeks with either
Lake Erie or Lake Ontario was an oft-discussed theme of the period.

To the north of the escarpment, the streams that drained north to Lake
Ontario in contrasting deep, steep-sided valleys either had a source below
its slope or crossed the escarpment by a series of waterfalls. As their lower
valleys across the plain next to Lake Ontario were oriented from north to
south, Merritt's concept was outwardly simple. As the Beaver Dams Creek
flowed east across DeCew Falls into the valley of Twelve Mile Creek, then
past his mill, why not cut a channel from this creek through the low intervening hills to the Welland River, the next major river to the south and one
with an abundant flow of water from its extensive upstream catchment
area? Nothing to it. The task was only the simple matter of digging a slender connecting trench across the intervening watershed – or so was the
thought of the day.

As the Merritt biography continues, the mill owner was 'determined to
make a rough survey of the ground ... He borrowed a water level ... [and
with neighbours] started on their tour of surveying.' The distance covered
was some 2 miles (3.2 km) south from the Beaver Dams branch of Twelve
Mile Creek. (Though these communities did not then exist, this was from
Allanburg to the Welland River at Port Robinson.) The height of the ridge
to be overcome was assessed at 30 feet (9.1 m). The reality of a more precise
measurement later virtually doubled Merritt's inaccurate assessment
(FAOF no. 2).

If the actual height difference to be dug through by hand labour using pick and shovel had been assessed more accurately, would the canal venture have been started? Probably not, because funds were scarce, and raising the increased finances for the additional works of construction would have posed an even more severe problem than already existed in the difficult financial climate of Upper Canada.

Much then happened, but as the political and economic aspects of the situation have been presented elsewhere (Aitken 1954; Heisler 1973; Upper Canada Assembly 1836–7), only certain aspects of an extremely complex situation will be presented here. Routes were surveyed to Burlington Bay, across the peninsula, and between the upper and lower lengths of the Niagara River. Niagara as the main town in the district wanted the terminal on Lake Ontario to be within their community. There were surveys and counter-surveys, arguments and counter-arguments, over the various possibilities that were deemed to exist to bypass The Falls. This debate culminated in a survey in 1823 by Hiram Tibbet, a civil engineer who had just completed a survey for a canal on the American side at Niagara Falls.

Starting 10 miles (16.1 km) upstream on the Welland River, Tibbet demonstrated that a cut was possible through the rolling topography north to the headwaters of Twelve Mile Creek, and that the steep slope of the escarpment could be offset by a railway incline. Encouraged by this report and the prospect that the costs could be undertaken by a private company without government aid, Merritt presented a petition for an act of incorporation to the House of Assembly. No work of comparable magnitude had previously been undertaken in the province.

Next year, in 1824, the formation of a joint stock company named the Welland Canal Company was approved. George Keefer was elected president and Merritt became its financial agent. This date was one year before the Erie Canal was to open in 1825. Indicative of Merritt's role is that, after Tibbet's survey, Merritt wrote 1000 letters, wholly in longhand, to every person of influence in the province. This persistent type of lobbying led Carter (1923) to describe the Welland as 'a one-man canal.'

The Welland Canal was named after the Welland Canal Company, and this from the intention to tap the Welland River as the source of water supply. This river, 82 miles (132 km) long, meandered east from its source in Ancaster Township north of Hamilton to the Niagara River at Chippawa. The name Welland Canal has survived to this day, but the Welland River has never been used to supply water to the canal as was intended. As the Welland River was crossed by an aqueduct, the name Welland Canal is a historical misnomer based on the expectation of an earlier but unfulfilled intention.

The Welland Canal Company in Action

By the act of incorporation passed in 1824, the Welland Canal Company was given powers to acquire land for purposes of the company, 'and required to make and complete the said canal, railway, towing paths, and other erections required for the navigation thereof from Lake Ontario to the River Welland within five years.' The canal was to have a depth of 4 feet (1.2 m). It would cross the high ground south of the Niagara Escarpment by means of a tunnel 9 feet (2.7 m) wide, which would involve about half the cost of an open cut and bring the canal down to a level of 4 feet (1.2 m) below that of the Welland River. To the north the canal would descend to Twelve Mile Creek below DeCew Falls by a double-track incline railway, and would then use the natural channel of Twelve Mile Creek to Port Dalhousie by canalizing this river with a series of locks. The charter also provided permission to extend the canal to Lake Erie and/or to the Grand River.

Before work commenced on these plans, the decision was made to widen the tunnel to 15 feet (4.6 m), increase the depth of water in the canal to 8 feet (2.4 m), and enlarge the locks. These changes had a twofold intent. The first was an improved ability to compete with the barge traffic expected on the Erie Canal. The Welland as a ship canal would have the ability to carry larger vessels and heavier cargoes through locks that would be 110 feet (33.5 m) long and 22 feet (6.7 m) wide, and thus avoid the trans-shipment of cargo that was required at the Buffalo terminus of the barge canal. The second factor was that the colonial authorities now wanted a canal that could carry gunboats between the two lakes, and not, as during the War of 1812, be limited for naval action to the vessels available on each lake. It was also thought that increasing the dimensions for the Welland would make it more profitable.

Contracts were let, and Merritt as agent for the company made an address at a ceremony in 1825 for turning the first sod at a location near present-day Allanburg. 'We are assembled here this day for the purpose of removing the first earth from a canal which will, with the least, and by the shortest distance, connect the greatest extent of inland waters, in the whole world' (Merritt 1875).

Three lengths of canal were envisaged, two canal cuts and an intervening length of canalized river. The river, the middle section of the Welland River

(Chippawa Creek), would combine the dual functions of feeder and summit level to the northern length of canal. It might here be noted that the Welland River was navigable downstream to Chippawa, where it connected with the Niagara River and joined the former Niagara Portage upstream along the left bank for traffic to Lake Erie, and that the Welland being broad, slow-flowing, and without falls was also navigable upstream and along its tributaries for a considerable inland distance. The Welland River, an important route of early travel, had been used, mapped, and named the Chenondac by the French at Fort Niagara.

The southern section of canal was to be a 12-mile (19.3-km) cut from the Grand River through the Wainfleet Marsh to the Welland River. To use the Welland and the Grand was to link the major and east–west waterways of the peninsula with the intervening north–south alignment of the canal. The intentions were on a grand scale. Regionally, they combined the envisaged canal with the potential that existed through the natural drainage system by incorporating the Niagara, Welland, and Grand Rivers into one united navigation system. In the North American context there was the importance of improving access to the Atlantic seaboard through improvements downstream along the St Lawrence River. Even though Merritt's majestic vision of a canal across the peninsula was not to be achieved in the form he anticipated, the Welland and the Grand were nevertheless to play a substantial role in the Welland Canal as achieved. Change was to be to matters of internal detail, not to the concept itself.

The idea of an incline railway at DeCew was dropped in favour of a series of wooden locks to cross the escarpment and, as the valley to the north was now considered too narrow to accommodate the larger size of vessel and the locks that were required, alternative routes were explored to cross the crest and link with Twelve Mile Creek. The alignment selected lay further east, and became what is now the Thorold-Merritton section of the First and Second Canal systems. This length required a series of locks and a man-made channel with storage reservoirs to cross the escarpment, then a canal cut with more locks and reservoirs as the land sloped north from the base of the escarpment. As described in 1825 by the engineer for the Welland Canal Company, 'Here the descent of the mountain commences, which is very easy and gradual ... Having descended the mountain the line is very straight with a gentle declivity, to the head of a ravine, which has very much the appearance of a wide Canal; along this ravine [Dick's Creek] the Canal is conducted with but little expense, except the necessary Locks and waste weirs, and clearing away the timber and constructing a towing path' (Upper Canada Assembly 1836–7).

Thus was introduced the canal along Dick's Creek next to the village of St Catharines, where it joined the originally intended route of Twelve Mile Creek past Merritt's mill. The canal, now in the form of a canalized river, then followed the drowned river valley of meandering Twelve Mile Creek for some 2 miles (3.2 km) through a ponded area next to the lake and into Lake Ontario across the sandbar that here formed the shoreline. A description that introduces Port Dalhousie reads: 'the creek runs through an opening in the beach continually ranging in width, and at times entirely shut, overflowing the meadows upstream, until the waters rise high enough to run over and cut away the bar of sand. At westernmost cape is the most favourable situation for a Lock' (Upper Canada Assembly, 1836–7).

Contracts were let in 1826. Oliver Phelps was given the contract for the thirty-four locks to be built of white oak and white pine between the summit and Lake Ontario. The contracts for the lock and harbour at Port Dalhousie were let out separately, but this work was unsatisfactory and had to be re-tendered.

Construction on the canal began at both ends of the tunnel intended to cut through the series of low hills that lay between Allanburg and the Welland River. Timber for shoring up the roof and the walls of the tunnel was delivered to the site, but the task was not easy. The excavation was into stiff clay with streaks of hard pan, large stones and rock had to be blasted, and the haul of spoil up the banks by horses and oxen required several yokes for each team. The work had to be done in the dry, as otherwise the banks were too slippery. Oxen died, the horses were hit by glanders, a contagious disease, ague hit the men, and many contractors unable to complete their work were replaced.

Now arose FAOF no. 3. While digging on the tunnel project, and after excavating the Deep Cut to some two-thirds of the required depth, a workman struck unsuspected underground water, which gushed out with such sudden and violent force that one labourer was carried to his death. As this flow could not be quenched, rather than pursue the original idea of a tunnel, the supply of water would be obtained from a channel cut down through the ridge to this depth. This required an excavation about 2 miles (3.2 km) long to a depth of from 30 to nearly 60 feet (9.1 to 18.2 m), a change in plan that would permit sailing vessels to pass between the two lakes, an omission from the earlier intentions.

FAOF no. 4 intervened next. In 1828, within a few days of the expected date for completion of the cut, the high banks collapsed into the deep trench that had already been dug. The excavated spoil had been deposited on the adjacent banks. It added to the weight that had to be supported, and the unstable mass was subject to movement during and after periods of rain, snow, and ice. The ridge, supposedly composed of clay throughout its depth, lay on very insecure foundations of sand. When this sand was exposed, the overweighted banks simply collapsed into the excavated area. The conundrum to be resolved was that the cut had to be shallower in order to leave the sand undisturbed, yet the depth of the cut had to be increased in order to secure the waters of the Welland River for the canal.

The Feeder Canal

The solution adopted was to return to an earlier concept of the route proposed in 1824 to connect the Grand River with the Welland River. Revived by James Geddes, this involved 'supplying the Canal with water from the Grand River instead of the Chippawa [Welland River] ... To raise the Grand River by a dam ... From this pond a supply from the canal to Lake Ontario is to be drawn' (Upper Canada Assembly 1836–7).

The new need was to provide an alternative source of water that was at a higher level so that the level of the canal through the Deep Cut could be raised by about 16 feet (4.9 m). Lake Erie could not be used because this would have meant an even deeper cut; the rock barrier of the Onondaga Escarpment also intervened. Nor was there a convenient mountain to hand from which to obtain water for a canal reservoir. The nearby high land of the Fonthill Kame did not generate sufficient water for canal purposes.

The canal company faced severe financial difficulties. It required a usable canal to provide revenue and to satisfy its shareholders, and this link also provided the opportunity for both extending the Welland system and linking it with the Grand River as a continuous through route of navigation. Engineers established that it was practical to obtain water from the Grand River, and in 1826 the Welland Canal Company was granted by the administration of Upper Canada 'all those lots or parcels of land ... situated, and being what is commonly called the Great Marsh in the Township of Wainfleet.'

Soon referred to as the Feeder Canal, the selected route was the cheapest and it offered the further advantage of draining some 30,000 acres (12,141 ha) of marshland. It involved almost 21 miles (33.8 km) of channel, 40 feet (12.2 m) wide and 5 feet (1.5 m) deep, sufficient for the supply of water and able to carry small boats but not of sloop and schooner size. An aqueduct would be constructed where the Feeder Canal crossed the Welland River and its slender valley (Paul 1983).

To raise the level of water in the Grand River to the required height, a dam was proposed close to its mouth at Port Maitland, but FAOF no. 5 now intervened. The commander of this naval base objected strongly, as a dam would restrict the use of the lower river as a naval base, a very signifi- cant consideration as the War of 1812 had finished only some ten years before and a British fleet on Lake Erie was deemed an imperative aspect of British defence policy.

As a compromise, and after the examination of various alternatives, the proposed location of the dam was moved upstream on the Grand River to above the Bear's Foot Rapids where this river crossed the Onondaga Escarpment, some 4.5 miles (7.2 km) inland from the mouth of the river. The commander had at first requested a distance of 18 miles (14 km) upstream. This dam extended the length of the Feeder Canal to 27 miles (43.5 km), and it provided yet another variant for the impact of the canal on the pattern of settlement, through transferring the potential of urban advance from Port Maitland upstream to this new location, later Dunnville.

The dam itself was 18 feet (5.5 m) wide at the base, 594 feet (181.1 m) long, and 7 feet (2.1 m) high. Construction was undertaken from both ends at the same time by laying trees in the river's bed, then adding cribs that were filled with stone to weight the substructure down. This barrier lasted for but a few years and had to be reconstructed, the first of many occasions, but the dam and its sluice gates raised the upstream water level by about 5 feet (1.5 m), which created a severe upstream flooding problem. The structure later car- ried a public road with bridges across the weirs at each end of the dam. It involved upstream embankments and a retaining wall to provide a sufficient volume of water to cover the sand at the Deep Cut, but the dam did not include a lock for the through passage of vessels on the Grand River.

The dam was criticized soon after its construction when examined by a commissioner. 'The erection and maintenance of this great dam across the Ouse [Grand] is unauthorised by any legislative act of this Province. The lands of individuals situated on the banks of the river, for a distance of about ten miles, are overflowed without the consent of the owners, and without recompense having been afforded them ...; fish are prevented from ascending the river, and neither ark gap, lock nor apron has been constructed or main- tained of a sufficient width and depth to admit boats, arks and rafts' (Randal 1831). Goods had either to be passed over the dam, or from upstream vessels were obliged to use the Feeder Canal to bypass the dam.

An important aspect of the situation was Merritt's conviction as MP for Haldimand that the dam would be of great advantage to the whole area. Instead of having a river that could be navigated only at certain periods of the year, if a series of short canal bypasses were constructed around the natural obstacles of the river, then the Grand River settlements, through the improved means of transport, would enjoy access to substantial new markets. No longer would goods have to be hauled expensively by wagon to Dundas to trade abroad. When in 1831 the British parliament granted a charter to the Grand River Navigation Company to construct such locks, the Welland Canal through the dam and its Feeder Canal now tapped into the vast new resources of the Grand River valley and its extensive hinterland area. The navigation company became an adjunct to the Welland Canal Company.

The Grand River valley also housed the Six Nations Indian Reserve. As a result of their loyalty to the British Crown, the Indians had been forced to leave their original homelands in the Mohawk valley of New York State and emigrate to what is now Canada after the American War of Independence (Hill 1971; Nelson and O'Neill 1989). They were granted a 6-mile (9.7-km) strip of land on both sides of the Grand River from source to mouth, some 768,000 acres (310,809.6 ha). This extent became attenuated through Indian sales to white settlers. The dam and its impact adversely affected these lands.

The Grand River Navigation Company was also unique in the history of Upper Canada in that it was the only major canal project to be financed by Indian funds. The sad side of this statement is that the Indians were not willing partners. It was an investment made without their knowledge or consent, and the Six Nations had no members on the board of directors. Money was taken from the Six Nations Trust fund on at least four occasions and transferred to the Welland Canal Company, an incident that is still under investigation by the Six Nations (Monture 1994).

The statute of 1824 that incorporated the Welland Canal Company provided that compensation would be paid to the Indians whose land was damaged by construction of the canal. Against the protest of the Six Nations, 2415.6 acres (977.6 ha) of land were flooded, but the Six Nations were not compensated for this loss. In 1895, arbitrators appointed by the Dominion government ruled that the Dominion of Canada was liable for claims that the Six Nations had against the Welland Canal Company. In 1988 the Six Nations Council formally filed a claim for compensation. In 1994 this claim was accepted as valid, and negotiations to achieve a settlement have been ongoing since then (Monture 1994).

To the east of the dam, the cutting for the Feeder Canal ran parallel to the Grand River to Broad Creek (now Stromness). A long straight cut 5 feet (1.5 m) deep and with a bottom width of 20 feet (6.1 m) then extended across the Wainfleet Marsh to the aqueduct across the Welland River. The marsh area had been granted to the canal company on the presumption that drainage through the canal would help to pay the cost of the canal works.

The completion of the dam and the Feeder Canal were delayed as the dam across the Grand River slumped and had to be repaired. Serious outbreaks of disease because of the insalubrious conditions throughout the marsh provided a second impediment. The water was not fit for drinking, and the thirst of the diggers was quenched by 'water-boys' who kept the men supplied with whiskey served from tin pails. Despite such precautions, 'Along the line of the works on the feeder the fever and ague was raging; strong men were wasted to skeletons, and the general feeling of despondency and discontent which all these vicissitudes bring in their train was felt in the ranks of the workmen who were there employed ... The season had been a dry one, and consequently the miasma from the stirred up earth was more severe in this section than usual; so severe had it been that the work was delayed in consequence' (Merritt 1875).

Despite such vicissitudes, the amazing fact is that the series of changes to the canal after the collapse at the Deep Cut were finished in less than a year and, although the canal works were not everywhere complete, water was let into the Feeder Canal in September 1829. A small vessel with Merritt on board then passed through the completed works from Dunnville to Port Robinson. Known locally with disrespect as the 'Big Ditch' or 'Merritt's Folly,' it is now referred to as the First Canal, the first of four to cross the Niagara landscape.

The Welland River was crossed by an aqueduct, 365 feet (111.3 m) long and 24 feet (7.3 m) wide. As described by Randal (1831), 'This is an excellent piece of workmanship and a monument of the superior skill and abilities of Mr. Marshall Lewis, the builder and contractor.' Space under the aqueduct did not impede the flow of boats and rafts on the Welland River below the arches. The two systems of navigation crossed without interfering with the flow of the other.

The Feeder Canal then ran parallel to the river (now Merritt Island) and to the southern end of the Deep Cut where, because of the now higher level, four additional locks were required. Two of these locks took the canal down into the Welland River at Port Robinson, and two lowered the now higher canal at Allanburg to the level at which the Niagara Escarpment was crossed.

The Completed Canal

A short but significant feature of the First Canal was that at Chippawa a narrow gravel peninsula projected north at the point where the Welland River entered the Niagara River. This effectively turned the Welland River into the Niagara River, but also posed the danger that any vessel leaving or entering the canal would be caught by the current and swept precipitously over The Falls to certain destruction. The task of navigation was made more difficult in that, to avoid this peninsula, vessels had to swing around the promontory into the river's strong current.

The Chippawa Cut, 8 feet (2.4 m) deep, constructed in 1829 across this spur of land (see map 7, page 105), permitted vessels to enter or leave the Niagara River safely and then proceed upstream along either the Niagara or the Welland Rivers without jeopardy. As reported by the engineer to the canal company in 1829, 'the cut to intersect the Niagara and Chippawa Rivers proved a far more formidable work than we anticipated, the earth very hard and difficult to excavate, particularly under the bed of the two rivers' (Carter 1923). Upstream along the Niagara River, capstans were placed to haul vessels around points where the current was strong, but the ascent had frequently to be delayed until there was a wind from the north.

The climactic event to all the canal construction occurred in 1829, when the first two vessels, again with Merritt on board, passed upstream through the canal system. Both suitably dressed with bunting and pennants, the schooners *Annie & Jane* of York (in the lead) and *R.H. Boughton* of Youngstown left Port Dalhousie on 27 November and arrived in Buffalo six days later, on 2 December. 'The banks of the canal were crowded with people, and the enthusiasm displayed on the occasion testified that those who witnessed the display were now fully satisfied as to the prospects of the great work which had so long occupied their attention' (Merritt 1875).

The record refers to an easy ascent through the locks of the escarpment but then to several difficulties that arose: breaking through ice that was in places 3 inches (7.6 cm) deep; a delay of some hours at the first lock of the Deep Cut because a chip had become lodged under the gate; then the passage through the Deep Cut, slowed by timber and ice; and, on the Welland River, the grounding of the boats on a sandbar, where they remained all day. Even so, the ever-optimistic Merritt noted in a letter from Chippawa to his wife: 'I have tested to my satisfaction that a vessel will pass on [through] the canal in twenty-four hours' (Merritt 1875).

The eastern terminal of the canal on the Niagara River had been reached. Passing through the canal cut, then 'stemming the broad and strong current of the Niagara, the vessels reached Black Rock then Buffalo,' where they were met with bursts of applause and honoured by a discharge of artillery from the Terrace. As stated in Merritt's biography, 'The Welland Canal was now an accomplished fact. The artificial wedding of the Great Lakes of the west and north, with the waters of the Ontario, and eventually with the St. Lawrence and the ocean, was complete.'

Delayed by settlement of the dam at Dunnville, the canal as an operating commercial route opened in mid-1830. Its first vessel was the schooner *Erie*, bound from Cleveland to Youngstown, New York. Rafts of staves were soon brought down the Feeder Canal from the Grand River, and lumber was brought from the same source to Buffalo and Lockport.

The schedule of toll rates 'will enable merchants in Ohio and other parts of the United States to make their calculations, and determine whether it will be to their interest to send through this channel such of their produce as might be destined for New York [via the Welland and the Oswego Canals to the Erie Canal and the Hudson River]. As to produce of every kind intended to be sent to Montreal [via Lake Ontario], there can be no doubt that it will pass through the Welland Canal.'

The savings in the expense and troubles of transportation over using the pre-existing Niagara Portage were calculated to be half or more of the previous costs. 'Vessels will convey produce from any part of Lake Erie to Prescott for less than the price heretofore charged for the two Lake Freights. As it is a mere continuation of the voyage, the grower will consequently save the charge heretofore made for storage at Fort Erie, Chippewa [sic] and Queenston, together with the boating and land carriage between these places, which hitherto caused their property to be removed six times; whereas, by this conveyance it will not require one trans-shipment.' The change from portage to canal was thus succinctly stated. Cargo would be handled on fewer occasions, and a longer distance could be achieved in a shorter period of time at a lower cost.

The First Canal had thirty-nine locks. Those 'below St. Catharines are 32 feet [9.8 m], and 130 feet [39.6 m] long ... all the others were originally 22 feet [6.7 m], and 110 feet [33.5 m]; but some of them have settled inwards so that they do not now exceed 20 feet 6 inches [6.2 m] in width' (Phillpotts 1842). Here was a foretaste of serious problems, but as noted by the same author, 'By constant watching and superintendence this canal has been kept in such a state of repair that no accident of any great consequence has lately occurred.'

The government of Upper Canada had purchased 42.9 per cent of the stock to bring about the canal. This was followed by individual contributions: 27.8 per cent from New York, 12 per cent from England, 10 per cent from Upper Canada, and 5.5 per cent from individuals in Lower Canada. 'Loans provided from 1826 to 1831 by Upper Canada were at twice the

level of those obtained from the British Government, and together represented a further 62 per cent over the funds that had been raised from the
sale of stock' (Aitken 1954).

American involvement played an important role. Contractors, engineers,
and skilled labour came from the United States, and Merritt was granted
remission from import duties on tools, wagons, and other equipment purchased in the United States, claiming correctly that this equipment was not
available in Upper Canada. There was reliance on American capital, later to
be reflected in naming Yates Street on the bank above the canal in St
Catharines after J.B. Yates, an American financier and a major shareholder
in the First Canal.

The funding for the canal was due largely to Merritt's enthusiastic
energy, but also noteworthy were the perseverance and personal influence
of the leading directors, support in the legislative assembly and newspapers, and the encouragement and patronage of the lieutenant-governor and
the governor-general. In addition, American assistance came by way of
Merritt's father-in-law, Dr Prendergast, who had lived close to St Catharines for a while before the War of 1812. He later resided in Mayville,
New York, where he became an influential member of the New York State
Legislature and introduced the canal concept to American capitalists.

The canal achievement was a combined operation that involved engineering, political, and financial inputs, plus the abilities of surveyors and
work gangs working under Merritt's guiding leadership to offset the physical barrier of the Niagara Escarpment that intervened between Lake
Ontario and Lake Erie. Incidental to the formulation of the canal proposals, a new city-forming element had been introduced across Niagara's landscape.

From north to south upstream, the canal left Lake Ontario at Port Dalhousie, where it followed the lower valley of Twelve Mile Creek to St
Catharines. It then diverted east towards the base of the Niagara Escarpment following the natural valley of Dick's Creek. The channel consisted
then of a cut with locks to the base of the escarpment, a closely spaced
flight of single locks with intervening ponds across its face to Thorold, then
a further excavation through the Deep Cut to reach the Lower Welland
River at Port Robinson. The passage was then downstream to Chippawa at
the head of navigation for the Upper Niagara River, through the Chippawa
Cut, and upstream against the flow of the river's current to Fort Erie, Buffalo, and Lake Erie.

Behind the dam at Dunnville lay the head pond for the whole canal.
When this dam leaked, as in 1831, the Feeder Canal was lowered in depth

by at least 18 inches (0.5 m), with disturbing consequences for navigation and water power to mills. As water might also be lost at the aqueduct, the canal was not perfect, but it had displaced the Niagara River and its portage on the west bank as the major route between Lake Ontario and Lake Erie. The estimated 300 c.f.s. (8.5 m³) that flowed through the canal, less the small proportion of this flow that was returned to the Lower Welland River at Welland and Port Robinson, was also diverted from The Falls on the Niagara River. This fact was not then an issue, but with other diversions from the Upper Lakes it later became a topic of serious concern to both nations.

William Hamilton Merritt in 1857 described the Welland Canal Company as Canada's first water-power company. The act of incorporation in 1824 stated that mill sites were to be surveyed for 'as many mills, manufactories, warehouses and other erections' as were required by the company for its purchase and use. The route selected for the canal maximized its hydraulic advantages and, after the canal company authorized the sale of its hydraulic privileges, in 1830 a second water-power company was formed. Purchased by John B. Yates and becoming the St Catharines Water Power Company, its purpose was to initiate industry through a cut known as the Hydraulic Raceway 2.5 miles (4 km) long from Merritton to St Catharines.

In 1830, the first year of active canal navigation, six mills existed, four were under construction, and fifteen applications to build mills had been made. By 1835, two years after the First Canal had opened to Port Colborne, some thirty-one mills existed. By the 1850s and 1860s over forty mills, as well as twelve or more driven by water subleased from the St Catharines Water Power Company, were listed in the public records. By 1882, the fifty water-powered plants along the canal and its feeder served twenty-one different industries. They included flour and grist mills (18), saw mills (8) and paper or pulp mills (4), and two each of machine shops and foundries, knife works, and cotton mills, plus other varied types of mills (Leung 1986). The First and Second Canals were industrial waterways par excellence, a trait that continued after their closure for navigation. The first canal constructed solely for purposes of navigation was the Third Canal.

In 1830 the canal company announced the sale of village lots at Allanburg, Port Robinson, and Marshville (Wainfleet), where the purchasers were required to erect a frame building within a year. Toll gates were erected at several places along the towpath and a rate was charged for the passage of vehicles and animals. Storehouses were erected at the dam on the Grand River, at Port Robinson for shipping freight to lake vessels, and at

Port Dalhousie on the west pier. To encourage the shipment of freight through the canal, four canal boats together with their horses and harness were contracted for the 1831 season. Each boat was 95 feet (29 m) long and 14 feet (4.3 m) beam, and a request was made to remove all restrictions on passing American produce through the canal that would be shipped to an American port. Two companies also chartered the steamer *Peacock* to run between Port Robinson and Buffalo, where it connected with their packet boats.

2

The First Canal Reaches Lake Erie, and a Second Canal Is Constructed

The First Canal was soon extended to Lake Erie. This new length opened in 1833, but the inadequacies of construction and the costs of improvement soon led to several important changes, the first of many to the canal. In 1837, the government authorized the purchase of the Welland Canal Company shares from its stockholders and, after the Union of Upper and Lower Canada in 1841, the Second Canal was constructed by the Province of Canada along broadly the same route as the first venture. This was essentially an enlarged and improved version of its predecessor, though stone locks replaced the now decaying timber locks of the First Canal and the new alignment differed slightly from that of the original route.

The Welland Canal Is Extended

The First Canal connected Lake Ontario and Lake Erie, but the route was circuitous via Twelve Mile Creek, the Welland River, and the Niagara River, where vessels faced the problems of its swift current, the rapids at the entrance, and the closure of this entrance by late-spring ice for several weeks after the start of lake navigation. The potential of trade to the developing American ports around Lake Erie and particularly to Buffalo had also to be taken into account, for this American city was expanding rapidly as a commercial port and trans-shipment centre at the head of the Erie Canal for barges and at the head of Lake Erie for sailing vessels.

Several alternatives for a more direct and safe route to Lake Erie were examined. These options included a series of different canal cuts from the Feeder Canal, ranging from short cuts in the vicinity of Port Maitland to longer canal cuts to different bays on Lake Erie as the southern length of the canal was moved to the east. Cuts to the west meant a longer journey by canal and lower costs for a canal cut, and those to the east the opposite of higher costs for the cut but a shorter distance for canal navigation. The aqueduct at Welland complicated this assessment. Should it remain or be replaced? A route due south to the lake from the aqueduct would require a longer cut than further east, whereas a longer cut south from Port Robinson would mean a shorter total length for the canal but would also require a costly new aqueduct.

When the issue was examined by a select committee of the House of Assembly, the Boulton Committee argued, 'completion of the work is, in the opinion of your committee, a matter no longer to be doubted ... Your committee is decidedly opposed ... to confine the [Welland Canal] company to the one channel by way of Niagara River ... The amount of expenditure within any reasonable bounds is not as important as that the most advantageous harbour should be selected' (Upper Canada Assembly 1836–7).

This paramount requirement to obtain the best potential harbour along a shoreline with numerous limestone headlands and intervening sandy bays introduces Gravelly Bay (now Port Colborne) to the Niagara scene. This locale was proved to be the most suitable, but Port Maitland at the mouth of the Grand River was also promoted because of its better access from the lake and the roomy harbour that already existed at the mouth of the river. However, Gravelly Bay had a sandy bottom that would hold ships' anchors, and the bay was protected from the prevailing westerly winds along Lake Erie by Sugar Loaf Point. As Robert Randall in 1831 reported to the lieutenant-governor, 'Gravelly Bay appears to me to possess very superior advantages for a harbour at which to terminate the ship canal; it is the best and deepest and also the nearest to the aqueduct where the feeder joins the canal' (Upper Canada Assembly 1836–7). One additional lock would be required to bring the water level in the canal down to the fluctuating level of Lake Erie.

The low Onondaga Escarpment had to be crossed by a canal route south from the aqueduct and also four distinct marshes: the Black Ash Swamp, the Cranberry or Wainfleet Marsh, the Hemlock Marsh, and another marsh between the junction with the Feeder Canal and the Onondaga Escarpment. What became Ramey's Bend on the canal can be explained through the canal taking advantage of a small creek that flowed west through these marshes. Less excavation would be required if the canal followed this channel.

Gravelly Bay, retention of the aqueduct, and a direct route to the south from the feeder junction were selected. Financial considerations were severe, excavation through the hard limestone of the Onondaga Escarp-

ment caused delays, and, as the marshy areas were unhealthy, cholera prevailed in 1832 to deplete the number of workers employed on the project.
Water also flowed into the works from underground springs; as contractors were unable to fulfil their commitments, the canal company had to re-
let the contracts and in some instances undertake the work on a day basis.
The lakeshore lock at what is now Port Colborne was situated at the end of
present-day Sugar Loaf Street, where a bridge crossed the canal. A small
harbour was provided west of the entrance where a protective breakwater
was constructed into the lake. A lighthouse and a second pier followed.

Through the accumulation of sand, the lock had frequently to be cleared
and dredged to maintain its depth. Merritt described its enlargement in
1835 from 110 feet (33.5 m) to 125 feet (38.1 m) to admit steamboats as 'one
of the most troublesome of any job on the whole canal' (Upper Canada
Assembly 1836–7). At the same time, the width of the harbour to the north
was also increased, thus confirming this location as the centre of business
activity. A stone grist mill and an associated sawmill were added on the
west side of the entrance to the canal, with a dock along its lake side where
grain from the lake was unloaded, ground into flour, pressed into barrels,
then shipped through the canal to trade on Lake Ontario or the St
Lawrence (Carter 1923).

In 1832 William Hamilton Merritt was nominated to parliament for
Haldimand County on the advice of friends that he would then be in a better position to advance the prospects of the canal and the Grand River
Navigation Company. The new length of canal to Lake Erie, then nearing
completion, opened in 1833. With the Grand River remaining the source of
supply, the flow of water along the now extended canal was south from
The Junction, where the Feeder Canal joined the main line of the canal
(Helmsport) and north through the Deep Cut across the Niagara Escarpment to Lake Ontario.

A through line of direct canal navigation had now been achieved
between Port Dalhousie on Lake Ontario and Port Colborne on Lake Erie
and, as Montreal and New York now each had direct access to the upper
Great Lakes, via the Welland Canal and the Erie Canal respectively, the
Niagara Peninsula had changed in its regional context to an area of intense
commercial rivalry between these two seaboard cities. Like York (now
Toronto) as the provincial capital, the canal zone now lay within the over-
lapping and competing hinterlands of both cities, a fact that was greatly to
flavour future development (Spelt 1973).

New York had the advantages of a year-round harbour, whereas Montreal was icebound throughout a long winter. For a continuous journey by

water to either Atlantic port, however, two further canals must be taken into account. On the American side, and wholly within New York State, the Oswego Canal was completed in 1828 to connect Lake Ontario with the Erie Canal at Syracuse. If used in conjunction with the Welland Canal and a passage free of locks along the south shore of Lake Ontario, this route provided a favoured alternative to the western section of the Erie Canal.

The Canadian side was less prepared to meet the future. Politically, the two competing provinces of Upper and Lower Canada (Ontario and Quebec) shared the outlet to the sea. Physically, where the rapids on the St Lawrence River presented their hazards to navigation, the military canals were increased to a depth 3.5 feet (1.1 m), but, as they could not take vessels with a beam larger than 12 feet (3.7 m), they were unsuitable for passage by either the larger vessels on Lake Ontario or those that passed through the Welland Canal. Further, because of the many shoals and rock obstacles to navigation in the St Lawrence, many vessels either had to be towed through their 6-mile (9.7-km) length or were obliged to unload their cargo for transportation by wagon. Improvement had to await the union of the two Canadas in 1841, a political process that also underlay the achievement of the Second Welland Canal.

The Second Welland Canal

The director's report for 1837 showed that the annual costs of administration, repairs, and improvements on the Welland were three times higher than the income from tolls and rents. The average loss was over £14,000 a year, and enlargement was also required to accommodate the larger ships that sailed on the Upper Lakes. It was increasingly evident that a private company could no longer finance the canal undertaking.

The company was in debt, the locks were deteriorating, and trade through the canal was hindered and sometimes stopped because of its dilapidated state. In 1834 the House of Assembly allowed the Welland Canal Company to increase its stock and this cost was covered by the government of Upper Canada. However, the actual takeover from the company was delayed through litigation against Merritt and his co-directors by William Lyon Mackenzie, and was followed by Mackenzie's unsuccessful but nevertheless challenging Rebellion of 1837.

The government took over effective control of the canal by appointing three of its directors, and the purchase of the private stock of its shareholders by public debentures was completed in 1841. The days of the initiating Welland Canal Company were over, and when the Act of Union of 1840 came into operation in 1841, the canal was placed under the Board of Works of the new Province of Canada. A reflection on these times is that Hamilton Killaly, the first chairman of this new board, is commemorated at Port Colborne by an east–west street centred on the Second Canal that is named after him.

During 1840, before work on an improved waterway could commence, a series of surveys, proposals, and counter-proposals were undertaken. In 1838 engineers N.H. Baird and H.H. Killaly had reported that a route from the Welland River to Queenston 'presented so many natural difficulties that ... we deemed it a waste of time to pursue it any further.' An alternative route from Thorold to the mouth of the Niagara River presented 'no difficulties and has the advantage of being 5¼ miles shorter,' but the harbour was deemed less acceptable than at Port Dalhousie because of its currents, storm conditions, and the late spring flow of ice on the Niagara River from Lake Erie. And the adverse frontier location of the canal was also again stressed: 'We wish particularly to record our decided objection to the selecting of a Harbor so perfectly under the control of our neighbours, as Niagara Harbor undoubtedly is, for the termination of the Welland Canal.'

The outcome of these deliberations was to 'recommend, without hesitation, that Port Dalhousie be adopted as the most suitable place for the construction of a good, commodious and extensive Harbor.' South from Port Dalhousie, where the channel of the First Canal had the disadvantage of following the meander channels of Twelve Mile Creek and Dick's Creek, the surveyors indicated various possibilities for new routes. These included a direct alignment between Thorold and Port Dalhousie that was later to be followed by the Third Canal, but this idea then received no support because of the additional costs that would result and the opposition of established business interests along the canal, as in St Catharines and Thorold, that would be adversely affected by this bypass channel.

The two canal communities preferred to retain the present route, but because of its deficiencies this choice required 'occasional departures therefrom.' These included dredging to below the water level and new piers at Port Dalhousie. Also, the line inland would be shortened and straightened across the meanders wherever possible, and two of the old locks could in places be combined into one larger-sized stone lock. The channel at the aqueduct would be widened by constructing a new lock, and at Port Colborne the harbour would be improved by upgrading the entrance and enlarging the channel to the aqueduct.

Work on this new canal commenced in 1841, and the new works came steadily into operation as they were completed. In 1848, the new harbour and its redesigned entrance at Port Dalhousie were brought into use. Now divided into an outer and an inner harbour through the construction of a weir, and with new protective breakwaters projecting into Lake Ontario, the old entrance lock through the sandbar was no longer required. This channel was filled and its navigation equipment was transferred from the old to the new piers.

The southern length presented more problems and several time delays. The Feeder Canal was enlarged by 1845, and the Port Maitland branch was built to provide a new outlet on Lake Erie at the mouth of the Grand River. This construction also enabled the route between the Feeder Canal and Port Colborne to be drained in order to permit its enlargement, but this task took longer than anticipated because of sickness among the labourers and the difficulties of excavation.

The intention had been to lower the summit level of the canal between The Junction and Port Colborne by 8 feet (2.4 m) to Lake Erie, but as this proved considerably more difficult than was anticipated because of having to blast into hard rock strata, the Feeder Canal and the Port Maitland branch were brought into full canal service for a time from 1845 to 1850 while the direct route was being deepened. The Feeder Canal offered the navigational advantage of an earlier spring opening period than at Port Colborne, but the disadvantage of a longer route between The Junction and Lake Erie. As noted by Bonnycastle (1846), 'The old miserable wooden locks and bargeway have been converted into splendid stone walls and a ship navigation.'

The source of water supply remained the Grand River. A portion of this was directed to Lake Erie via the Stromness–Port Maitland Branch, and some as before was returned to the Lower Welland River via the industries at Welland and Port Robinson. The remainder of the larger flow caused by deepening the canal continued to bypass The Falls.

The completed Second Canal received a variety of different names: the 'stone canal,' because the earlier wooden locks had now been replaced by more permanent stone structures; the 'new canal,' because it displaced the older canal; and the 'government canal,' because of its new management. Each was correct and each was used, but with the addition of a later Third Canal, 'Second Canal' now prevails.

The more practical route that had been selected was rebuilt along broadly the same route between Port Dalhousie and the Feeder Canal, but the canal dimensions were expanded, new locks added, and all the former wooden locks were replaced by cut-stone locks. In total the locks were reduced from forty (thirty-nine, plus one at Port Colborne) to twenty-seven, which in turn meant slender changes in the location of the channel. Previously, the locks were located at the lowest points in the valley of Dick's Creek south from St Catharines, but as the old channel had to remain open for shipping until its locks were replaced, the new stone locks were constructed higher up the bank on the west side of the First Canal. Water was then introduced to the new channel upon completion of the canal works.

The canal length between Port Dalhousie and Thorold received twenty-five new lift locks of limestone that was obtained from the quarry on Queenston Heights, brought down to docks at Queenston, loaded onto barges for transit down the Niagara River, then along Lake Ontario, up the First Canal, off-loaded, then hauled the short distance to the site of each lock. Each stone was then dressed, shaped, and put into place along the walls by the skilled hands of imported Scottish stonemasons.

The size of the stone locks caused considerable discussion, for they were the controlling factor that determined the size of ships that could make use of the canal. In this period the sailing vessel was giving way to the steamship on the lakes, and the latter had substantially larger dimensions. The debate was between those who thought that freight movement was the vital factor and those concerned with the high cost of larger vessels and an improved canal. Against this, the cost of designing to such high standards throughout the canal was much higher.

The compromise solution was to construct locks of different size, with locks of larger dimension at the canal entry points to serve the ports and smaller locks along the line of the canal. Their size was increased in 1842, a decision that resulted in entry locks 200 feet (61 m) long, 45 feet (13.7 m) wide, and 9 feet (2.7 m) deep at Port Dalhousie, below St Catharines, Port Maitland, and Port Colborne. Elsewhere the locks were 150 feet long (45.7 m), 26.5 feet (8.1 m) wide, and 8.5 feet (2.6 m) deep; the width of the bed along the main line of the canal was increased to 26 feet (7.9 m) at the bottom.

In addition to the replacement locks across the escarpment, the two locks at Port Robinson and Allanburg were each replaced by a single lock. A lock at Welland connected the canal upstream with the Welland River. Although contemplated, a downstream lock was not constructed. At Dunnville, where the dam across the Grand River blocked navigation and had no lock, the nearly 2 miles (3.2 km) of canal that opened for traffic in 1845 between Stromness and Port Maitland provided the only alternative for lakebound traffic, and a new large-sized lock was constructed at the Lake Erie entrance to this canal.

Further important new features included the replacement of the aqueduct at Welland by a larger stone version 45 feet (13.7 m) wide, 316 feet (96.3 m) long, and 10.7 feet (3.3 m) deep. This work was slow and the wooden predecessor was not phased out until the early 1850s.

The harbours at Port Dalhousie and Port Colborne were greatly extended with improved facilities. At Port Dalhousie, a weir constructed across the river channel of Twelve Mile Creek created an upper and a lower harbour that were connected by Lock 1 of the Second Canal. At Port Colborne, where the lock at the canal entrance from Lake Erie was moved inland from the lake by nearly 0.5 miles (0.8 km) to Clarence Street, a wide basin was created on each side of the outer channel to accommodate ships.

More changes during the 1850s included the deepening of the canal to 10 feet (3 m) and increasing the bottom width of the summit level. This redesigned canal was constructed at the end of the pioneer period with its close dependence on inland waterways and river navigation. As the first railway arrived on the scene in 1853, the opening of the Second Canal marks the start of a new era, with an improved canal system soon to be subject to intense rivalry from the newfangled railway locomotive that ran on iron rails.

As an indication of the volume and type of traffic carried on the canal, the Welland Canal office in 1847 announced that, from the start of navigation to the end of April, 185 vessels passed through the canal. Of these, 127 were upbound, 84 from Oswego, and 43 from Canadian ports, of which 23 went to American ports. Of the 58 downbound vessels, 27 were destined for Oswego. The American cargoes were nearly all grain, and on Canadian vessels they were principally flour. Over the same period 47 scows passed up or down, carrying principally stone, gravel, staves, and hoops. One scow even carried the distinctive load of a house from Port Robinson to St Catharines on top of its load of staves and wood (St Catharines *Journal*, 13 May 1847).

Ships passed through the canal on Sundays until 1845, but the Sabbath Observance Laws then prohibited the passage of vessels through the canal between 6 a.m. and 8 p.m. on Sundays, a proviso that lasted until 1876.

The excitement of a vessel being towed through the escarpment series of locks did not pass unnoticed. The vista upon leaving Thorold has indeed been described for its powerful scenic grandeur: 'we rounded the corner of the height above the mountain tier of locks. It was a wondrous sight to see laid out before us the wide landscape of tableland and valley spread out below, through which we were to navigate and drop down 340 feet [103.6 m] on the next four and one-quarter miles [6.8 km]. To the left was the series of locks which circled, in gray stone structures, like a succession of great steps, down the mountain side. These were separated one from the other by small ponds and reservoirs with waste weirs, whose little water falls tinkled, foaming and glinting in the sun. Directly in front of us, were the houses and factories of Merritton' (Cumberland 1913).

Access to the Atlantic Seaboard

The First and then the Second Welland Canals provided a direct link by water for vessels sailing between Lake Erie and Lake Ontario, but this initial progress and the need to provide a serious competitor to the Erie Canal was insufficient by itself. William Hamilton Merritt had a far wider vision in mind, namely a direct connection from the Upper Lakes to the Atlantic Ocean via the Welland Canal, Lake Ontario, and the St Lawrence River. As explained by him at the inaugural sod-turning ceremony in 1824: 'This canal is the commencement of a similar undertaking; it is the most important link in that chain of communication – we hope to see [it] effected within three years. We remove the only natural barrier of importance – the Falls of Niagara. The rapids between Prescott and Lachine commands the next consideration. If the subject is properly before the Legislature of the two Provinces this winter it can be commenced the year following ... When we contemplate the natural advantages we possess over the Americans in our water communication, it is astonishing to think of the apathy and indifference that has hitherto prevailed amongst us on this subject' (Merritt 1875).

Merritt, however, was over-optimistic about the time required to achieve this further venture, which anticipates the opening of the St Lawrence Seaway in 1959. By 1781 four small military canals had been constructed to avoid the rapids of the Soulange section of the St Lawrence between Lake St Louis and Lake St Francis. They together had five locks. With a depth of only 2.5 feet (0.8 m) over the sills, and 6 to 7 feet (1.8–2.1 m) wide, they took small boats only. They were replaced in 1805 by enlarged locks with a depth of 4 feet (1.2 m). Though by 1814 some 15,000 to 18,000 public loads had been carried, the locks could not receive the commercial vessels that used the First Welland Canal. They were not replaced until 1845, the period of the Second Welland Canal.

At the Lachine Rapids closer to Montreal, an effective portage road existed that carried loads on wagons. This became deficient only when the advent of steam vessels led to a regular service between Quebec City and Montreal, with an increase in trading volumes that provided a strong impetus to the demand for a canal between Montreal and Lachine. This opened in 1825, with six lift locks and a five-foot (1.5-m) depth of water over the sills – again inadequate to accommodate the larger size of vessel that passed through the Welland Canal.

An alternative route to the sea came about after the War of 1812 – a safer inland route between Kingston and Montreal that avoided the St Lawrence, its rapids, and the internal section of the river route where the Americans had jurisdiction and presented a constant threat of attack. This route involved a channel via the Ottawa River and the Rideau Canal between Kingston and Ottawa. Surveys were ordered in 1816. Work started on the Grenville Canal three years later, and navigation opened there and through the Carillon Canal in 1834. The Rideau Canal, which rose from the Ottawa River at Bytown (Ottawa) to a summit at Upper Rideau Lake, then dropped to the Cataraqui River and Kingston harbour, opened in 1832. With lock dimensions equivalent to the Lachine Canal, the Rideau-Ottawa line was used for almost all through trade between Lake Ontario and Montreal from 1834 to 1851 (Heisler 1973).

It was not until the latter date that the last of the locks and cuts of the St Lawrence Canals were completed, and vessels were able to sail directly between Port Dalhousie and Montreal. Over the period from 1834 to 1851 the military and commercial route through the Rideau Canal and along the Ottawa River carried regular passenger and freight services, military supplies, naval vessels that were toll-free, and a few small cargo ships. As the locks were too short and shallow for most merchant vessels, this traffic had to await the completion of the St Lawrence system in 1851 to a depth of 9 feet (2.7 m), with locks about 200 feet (61 m) in length to match those that had been constructed on the Second Welland Canal.

To these canal considerations must be added the political factor of British-American rivalry. In a period of great internal expansion, both sides wanted to enlarge their transportation network and tap the ever-increasing trade from the West. The Erie Canal alone could not cope with this expanding flood, and American and Canadian merchants both wanted the additional facilities of a St Lawrence–Great Lakes commercial system. The Welland Canal, used extensively by American vessels, was successful because it contributed materially to these growing demands, but then other international considerations downstream from the canal also rose to importance.

American access to the St Lawrence became a leading matter of concern after the War of 1812. Even after the completion of the Erie Canal, the St Lawrence was perceived as the most efficient and natural trade route to the sea. Yet the international boundary line that divided the Niagara River and the Great Lakes between the two nations had also placed much of the St Lawrence River under British control. The international section was through Lake Ontario then along the St Lawrence River between Kingston and Cornwall, but the length between Montreal and Quebec City and the Gulf of St Lawrence was wholly British, which created tensions and animosities between Great Britain and the United States until the Reciprocity Treaty of 1854.

In that year the right was given for American vessels to freely navigate the St Lawrence within British territory. Previously there had been heavy duties and the possible confiscation of property. Under the Navigation Acts, American vessels were allowed to sail down the river to Montreal and Quebec and to sail the ocean up to Quebec, but an unbroken voyage from the ocean to the Great Lakes was not permitted.

Great Britain by the Canada Trade Act of 1822 restricted the freedom of trade across the boundary by placing duties on goods entering the two provinces of Upper and Lower Canada by land and inland navigation (Brown 1926). Until then Americans had been able to ship their goods in bond via the St Lawrence to foreign markets free of duty. This act of 1822 was attacked by American interests as prohibitory. Americans argued that the St Lawrence was the gift of God, and that the American people had the inherent right to navigate the St Lawrence River to the Atlantic.

William Hamilton Merritt, a staunch proponent of free trade, was a leading Canadian spokesman who also argued for the unrestricted use of the St Lawrence by all nations. The *Niagara Gleaner* (22 February 1823) urged the same approach: 'The produce that goes down the St. Lawrence from the United States, particularly lumber, creates employment for a number of hands, and being exported all by British shipping, gives employment to a number of seamen and carpenters.'

Concessions started in 1826 when American flour destined to the West Indies, and American salt, beef, and pork for Newfoundland, were allowed to pass through Canada duty-free. In 1831, after the opening of the First Welland Canal, further duties were removed, which suggests the growing importance of the Canada trade for the United States. In 1843 American wheat could be processed in Canada and re-exported as Canadian flour, which led to the construction of mills along the Welland Canal across the Niagara Peninsula. This new prosperity was short-lived for, with the repeal of the Corn Laws along with American drawback regulations, Canadian goods could transit through the United States without paying duties (see Easterbrook and Aitken 1958).

The commitment of the British government to free trade began in 1846 with the pledge to repeal the Corn Laws. This removed trade protection from the provinces. They no longer had imperial protection, and this in turn led to the search for new outlets. The obvious choice was for the Province of Canada to establish closer trade relationships with the United States. Because of their proximity to the international boundary, both the Welland Canal and the Niagara Peninsula were well situated to take advantage of this new trading situation.

The Transition from Sail to Steam

The era of the First and Second Welland Canals was also transitional in the types of vessels using the canals, and their significance for the character and economy of the canal communities. The change was from sailing vessels that were towed through the canal by horses or oxen that required a towpath as part of the canal design, to small steam vessels that could navigate the canal more speedily and under their own power. This change was not only from sail to steam, but also from wooden to metal hulls, and from towpaths, tow boys, and tow animals to the towing of vessels by tugs. The steamers carried more cargo, but their higher speeds and greater bulk increased the wake and had an erosive impact on unlined and often unstable banks.

There was also the personal factor. The towing of sailing vessels through the canal has been compared with the miseries of human slaves on the Atlantic passage. 'It is doubtful if they were in any way worse than those of the miserable beings then struggling on the canal passage between Lakes Erie and Ontario. The canal banks and towpaths were a sticky mush, which in those autumn months was churned and stamped into a continuous condition of soft mud and splashing pools. From two to six double teams were employed to haul each passing vessel, dependent on whether it was light or was loaded, but in either case there was the same dull, heavy, continuous pull against the slow moving mass, a hopeless constant tug into the collars, bringing raw and calloused shoulders' (Cumberland 1913).

When horses did the towing, one was usually hitched to the end of the vessel being towed and another was hitched a few yards back, while the driver controlling the horses walked along the towpath between the two horses. The man who operated the towing business kept a number of horses and contracted with the canal company to undertake this work. Such conditions changed as the steam-powered propeller vessel came gradually into vogue, but conditions in the engine room could be hot, dirty, and unsafe. Inefficient steam boilers might explode and the vessel might sink, and there was the ever-present danger of human contact with hot pipes and scalding water.

That the Welland Canal drastically changed the design of vessels has been stressed by Cuthbertson (1931): 'The construction of the Welland and St. Lawrence Canals undoubtedly has had the greatest single influence on all lake construction, either oil or steam, up to the present.' Previously, if the shipowner wished to build a vessel, he could build to any length, beam, or tonnage to sail on the lakes. Now, if he wished the vessel to be sent from Lake Ontario to the sea or between Lake Ontario and Lake Erie, he had to conform to the restricted dimensions of the locks, which were smaller on the St Lawrence Canals than those on the Welland waterway.

As every inch of space had now to be used to advantage. 'Vessels built for the canal trade had to sacrifice many things which hitherto had been the pride and joy of shipbuilders. A straight stem took the place of the nicely curved cut-water. The bowsprit had to be so short as to be practically negligible, and only served as a fastening for the jibboom which, in order not to foul the lock gates, was made to cant up almost vertically. On the bottom, only the smallest curve was permissible, and the stems were very flat with almost no cut away aft whatever. Indeed, these old canallers were little more than boxes, with their beam carried almost to the bows giving them ... anything but [a] graceful appearance either loaded or when running light.

'To avoid leeway with such a restricted draft, centreboards or drop keels had to be built into them. As many as three and four centreboards were used on some ships' (Cuthbertson 1931). Leeboards as used were impractical. They sacrificed valuable width and were liable to be carried away during lake storms. With vessel form changed so remarkably by the new type of vessel, shipyards around the lakes and along the canal had inevitably to bow to the new trends of ship design and produce in sail then steam what became known as 'canallers.'

Despite these disadvantages of design, the Welland Canal, always a ship canal in contrast with the Erie Canal, designed to carry only barge traffic, offered a considerable advantage over the competing American waterway. As an indication of this varied yet rapidly changing situation, 'in 1845 all the vessels engaged in the grain trade on the upper lakes could pass through this [the Welland] canal, but in 1855 there were at least twenty propellers that could not use the canal on account of their size' (Mansfield 1899). By this criterion of usage by lake vessels, the Second Canal became obsolete soon after construction, but at first it could accommodate the vessels that might be expected to use its now improved facilities.

Steamboats arrived on the Great Lakes during the first decade of the nineteenth century. According to Mansfield, the steamer *Dalhousie* was built at Prescott in 1809, and the steamboat *Accommodation*, described as 'the first vessel of the kind that ever appeared in this harbor,' made the journey from Montreal to Quebec the same year. 'The great advantage attending a vessel so constructed is that a passage may be calculated on to a degree of certainty in point of time, which cannot be the case in any vessel propelled by sails only.'

The steamers *Ontario* and *Frontenac* were constructed on Lake Ontario in 1816, and the *Walk-in-the-Water* on Lake Erie in 1818. By 1820 there were four steamers on the Great Lakes. It was the recognition of this new and more powerful trading force that led to the provision of an increased depth during the construction period for the First Canal and provided it with a considerable commercial edge over the smaller, and then unimproved, Erie Canal.

The End of the First Canal

As the Second Canal gradually opened, the First Canal became an unwanted feature of the past. South from the escarpment to Lake Erie and along the Feeder Canal, it was generally widened and became part of the Second Canal. Along other sections, such as across the Niagara Escarpment, the reservoirs were reconstituted to suit the location of the new locks. The disused channels were usually filled, either gradually as suitable locations for local domestic and industrial garbage or with material from the newly excavated channel after it became available for navigation.

Occasionally the old channel might survive, but with a diminished flow of water, as when the new channel was cut across a river meander at Welland Vale on Twelve Mile Creek. The lock sites might be sold for industrial purposes, used as part of a drain, or harvested for their timber. More generally they were filled with spoil or garbage like the channels. No thought about historical preservation existed. Given this type of treatment, as the First Canal was displaced by the Second Canal, little evidence has survived on the ground of the original channel. An exception is the weed-filled, former stream bed at Welland Vale, where the historically important initiating Merritt mill was located (see map 2, page 31).

3

The Regional Significance of the Canal for Pioneer Life

Many drastic and broad-ranging changes were introduced to the pioneer world through the course of canal construction and again when the canal opened. The workers on the project had to be housed and fed, and the presence of Irish labourers contributed strongly to the advance of the Roman Catholic church in a former Protestant locality. The canal had to be administered, repaired, and maintained. The pattern of roads was augmented, and bridges were added to the pre-existing system. Agriculture was improved and advanced, and new sites with water-power opportunities were created in locations other than within the natural drainage areas where mills had previously developed. The Niagara Peninsula was invigorated through the expanding period of development that prevailed after 1825.

A Military Appreciation

The date and inland location of the First Canal, the dam across the Grand River, and the inability to use the northern length of the Niagara River as an entrance to the canal were each influenced strongly by military considerations. Then, after its construction, the canal was viewed as an important line of defence across the Niagara Peninsula.

Memories of the American invasion in 1812, their takeover of the frontier, and the tribulations of war were etched into the minds of many settlers, several of whom had served with the military. And no longer were the two lakes isolated, with each requiring its own naval fleet. They were now connected into one system by the canal, which could be used to convey vessels and supplies between the two lakes.

During the 1820s and again during the 1840s, plans were prepared for a major fortress to be constructed just west of the canal on the crest of the Fonthill Kame as the first continuous line of defence inland from the frontier. Straight roads for rapid military movement were then intended – to the canal, where forts would be located at Port Dalhousie and Port Colborne, and to the frontier, along the Niagara River at Fort George and Fort Erie. These ideas never reached the stage of construction, but they were revived in 1837 during the Mackenzie Rebellion and again during the Fenian raids of 1865, when armed insurgents invaded Canada from the United States and the local militia was called to arms.

This potential military role of the canal reflects both the temper of the times and the close proximity of the canal zone to the Canada–United States boundary along the Niagara River as a significant and ongoing feature of importance in the history of the canal, its industries, and the communities that had arisen along its banks (Hill 1918). This military role is also reflected in the late building of the Welland Canal, its private rather than government achievement, and its inadequate wooden locks when compared to the Rideau Canal, where construction started in 1826 to link the Ottawa River at Ottawa with Lake Ontario at Kingston as a government venture under the control of the Royal Engineers.

The Canal Is Open for Business

After the symbolic and ceremonial opening in 1829 of the Welland Feeder Canal and then of the First Canal to Chippawa and the Niagara River, substantial and increasing volumes of freight soon moved through the completed waterway (Kingsford 1865). Rafts of staves were brought down from the Grand River, flour was received at Port Dalhousie from Brantford and exported to Prescott – all within a few days of opening the canal in 1830. By 1831 packet vessels owned by the Welland Canal Company served between Port Dalhousie and the Grand River. And in 1833, the first official year for a through lake-to-lake canal, the Merritt biography records that 'the canal was now rapidly coming into public use ... From the 1st to the 20th of June in this year, 34 vessels had passed up the canal and 20 went down, and in the following month 219 schooners, 138 boats and scows, and 30 rafts, loaded with produce of all descriptions.'

Every passing vessel, scow, and raft presented new opportunities to each community along the canal. The vessels had to be manned with a crew, supplied with stores and supplies, and hauled through the canal by tow boys and tow horses. The vessel, barge, or raft may well have been constructed locally, or perhaps repairs or new supplies were required along the canal. The banks of the canal, its locks and bridges had each to be manned and kept in a reliable state of repair. Tolls and customs had to be levied and tallied.

The range of such service elements, with their attendant employment, then the demand for houses and other community requirements, has been indicated in the introduction. These service demands stemming from the canal system were in addition to the cargo and trading importance of the canal, to its potential catchment area stretching to the expanding ports around the Upper and Lower Lakes, and to the canal's potential as a body of water that might be tapped for mill purposes or used for the benefit of agricultural land.

The usual way for a sailing vessel to enter the canal was to await the arrival of a tug, and then be towed to the entrance lock (Cuthbertson 1931). When leaving the canal, many vessels might club together and form a tow of sometimes six ships until the crew of each vessel cast off and set sail. Towing on the canal was also done by tugs that could handle several vessels at a time, with teams of horses or oxen, or even the crew when necessary, being used when the tugs were busy.

Upbound cargoes (towards Lake Erie) of salt increased by 23 per cent, and merchandise by 8 per cent during the 1830s, and the downbound freight of square timber by 35 per cent, wheat by 15 per cent, and pork and beef by 8 per cent. Schooners passing through the canal increased from 240 in 1832 to almost 2000 by 1840.

Much of the traffic that passed through the Welland Canal came from American ports. In 1834 the bushels of wheat carried from Canadian ports constituted but 7.5 per cent of the total, and 95 per cent of this was consigned via the Oswego Canal to the New York market. The Welland Canal had become a feeder to the eastern part of the Erie Canal system, and could compete as an alternative route because it had fewer locks, carried a greater depth of water, and could accommodate larger vessels carrying over twice the tonnage. On the Erie Canal, a vessel required a master and four hands, plus relays of horses for towing day and night, and could not exceed 50 miles (80.5 km) in 24 hours. Overall, 'the difference between the two in daily expense and quantity of freight will reduce the whole expense of transportation by a lake vessel to one fourth of that necessarily incurred by a canal-boat' (Creighton 1830).

In addition, the American destination of cargoes was influenced strongly by the growing strength of New York City relative to Montreal. American merchants now exercised far stronger control over Lake Ontario than when the Erie Canal was constructed using an inland course through New York State. Two other factors influencing the flow of traffic through the Welland Canal were the inadequate depth past the rapids of the St Lawrence River and the trading controls exercised by Britain. Even so, by 1847 the trade through the Welland Canal had reached nearly 256,000 tons (235,872 tonnes), of which 75 per cent was downbound and 25 per cent upbound. By 1849, when this total had increased to 265,000 tons (240,408), the figure included a higher up-trade and a lower down-trade.

After the Feeder Canal from Dunnville and the Grand River had been materially improved, an account of the early 1840s reads: 'Immense quantities of pork, flour, and almost every other description of produce, are shipped through the medium of this work, to every port on the lakes. I have seen strings of schooners wending their way from lock to lock, resembling a floating forest, at a distance. Here or there a raft or scow crawling sluggishly along, hugging the heel-path, to make room for the swifter packet boat' (St Catharines *Journal*, 2 June 1842).' The south side of the Feeder Canal was the towpath for the ladened downstream journey, and the north side or heelpath was normally used for the empty return journey. Timber rafts were used extensively. They could carry eighty barrels of flour, pork, ashes, or other produce, and square sails might be used to aid the oarsmen (Guillet 1933).

By 1850 these exports from the Grand River valley through Dunnville and the Feeder Canal to American and British ports included over 30,500 barrels of flour, nearly 230,000 bushels of wheat, almost 190,000 bushels of oats and barley, 2239 tons of gypsum, over 16 million feet of lumber, and nearly 560,000 cubic feet of square timber in rafts. The sawmills along the Grand also supplied nearly a further 13,500 cubic feet of square timber, together with laths, hoop for barrels, and fence pickets. Pine lumber together with flatted and round timber were tallied at nearly 16.4 million feet. Pipe barrel and West Indian staves were numbered at 600,000 and shingles at over 1500 million. Sawn logs, empty flour barrels, and cordwood each received further mention. (These measurements have not been translated into their metric equivalents because of the differences that exist between weight, mass, volume, and capacity and the problems of their inter-comparability.)

The flow along the canalized Upper Welland River complemented that along the Feeder Canal. It began when large numbers of men were employed to remove timber. The trunks and logs were hauled to the banks, where they might be lashed together and rafted downstream. Cordwood was stacked ready for shipment. Other workers made shingles for the roofs and sides of houses and staves for the barrel factories, either on the site where the trees were felled or in the downstream canal communities.

Scows and barges were built to carry these products from the Grand and Welland Rivers to shipping and processing points, the markets being along the canal in Welland, Port Robinson, Allanburg, Thorold, and St Catharines, and in the United States at Tonawanda and Buffalo. Both American cities became major lumber ports, based to a considerable extent on Canadian supplies of timber and with markets downstream along the Erie Canal, including at Boston and New York, and also at the ports around the Upper Lakes.

Some sixty scows and barges have been recorded as working at one time on the Welland River, and then the process gradually changed. As the land was cleared of its timber resources and converted to settled agriculture, farm products such as wheat, other grains, and animal products took over from timber, and rafts poled slowly downstream were displaced by tugs, barges, and cargoes reflecting the barrel economy. The entrepreneurs of the day were those who organized this marine traffic and who established the mills, wharves, and stores, often as an integrated enterprise under single ownership.

The resources of the land were being exported on a grand scale, with the Welland Canal and its related waterways contributing in an important way to this outflow of materials. Exports from Dunnville were consigned to Port Dalhousie for Montreal and the British market, to Buffalo and Tonawanda via the Lower Niagara River and to the Upper Lakes, including the Chicago area, via either Port Colborne or Port Maitland.

Coal from recently discovered deposits in Pennsylvania was added as a new but valuable import to the Niagara Peninsula, and coal yards next to the canal soon became a feature in all the main centres. Coal was also brought through the Welland Canal to the markets of southern Ontario in rapidly increasing volumes: the 3486 tons (3162.5 tonnes) in 1847 had increased to 5533 tons (5019.5) two years later. This access to American mines via the Welland Canal soon brought coal into regular use for domestic heating and for driving steam engines in a range of manufacturing enterprises. The industrial revolution had reached the Niagara area through coal imports to a province that had no coal deposits. The earlier days of pioneer dependence on water power were now numbered, but water power remained important into the period of the Third Canal and until the end of the nineteenth century.

The canal was also important for its passenger traffic. Port Dalhousie and Port Colborne had regular service access to the ports around their respective lakes by sailing vessel and then by steamship. The Grand and the Niagara Rivers were accessible from the canal communities on regular schedules. A through passenger movement by steamer took place along the Feeder Canal and the Grand River between Buffalo and Brantford and between Dunnville and Port Dalhousie. In an era of poor roads, the canal was an important transportation artery, cheaper and often more direct than the slow and uncomfortable journey by road.

In 1851 the fare by boat was the same, at 3s.9d., from Port Dalhousie to Hamilton as from Port Colborne to Buffalo, and higher, at 4s.6d., from Port Robinson to this American terminal. To journey by steamboat per unit distance of travel was cheaper than by stagecoach on rutted and dusty roads. The stage fare between Port Dalhousie and St Catharines was 1s.3d., and 2s.3d. from Port Robinson to the same location (MacKay 1851). The journey by boat was also more direct and faster than the circuitous journey by land around the head of Lake Ontario from St Catharines to Toronto.

Of the major competitors to the Welland Canal, the Niagara Portage as the former prime route dwindled significantly and was soon abandoned as a through route between the lakes. The Ohio Canal, with its route via the Mississippi to the Gulf of Mexico, was oriented to neither the Atlantic Ocean nor to trade with western Europe. The Erie Canal, whether used entirely from Buffalo to the Atlantic or along with Lake Ontario and the Oswego Canal, remained dominant, enjoying more trade and traffic than the Welland facility. Its number of vessels and annual revenue from tolls were both much higher. The Welland was definitely now the baby of the two canals (Keefer 1850).

Housing the Workers

The change was drastic, from isolated farms and small communities to bustling construction sites at the locks, the aqueduct, and along the line of the First Canal – a process to be repeated within a decade for the Feeder Canal, the extension to Port Colborne, and then the Second Canal. As each canal involved much human toil and sweat, disease, and deprivation, a human interpretation is that it is the pick-and-shovel labourers who have left their indelible mark on the landscape across the Niagara Peninsula.

As labourers and contractors with their demands for food, equipment, shanties, and boarding houses were attracted into the thinly populated townships, 'clusters of tents and shanties sprang up all along the works. Soon Slabtown at the foot of the mountain [the Niagara Escarpment], Stumptown on its brow, Allanburg at the Holland Road crossing, and Port

Beverley in the hickory bush were recognised settlements. A little later the Aqueduct, Stonebridge and Gravelly Bay were added to the list of local place names ... Traders and liquor vendors followed the navvies; shoemakers, tailors and other craftsmen set up little shops where there were prospects of custom ... Soon an air of permanency began to pervade the mushroom villages. Buildings of some permanent character were erected, merchants of the substantial sort located "on the line" ... Slabtown became Merritton, Stumptown took the name of Thorold, Port Beverley changed to Port Robinson, the Aqueduct was Merrittsville (later Welland), and Gravelly Bay was dubbed Port Colborne to balance Port Dalhousie, that had sprung up at the Lake Ontario portal of the canal' (Green 1930).

It was from Centreville (Merritton) that Oliver Phelps advertised in 1827 for 'all classes of Labourers' to complete the Deep Cut and to build all the locks. The relative rates of pay for different activities are interesting: $12 a month for common shovellers, teamsters and men to hold the plough from $15 to $18 a month. Two good yokes of oxen and a good stout cart were assessed at $28 a month. 'Any person employing and bringing on fifteen good shovellers, shall be entitled to the wages of an overseer, and hold that station, and may draw for his men the wages above stipulated; or *seven cents* per yard for shovelling into carts or wagons, (after the earth is ploughed up,) and his own pay as an overseer.' Further, 'Cash will be paid for 100 yokes of good, young working *Oxen*, and for all kinds of *Grain*' (*Farmers' Journal and Welland Canal Intelligencer*, 28 May 1827).

The construction difficulties were many. The land had first to be cleared of all vegetation including its forest cover, and then ploughed by four or six yoke of oxen per plough. The dirt was carried away in sacks on men's backs or in wheelbarrows and then in carts and wagons (Johnson 1926). At the Deep Cut, 'An Improved Machine for removing Earth in deep Cutting,' invented by Oliver Phelps the American contractor, involved a wagon wheel on top of a 7-foot (2.1-m) pole fixed firmly into the ground at the top of the bank. A rope and hook were attached to an empty cart. Its oxen going down helped to bring up the loaded cart and its yoked oxen at the other end of this rope. Fifty of these contrivances were at one time stated to be in use on the Deep Cut.

The landscape was transformed by the work project. 'Small shacks and shanties sprouted along the proposed route of the deep cut and small settlements were begun at both Allanburg and Beverley as Port Robinson was first called ... Scores of workers swarmed into the area seeking work on the digging of the Welland Canal or "Deep Cut" as it was called from Port Robinson to Allanburg. Shops and lodging houses followed in the wake of the labourers who hastily erected shacks and shanties along the proposed route of construction. The Welland Canal Company opened a store to sell goods to the workers who were also able to buy goods from the established shops ... Local farmers had a ready sale for surplus produce at a market that was held at the Deep Cut. Many farmers sent their wagons twice a day to meet the demand for fruit, eggs, vegetables, grains and butter' (Michael 1967).

At the edge of the Niagara Escarpment, 'in 1824, a large number of Irish and English immigrants employed on the canal made their houses at Thorold, and shops, inns and houses had to be built for the accommodation of the newcomers ... During the building of the old canal, the new residents lived near the works, and Front and Pine Streets became the centres of business ... [There was] a large general store ... and many smaller shops were afterwards opened' (Thompson 1897–8).

By 1840, some seventy shanties were enumerated along the canal between St Catharines and the Mountain Locks. These poor-quality, substandard, and often short-lived types of accommodation arose where the work was most intense and concentrated, including at the lock sites between St Catharines and Thorold, at both ends of the Deep Cut, and where the aqueduct crossed the Welland River. Where the essential work was to cut a canal channel, as between Welland and Port Colborne and along the Feeder Canal between the Junction and the Grand River, labourers moved progressively from one contract site to the next.

Violence in the Community

The labourers worked long hours for subsistence wages, sometimes having to wait several weeks or even months to be paid. Often there was insufficient work to go round, and some men were laid off as a contract was completed. The workers endured wretched conditions. They were at the mercy of the weather, insects, snakes in the marsh, mud, dust, the heat of summer, winter cold, and difficult terrain in which to work. They suffered from fever, ague, cholera, malaria, and typhoid, as well as the frequent risk of inopportune accidents and unscrupulous contractors. In the best of circumstances, the worker lived in dire poverty, in a constant state of insecurity, and often with either the contractors or the Welland Canal Company having insufficient funds to meet the wage demands. Not unnaturally, in periods of special duress the workers resorted to violence in order to improve their situation, as when in the summer of 1842 there was a serious work stoppage on the canal and 2000 labourers marched into St Catharines.

They plundered stores and mills, raided a schooner on the canal for food, and endangered the safety and the security of the local community.

Problems arose because of poor working conditions, rivalry among the workers for the jobs available, and the religious differences between the various gangs of canal labourers. Some were Protestant Orangemen from the North of Ireland and others were Roman Catholics from the South (Runnalls 1973). In the early 1840s, these two groups 'reduced canal areas to virtual war zones. Parties of armed men constantly assembled and paraded, planning attack and counter-attack, at times fighting the issue out in the streets of St. Catharines' (Bleasdale 1978). Tensions were high in a double sense: between the various labour groups as they competed for the available jobs, and between the labourers and the established communities because of their different economic and social backgrounds. The strong brogue of the Irish workers' accents was unintelligible and foreign to many settlers, and their generally Catholic religion was alien to the mores of the community as a whole. The Irish wanted work and food, the established community feared for their property and safety, and neither really understood the problems and the attitudes of the other.

As faction fights were frequent, during the 1840s detachments composed of 'men of colour' were at various times placed at St Catharines, Stonebridge (now Humberstone), Slabtown (Thorold), The Aqueduct (Welland), and Port Robinson, their headquarters, to maintain law and order. In the 'coloured corps,' with some four sergeants, four corporals, one drummer, and eighty privates, the commissioned and non-commissioned officers were white men; the Black privates were former escaped slaves from the United States (Groh 1969). 'To the Irish the sight of Negro soldiers wearing the red coat of authority brought out further resentments. The matter came to a head one day when the Irishmen dropped their shovels and started to march to Port Robinson where the company was stationed. They were joined by fellow workers from Thorold, Allanburg and the Deep Cut' (Michael 1967). Orders were given to fire upon the rioters, but the tense situation was relieved when a Catholic priest known to the rioters for settling earlier disputes arrived from St Catharines and forbade any further advance towards the armed soldiers.

One severe incident that broke out in 1849 has become known as the Battle of Slabtown (now Merritton). The Orangemen planned to celebrate their 12th of July with a dinner at Duffin's Inn. The men on the canal downed tools and several hundred besieged the inn and fired shots through the window. The coloured corps were summoned from their headquarters at Port Robinson to restore order. Confusing reports list two to five dead and several wounded. The corps was disbanded in 1851 after completion of the canal and the dispersal of its labour force. Clearly, with epidemics and group tensions, the task of canal construction left bitter memories.

The Roman Catholic Church

In St Catharines, land for the first Roman Catholic Church was purchased in 1832. With money from both the permanent residents and the Irish labour force along the Welland Canal with whom the church was closely associated, a frame wooden building dedicated to St John was built on Church Street. After this church burned down in 1842, a new building dedicated to St Catherine of Alexandria was constructed. (Except for the confusion of two similar-sounding names and a different spelling, no connection exists between the name of this saint and the city's name.) Through time this church became the Catholic cathedral, and St Catharines a Catholic diocesan city (Harris 1895). As stated on a commemorative tablet in this church building, in its English translation from the Latin, 'the Irish working on the Welland Canal built this monument of faith and piety.'

Thorold also developed as a Catholic centre. The first mass was held in 1841 in the home of a Catholic named Thomas O'Brien, a frame church was erected in 1843, and the first separate school, a frame school known as St Joseph's School, followed in 1854. These developments can be attributed to the faith of Catholic workers along the canal, a new religious presence that supplemented the earlier Episcopal and Methodist congregations.

When the first official census was taken in 1851, at St Catharines one-quarter of the population (1093 out of 4368) were of Irish origin, and by religion, 26.8 per cent were classified as adhering to the Church of Rome. In the village of Thorold, with a population of 1091, the proportion of Irish origin was 24.7 per cent and adherents to the Church of Rome were 22.4 per cent of the total population. Although some were immigrants and not all were former canal labourers, many from the latter group had settled in these two communities.

The Catholic proportion was lower but still significant in the townships along the canal. In Crowland, 9.2 per cent of the population belonged to the Church of Rome and 8.4 per cent had Irish origins. Certainly the Roman Catholic religion had now penetrated a regional area that had been essentially a Protestant stronghold after its settlement by the first pioneer immigrants and later settlers from the United States and the British Isles.

The Development of Service Facilities

The selection and determination of canal routes, with the many necessary letters, contracts, and deeds, was conducted by William Hamilton Merritt from his home in St Catharines that overlooked Twelve Mile Creek. As stated in his personal diary, 'I write much, and employ myself incessantly in some things ... [And later that year], I have the camp iron-bedstead placed in the corner of my office, where I keep a good fire, and write till 9 or 10 P.M.' (Merritt 1875).

Much has followed from such slender beginnings. As staff such as surveyors and then engineers, superintendents, clerks, treasurers, and draughtsmen were appointed to promote the canal business and to organize the maintenance, repair, and reconstruction of the channel, administrative offices for the Welland Canal Company were built in 1855 at the west end of St Paul Street, close to the canal and next to the growing commercial hub of St Catharines. As this office building, which also housed the Collector of Customs, survived for some 125 years until the St Lawrence Seaway took over the Fourth Canal, the role of St Catharines as the regional administrative centre for canal affairs became one important service element that helped to maintain the city's regional supremacy over the line of canal communities through the decades.

Other local centres of canal administration arose when toll collection located at Chippawa, Port Robinson, and Dunnville, contributing to the importance and centrality of these canal villages. Also, as two lock tenders were employed at every six locks under one superintendent and the money so obtained was used to pay labour for repairs to that part of the canal work, the locks became local centres of employment.

Not unnaturally with the ebb and flow of passing vessels, their mooring at the wharves and in the harbours along the canal, and their winter berthing when the canal was closed, yards for the construction, repair, and maintenance of vessels arose on the banks in many canal communities. These vessels ranged from sailing ships to steam-powered ships and included tugs, scows, and barges for service on the Welland Canal, the Grand, Niagara, and Welland Rivers, and the Upper and Lower Lakes. Some even crossed the Atlantic to ports such as Liverpool. Noted shipyards developed in conjunction with the shipping on Lake Ontario at St Catharines and Port Dalhousie, centrally within the canal system at Port Robinson and Welland, and for the Niagara River–Lake Erie trade at Chippawa.

Post offices followed the canal from its inception and provided a rough index for the emerging importance of the canal communities. A post office had existed at St Catharines since 1817, but the development of the canal sequence is recognized by a post office established at Thorold in 1826, followed by Dunnville in 1830, Port Dalhousie the next year, Port Robinson in 1835, and Port Colborne and Marshville (Wainfleet) a year later. The period of the Second Canal brought post offices to Wellandport and Port Maitland during 1841, Allanburg in 1846, then Merrittsville (Welland) in 1849, and Humberstone (Petersburg) in 1851 (Campbell 1972).

Newspapers were more a reflection of community size, but they also played a powerful role in community development and identity. They started in St Catharines in 1826 with the weekly *Farmer's Journal and Welland Canals Intelligencer*, which reflected both the canal and its rural setting in its formidable title. This newspaper became the *British American Journal* and then the *St Catharines Journal*. A second weekly, the *Constitutional* (added in 1850) gave St Catharines its second newspaper. The *Thorold News* also started at mid-century, and newspapers at Chippawa the next year and at Welland in 1854.

Various mills, schools and churches, carpenters and wagon-makers, shoemakers, and the other trades and professions of the day were soon added in each community, together with some breweries and distilleries, taverns, and inns. Groceries, butchers, and a variety of stores supplied the needs of vessels, those who lived in the canal settlements, and those who farmed the neighbouring land. A 'grocery boat' plied the Feeder Canal to serve the farms along its route.

The towing of vessels brought in teams of horses and mules, tow boys, stables, teamsters, and blacksmiths. Service activities that emerged as a direct consequence of the new waterway included custom houses, toll-keepers, lighthouse-keepers, and bridge masters. Although the passage of a vessel through the hand-operated locks was undertaken by the ships' crews, the locks required lock tenders appointed by the Welland Canal Company to supervise this operation. Housed next to the locks, their buildings prevailed across the Niagara Escarpment and the Ontario Plain where the locks were most closely spaced.

Whether engaged in the service activities or working on the canal, all those employed in any aspect of the canal had to be housed somewhere nearby. It was not only the mills, trade, and the shipping activities that caused urban expansion. The trades and professions of the era, the churches and schools, inns and taverns, merchants and storekeepers were each added steadily to the communities initiated and promoted by the canal. The service activities are recorded in the various directories of the period such as *Smith's Canadian Gazetteer* and in newspaper advertisements, but the continuity of data is uncertain as not all years are covered, and descriptions vary by place and in the detail contained in each directory.

The Road Network

As Merritt's waterway crossed land that had been settled since the American Revolution, its vicinity reflected the eras of native occupancy and some fifty years of active pioneer settlement. That the land was farmed and settled had indeed helped considerably towards the achievement of the canal, and many of the area's local characteristics were to be used and amended as the canal began to exercise its sway over the inherited environment.

The roads that preceded the canal, and that later provided the setting within which the canal communities developed, had two primary elements (Burghardt 1969). The older system of former Indian trails that had been used as pioneer routes of entry were now entrenched into the landscape through their frequent use by settlers. They stressed the east–west orientation of major physical features such as the shorelines of Lake Ontario and Lake Erie, the Welland River, the Niagara Escarpment, and the former shoreline of Lake Iroquois (an earlier form of Lake Ontario along which Regional Highway no. 81, former Highway 8, is now aligned). These established east–west roads, including the Lundy's Lane–Canboro Road west from The Falls, had necessarily to be crossed by the Welland Canal on its north–south alignment. They became bridge crossing points of the new waterway, which perpetuated their location and their importance as roads.

These earlier roads were supplemented by the rectangular imprint of the surveyor's grid of lot and concession roads across the land. Developed during the 1780s, these straight lines provided the basis for agricultural settlement within each township. Their regular pattern used the front and rear system, under which the base and the side lines of the lots were surveyed. Each lot was 20 chains (1320 feet, 402.3 m) by 50 chains (3300 feet, 1005.8 m), and contained 100 acres (40.5 ha). A road allowance of one chain (66 feet, 20.1 m) was left between each concession, and an allowance of half a chain (33 feet, 10.1 m) between every other lot. This survey system applied in the south along the canal through Thorold, Crowland, and Humberstone Townships, but changed in the north across Grantham Township, where the 'east–west' roads were drawn parallel to Lake Ontario to create a parallelogram pattern of streets within which the village, and then the town, of St Catharines was obliged to develop.

The canal amended this dual pattern of inherited Indian routes and survey roads, for it was promptly accompanied by roads in the new canal settlements that were laid out parallel to the waterway. A typical feature of the canal communities, but not of St Catharines which preceded the canal, therefore involved a change in road direction upon approaching the canal settlement, a change necessary to link the regular pattern of the township's survey grid with that of the road pattern created by the canal.

At the regional level, a 'canal road' also followed the length of the canal. This came into operation in conjunction with the construction works, and was used to bring men, equipment, and supplies to where the works were under way. After the canal opened, it connected 'the urban places that were growing rapidly along the course of the waterway. The canal road, thus, became the transverse road of the eastern peninsula, and replaced the Portage Road for this function' (Burghardt 1969).

Bridges across the Canal

Twelve swing-bridges were constructed by the Welland Canal Company to carry roads across the new water channel. They were necessary to retain or provide land communications across the canal for local purposes between the two banks, and for long-distance stagecoach travel on the main roads of the peninsula. Their significance was to confirm the location of the bridge, the highway, and the associated settlement at crossing points of the new canal system.

An interesting point is that railways and then roads competed with the canal for traffic. These land routes crossed the canal by swing- or lift-bridges. Ironically, the canal administration had to accept responsibility for constructing and operating these crossing bridges for competing transportation systems (Parks Canada 1980). The provision, form, structure, and location of these bridges were each dictated by the canal administration as part of their decision-making process.

As the canal followed natural waterways in the north and was a new body of water only between the Niagara Escarpment and Lake Erie, the only bridge of importance that preceded the canal lay at the west end of St Paul Street in St Catharines. The remainder, in new locations, were located and built during the process of canal construction. These swing-bridges became the economic focus of each canal community, and the community business centres developed nearby on the approaching roads, usually at first on one side of the canal but with growth later extending the settlement across the canal to the other bank.

In between the bridges, the sometimes lengthy intermediate sections of canal functioned as a water barrier to movement on the land. Some farmers objected to this unanticipated break across their property and to the circuitous journey then required to get to parts of their land, but mostly the canal was endorsed and supported for its greater advantages of trade. In those instances where the canal and its locks caused flooding upstream, an even greater hindrance and a potential hazard to people and animals was caused to local farmers.

If the canal had been designed for barge traffic as on the American Erie Canal, then fixed arched bridges to take horse-drawn traffic across the channel and its low passing vessels might have been constructed. However, Canadian circumstances differed from this American precedent. Sailing vessels using the canal required swing-bridges across the channel to permit the passage of vessels with tall masts. Without bridges for the tow animals to cross the canal, the towpath on the Welland remained on the east side of the canal.

Marine traffic on the canal was thus provided with priority in the design of the canal works and in the later operation of the canal. This arrangement was not of great importance until the volume of either land or water traffic increased, but throughout the history of the canal, until high-level bridges and tunnels under the waterway removed this dilemma, it established that road traffic crossing the canal had to be delayed for passing vessels and priority given to the movement of vessels through the canal.

Where locks existed, it was possible for a pedestrian to cross the canal at one or other gate. This offered the possibility of limited subsidiary crossing points, but it did not affect the primary circumstance that where no bridges existed the canal introduced a considerable barrier to cross movements. To cross at a lock was a dangerous practice. According to the obituary notice of Thomas Duffin, lock tender at Lock 2 of the First Canal who was accidentally drowned when slipping off a lock gate into the canal, the gates were not provided with foot bridges and 'no less than eleven deaths have occurred in consequence, it is believed, of this very reprehensible omission. This neglect in the *old* lock gates, will, we are happy to know, be remedied in the *new*; as there is to be attached to each pair of gates [on the Second Canal] a substantial foot bridge, for the convenience and safety of those employed about the locks' (St Catharines *Journal*, 16 November 1845).

The canal settlements developed at the points where the canal could be crossed by one of its road bridges. As the communities developed differently in terms of their manufacturing, service, and residential activities on the two sides of this crossing point, not only the location but also the form of each emerging settlement was thereby influenced strongly by the canal.

This statement applies to all the canal communities. Their central areas lie on one or other side of the canal. The pattern of their activities is asymmetrical, being greater on one side rather than being evenly balanced on both sides.

In the history of world cities, the association is close between the growth and character of cities and their physical location as sea or river ports. Quebec and Montreal, Ottawa and Toronto have this important paradigm in their character. Stourport has been mentioned as a British canal town, but the Rivers Severn and Stour also had significant roles to play in the evolution and character of its urban environment.

Along the Welland Canal, Allanburg, Humberstone, Merritton, Stromness, Thorold, and Wainfleet had no important prior natural watercourse to shape their settlement patterns. Elsewhere, the nuances of urban form and character are indebted to lakes and rivers. Port Dalhousie and Port Colborne grew on lake shores; Chippawa and Welland had the Welland River, Port Maitland and Dunnville the Grand River, and St Catharines and Port Dalhousie Twelve Mile Creek to flavour their form and growth.

The emerging nodes in the landscape to which the development of routes had to conform were created by engineering decisions in that landscape. As the canal, its locks, and bridges were each changed to new locations as each successive canal was constructed, the community and the businesses that had grown in response to previous circumstances might now be adversely affected. To respond to the new situation might require a considerable switch in emphasis, especially when the canal alignment was changed drastically to some new and perhaps distant route, as has happened to the canal communities at the northern end of the canal.

Examples of the canal's direct impact on urban form and character include the one-sided growth of St Catharines to the north and northeast away from the generating impact of the First and Second Canals. Welland grew around an aqueduct that varied in location as the canal alignments changed with each successive canal. Port Colborne, Allanburg, Port Robinson, and Humberstone each grew around their respective bridges and locks, but as these locations changed, the established form of each settlement had to adjust to the new circumstances. Dunnville, which grew at the eastern end of its dam, provides a further illustrative example.

At Thorold, Merritton, and St Catharines, and later at Welland, the canal has even been removed from the centre of the settlements that it so earnestly created. In many centres, and especially around the harbour at Port Colborne, and to a lesser extent at the heart of Welland and next to Port Robinson, new canals designed with engineering criteria to the forefront have slashed through the very community that over the years an earlier canal or canals had helped so considerably to create.

The canal engineers when locating their channels, reservoirs, ponds, locks, and bridges had unwittingly created a new context of urban circumstances within which the canal settlements had to grow. When these canal circumstances were changed in response to new navigational needs to accommodate vessels of increased size, then too the communities created by the earlier set of canal conditions might be adversely affected and have to take on new directions of growth. Renewal and redevelopment because of changed canal circumstances in addition to redevelopment because of technological and social change became unenviable aspects of the canal scene. When the canal changed, as it did on three successive occasions as it advanced into the Second, Third, and Fourth Canal periods, and then again with the arrival of the St Lawrence Seaway Authority, so also did the urban relationships that were involved.

The Agricultural Land Base

Within the townships crossed by the canal, a settled agricultural landscape prevailed. From north to south, the canal passed through Grantham, where the Niagara Fruit Belt later emerged. As reported by Gourlay (1822), 'The soil in its natural state is covered with a black loam, from three to nine inches deep; is of two kinds; the northern part, a sandy loam; the other a brown clay, intermixed with marl, generally rich and productive. The lands are heavily timbered with white and red oak; white pine; beech; sugar and white maple; red and white elm; black and white ash ... White clover, red top, and spear grass, natural to the soil ... white clover, best feeding pasture. A good four year old ox will gain, if attended, from 200 to 250 lb. [90.7 to 113.4 kg]; by running on the commons, or in the woods, will gain 150 to 170 lb. [68.0 to 77.1 kg]. A milk cow will produce, (well kept), 8 lb. [3.6 kg] butter, or 14 lb. [6.4 kg] cheese per week ... Manure is applied for flax, potatoes, oats, Indian corn, wheat, and rye.'

In Thorold, south of the Niagara Escarpment, 'the face of the land is level: the chief part of the timber beech and sugar maple, with plenty of white pine and oak; black walnut and a variety of other timber. The soil chiefly clay and loam; produces, besides wheat, pease, good oats, barley, rye, Indian corn, and buckwheat. Our meadows generally yield from one to three tons of timothy and clover hay per acre; and our fields afford good pasture from the 1st of May to the 1st of December, four months being the ordinary time for feeding cattle in the winter.'

At Crowland, 'the soil is various and much given to grass,' and 'clay is most prevalent.' To the north in Thorold, 'the soil chiefly clay and loam; produces, besides wheat, peas, good oats, barley, rye, Indian corn, and buckwheat. Our meadows generally yield from one to three tons of timothy and clover hay per acre; and our fields afford good pasture from the 1st of May to the 1st of December, four months being the ordinary time for feeding cattle in winter.'

At Humberstone, by 1818, 'much of the soil is rich black loan; some of a yellowish cast and poorer, and a smaller proportion clay. There is a considerable extent of marsh ... Timber abounds in the following order; Oak, pine, hickory, beech, maple, walnut, ash, elm ... Beasts are turned out to pasture about the beginning of May, and taken home during the beginning of November. Wheat is sown in September, and reaped the beginning of August. The pasture is capital. Cheese is seldom taken to the market. After clearing the land, wheat is the first crop, and is often sown the second year, when it is sown down with timothy and clover ... Sometimes the succession is wheat, Indian corn, wheat and grass. On the best spots, Indian corn is grown several years in succession.'

These four townships contained an estimated population of some 3200 by the late 1810s. They had been occupied since the 1780s by United Empire Loyalists, by Mennonite groups, by retired military personnel from the Butler's Rangers, and by Americans seeking cheap or free land. After the War of 1812 immigrants arrived from the British Isles and American immigration was halted.

Niagara (now Niagara-on-the-Lake) was the county town, the principal market, the centre of the import and export trades, and the community with the most stores and services. The settlement pattern elsewhere emphasized the portage communities of Queenston and Chippawa, which fulfilled these functions but to a lesser extent. Inland, St Catharines on Twelve Mile Creek, navigable up to the village and to the base of the escarpment by boats carrying 10 tons (9.1 tonnes) of cargo, and also with flour mills and a salt works, had developed into a rural and agricultural service centre of local importance.

Across this farming landscape, more intensely settled along the Niagara River and next to Lake Ontario than on the heavier clay lands closer to Lake Erie, mills had clustered along Twelve Mile Creek and in the Short Hills, and at selected points of advantage such as at Cook's Mills, DeCew, Beaver Dams, and the Sugar Loaf. Incipient villages with a church, store, or other local service trades had developed at these places. The landscape was settled, and the desire to export the surplus produce from these well-

cultivated lands and their associated rural centres provided one serious justification for the First Canal and then for the improved Second Canal.

The farming community along the line of the canal gained immediately
and considerably from the canal works. Farm animals and wagons might be
used in the construction projects. The labourers and their families had to be
supplied with food and the animals with fodder. Moreover, in a time of
barter, when farmers sold produce to the merchants in exchange for
imports, purchases by the Welland Canal Company and its workforce had
the advantages of payment on a cash rather than a credit basis. Local sales
also avoided the need for an intermediary role by a merchant. One illustration of this impact of labourers on farming is that in the Port Colborne
area, 'a number of hands being employed on the canal during the last season, they consumed all the surplus produce of the neighbourhood, leaving
nothing to be exported' (Smith 1851).

Farms could increase their output for a local market, and exports were
encouraged by the canal. By the mid-1830s, 'it was clear that the Welland
Canal stimulated the development of the country adjacent to it. The canal
spurred on the inhabitants in the more remote areas to improve the rivers
and streams flowing into Lake Erie in order to avail themselves of the
advantages afforded by the canal. This improvement in the more remote
waterways tended to new sources of trade' (Heisler 1973).

New Agricultural Lands

Three areas in the early process of settlement received particular advantages
because of the canal. In the first of these locations, where the canal crossed
the Welland River at The Aqueduct (now Welland), the upstream resources
along the Welland River obtained a new degree of accessibility. The
exports included agricultural produce to Buffalo, timber to this city and to
Tonawanda, and timber downbound to the St Catharines market. This
trading movement encouraged the rapid clearance of the land, the development of surplus products for export, and the expansion of settlement.

Smith (1851) noted that Canboro (now Canborough) on Oswego Creek,
a tributary of the Welland, 'is a pleasant little village, containing about 100
inhabitants, a saw mill, post-office, and two churches, Methodist and Baptist.' And at Wellandport (The Narrows), 10 miles (16.1 km) downstream
between the Welland River and Beaver Creek, 'is a small village, containing
about a hundred and fifty inhabitants, and a steam saw mill.'

In the second location, 13,000 acres (5261 ha) of Wainfleet Marsh on the
line of the Feeder Canal were granted to the Welland Canal Company by

the provincial legislature in 1826. The Gourlay survey of 1817 had recognized the innate potential that existed. 'There are 22 square miles [57 km²] of marsh land owned by the government, which if drained, would be preferable to any other land in the province for hemp, &c; its soil or surface is three foot [9.0 m] deep.' The directors of the canal company received many applications for lots of land on Wainfleet Marsh and in 1829 resolved to lease lots of 50 acres (20.2 ha) adjoining the canal for periods of five or ten years, with the tenant then having the option to buy.

Known also as the great Cranberry Marsh, the reclamation of this important extent of high-quality land centred on the canal village of Marshville (now Wainfleet). The marsh was purchased back from the canal company by the government in 1854 and then by the provisional Welland County Council. Gradually drained over the next few decades, tracts were sold to settlers for their high arable potential.

The most extensive of the three agricultural areas to be improved by the canal was the Grand River valley, which extended for some 130 miles (209.2 km) upstream from Dunnville. The construction of the Welland Canal, and in particular its Feeder Canal from the Grand River, led to the canalization of this river, which should be viewed as an extension of the Welland Canal. The Grand River Navigation Company, promoted as a necessary link for the Welland Canal by William Hamilton Merritt, was chartered in 1831, and work started in 1832. Eight locks were constructed to overcome the falls on the river, one at Indiana, two at each of York and Caledonia, and three at Brantford. Each created milling complexes, and in 1850 Cayuga was selected as the county seat for the newly created County of Haldimand.

Mostly on Indian land granted to the Five Nations, the valley of the lower Grand contained valuable deposits of gypsum to the south of Cayuga from which plaster of Paris, plasterboard, and fertilizer were made. Adit shafts were cut in from the river bank and a tributary of the Grand. The gypsum was ground in a nearby mill, then exported by boat and later by railway. A new local resource had entered the trading scene of the Welland Canal. Pine and hardwoods such as white oak, beech, and maple provided another substantial export. The sawmills along the Grand produced boards, shingles, laths, and barrel staves in large quantities. Some of these products were moved by boat, and rafts of timber were sent downstream through Dunnville and the Feeder Canal until this resource became exhausted.

These developments were anticipated by Merritt, who, at the canal's inaugural ceremony in 1825, observed that 'the banks of the River Welland

and the Grand River abound with an almost inexhaustible supply for fine timber, now useless, which will be floated down to our establishments, converted into lumber, and transported to the entrance of the American canal at Tonawanda, where it must ever find a constant and ready demand, as their borders are destitute of the article. There are likewise quarries of the finest white gypsum, or plaster, ... which will soon become a profitable article of commerce' (Merritt 1875).

An adverse impact of the Welland Canal was that its north-flowing course decapitated the headwaters of several east- or west-flowing creeks and denuded the water supply of their lower lengths. This diminution of flow happened across the upper waters of the Beaver Dams Creeks. The same features of attrition and water deprivation to downstream mills occurred across the upper waters of Lyons Creek in the southern part of the peninsula. And on the Grand River upstream from the dam, fertile Indian lands were flooded and their production lost.

The Canal as a New Source of Water Power

By its act of incorporation in 1824 the Welland Canal Company was granted the power 'to take and appropriate for the use of the said Canal as much water as they may find necessary from out the Niagara River, the said Grand River, and the River Welland.' In the mid-1840s these rights were transferred to the Board of Works for Upper and Lower Canada, then through the Department of Railways and Canals to the St Lawrence Seaway Authority. Here lay an exceptionally important link between the canals as engineering and as a marine highway and the attendant development of land and the advances of urbanization and industrialization along the canal banks.

A major and ubiquitous feature of the canal has always been that its navigable channel provided a regular supply of water, which from the start of the waterway and during its construction encouraged the substantial promotion of mill developments. In an era during which the power of falling water turned the wheels that drove machinery, the Welland Canal performed this function with noted ability (table 1). As Smith stated in 1851, 'a fall of more than three hundred feet ... is such as probably no country in the world can equal in a similar space. And there is no doubt that considerable manufacturing towns will eventually spring up on the canal. The unlimited supply of water power for turning machinery ... offer advantages such as few places in the Province possess for similar undertakings.'

Table 1 The population, mills, and churches in the canal communities at 1851

Mills and industries	Port Dalhousie	St Catharines	Thorold	Allanburg	Port Robinson	Welland (Merrittsville)	Humberstone (Petersburg or Stonebridge)	Port Colborne	Dunnville	Chippawa
Population	200	3400	1200	300	400	1500	200	160	1000	1000
Grist mills	1	6	5	2	1	2	–	–	3	–
Sawmills	–	1	1*	2	1	2	–	–	4	1
Woollen factories	–	1	–	2	1	–	–	–	1	–
Tanneries	–	–	1	1	–	–	–	–	1	2
Machine shops	–	5	–	–	–	–	–	–	–	–
Axe and edge tool	–	1	–	–	–	–	–	–	–	–
Carding machine	–	–	1	–	–	–	–	–	–	–
Cotton factory	–	–	–	–	–	–	–	–	–	–
Broom factory	–	–	1	–	–	–	–	–	–	–
Plough factory	–	–	1	–	–	–	–	–	–	–
Soap and candle factory	–	–	1	–	–	–	–	–	–	–
Pail factory	–	1	–	–	–	–	–	–	–	–
Pottery	–	–	2	–	–	–	–	–	–	–
Plaster mill	–	–	1	–	–	–	–	–	–	–
Last factory	–	1	–	–	–	–	–	–	–	–
Ashery	–	–	–	–	1	1	–	–	–	–
Cloth factory	–	1	–	–	–	–	–	–	–	–
Shipyard	1**	1	–	–	1	–	1	–	1	–
Foundry	–	2	–	–	–	–	1	–	–	1
Brewery	–	–	–	–	–	–	–	–	1	–
Distillery	–	2	–	–	–	–	–	–	1	–
Marble factory	–	1	–	–	–	–	–	–	1	–
Churches†	1(E)	7(2B, 1C, E, F, M, P)	3(C, E, M)	1(M)	2(E, P)	–	–	–	2(E, P)	1(P)

Source: Based on Smith 1851.
*Includes a carding machine
**Smith 1846
†B = Baptist, C = Catholic, E = Episcopal, F = Free Church, M = Methodist, P = Presbyterian

Many sites along the canal offered this advantage of water power. Regulations and terms laid down by the Welland Canal Company for the occupation of sites and the consumption of water, the so-called water privileges, gave first priority to factories requiring expensive machinery and employing a large labour force. Then came grist mills, after them the carding and fulling mills, then planing, turning, pail, last, wainscot and sash making, and finally sawmills.

The locations with hydraulic privileges were soon recognized for their great industrial potential, greater than that which prevailed on the pre-existing waterways and available through the construction of short hydraulic raceways from the canal. Even by 1830, within a year of the canal's opening, an advertisement placed extensively in Canadian and American newspapers by the Welland Canal Company listed 4 acres (1.6 ha) at Dunnville with a grist mill, carding machine, and three sawmills; a mill at Wainfleet; 5 acres (2.0 ha) at the Aqueduct (Welland); a town plot of 75 acres (30.6 ha) at Port Robinson; 70 acres (28.3 ha) at Allanburg including a sawmill; 100 acres (40.5 ha) at Thorold with a grist mill, two sawmills under construction, and one in operation; four mills or factories including a furnace at St Catharines; and a sawmill and a dry dock at Port Dalhousie.

The types and location of canal sites with industrial water opportunities varied considerably. Of prime importance were the canal locks, where water might be diverted from upstream around the lock chamber to provide a fall equivalent to the depth of the lock. These points of power advantage existed at both ends of the Deep Cut, at Port Robinson where the canal was taken down into the level of the Welland River, and at Allanburg where the canal was lowered to the level at which it crossed the escarpment. They occurred at Port Colborne where the canal flowed into Lake Erie, and along the northern length of the canal between Merritton and Port Dalhousie where it crossed the gently sloping Ontario Plain between the Niagara Escarpment and Lake Ontario. The most extensive potential of all lay between Thorold and Merritton, where there was the greatest change of levels and the greatest concentration of locks on the whole system as the canal crossed the formidable slopes of the Niagara Escarpment.

A second opportunity for water privileges existed where dams or weirs were constructed across the water system. This existed where the Grand River was impounded by the dam at Dunnville. Here, water at a higher level upstream from the dam was diverted to the east bank of the river. Mills located between the canal and the river below the dam could use the water diverted into the Feeder Canal to drive their machinery, and a short tail race then returned this water to the Grand. After the weir for the Second Canal was constructed across Twelve Mile Creek at Port Dalhousie, this dam could also be used for the generation of water power.

A third set of opportunities occurred where the canal crossed the natural drainage system at a higher level. At each of the four quadrants, the potential existed to construct a mill race and locate a mill. At Welland the main line of the canal crossed the Welland River, and at Wainfleet the Feeder Canal was at a slightly higher level than the small nearby tributary of Big Forks Creek.

The most ingenious solution of all was to construct a mill flume along the top of the canal bank. This practical solution was achieved most proficiently and extensively along the eastern side of the canal to carry water from Merritton to St Catharines. As the canal gradually dropped to a lower level through a series of locks along this route, the top of the bank at St Catharines immediately to the south of the town's main street (St Paul) was some 70 feet (21.3 m) above the valley bottom. Moreover, after the water was used by a mill, rather than let it cascade uselessly down the canal bank, a series of parallel raceways were constructed to follow the contours along lower levels of the bank.

The availability of water power at selected points along the canal system soon introduced a differential element in both the number and the category of mills located in each place along the canal, and in the population size of the associated canal communities. In sequence by size in 1847 were St Catharines with twelve mills and Dunnville with ten. Of the smaller mill centres Allanburg had four, Thorold three, Marshville two, and Port Robinson one mill (table 1).

The grist and saw mills were ubiquitous. St Catharines, Dunnville, and Allanburg had carding, wool, or fulling mills, and St Catharines had a further variety that included a turning and a cement mill, a foundry, a tannery, and a pump and axe factory. Even so, only about 23 per cent of the available horsepower capacity was used, and the total of used and available water could be almost doubled if the velocity was increased by keeping the levels 6 inches (15 cm) higher. The line of the canal across the Niagara Escarpment from Thorold to St Catharines had the largest reserves and the greatest availability of additional potential water, whether or not the velocity of flow was increased.

The Milling Industry

The mills introduced employment and, thereby, a parallel demand for service activities as the local population increased. The mills also promoted local agriculture and encouraged trade on the canal through the inflow of raw materials such as wheat and timber, and the export of processed commodities that included barrels of flour, meat, timber, and other agricultural products.

As the mill sites along the canal offered the advantage of a more regular water supply than existed along the natural, now often intermittent, local streams, a greater volume of flow, and the transportation advantages of the canal, these canalside mills were soon the largest milling units of the Niagara Peninsula. They had a greater run of millstones, a higher output of barrels of flour, and employed more people than elsewhere. The canal introduced not only a change in distribution, but also a more concentrated output in a few selected centres.

Even so, the presence of the new mills was not always advantageous, and conflict situations arose between the mill owners and canal interests. Only a certain amount of water was available through the Feeder Canal, and any flow that was diverted and not returned to the canal meant that less was available downstream to float ships. The clash was severe, in that the mills served important needs required by the community and also contributed steadily to the traffic carried by the canal. To close down a mill because of a low water supply in the canal was to reduce the tonnage carried through the canal.

Grist mills to grind flour or meal were the most favoured. They also used less water than sawmills for 'one saw-mill with one saw will consume in one day as much water as it would require to keep a grist-mill with four run of stones in operation for six days' (Upper Canada Assembly 1836–7). Even so, sawmills were essential to the pioneer economy. Their timber was in high demand for a multitude of local, domestic, agricultural, building, and manufacturing purposes, and nearby export markets that included the expanding American centres of Buffalo and Tonawanda, the Upper Lakes, and the ports of Lake Ontario, as well as across the Atlantic to England.

A worrisome issue was that waste from sawmills was often deposited in the tail race, then returned to the canal. Pollution of the new water resource had started, and a common complaint from the masters of vessels was that it was more a canal for mills than for vessels. When the slow-flowing channel was used for effluent purposes, bars might be created that reduced the water level and led to demands for dredging below the mill sites. In 1835, a one-third remission of rents was granted to the owners of sawmills who bore a proportion of the downstream dredging expenses.

When water was diverted from the canal and not returned to it, downstream sites were deprived of their envisaged water-power potential and the canal's use for navigation was also diminished. This loss occurred especially in summer during periods of low water flow. The worst culprits were located along the Feeder Canal at Dunnville and Marshville (Wainfleet), and then at Welland and Port Robinson, where in each instance the canal was at a higher level than the local river system. This loss of water did not occur north from Thorold across the Niagara Escarpment where the majority of mills had located, because the canal was here in a valley and water diverted from the channel at a higher level to serve the mill wheels flowed back into the canal down the side of the valley.

The most serious loss of water was along the Feeder Canal, for if this was diverted away from the canal, all the lower lengths of the canal would be deprived of water. In 1847 the letter books of the Welland Canal Company stated that the supply of water to the mill flumes at Dunnville was stopped for 'many weeks together in order that the navigation may not be deprived of its proper supply.' Rent abatement for such stoppages was permitted only if the supply of water was discontinued over a six-month period.

These problems of conflict and the fluctuating supplies of water from the Grand River were not resolved when the channel was widened and deepened for the Second Canal. These works admitted more water, but the source of supply remained the same and, as more mills were added to the water system, the demand for mill water also increased steadily.

A more extensive and assured water supply had to await the deepening of the canal to the level of Lake Erie in 1881, but even here lake levels fluctuated. A safeguarding factor was that by this time water power to supply mill enterprises no longer occupied the prime place of importance that it had held over the earlier decades of the century. In marked contrast with its predecessors, no line of mills or industry followed the Third Canal across the brow of the Niagara Escarpment. The technological means of factory operation had changed to coal and steam rather than water power.

Apart from vagaries in the flow of the Grand River as the head pond for the flow of water in the canal and its water-power capability, two further adverse factors affecting mill operations by water power were the winter closure of the canal because of ice and below-freezing temperatures and the need to de-water sections of the canal for its annual maintenance and cleaning operations. Both items might be offset if the mills and factories that relied upon the canal for water made temporary use of coal-powered machinery during that period, or if they devoted the same period to their own maintenance and repair schedules.

A late example of this variation in water power is that when the electric streetcar system was supplied with power from the locks of the canal, closing down the supply of water in the canal meant that horses had temporarily to be used to tow the electric vehicles. But despite these difficulties in the supply and use of water for mill and other operations, the water-powered mill remained the economic lifeblood of the canal settlements until well into the twentieth century. An illustration of this late use of water power is provided by the mill at DeCew Falls (Robson 1994). Built in 1872 and using water diverted from the canal at Allanburg, it was taken over and operated as a grist and flour mill by Wilson Morningstar and survived in active production until his death in 1933.

The Canal as an Urban Catalyst

When these several elements of trading flow and the shipping industry along the canal are combined with agricultural progress, the mills, and service developments, together with the new components such as the novel flow of water and the bridges and roads at strategic locations in the landscape, then it is clear that the Welland Canal from the 1820s onwards had created an innovative and vital force for urban formation and growth.

The overall pattern of these canal settlements was that of a large cross. The centre piece from north to south extended along the main line of the canal between Port Dalhousie and Port Colborne, and the transverse arms east along the Lower Welland River (Chippawa Creek) to Chippawa on the Niagara River and west along the Feeder Canal to Dunnville and Port Maitland on the Grand River. The area of overlap served by both systems and with the most intense shipping activity lay between Port Robinson and The Junction (Helmsport), south of Welland.

Within this cruciform pattern, the analogy of beads along a string is not inappropriate. Most of these 'settlement beads' made their first appearance in the landscape during the period when the works of the First Canal were under construction, then expanded with mills, canal, and service activities when the waterway opened. This youthfulness of settlement applied along the main line of the canal from Port Dalhousie to Port Colborne and along the Feeder Canal, whereas other centres, including St Catharines, Humberstone, Chippawa, and Port Maitland, had reached an earlier degree of maturity.

At St Catharines, the canal caused this previous village to balloon from a small but growing rural-agricultural centre into a new character as an industrial town of increasing importance. The small pre-existing village of Humberstone inland from Lake Erie and its marshy fringe was encouraged to grow because of the canal. At Chippawa the declining portage centre was reinvigorated into active community life, and Port Maitland was renewed with expanded port facilities. Through the canal endeavour, these centres each obtained an extended lease of life that included a greater marine commitment and a much larger trading catchment area.

Pioneer names that preceded the First Canal often reflected an important feature in the landscape such as the descriptive Sugar Loaf Hill or Beaver Dams. Names might also have identified a feature through an associated pioneer settler or family, as at Cook's Mills or Brown's Bridge. St Catharines and St Davids were both named after prominent local individuals of the period. Stonebridge, which became Petersburg after Peter Neff, a pioneer of German descent who arrived in Upper Canada from Pennsylvania, provided an interesting combination of these two approaches. In addition, the Welland River and townships such as Grantham, Wainfleet, and Humberstone had been named by John Graves Simcoe, the first lieutenant-governor, to replicate the British scene in Upper Canada as an encouragement for Loyalist settlers.

A different policy was pursued when the new canal communities arrived on the scene (Jackson 1989). Port Dalhousie, previously simply Welland Canal Harbour, was named after Sir George Ramsay, the 9th Earl of Dalhousie, governor-general of Canada from 1819 to 1828. Former Gravelly Bay, named in the pioneer tradition after its sandy beach, was redesignated Port Colborne after Sir John Colborne, 1st Baron Seaton, lieutenant-governor of Upper Canada from 1828 to 1836. Along the middle section of the canal, Port Beverley, which became Port Robinson, used first the Christian and then the family name of Sir John Beverley Robinson, chief justice of Upper Canada in 1829, speaker of the legislative assembly, and president of the executive council.

Allanburg recalled William Allen, president of the Bank of Upper Canada, vice-president of the Welland Canal Company, and a member of the legislative assembly. A note of relevance may be that the Welland Canal Company in 1836 issued company bills in $1, $5, and $10 denominations. On the Feeder Canal, Dunnville obtained its name from the Honourable John Henry Dunn, receiver-general of Upper Canada in 1824, president of the Welland Canal Company, and a major shareholder of this company. Port Maitland honoured Sir Peregrine Maitland, who was appointed lieutenant-governor of Upper Canada in 1818. The Welland Canal through its naming policy thereby became an important biographical reminder of the Family Compact, the oligarchy of leading families that controlled public expenditures and directed public works.

At Thorold, first known as Stumptown after the clearance of its forested environment, the name St George after George Keefer, a major landowner and promoter of the town, was considered, but Thorold (after the name of the township) was selected when Beaver Dams was ousted by the canal from its earlier status as the most important centre in the township. Thorold in turn was an English introduction after Sir John Thorold, an MP for Lincolnshire interested in colonial policy.

The canal was distinctive in many respects, including its names. The above account has recognized its benefactors, but 'he who was justly called the Father of the Canal had not received the honour due him, insomuch as his name has not found a place in any of those villages which had arisen from his wisdom and energies and in order to perpetuate his name, it was resolved that the place [Aqueduct] should be called Merrittville' (St Catharines *Journal*, 1850). Thus was William Hamilton Merritt at last given due local recognition through a name on the land for a community that he had instigated. Later transferred to Merritton when Merrittville (Merrittsville) became Welland, now only this canal centre of all those along the canal recalls its prime promoter.

4

The Canal Settlements by the Early 1850s

By 1850 the line of canal settlements between Lake Ontario and Lake Erie had replaced the former pioneer concentration along the Niagara River. William Hamilton Merritt was correct when, at the sod-turning ceremony in 1824, he.avowed: 'This canal ... will afford the best and the most numerous situations for machinery, within the same distance in America; wet or dry, warm or cold, we always have the same abundant and steady supply of water ... We will mingle in the bustle and active scenes of business; our commodities will be enhanced in value, and a general tide of prosperity will be witnessed on the whole line and surrounding country' (Merritt 1875).

Some twenty communities that had obtained their activities, form, and character from the canal can be identified by the early 1850s. They included lake and river ports, small villages along the line of the canal, and industrial towns with a greater range of manufacturing, commercial, and service facilities. The canal had become the major instigator of a new settlement pattern inland from the frontier and across the Niagara Peninsula from Port Dalhousie to Port Colborne (Watson 1945 a, b).

The transition in this discussion from the canals as engineering to the examination of the associated canal communities in terms of form, function, and character involves a substantial change in the availability and use of source materials. Government reports concerning the canals are numerous and well documented. The switch in source material is to local authors, directories, and newspaper accounts. Not all communities and certainly very few industries have histories, and the few available studies are dispersed over the libraries, museums, and private collections along the canal. Outside the provincial and federal archives, the single best source within the Niagara Region is the Special Collections Division of the Brock University Library.

The Canal and Canal River Ports

Four centres designated as ports by their prefix emerged by the 1850s: Port Dalhousie at the northern entrance on Lake Ontario; Port Colborne at the southern entrance on Lake Erie; Port Maitland, which preceded the canal at the mouth of the Grand River and the second point of entry from Lake Erie; and Port Robinson at the mid position of the canal, where boats were locked into the Lower Welland River. Chippawa, Dunnville, and Welland, though not designated as ports also functioned in this capacity where the canal interacted with traffic on the Niagara, Grand, and Welland Rivers respectively.

At each of these seven ports, through and local shipping was serviced from chandlers and other stores. Here vessels were repaired and equipment replaced, their crews had to be victualled, and the tow animals stabled and fed. Cargoes had to be stored on the wharves, in sheds, warehouses, or elevators, and trans-shipment to vessels of different size was undertaken. As during the winter months the harbours might be filled with vessels awaiting spring and the end of ice, the ports were busy year-round, though the navigation season was their busiest period.

Port Dalhousie

Known locally as 'the peninsula within the peninsula,' Port Dalhousie developed at the eastern end of a narrow peninsula of land that rose steeply to its crest above the bluffs of Lake Ontario (map 3). This promontory, described as a cape and forested with pine trees, was bounded on the north by Lake Ontario and to the south by the impounded waters of Twelve Mile Creek. At its eastern end, a sandbar created by the currents of Lake Ontario crossed the mouth of Twelve Mile Creek, its open but fluctuating entrance being on the western side across the creek and away from the promontory. Here, before the days of the canal, lay a small natural harbour protected by bluffs from the westerly winds and surrounded by marsh, but of sufficient depth and size to accommodate small vessels that could travel upstream to as far as the base of the Niagara Escarpment.

The route selected for the First Canal cut across the head of this promontory. A lock with protective breakwaters took the canal through the sandbar into Lake Ontario, and a floating towpath was used to tow vessels through the ponded area of Twelve Mile Creek. The settlement began at the end of the promontory, where a number of shanties were constructed to house workers on the harbour, the lock, and the breakwaters. Boarding

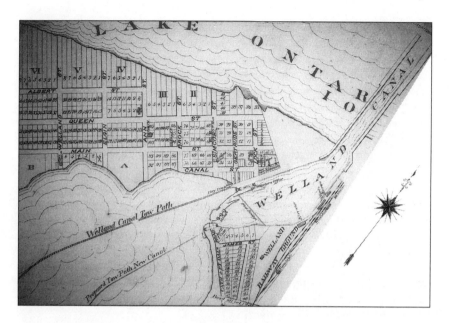

Map 3. *Port Dalhousie: The northern port of entry.* The First Canal entered Lake Ontario through a lock and breakwaters located at the western end of the sandbar across the mouth of Twelve Mile Creek, a location indicated by the narrow lots at the eastern end of Queen, Main, and Canal Streets. The Second Canal was located east of the former channel, its entry lock being inland where the weir constructed across Twelve Mile Creek created an Upper and a Lower Harbour across the river channel. (1876)

houses, stables, a stone house, and small service establishments followed as the canal opened for traffic.

In early 1826 Nathan Pawling advertised: 'The subscriber, having laid out a *Town Plot* on an extensive scale, with regular and extensive streets, ... now offers BUILDING LOTS For sale, to *Actual Settlers*, on moderate terms. Located on a Peninsula at the confluence of the Welland Canal with Lake Ontario, with a fine commanding view of a spacious Harbour (which will contain about 280 acres [113.3 ha]), on one side, and the Lake on the other – a high, dry, and healthy situation ... On the completion of the Welland canal ... numerous commercial advantages will immediately present themselves' (*Farmer's Journal and Welland Canal Intelligencer,* 5 April 1826).

This physically determined layout centred on Main Street followed the spine of the peninsula. It did not conform with the standard survey grid of Grantham township. The village alignment then changed direction to enter Louth township at the western end of the peninsula. This road layout was restricted to the western bank of Twelve Mile Creek, and did not cross to the east bank, which was separated from the village by the extensive ponding that existed behind the lakeshore bar. The main road to St Catharines followed the crest of the promontory, then turned south along Twelve Mile Creek, crossing the creek at Welland Vale some 2 miles (3.2 km) inland to enter St Catharines via Ontario Street. As Twelve Mile Creek, a meandering river set within steep banks, provided a sharp barrier to east–west movement on land across the Ontario Plain, Port Dalhousie offered a suitable harbour but an isolated situation. The first inland crossing of the creek was inland at Welland Vale, now St Catharines.

A major advance to the progress of this incipient settlement occurred in 1837, when Robert Abbey established a shipyard next to the canal in order to build various sailing and steam vessels. Followed in 1850 by Alexander Muir's addition of a ship repair business, which was later greatly to expand, these marine activities added to the service character of the village and were largely responsible for its growth.

With the advent of the Second Canal, new piers and a new canal exit to the lake were added at the east end of the sandbar, and the earlier route and its piers were abandoned. A dam or weir was added across Twelve Mile Creek that effectively divided its outlet into two units, an Upper or Inner Harbour, which became known as Martindale Pond after a settler of that name, and a Lower or Outer Harbour next to the lake. The floating towpath now crossed the Upper Harbour only.

The dam introduced a source of potential water power equal to the height of this embankment. The first lock of the Second Canal was placed at its western end, that is, next to the promontory but to the south and around its edge from its predecessor, and the waste weir was constructed at the opposite end. A swing bridge at the lock and then the dam and weir across the harbour entrance were followed by a road (now Lakeport Road), which crossed the harbour entrance of Twelve Mile Creek, connected the two banks, and relieved the earlier isolation of the village. Here now lay an alternative route on the east side of Twelve Mile Creek to St Catharines.

The focal point of village development had changed from the end of the peninsula's high land above the entry lock of the First Canal to a position next to the first lock of the Second Canal. It now lay on the west bank of the greatly extended harbour and at the point where the harbour was crossed by the weir and its road. Main Street along the peninsula now turned into Lock Street at right angles to the canal and its first lock. As described by Smith in 1846, 'A harbour has been formed, having a basin of 500 acres [202.4 ha] in extent, with a depth of water of from twelve to sixteen feet [3.7–4.9 m]. There is a small village on the east side of the canal, in the township of Grantham, five miles from St. Catharines where is a ship yard. Port Dalhousie contains about 200 inhabitants, two stores, one tavern, two blacksmiths.'

Five years later, doubts were expressed about further expansion through the close proximity upstream of St Catharines as a larger, competitive centre. 'The village, however, does not grow much, St. Catharines absorbing the business of this end of the canal' (Smith 1851). The first two locks of the Second Canal had been constructed to larger than the standard size along the canal in order to enable lake vessels to move easily upstream to this expanding industrial town. The history of Port Dalhousie was changed by this engineering decision: from being a small independent entity on the First Canal, it became, with the Second Canal, a village that developed as a port in conjunction with its bigger, upstream neighbour.

Port Colborne

As described in 1832 by Picken, Humberstone township, in which Port Colborne was located, had 'soil, clay and black mould. Advantages, bounded on Lake Erie, and the dry parts of the township well settled. Disadvantages, a great part is Tamarack and Cranberry marsh, the land generally low and flat, the front of the township thinly settled, and not mill streams.' Here, by contrast with the natural and then the later engineered expansion to the harbour at Port Dalhousie, the harbour of Port Colborne at the southern end of the canal was a man-made creation from its inception (Koabel 1970; White 1969).

Physically, the site lay behind the sanddunes and marshes that lined the shore of Lake Erie, and inland at the edge of the limestone highland lay the small village of Petersburg (later Humberstone). When in 1833 this newborn centre on Gravelly Bay was named Port Colborne and established as a port of entry, Mrs Hamilton Merritt described the scene in her diary. 'The ground here is all marsh on both sides of the canal. There is a small ridge, with one white house and a store and some shanties.' The ridge on the western side of the canal is presumed to be the earth excavated from the canal and placed between the canal and the ditch that followed it. On the eastern bank, not mentioned by Mrs Merritt, the ridge was flatter and some 40 feet (12.2 m) wide to carry the towpath and a roadway. As Merritt continues, 'Eagles' nests were viewed along the lakeshore road to Sugar Loaf Hill, which wound between trees, and had a large marsh or pond to the north.'

Before the days of the canal two small mills operated. Two miles (3.2 km) east on Oil Creek, a mill ground flax seed into meal and linseed oil, and a sawmill operated at Gravelly Bay. Both were operated by marsh water drawn to the lower level of the lakeshore by slender channels, and at the sawmill logs were stored behind a minor breakwater in the lake. The marsh and its inhospitable conditions, causing diseases such as cholera and typhoid, and the difficulties of building on saturated ground drew the first land grants south to Petersburg (later Humberstone) in 1796, six years earlier than the first grant next to the bay.

It was solely the canal that formed then shaped the place soon to be known as Port Colborne. Shanties and boarding houses were erected next to Lake Erie close to the lock constructed at the southern entrance to the canal. This site became a local market centre in conjunction with Petersburg for the surrounding farmers. In 1834, when William Hamilton Merritt purchased 100 acres (40.5 ha) of land across the canal, there were six houses and a store.

Merritt, the canal entrepreneur, established a post office, which was located in the small general store on the east bank of the canal, and provided the community with an advantage over Petersburg, which did not obtain a post office until mid-century. Together with the few homes, this canal development clustered next to the lock at the lakeshore, where the canal dropped some six feet (1.8 m) into Lake Erie. The streets of this survey, which determined the form of the new development, had two major streets (East and West Streets) parallel to the canal and along the banks, as at Wainfleet on the Feeder Canal, and other streets were drawn at right angles to the canal on an alignment some seven degrees west of south (map 4). They met with the earlier, straight east–west and north–south survey lines of Humberstone township, with the requisite change of direction at Fares and Elm Streets. To the north, east, and west these streets in a checkerboard fashion extended into marshland, and in the south a lakeshore road that followed the sand hills crossed the canal at its entry lock.

The various streets of the Merritt survey were named patriotically after members of the British royal family. Adelaide commemorated Queen Adelaide, the wife of William IV; Victoria, the queen at the time of the land survey; Kent, the Duke of Kent, the father of Queen Victoria; Charlotte, Queen Charlotte, the mother of William IV; and Clarence, the duke of

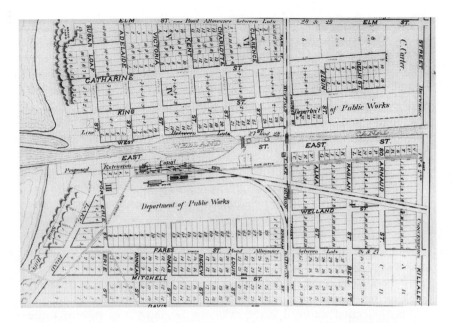

Map 4. *Port Colborne: The southern port of entry.* The First Canal entered Lake Erie by a lock constructed through the lakeshore sand dunes. The artificial harbour behind the dunes was widened for the Second Canal, and a new lock and bridge were then constructed inland to provide the bar of an H-shaped pivot around which the community developed on both sides of the artificial harbour. (1876)

Clarence, later William IV. In progression inland from the earlier Sugar Loaf Street on the west side of the canal, Catharine Street was named after Merritt's wife, and King was named in the popular fashion of the period. The survey stressed land west of the canal for development, but growth trends were for the village to follow the canal to the north rather than spread east–west, which would have removed development from the canal, its lock, and bridge as the focus of activity.

With the arrival of the Second Canal, the point of emphasis in the growth of Port Colborne was transferred inland from the lakeshore to the first lock of the Second Canal on Clarence Street at the head of the harbour, where an accompanying swing-bridge crossed the canal. The earlier lock and bridge at the entrance to the canal were replaced by a government ferry. Smith (1846) indicated that the village was smaller than nearby Humberstone, but with a population of 150 it had attracted a post office, with mail three times a week, and a resident collector of customs. Its professions and trades included a 'steam grist mill (not at present in operation), one store, three taverns, one baker, one grocery, one shoemaker.' The reference to location, 'situated on Lake Erie, at the mouth of the feeder of the Welland Canal,' would seem to indicate that the southern extension of the canal was still not accepted as the major port of entry. This may have led to 'skepticism concerning the need for a Port Colborne harbour,' and may also have stunted early commercial growth as potential businessmen hesitated, waiting to see if the works would be abandoned or the route changed (Koabel 1970).

Smith (1851) noted that, through the Second Canal, the size and potential of the village had increased greatly. 'A large basin has been formed a short distance from the entrance, capable of holding two hundred vessels; as many as one hundred and eighty sail have been in it at one time.' Even so, the population was stated to be about 160, a figure that probably excluded the hands working on vessels in the canal. The land in the vicinity was still extensively forested, for, as the author continued, 'After passing the Junction [with the Feeder Canal] most of the timber was still standing, the clearings being few and far between.'

This extended artificial harbour of the Second Canal proved to be a major factor in the subsequent growth and development of the port. Port Colborne now had sufficient merit to resist competing claims from Port Maitland at the mouth of the Grand River, where the natural harbour was stated to be safer and more spacious and to provide easier access for vessels, but when it was later argued that the envisaged Third Canal should be diverted from the Feeder Junction at Welland south to the mouth of the Grand River, the decision of the 1830s to develop a harbour at Port Colborne was not endangered.

Port Dalhousie and Port Colborne had marine and inland catchment areas of different calibre. Port Dalhousie was under the shadow of the nearby larger centre of St Catharines, its lake connections were across the lake to Toronto, the provincial capital, and its export trade was to either Montreal or the Oswego Canal. Port Colborne had no large-size urban neighbour, its trade from the Upper Lakes included the downbound grain trade, and competition for growth and trade was with the American port and industrial city of Buffalo, which commanded the shipping trade on Lake Erie.

With settlement around each lake expanding, and with both lakes being divided about equally between Canada and the United States, the densest clusters of settlement lay on the Canadian side of Lake Ontario and on the American side of Lake Erie, a regional distinction that favoured Port Dalhousie rather than Port Colborne. The urban centres along the canal had access through the two ports to each of the Lower and the Upper Lakes, but the presence of locks somewhat mitigated these trading advantages through adding the burden of tolls to the costs of transportation. This became a differential element in industrial location, depending upon the number of locks that had to be traversed for the import of raw materials and the export of processed and finished products.

Port Robinson

Inland at the centre of the peninsula, on the east side of the main-line canal system and the connecting port between the canal and the Lower Welland River, Port Robinson (at first Port Beverley) developed in the northeast quadrant of the angle between the First Canal and the Welland River (map 5). Located at the southern end of the Deep Cut, it soon became the largest port on the canal in the terms of population size, industry, and services. Work gangs and their shanties initiated the community within an area of pioneer agricultural settlement.

Upon completion of the First Canal, two lift locks linked the canal with the Lower Welland River. The intention for the original canal had been to dig down to the level of the Welland River, but the accidents of achievement in the Deep Cut prejudiced a direct link in favour of a canal at a higher level. Port Robinson therefore arose as an FAOF community, but the village soon became a port of entry. By 1846 (Smith), with a population of some three hundred, it housed 'the "headquarters" of the coloured company employed for the maintenance of order on the canal.' The village had a grist mill, a post office, an Episcopal and a Presbyterian church, and a collector of canal tolls. The trades and professions, more substantial than at the two lake ports, included three each of stores, taverns, groceries, and tailors. There were two each of wagon-makers, blacksmiths, and shoemakers, and one saddler, baker, watchmaker, and tinsmith.

With the arrival of the Second Canal, the wooden locks of the First Canal were converted into a dry dock, a major shipbuilding and repair enterprise run by the brothers John and James Abbey. This business took advantage not only of the vacant locks and their water power, but also of Port Robinson's central location on the canal, its associated systems of water navigation, and the availability of timber that could be brought downstream along the Welland River. Smith (1851) recorded a population increase in the growing centre to about four hundred inhabitants.

Port Robinson was shaped and defined by its dual water systems: the

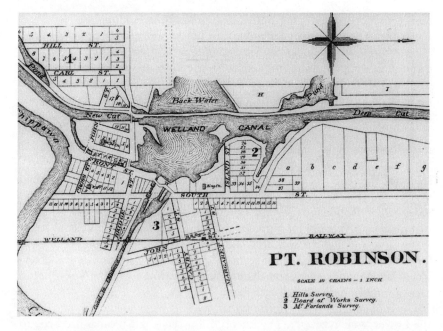

Map 5. *Port Robinson at the southern end of the Deep Cut*. A Feeder Canal constructed from the Grand River at Dunnville brought water along Front Street to a canal basin, where it was fed through the Deep Cut to the northern length of canal and through a canal cut with locks to the Welland River (Chippawa Creek). (1876)

Welland River and the Feeder Canal of 1829, which became the through line of the canal in 1833. The village developed on the outside curve of a sharp bend in the Welland River where the Feeder Canal flowed in from the south to a large reservoir that bounded the north of the village at the entrance to the Deep Cut and its canal channel. This reservoir supplied water to the canal north towards Lake Ontario and to the two canal locks for descent into the Welland River. Because of this physical situation, only Port Robinson of all the canal communities received all through vessels on the canal system. The village had more canal traffic than any other location along the waterway.

Two opportunities for water power existed: the first at the locks between the reservoir and the Welland River, a site that soon attracted a sawmill and the shipbuilding enterprise; and the second where the Feeder Canal ran close to the Welland River but at a higher level. The latter by 1834 had attracted a plaster mill with supplies of gypsum from the Grand River, and grist mills were soon to follow.

The village grew, contained by the First Canal to the west and the cut to the Welland River, its locks, and the mill pond to the north. The main street (then Front, now River Street) ran parallel to and along the east side of the Feeder Canal. A swing-bridge crossed the canal, and a second swing-bridge the Welland River at the south end of the village. As Bridge Street at right angles to Front Street also crossed the First Canal to the west of the village, and another road (now Canby Street) served the agricultural area to the east, the village combined centrality on the water system with being a centre of road communications for the surrounding agricultural areas.

When the Second Canal was constructed, this New Cut (so named) took the form of a wider channel across the village along its western side. A section of the old First Canal was retained to provide a supply of water to the mill complex next to the Welland River, but the northern length was filled to provide new sites for development. Within twenty years of its founding, the village was having to adapt its physical structure to changes introduced by the canal.

Welland and Helmsport

Known first as The Aqueduct (then Merrittville or Merrittsville, now Welland), where the Feeder Canal then the First Canal crossed the Welland River, this location offered a strong impetus for growth through the potential of water power at the aqueduct where the canal crossed the Welland River at a higher level (map 6). There was also access to the considerable catchment areas of the Upper Welland River and the Grand River and the opportunity to manufacture and service vessels plying these waterways, the Feeder Canal, and the Canal, circumstances not too dissimilar from those at Port Robinson.

The aqueduct was constructed of local white pine at approximately the end of Seeley Street. As its flow of water was at a higher level than the Welland River, a space of about 10 feet (3.0 m) existed between its base and the height of the river (Michener 1973). This difference was sufficient to permit the passage of small boats, barges, and rafts in the river below. Whether or not a lock connected the aqueduct downstream to the Welland River is uncertain. One is shown in Michener's later interpretation, but this is not depicted on the map series available through the Commissioners of Public Works (Brock University Library, 1841–80). It might of course have been removed in the intermediate period, but this author suspects that no

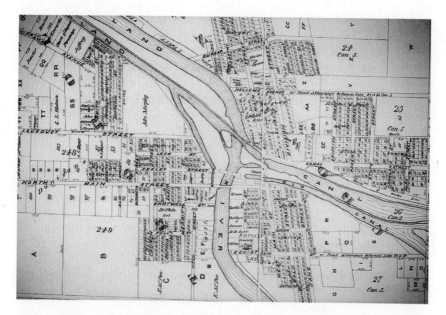

Map 6. *Welland and its aqueduct.* Welland arose where the Feeder Canal and later the First Canal crossed the Welland River by an aqueduct. The commercial centre grew where Main Street crossed the canal and North Main Street (now Niagara Street) crossed the river. (1876)

connection between canal and river existed until the Second Canal was constructed. There was little need for a lock as the trading flow downstream could pass under the aqueduct, and canal and river were connected downstream at Port Robinson.

Upstream the Welland River had extensive stands of timber, and several small ephemeral ports developed where barges could be loaded with local agricultural and forest products. Port Davidson, The Narrows (Wellandport), Beckett's Bridge, and Misener's Bank provide examples. Almost every farm along the river had its own private wharf where timber and cattle were transported from the upper reaches, and also grain, shingles, cheese, and other farm products. In addition to farm work and the extraction of timber, men were employed to raft the timber downstream, to make shingles, and cut cordwood. The shipping points were from Welland to the centres north along the canal, and to Chippawa, with exports to serve the Buffalo and Tonawanda markets.

Earlier crossings of the Welland River were upstream at Brown's Bridge, at the southern end of South Pelham Street, and downstream at Misener's Bridge on Quaker Road. The construction of the aqueduct led to a towpath and pedestrian paths across the Welland River. At this early period, most inhabitants lived west of the canal and north of the river. Welland was confirmed as a route centre where the canal was crossed by a swing-bridge at the western end of Main Street (then a track along the river bank from Cook's Mills), and where either Niagara Street or Aqueduct Street crossed the river.

This new centre ended the importance of Brown's Bridge, which became unused and decayed. In 1867 the county's Road and Bridges Committee reported that they 'had the Bridge over the River in the Town of Welland put in a complete state of repair.' Brown's Bridge was then 'in a very dilapidated state and altogether unsafe and dangerous ... Several parties say that the said bridge is very little if at all required for public travel ... [and] can be dispensed with without any serious detriment to public travel' (Report of Committee 1867). Next year its removal was authorized.

As the First Canal and the aqueduct had each to remain in operation during the period of construction for the Second Canal, the new route followed an alignment that lay some distance east of its predecessor (map 6). Now of stone, the new aqueduct was constructed at approximately the base of Aqueduct Street. The top of its arches were now closer to the river. The maximum clearance varied with the level of water in the Welland River, but, reduced to perhaps 6 feet (1.8 m), rafts of timber and barges could still pass under them.

A lock was constructed to provide a link between the canal and the upstream length of the Welland River. Located in the north-east quadrant of the aqueduct between the routes of the First and Second Canals, the lock and the river basin in its vicinity became known as the Lock Pit. As manpower and the poling of rafts were displaced by tugs, it now became the practice to push the load being hauled under the arches of the aqueduct. The tug would then lock up to the canal, follow the canal to Port Robinson, then lock down into the Welland River and return upstream along the river to Welland to reclaim its haulage load. The aqueduct, with its lower height over the Welland River, had now become a barrier to river traffic, but more suitable for canal traffic as engineering technolology advanced.

The hinterland of the area served from the aqueduct and the Lock Pit at the lower level was via the Upper Welland River. At the upper level access was via the canal in both directions and along the Feeder Canal to the Grand River valley with its abundance of resources. Wharves soon located near the aqueduct along both the canal and the river. Timber was obtained from the Grand and the Welland River valleys and, as the lumber trade became important, Sherwood's Warehouse and Wharf next to the canal became an active centre of business. Cordwood, stave bolts, and shingles were the main products, then grain and other agricultural products as the land was cleared and cultivated. When wood-burning vessels came into operation, this wharf supplied the cordwood required for the daily passage of vessels that included the Northern Transportation Company, an American line of passenger and freight vessels that traded between American ports on the Upper Lakes and Kingston. Usually two of these vessels passed through the canal every day. They burned wood to fuel their boilers and stopped regularly at Welland for new supplies of cordwood.

The first building in the area that ultimately became downtown Welland was probably a blacksmith shop in the north-western quadrant next to the lock and the aqueduct. Saloons and stores soon followed, but development was at first slow to follow the construction works on the canal and its aqueduct. Smith by 1846 recorded only 'about 100 inhabitants, five stores, three taverns, two tailors, two shoemakers.'

A postal designation was obtained in 1851, and a small industrial village community developed as water power at the aqueduct was used 'to turn the machinery of two grist mills, one of which has three run of stones; two saw mills, one containing two single saws, three circular saws, and a planing machine, and a small cloth factory. The resident population scarcely numbers one hundred and fifty, but the labourers casually employed on the canal will raise the numbers to nearly three hundred and fifty' (Smith 1851).

To the south at The Junction (Junction Village then Helmsport), where the western end of the Feeder Canal from the Grand River met the main line of the canal, less development had taken place. Smith (1851) disparagingly states that 'a few houses, taverns, &c, have been erected, but there is nothing here that can be called a village.'

Physically, Welland's setting lay in a marshy environment on the edge of Wainfleet Marsh, a situation aggravated by ponds and islands as the Feeder Canal and then the First Canal both impeded the drainage of small streams to the slow-flowing Welland River. Welland in 1829 has indeed been described miserably as 'a swamp with a ditch running through it' (Michener 1973). Economic activities had to develop against such unpropitious circumstances.

With mills at the aqueduct and upstream on the Welland River, the village grew along the primary axis of Main Street, which crossed the canal south of the aqueduct and the Welland River. This highway (now East and

West Main Street) tended from north-north-east to south-south-west, with parallel streets (Division to the east and Bald to the west) to the south. As the north–south connecting streets were drawn along the cardinal orientation, the business centre took on the form of a parallelogram, less sharply defined than in St Catharines, but different from the regular north–south, east–west grid that prevailed in Crowland township. The transition from canal-oriented streets to the survey pattern of the township streets is now reflected east of the town centre, where East Main Street bends to meet the township alignment after it crosses the former Welland Railway tracks.

The flat and marshy setting together with the canal also created a series of severe physical barriers to development. The canal had a north–south alignment and, as the nearby river flowed on a parallel course north of the aqueduct, the two water bodies were separated only by a narrow peninsula known locally as The Island (now Merritt Island). As the two water barriers were not crossed by road bridges, the north–east and the north–west quadrants of Welland's growth were isolated from each other. South of the aqueduct the canal and the river diverged at an angle of some 30 degrees, which isolated the south–west quadrant and separated the area between the two water bodies from the more extensive south–east quadrant.

An administrative complication was that, as the Welland River was also the township boundary between Thorold Township to the north and Crowland Township to the south, the north–south survey grids of their respective rectangular road systems did not line up with each other. For example, when approaching Welland from the north (now Niagara Street), after crossing the Welland River, the jog is east onto East Main Street to cross the canal then south onto Canal (now King) Street. The canal-river environments have provided a confining obstacle to land movement since the earliest days of the Welland community.

When the Second Canal replaced the First, two bridges were required to cross the canal system of waterways, a fixed structure to the west over the First Canal and a narrower swing-bridge over the Second Canal. Several long but narrow islands were created between the two canals or its remnant channels, and between these channels and the river. Ponds had also been formed in shallow depressions where drainage had been blocked. These ponds and the islands along the canal banks to both the north and to the south of the bridge crossings provided a firm boundary to development and a series of severe physical impediments to the outward growth of the emerging business centre along its western side. The village had necessarily to grow in three distinct sections, each defined by water and linked by only the bridges between East and West Main Streets.

Perceived as a barrier to movement on the land, the canal was more inhibiting than the river. The river though wider could be spanned by conventional bridges. On the canal for eight months of the year the steady flow of marine traffic restricted vehicle and pedestrian traffic across the canal, and the bridges had to be continually opened to permit the passage of slow-moving, hauled or towed, vessels.

Chippawa

Before the First Canal, Chippawa (Chippewa) had grown with warehouses, the palisaded Fort Chippawa, and a bridge across the Welland River. The key factor was its location on the frontier at the southern end of the Niagara Portage and at the head of navigation for the Upper Niagara River, where the slow-flowing Welland River turned north towards The Falls of the Niagara River.

When the First Welland Canal bypassed the Niagara Portage in 1829, the portage trade almost immediately transferred inland to the new waterway and the formerly important river communities of Niagara (now Niagara-on-the Lake), Queenston, and Fort Erie went into steady decline. But not so Chippawa, which became the southern terminal of the First Canal (map 7). Further, when the canal was extended to Port Colborne in 1833, Chippawa remained as the canal's eastern terminal and port on the Niagara River, but with diminishing traffic into the latter part of the nineteenth century.

An immediate consequence of the canal was that the Lower Welland River was dredged frequently to remove obstructions from the channel, a towpath was provided along its northern bank, and the Chippawa Cut afforded access to the Niagara River. It also created an island where the spit and its military reserve had previously existed, which was known later as Hog(g) Island, perhaps because the logs brought down the Welland River when stored behind booms looked like hogs in a pen. Tolls were levied at the cut, but objections arose as no locks were involved. Now as a river and a canal port, the harbour attracted docks and wharves along the banks of the Welland River, the village also became a customs port, and the slender King's Bridge was replaced by a sturdier wooden structure with a central swing span to permit the passage of canal vessels.

Upstream, barges and rafts of timber from the Upper Welland River could pass under the aqueduct at Welland to Chippawa, and access was available to the main line of the canal at Port Robinson for northbound and southbound traffic, and to the Feeder Canal for links with the Grand River.

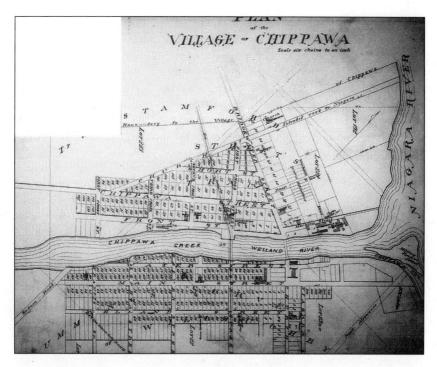

Map 7. *Chippawa, where the First Canal entered the Niagara River.* The earlier
village was located at the southern end of the Niagara Portage and at the northern
end of navigation on the Niagara River. When the Welland River was canalized for
the First Canal, the Chippawa Cut provided safe nagivation between the Welland
River and the fast-flowing Niagara River. (1876)

Chippawa, unlike Queenston, its companion on the portage, had substantially extended its catchment area and its potential for growth through the
canal venture. As stated by Bond (1964), 'The First Welland Canal which
had changed the destiny of the lower river settlements had been beneficial
to Chippawa. As a busy canal port she was still important.'

Vessels under tow on the canalized Lower Welland River, its scows, and
rafts stopped at Chippawa so that the crews, the teamsters, and the tow
boys might rest and be refreshed before their return journey. There were
steamer links across the Niagara River to Buffalo, Tonawanda, and LaSalle,
and via Fort Erie to Lake Erie for lake commerce. A line of steamboats also
made regular daily trips between Buffalo, Port Robinson, and Dunnville
then via the Grand River to Brantford. This passenger service with some

freight was inaugurated by stern-propelled steamers that took some twenty-four hours to make this journey, but it became so popular that in 1843 over a hundred crowded steamers were reported on this route (Ott 1967).

At Chippawa, the land was essentially flat. It sloped gently to the Welland and Niagara Rivers, but as water could not be diverted from the slow-moving Welland River, water power was not available and the milling centre for the community lay downstream on the Niagara River at Bridgewater, where Dufferin Islands are now located.

Coach travel along the Portage Road from Buffalo to The Falls brought in some through tourist trade. As steamboat travel became more proficient and with the improved facilities of the Chippawa Cut, this journey could more comfortably be undertaken after the mid-1820s by boat to Chippawa, then by coach for the rest of the journey. As some travellers rested or remained at the village to enjoy the scenic attractions of two rivers and the proximity of The Falls by a gentle drive, tourism also became a part of the village's character.

A shipbuilding industry along the banks of the Welland River was introduced by John Lovering, who launched a vessel in 1832 that sailed from Buffalo to Detroit. More important was the building of boats for the Niagara Harbour and Dock Company on the south bank. Vessels built at the company's Niagara site could sail on Lake Ontario, but as they were too broad to sail through the locks of the First Canal, materials and components were sent to Chippawa for assembly. The boats could then be used on either the Buffalo–Chippawa run or in the expanding trade on Lake Erie.

Engines and boilers were manufactured at the Macklem Foundry on the north bank. This foundry, destroyed by fire in 1842 but then immediately replaced, was multi-purpose. In the words of an advertisement placed by Oliver T. Macklem: 'The undersigned is now prepared to do all kinds of work in the best manner and at short notice.' The foundry produced not only boilers and engines that were custom-built for steamers on the lakes, but also gearing and shafting for grist and saw mills, stationary fire engines, iron planing, turning-screw cuttings, brass casting, and finishing. The iron works also produced cooking ranges and stoves, one record being an output of forty stoves a day.

Activities other than shipbuilding included a distillery that could produce 1200 gallons (4542.5 litres) of whiskey a day, in reputation the largest in Upper Canada. A nearby grist mill supplied the grain used in the distilling process. A tannery incorporated in 1846 reportedly had an annual production by 1851 (Smith) of 22,000 sides of leather, 4000 calf skins, and 2000 sides of upper leather. Chemicals for use in the manufacture of glass and soap were also produced at potasheries for export via the Welland Canal to supply the English market.

Smith (1846) recorded a 'large village' with a population of about a thousand. Shipbuilding for the upper lakes was an important activity, 'many vessels have been built here, and a fine steamboat of 800 tons [725.8 tonnes] is now on the stocks, intended to be ready for service during the present season.' The associated trades and professions included an agency of the Upper Canada Bank, a steam grist and a steam saw mill, three distilleries, two tanneries, an iron and brass foundry, a tin and sheet-iron foundry, seven stores, six groceries, and six taverns, together with blacksmiths, saddlers, tailors, bakers, shoemakers, cabinetmakers, and two physicians and surgeons, inter alia.

Here was now the largest canal port, larger than any that had accrued along the main line of the canal. The village grew, centred on the Welland River, which was also the boundary between two townships. The northern section in Stamford Township was laid out in 1816, and the southern section in Willoughby Township followed after a decade. Streets appropriately named Front Street and Water (now Bridgewater) after their setting ran parallel to the river along its two banks, a pattern in accord with neither of the regular township surveys but determined by the alignment of the Welland River. Examples of the resultant sudden change in direction where the village grid met the township system of survey roads include the places where Main Street now becomes Sodom Road and Niagara Street becomes Willoughby Drive.

The two sections, each with rectangular layouts, were linked by a wooden swing-bridge. The northern bank carried the foundry, tannery, post office, and a store. The south bank supported the shipyards, further homes and stores, and the square (now Cummington Square) next to the bridge, which was intended for the town hall and market place when the village was incorporated. Since civic squares as a traditional feature of urban design are not a common feature of Ontario's towns, this small area of municipal open space provided an unusual feature in the urban structure.

Dunnville and Byng

The Townships of Moulton and Sherbrooke within which Dunnville and Stromness grew had been granted to the Six Nations under the Mohawk

War Chief, Captain Joseph Brant (Thayendanegea), at the end of the American Revolution for services rendered to the British Crown. The original extent of this land grant, 6 miles (9.7 km) in width on both sides of the Grand River upstream from its mouth, was too small in area to sustain hunting on a permanent basis, and land was sold or leased to white settlers in order to provide an income. The original Indian area was reduced considerably in extent. Sherbrooke Township was patented in 1820 and Moulton Township was opened for settlement just before construction of the dam at Dunnville commenced.

Communities with accidental origins emerged at each end of this dam: Dunnville on the east bank of the Grand River, where water was diverted into the Feeder Canal, and Haldimand Village (later Byng) on the opposite bank (map 8). Both were FAOF communities, for the dam had originally been intended at Port Maitland at the mouth of the Grand River. The dam itself, named T'Kane Khodh, meaning the big dam by the Indians of the Grand River, became the fulcrum of development for its two associated communities. Previously the most important settlement was Anthony's Mill on the north shore of the river about one mile (1.6 km) below Dunnville.

The dam supplied water to feed the canal. It also separated the Grand River into two navigable lengths. Upstream next to the river, Indian corn-growing land was flooded over the banks for several miles upstream in order to provide the mill pond that was needed to supply the canal with water. As no provision was made for a lock on either side at the dam, the downstream through movement of vessels along the Grand River was directed through the Feeder Canal into the Welland Canal system.

As vessels that entered the Grand River could either sail or be towed upstream to Dunnville, wharves and storehouses developed on both sides at the dam at the entrance to the Feeder Canal. The canal company had built a towpath along the west side of the Grand from Dunnville to Lake Erie. At its upper end cargoes might be transferred through a storehouse between vessels on the Grand and the feeder, unloading at one door and receiving at the other. This operation had the advantage that in late spring it avoided the ice that still blocked the harbour at Port Colborne and the entrance to the Erie Canal at Buffalo. Dunnville therefore functioned as a port in a triple sense: at the western entrance to the Feeder Canal; as the terminus for navigation on the Upper Grand River; and as the head of navigation on the Lower Grand River. A lock connection between the Feeder Canal and the Grand River was started but abandoned, probably because of site conditions.

This situation prevailed until the period of the Second Canal, when the

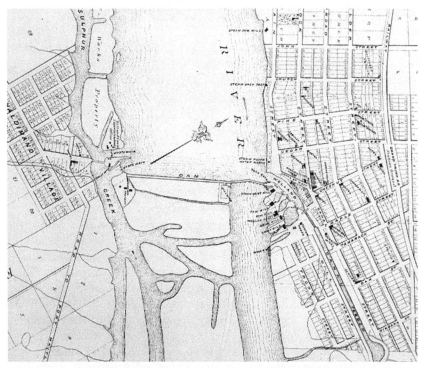

Map 8. *The dam and feeder canal at Dunnville.* The dam across the Grand River raised the upstream water level, from which the Feeder Canal was supplied. Its reservoirs provided the head ponds for mills between the canal and the river. Facing Dunnville on the opposite bank, Haldimand Village (now Byng) developed close to the sluices at the southern end of this dam. (c. 1877)

length between Stromness and Port Maitland opened with a lock to connect the Feeder Canal at Stomness and the Grand River at Port Maitland. In addition to providing a second outlet to Lake Erie for the canal system and a connection around the dam, the canal between Dunnville and Port Maitland encouraged the trade from the Grand River to Buffalo, Lake Erie, and the Upper Great Lakes. The canal bisected the town, and when it was enlarged for the construction of the Second Canal, shipping on the feeder increased tremendously and the port greatly expanded its activities.

At the dam, with the inflow to the canal system on the Dunnville side and the control weirs at Byng, the Byng side attracted the office of the canal superintendent, which controlled the waste weirs. Mill sites were available between the Grand River and Sulphur Creek, which ran parallel to the Grand at the western end, and a fish ladder later connected the now divided sections of the Grand River.

Bridges at both ends of the dam across the Grand River and Sulphur Creek connected the two riparian communities. Constructed in 1834, maintained as a toll bridge for pedestrian, horse, and wagon traffic, and, with the dam, referred to as the Long Bridge, this road bridge now provided the shortest route between Buffalo–Fort Erie and Windsor-Detroit along the north shore of Lake Erie. Earlier, Gifford's ferry upstream between Dunnville and Cayuga provided the only place where teams could cross the Grand; downstream a ferry crossed at Port Maitland. As the dam was the first bridging point on the Grand River upstream from Lake Erie, Dunnville soon became an important strategic position on the regional highway network that emerged after the dam was constructed.

This new highway was more than a link across the river. As Dunnville lay in Moulton Township and Byng in Dunn Township, the bridge crossed an administrative divide between these two previously separated and disassociated communities and their disconnected farming settlements. The dam had both local and regional significance for the road communications of the period, quite apart from its importance for canal and river navigation. Previously, the area was isolated. 'There were no roads in the adjacent townships communicating with Dunnville, only a path [an Indian Trail] along the river bank. The Robinson Road was chopped out in 1833 and the Diltz Road the following year, but neither were fit for travel for a couple of years afterwards, except in winter or late in summer' (Page 1877). The construction of the dam also meant that the former road upstream along the river had often to be rerouted inland from the river as the side streams were pushed back by the dam and the greater depth of water in the Grand River.

The dam also brought in water rights and water power, and their high potential on both banks of the river at the dam site. The Welland Canal Company had announced free water rights in perpetuity for mills completed before the Feeder Canal opened. Oliver Phelps of Dunnville – a contractor on the Erie Canal who had designed the machinery for excavating the Deep Cut and was later the owner of several mills along the canal – was successful in winning this prize. Other mills soon followed, both at Dunnville and at Byng. Further inducements to development included locks on the Feeder Canal, a wharf that extended from the Grand River to the first lock, and a turning basin at Niagara Street.

The construction of the dam introduced Slabtown (Scottsville) west of the river to initiate Dunnville. Here, shanties and a communal boarding house housed the Irish and other workers brought in for the construction project. The first street, before the present business centre arose, followed the river facing the dam. As soon as the dam was announced, Oliver Phelps bought land and laid out village lots. These were soon sold and buildings erected on them.

The Feeder Canal, at first only a raceway to supply water to the canal, but widened and deepened in 1842 to a fully-fledged canal, attracted water-side mills to the strip of land that lay between the Feeder Canal and the Grand River. Water diverted from the river at the dam fed into the canal, which in turn fed a number of mill ponds. The mills after using water from this source of supply simply returned it back down the banks into the river.

A further bridge was required on the Dunnville side to cross the Feeder Canal. Long Bridge and the waterside mills lay to the west, whereas Front and Mill Streets as the business centre and the adjacent residential streets had developed in the less confined area east from the canal. First one and then two swing-bridges crossed the feeder. As a measure of the trade involved through the canal, 575 vessels (including scows) were recorded as loading on the Grand River at Dunnville by Smith in 1851: 216 were British vessels bound for British ports, and 208 British vessels and 75 American vessels were bound for American ports.

An interesting comment is that, in the early days of Dunnville, three ships were loaded for the place later to be known as Chicago, 'one with timber and lumber cut in a sawmill located at the Byng end of the dam; one with meat of all kinds and cattle; one with grain and flour ground in local mills ... Thus it happened that lumber from Dunnville was used to build Chicago's first permanent house' (Sorge 1977–89).

Because of the dam, nodality at a prime river crossing point, water power, and the flow of trade from upstream along the Grand River valley, Dunnville grew rapidly. After only some twenty years of existence, it is described by Smith (1851) as 'a place of considerable business ... Large quantities of produce coming down the Grand River are shipped from it. Goods intended for places on the upper portion of the river are generally reshipped here into small vessels.' With a population of about 1000, the village had 'three grist mills to one of which a plaster mill is attached, four saw mills, a foundry, woollen factory, brewery, distillery, and tannery; two churches, Episcopal and Presbyterian; post office, collector of canal tolls, and collector of customs.'

As the town was linked by trade and a regular passenger steamboat service along the Feeder Canal–Welland River route to Buffalo, these marine connections helped to extend the influence of that rapidly expanding American city west through Port Colborne across the northern part of the Niagara Peninsula and up the Grand River valley. The same route also linked the pine stands of the Grand River valley to Tonawanda and encouraged the export of processed timber via the Erie Canal to the eastern American states.

Byng on the opposite bank of the Grand River had a comparable water-power potential, but it lacked the shipping-river-canal traffic of its neighbour. It grew more discretely as a local service centre with taverns, blacksmiths' shops, stores, a large grist mill, and a distillery. Dating to the early 1830s, a carding mill also processed the fleece of local sheep.

Port Maitland

Situated at the mouth of the Grand River and one of the best natural harbours on Lake Erie, here behind the sand dunes in 1825 nestled a naval reserve with a house for the officers and barracks for the soldiers and sailors. After the Rush-Bagot Agreement of 1817, which limited naval power on the Upper Lakes to two American and two British ships, the naval base declined in importance. Even so, the port remained as a refuge for ships on Lake Erie, for vessels using the Grand River or transferring to the Feeder Canal, and as a lake and river fishing port. The bar at the entrance to the river, which fluctuated in size but allowed for the entry of vessels with a draught of some 7 feet (2.1 m), was deepened. The port had a pier by the 1820s and soon a lighthouse to assist navigation. Further improvements followed the arrival of the Second Canal and the additional traffic it carried.

Described as a village built on sand, Port Maitland was described by Smith (1846) as follows: 'A settlement and Shipping-place at the mouth of the Grand River ... It contains about 50 inhabitants, and an Episcopal Church, two stores, two taverns, one tailor, one blacksmith.' The canal lift lock at the entry point to the Second Canal, larger than the norm as at Port Dalhousie, with a lock tender's house, was situated on the east bank facing the fishing village and the extensive harbour at the entrance to the Grand River. A ferry across the river linked the two communities and carried the small volumes of local highway and agricultural traffic that existed. By 1851, according to Smith, the 'settlement has but a small resident population; and since the completion of the new cut of the Welland Canal by Port Colborne, its business, even during the season of navigation, is said to have fallen off considerably.'

The Inland Canal Villages

In addition to the settlements that have been classified as ports, several small canalside communities emerged along the main line of the canal and its tributary Feeder Canal. They each reflected some combination of the early impact of canal labourers during the period of canal construction, the service needs of vessels passing through the canal after it opened, the potential of new hydraulic situations, and the growth of service activities as the adjacent agricultural land was farmed more intensively and the population increased.

Merritton

As the initial idea of a railway incline for the First Canal to cross the Niagara Escarpment changed to the provision of canal locks, small shack communities developed at work sites along the canal. Moving upstream these included Centreville, Westport at Lock 15, Protestant Hill, named after the religion of its inhabitants, and Slabtown, named after the outside piece cut from a log when squaring it for lumber and used to produce makeshift housing. When a labour force had again to be accommodated during the construction works for the Second Canal, seventy shanties were recorded in 1842 between St Catharines and the mountain locks (St Catharines *Journal*, 18 August 1842). These shanty settlements merged in 1850 and took on the communal name of Welland City after the canal. This cluster of small settlements was renamed Merritton after the founder of the canal in 1858, an exchange with Merrittsville, which became Welland. The reasons for this exchange of names are unknown, but Merritton was the more important settlement and closer to Merritt's home town.

Merritton's outstanding characteristic was its exceedingly complex water system (map 9). Attached to the First Canal were the shipping channel and some fifteen locks and weirs, the ponds along the channel where ships awaited oncoming traffic or passed each other, and the turning basins where vessels waited for their turn at climbing or descending the escarpment. This system was supplied from a series of weir ponds east of the canal, and these had interconnecting channels to supply the next weir pond, and so on down the system. Then there were the mill races, the mill ponds for the local storage of water, and the tail races. This water system was continuous over a length of some 3 miles (4.8 km) across Merritton and Thorold. Here undoubtedly was the greatest head of available power anywhere along the whole canal system, if not in Upper Canada and North America of the day.

This grand availability and potential of an abundant supply of falling water soon attracted numerous flour and saw mills and other establishments, but their number and sequence is now impossible to determine.

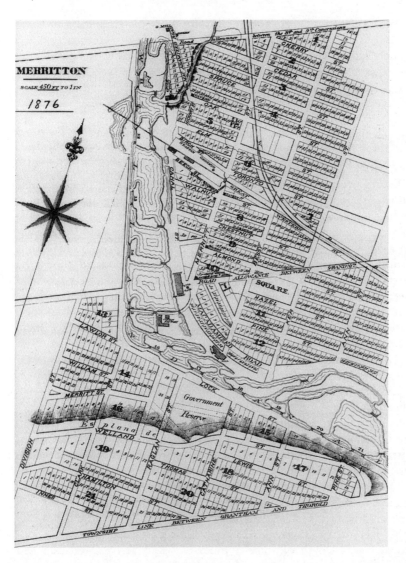

Map 9. *Merritton, where the Canal crossed the Niagara Escarpment.* The First and Second Canals with their continuous sequence of locks, basins, storage reservoirs, and mill races together provided a pronounced barrier to land communications that was crossed by only one road-allowance street (now Glendale Avenue). (1876)

'The village's heyday was short but complicated. Businesses changed hands frequently. Mill structures burned ... Buildings devoted to producing one sort of product were refitted to produce a different sort. One structure would house two enterprises. Partnerships were formed and dissolved quickly. The surviving records are confusing, as well as sparse' (Taylor 1992a). Among the more important enterprises, however, were Oliver Phelps's sawmill at Lock 7 in the 1830s, the Smyth and Smyth Flour Mill at Lock 10 in the 1840s, and a distillery at Clifford's Creek in the 1840s.

The water system was augmented considerably in width and in its potential for industrial development with the arrival of the Second Canal. The channel was moved slightly west to higher ground to enable the First Canal to remain in full operation during the construction works. Most of the earlier system was retained, with the old being used to help supply the new channel, its ponds and weirs; new ponds, weirs, and channels were also created and added to the old.

In illustration of these changes, whereas the old canal swung east, the Second Canal followed a more direct route. Parts of Merritt Street ascending the escarpment are presumed to have evolved from the towpath of the First Canal (Duquemin 1979), and the entrance to the Hydraulic Raceway was moved downhill to a new location. Locktenders' cottages were added to the Second Canal, being added for safety reasons after the Battle of Slabtown, when the Coloured Corps was brought in to restore order, and to protect government property, the canal, its vessels, and their cargo.

The new water potential added considerably to the industrial prowess of Merritton, which became known as The Factory Town. Industry escalated, including William McCleary and John Maclean's lumber mill, founded in the late 1840s at Lock 21, and John Brown's Cement and Plaster Mill, founded in 1853 near Lock 19. As the water complex presented an increased barrier to east-west movement on land by road across the canal, only one concession road (now Glendale Avenue) crossed the canal in Merritton. This road met the north–south road (now Oakdale Avenue, formerly Thorold Road then Merritt Street) that followed the canal at Tenbroek's Corner, where a toll gate and then an inn were established as roads were improved. The village centre developed to the north after the arrival of the railway, but Welland City as then named was not of sufficient importance to be included in Smith's Directory in either 1846 or 1851.

Allanburg

A considerable new community developed on the flat land at the north end of the Deep Cut, where Merritt's inaccurate survey had commenced (map 10). Starting with the labour gangs, then water power at the locks, development was also influenced by a well-settled agricultural area in the vicinity and by the addition of a swing-bridge where the coach road between The Falls and Ancaster crossed the canal. Important precedents at the village included Holland Road and Black Horse Inn to the east, where the dignitaries repaired for a meal after the sod-turning ceremony that initiated the canal excavations. The canal village that arose (map 10) was known variously as Allanburg, Allenburg, or Allanburgh, the former now being the accepted name (Vanderburgh 1967).

This pioneer progression from a close pattern of rural settlements to a village began when shanties were erected to house workers on the Deep Cut. 'The Canal Company had a general store, and there were three or four other shops on the bank that sold goods of all types. There was also a market ..., and so great was the demand for farm produce that some farmers sent wagons there twice a day, with fruit, vegetables, eggs and butter' (Thompson 1897–8).

The village nucleus with docking facilities on the canal developed between where Holland Road and Canborough Road crossed the canal. In between were the two locks of the First Canal, the mill races, and ponds that fed from the canal. Laid out in 1832, the road pattern included Canal Street next to the canal and Centre Street as the spine of the community on an alignment in accord with the survey system for Thorold Township. Village form centred on the canal, these two roads, and the mills at the two locks. By 1835 there were a grist mill, a sawmill, and a shingle factory. These were destroyed then replaced when the First Canal became the Second Canal, and the wooden locks replaced by one stone lock. The new canal, on a parallel route close to its predecessor, was crossed by the two swing-bridges. The water system that confined the village to the west with narrow islands between the two canals had about quadrupled in width.

The insecurity of the era involved an attempt in 1841 to blow up a lock at Allanburg by a keg of gunpowder, one year after the Brock monument at Queenston Heights was destroyed. The reasons were Irish grievances against the British occupation in Ireland, the failure of the Mackenzie Rebellion in 1837, and dissent among the Irish canal workers. The temper of the times was such that the frontier was still garrisoned and patrolled by regular troops. As the canal had adversely affected the trade of Buffalo, Irish insurgents from that city were suspected when the explosion took place.

By 1846 (Smith), the Allanburg twin to Port Robinson at the north end

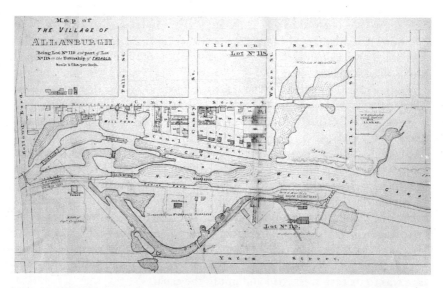

Map 10. *Allanburg, at the northern end of the Deep Cut.* The town was located where locks stepped the First Canal down from the Deep Cut to the escarpment level and where the canal crossed and impounded a small stream. Water was typically diverted from two locations above the lock to serve mills, a situation extended considerably with the Second Canal. Canby Street is the modern crossing point on Highway 20. (c. 1850)

of the Deep Cut (but lacking its port facilities) was a village with about five hundred inhabitants that possessed a town hall for public meetings. Industry included a grist mill, a sawmill, a carding machine and a cloth factory, a candle factory, and a pipe factory. There were four stores, two taverns, and one each of wagon-maker, cabinetmaker, blacksmith, and baker. It was a growing entity that promised future growth and expansion. Smith by 1851 noted that the economic base had expanded to include two each of grist and saw mills and woollen factories, a tannery, and a Methodist church.

Humberstone

Known first as Stonebridge from the stones sunk across a shallow stream flowing south to Lake Erie, but named Petersburg after Peter Neff, its first settler, and then Humberstone after the township and the name of the post office established in 1851, this roadside hamlet preceded the canal and at first grew faster than Port Colborne, which lay about one mile (1.6 km) to the north. The land was higher and drier, it lay inland from the marshy lakeshore, and a small cluster of houses, services, and a hotel had developed along the lakeshore road (now Highway 3) west from Fort Erie and the ferry crossing from Black Rock. This route followed the dry edge of the Onondaga Escarpment rather than the marshy land next to Lake Erie, and a swing-bridge across the canal replaced the former stone crossing of the creek.

By 1846 (Smith) the village had a population of 200 against 150 at Port Colborne, more services, and a stronger industrial base. 'It is supported almost entirely by the works on the canal. A detachment of the Coloured Company is quartered there ... [Then listed were] One physician and surgeon, one distillery, one foundry, seven stores, one druggist, three taverns, two wagon makers, three blacksmiths, three butchers, four shoemakers, two saddlers, three tailors, one tinsmith.' As Humberstone was a local market place with no lock and only passing vessels, the orientation of the village was to its east–west road and not to the canal.

By 1851 the relative position of Port Colborne and Humberstone had started to change. Humberstone as the market centre for the agricultural area at the southern end of the canal could no longer compete with the improved harbour and service facilities at Port Colborne. In the words of Smith's 1851 description, 'this village has hitherto generally done the business of this end of the canal. It contains about 200 inhabitants, a small foundry and brewery.'

Marshville (Wainfleet)

Marshville, renamed Wainfleet after the township in 1923 as its original discouraging name was no longer acceptable after the marsh had been drained and converted to rich farmland, was described as a 'small Village' by Smith (1846), who noted sixty inhabitants, a grist mill, two stores, a tavern, and a blacksmith. The village came into being as a construction camp to house workers on the Feeder Canal. It had an engineer's office of the Welland Canal Company to control these works, a store established by Edward Lee supplied food to the workers, and a hotel followed. Each clustered next to the other at the centre of the settlement.

The canal had a towpath to the south and a heelpath to the north, both located on the raised banks of material excavated from the canal. Higher than the adjacent flat land, together with the canal they formed the centrepiece of the village. Its centre lay where a north–south road crossed the

Feeder Canal at a swing-bridge. As at Stromness, Wainfleet was a relay station where horses and mules were changed during the towing of barges
through the canal. Lee's store was the emporium that handled this business,
victualling and serving the boats, barges, and rafts as required.

When the Feeder Canal was completed and into the period of the Second
Canal, the canal banks were lined with piles of cordwood ready to be
shipped to Buffalo, and timber was available to be made into rafts and sent
to Buffalo, Tonawanda, and local destinations including Allanburg,
Thorold, and St Catharines. A flour and a grist mill and then a sawmill
located on the north bank of the canal. Powered by water diverted from the
feeder, the combined mill race used a deepened tributary of Forks Creek
that emptied into the Welland River and was part of the contentious loss of
water from the canal system for navigation purposes. Before the canal, and
then the draining of the marsh and its conversion to highly productive agricultural land, the township had been settled only to the north and south of
the marsh. Wainfleet at the uninhabited centre of the marsh now attracted
the township offices, and the village prospered because of the canal, its
centrality, and its close associations with the new but expanding farming
community.

Broad Creek (Stromness)

Smaller than Wainfleet, Broad Creek (later Stromness) was located on the
Feeder Canal between Wainfleet and Dunnville. It gravitated to greater
importance with the Second Canal, when the canal link to Port Maitland, a
turning basin, and two swing-bridges across the two branches of the waterway were added to the scene (Sorge 1977–89). The village grew around this
basin, under the leadership of Lachlan McCallum who arrived from Scotland in 1842. He purchased timbered land with the money saved when
employed on building the pier at Port Maitland. He then supplied cordwood to wood-burning vessels on the canal. Later an MP and then a sena-
tor, McCallum renamed Broad Creek Stromness in remembrance of that
town in the Orkney Islands, Scotland, in 1858, when a post office opened
in the village.

The Two Industrial Towns

Thorold and St Catharines each eclipsed the ports and the smaller centres
in population size and industrial importance. Thorold had an abundance of
water power where the canal crossed the crest of the Niagara Escarpment,

and St Catharines supplemented the availability of power at the canal locks by a tier of hydraulic raceways along the bank above the canal. The two towns, respectively in Thorold and Grantham Townships and in both instances their major centres, awarded urban dominance to the northern length of the canal. This reinforced the earlier natural circumstances of more productive soils, a more favoured climate, and a higher density of farm settlements than south of the Niagara Escarpment.

Thorold

Had a railway incline been constructed for the canal across the escarpment as was the intention of the Welland Canal Company, then 'Thorold' would presumably have developed at DeCew Falls. It is another FAOF town that enjoyed its first industrial advance when a stone grist mill was built on the west bank of the canal by George Keefer in 1827. As recorded in the family memoirs, Keefer 'built the mills in the woods at a spot where the canal was surveyed to pass. He said he felt like Noah, when his neighbours derided him for building the Ark ... So George Keefer, feeling certain the waters of the canal would come, intended to have the mill in readiness' (Keefer 1935). Together with Phelps's mill at Dunnville, the Keefer Mill at Thorold enjoyed free water rights in perpetuity because the building was constructed with four run of stones and available for grinding before water was let into the canal.

The village grew with an influx of English workers and Irish labourers from the Erie Canal and Ireland when work commenced on the Welland Canal in 1824 (Orr 1978). New enterprises taking advantage of the high water-power capability soon included a tannery, two potteries, and a plaster mill, together with factories producing brooms, ploughs, soap, and candles. Catholic, Episcopal, and Methodist churches were founded. Shops, taverns, the service professions, and the trades were also each numerous. Smith (1846) recorded a population of about a thousand inhabitants, and the vicinity was described as 'one of the best settled townships in the Niagara District, containing a great number of excellent, well cleared farms.' The arrival of mills and services from 1826 on was the start of intense urbanization in an otherwise pioneer agricultural township.

By 1849, four flour mills operated. The most imposing, the Welland Mills built by Jacob Keefer, the son of George Keefer, and described as the largest and best equipped in the province, could manufacture up to three hundred barrels of flour a day, but the others had a daily capacity of over one hundred barrels. As the staves and the barrels were also produced

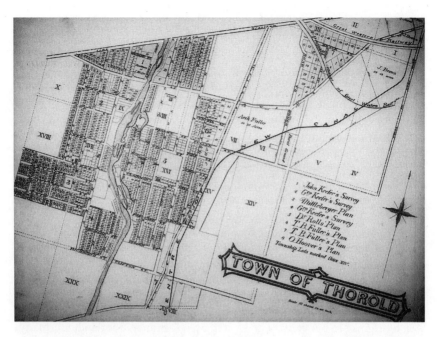

Map 11. *Thorold, at the crest of the Niagara Escarpment.* As mills developed along the canal, Front Street developed east of the canal as the main commercial street. Note how the Third Canal stringently confined this growing community and curtailed its eastern growth. (1876)

nearby, the economy was self-contained. The industries depended on each other, and then on the canal for the export of their products. A setting on the Niagara Escarpment is emphasized by the fact that limestone blocks from a nearby quarry were used for the construction of the Keefer Mill. The escarpment also provided a source for lime and cement for use at the Thorold Cement Works next to Keefer's Mill, and at Brown's Cement and Plaster Mill. Both mills had canalside locations at lock sites, as had a pottery that made use of local clay deposits.

A cotton mill owned by Keefer wove cotton in 1847. Some twenty looms driven by water power produced unbleached sheetings and cotton batting on a mill race between Front Street and the canal. This venture, the first in the province and the second in Canada after a Quebec enterprise, was short-lived and soon offered vacant space.

The village was centred on the canal, easily crossed on foot at the series of locks and the swing-bridges (map 11). The two halves, known respectively as the East and West Sides, yielded a total population of 370 in a census taken during 1835 (Thompson 1897–8). The canal was straddled by homes; 226 people lived on the East Side and 144 on the West Side. Ten families recorded 10 or more people and the overall average family size was 6.5. Three churches, Episcopal, Catholic, and Methodist, existed by 1847, and advances in the growth of mills, population, and its functioning as a local service centre led to the incorporation of Thorold as a village in 1850.

As described the next year, Thorold 'has greatly increased in size within the last few years, and now contains a population of about twelve hundred. The hydraulic powers of the canal have been taken advantage of to a considerable extent, and five grist mills have been erected having an aggregate of fifteen run of stones; also a saw mill, containing one upright and two circular saws, and two planing machines. There are also attached to the establishment a machine shop and carding machine. A cotton factory was in operation here for some time [since 1847], but has ceased working, and the establishment, with machinery, &c, is for sale' (Smith 1851).

By the census of 1851, the greatest percentage of the population in both the village and the township were natives of Canadian not French origin. The village had a large population of Irish (25 per cent) and of English and Welsh origins (10 per cent). The United States, France, Scotland, Germany, and Holland were also represented. By religious affiliation, 30 per cent were Church of England, followed by Methodists (25 per cent), and Roman Catholic and Presbyterians (8.5 per cent each).

Thorold was not centrally located with the township after which it was named, but lay on the northern edge next to the township boundary, which was also the boundary between Welland and Lincoln Counties. As the flow of water in the canal also crossed this county divide, the industrial and urban continuity that developed north from Thorold along the canal through Merritton was divided between these two centres and their respective counties. By contrast, expansion to the south lay within the same county and within Thorold township. This diversified administrative control over a unitary canal system created problems that have remained to this day.

St Catharines

Several important events had taken place at St Catharines before the First Canal used, then amended, the previous set of circumstances. These antecedents included a road pattern within which the industrial town later developed, a central location within a settled farming community on mostly cleared land of high productivity, the local provision of services as a rural service centre, and an industrial base of several mills along Twelve Mile Creek. The village even had earlier names, The Twelve to denote its situation on this creek and Shipman's Corner(s) from Paul Shipman, a tavern owner after whom the main street, St Paul, was named.

The pattern of roads included a radial pattern of former Indian trails: St Paul Street as the curving main street, Queenston and Niagara Streets to Queenston and Niagara (now Niagara-on-the-Lake), Pelham Road across the brow of the Niagara Escarpment, St Paul Street West to Ancaster and York (now Toronto). The village had grown at the meeting point where these former Indian trails, then pioneer routes of entry and travel, converged to cross the deeply incised valley of Twelve Mile Creek by a log bridge.

This radial pattern of earlier Indian trails that became the main roads of the village was overlaid by the regular pattern of the survey grid for Grantham Township (map 12). This grid had the conventional north–south lines as elsewhere in the Niagara Peninsula, but these were crossed by lines drawn parallel to Lake Ontario at the acute angle of north 65 degrees east. Welland Avenue and Geneva Street as survey roads defined the northern and eastern edges of the growing village, and Dick's Creek and Twelve Mile Creek, as part of the First Canal, its southern and western limits. The village itself had grown along the former Indian trails of St Paul and Ontario Streets, and up the bank from where Twelve Mile Creek was crossed by the slender bridge that preceded the canal.

By the 1790s five mills had developed along Twelve Mile Creek, the most intense milling locality in the peninsula, and The Twelve was crossed by the log bridge on the stagecoach route between Niagara, Ancaster, and York. A storehouse for goods where Indians traded furs and pioneer settlers obtained supplies located next to this crossing point. A church, a tavern, and a schoolhouse followed by the turn of the century, emphasizing the bank up to St Paul Street and its junction with Ontario Street as the core of the emerging village. The vicinity had been occupied by disbanded members of the Butler's Rangers, and the pioneer pattern of settlement expanded rapidly as agricultural land on fertile soils replaced the indigenous forest cover.

The War of 1812 brought damage and destruction, but after hostilities ended, an expanding situation resumed. Of particular importance to this narrative was the return of William Hamilton Merritt from war service,

Map 12. *St Catharines: Queen City of the Canal.* The village of St Catharines took shape where an earlier St Paul Street crossed Twelve Mile Creek by a crude wooden bridge. It became an industrial town when Dick's Creek immediately north of St Paul Street became the First and then the Second Canal. (1875)

his purchase of a store that sold numerous items by barter in exchange for country produce destined to the Montreal market, and then his purchase of a sawmill downstream from the village on Twelve Mile Creek. It was this mill at Welland Vale that, running dry at the end of the season, led to his ambitious idea of diverting water from the Welland River to drive his mill wheels and, eventually, to the canal as the centrepiece of this account.

The road between St Catharines and Port Dalhousie crossed Twelve Mile Creek by a swing-bridge at this mill, and a small village with a tavern and a store developed. The arrival of the Second Canal with a lock at this point created an island between the new channel and the meandering First Canal. This lock, as at Port Dalhousie and Port Maitland larger than the norm along the canal, permitted sailing vessels to make the journey upstream from Lake Ontario to St Catharines.

In the meantime the valuable pioneer commodity of salt for preserving and seasoning food had been discovered on Merritt's property. This added centrality and increased the service character of the village. When a farmer brought wheat to a mill or some agricultural product to barter at a store, he could now return home with a bag of salt and no longer had to travel further to Niagara to purchase this expensive commodity.

By 1827, when the canal was under construction, the village recorded nearly four hundred inhabitants and a goodly array of local service activities that ranged through a barrister and a printer, saddlers, a hatter and milliner, cabinet and shoe makers, blacksmiths, coopers, and a wagon-maker. The subdivision of land next to St Paul Street had begun by about 1817, and this road was turnpiked in 1827. St Catharines was already a considerable village that could be described as 'a town growing into note' before the canal came into being (Jackson and Wilson 1992).

The transformation from an expanding village and local service centre through the canal occurred not only because of the water power available at the canal locks, a typical feature of the canal scene, but also through a lengthy mill flume known as the Hydraulic Raceway that was constructed from Merritton in the 1830s by the Welland Canal Company along the top of the canal bank on its eastern and southern side. At the St Catharines end of this noteworthy feature for urban advancement, a series of levels were constructed along the bank of the canal immediately to the south of St Paul Street. One channel, still clearly visible, is situated between the Canada Hair Cloth mill and St Paul Street. The name Race Street and the sinuosities of Gale Crescent also recall this raceway system.

Water power at St Catharines was therefore available both in the valley below at the locks and in a linear fashion along the bank above the canal, and the rapid development of a strong industrial base testifies to the significance of this dual commitment. By 1830 a description of this multilevel Hydraulic Raceway noted that 'the Water Company have constructed at their expense, three separate races ... and have placed thereon the following mills and machinery ... Upon the higher level an extensive grist mill ... Surplus water upon a lower level moves machinery for a pail factory, a carding mill, brewery, and tannery; also a saw mill in progress of building ... All the surplus from these mills and levels is then carried upon the lower level to Mr. Merritt's establishment, now building to consist of saw and grist mills, besides other machinery' (Upper Canada Assembly, 1836–7). By 1847, of the thiry-two mills located along the canal, twelve were supplied with water from this raceway.

St Catharines had a more favoured location than elsewhere along the canal. Its main street, the foremost through route of the peninsula, and its growing array of service establishments were located on high ground above the canal. This was a dry site on a sandbar of an earlier Lake Ontario. Water was available at the base of these deposits, springs occurred round the perimeter, and underground saline deposits capable of providing good supplies of salt had been discovered at the bottom of deep wells.

The new offices of the Welland Canal Company located here in 1855, and in the valley below were the locks, ponds, and wharves where ships might load and unload their cargoes. And the mill sites were capable of being supplied with power from mill races along the bank, from water diverted around the locks along the canal valley, and from the upstream waters of Twelve Mile Creek. The circumstances were ripe for substantial industrial and urban advance because of the many resources that existed and also because of the village's start before the successful achievement of the canal. Named after the wife of an important Queenston merchant, Robert Hamilton, who owned extensive tracts of land in the vicinity, doubtless the settlers' social connections with Detroit, Montreal, and the British civil and military authorities had also helped considerably to foster the fledgling settlement.

With the arrival of the First Canal, there soon arose several flour and saw mills together with a tannery and a wool-carding and cloth-dressing factory. These structures included a second mill enterprise for William Hamilton Merritt, this time next to the village at the confluence of the First Canal with Twelve Mile Creek.

In the metal industries four foundries were recorded in the census of 1851: the St Catharines Furnace Company, founded in 1828, produced all sorts of mill machinery and castings; Chauncey Yale's Tin and Sheet Factory, founded in 1831, manufactured items that ranged from tin whistles to tin kettles; the Union Foundry produced a versatile range of agricultural implements and domestic products; and the St Catharines Foundry and Plough Factory advertised that it produced castings of every description.

Other shops produced well and cistern pumps, hot-air furnaces, and chain and iron of all kinds. Thomas Towers made engines for canal and lake vessels, and to drive planing machinery and horse-power threshing machinery. Gilbert Samson, in a large plant with some fifty mechanics and labourers, produced agricultural machinery. An important factor in this extensive range of enterprises was the extensive hinterland served by the canal upstream to the Upper Lakes, the Grand, and the Welland River valleys and downstream to the ports around Lake Ontario, including in each instance their expanding agricultural and settlement economies.

A series of other establishments produced wooden items such as pails, doors, and window sashes for domestic and farm use. There were a soap factory and a potashery, two cement mills, a distillery, and a brewery. The salt works had expanded to 58 kettles, each holding 110 gallons (416.4 litres). Brickyards took advantage of the clay deposits in an earlier lake bed, and limestone was quarried from the Niagara Escarpment for lime, road construction, building foundations, and tombstones.

Major shipbuilding activities began when Russell Armington, a master builder, was persuaded by William Hamilton Merritt to enter into the business of repairing and building ships. Three schooners were built on Twelve Mile Creek below the village. The first, with the appropriate but anticipatory name of *Welland Canal*, made its maiden voyage in 1827, when it carried a thousand barrels of flour to Prescott. The captain and part owner, Job Northrup, who had established a forwarding business, superintended the construction of the vessel. The following year consideration was given to constructing a packet boat that would run between the village and Port Dalhousie.

A further significant impetus to shipbuilding occurred in 1830, when the Welland Canal Company authorized the construction of a dry dock in the valley below St Paul Street. A schooner with the now familiar name of *W.H. Merritt* was launched in 1832 by Armington, and two more such vessels were launched in 1834. Although primarily a repair yard, at least four other schooners were built there. St Catharines had become a major centre for shipbuilding on the Great Lakes, and as its vessels fitted the size of the locks, they had the advantage of being able to carry maximum cargoes through the canal. They also served to advertise the canal in the ports around the lakes and influenced the design and construction of ships, as sailing vessels had now to be constructed to meet the lock dimensions of the canal.

When Armington died in 1837, his shipyard and its effects were purchased by Louis Schickluna, a shipbuilder of Maltese extraction who became one of the most prolific and competent of all the canal and indeed of the Great Lakes' yards. He operated first from the former Armington premises, repairing and refitting the sailing vessels of the day, but with the coming of the Second Canal he established a new yard on the west bank of Twelve Mile Creek between Locks 2 and 3 (just north of the present Burgoyne Bridge) and there constructed schooners and steamships. The first vessel to be constructed was the *Chief Justice Robinson* in 1841, and many other names followed. The story of this yard will be resumed in a later chapter, but it should be noted here that the foundations of a major shipbuilding industry existed at both St Catharines and Port Dalhousie by the mid-nineteenth century.

To this expanding industrial base as a creature of the canal were steadily added the social and cultural conveniences of schools including Grantham Academy, which opened as a grammar school in 1829, two weekly newspapers by 1850, and bank, insurance, and fire agencies. The churches served six different faiths. The temperance movement became important, as did amateur theatre, a debating society, a circulating library, and a Mechanic's Institute. In 1826 Dr William Chase fitted up a bathing establishment with hot- and cold-water baths at the salt works, and bottling the mineral waters started in the late 1840s.

Taking advantage of the soils and favoured climate, Dr Chauncey Beadle founded a large nursery in 1830 on a 100-acre (40.5-ha) site on Geneva Street near Russell Avenue. Its catalogue of 1841 listed 100 varieties of apple trees, 26 peach, 23 cherry, and an assorted array of other trees and plants.

Incorporated as a town with a population of 3400 in 1845, this now urban-manufacturing centre had grown to a population size of 4358 by the first census in 1851, four times larger than Thorold, which had 1091 residents on that date. Smith (1851) noted that 'St. Catharines, which aspires to rival Hamilton, is fast rising into a place of importance.' He listed 'six grist mills having an aggregate of twenty-eight run of stones,' with three of these producing above 72,000 barrels of flour during the season, and also noted 'a woollen factory, two foundries, five machine shops, an axe and edge tool factory, a saw mill, a pail and a last factory, ship yard and dry dock, telegraph office, marble factory, brewery and two distilleries; a nursery [horticultural] and two newspapers, the *Constitutional* and *Journal*.' And although the town was not officially declared a port, it functioned in that capacity. Six vessels were owned there, and three were under construction. Three banks had agencies, and the town with its Welland Canal offices was the centre of canal administration.

St Catharines had grown from its point of initiation at the junction of St Paul and Ontario Streets by commercial and service expansion east along St Paul Street. Church then King Streets were added to the north, followed by linking cross streets to accommodate this growth and its demands for housing and service provision. Meantime, industrial enterprises followed in the canal valley and along the raceway system.

Here by the early 1850s was the expanding lead city of the whole canal system. St Catharines had also taken over from Niagara as the most important centre in the Niagara Peninsula. The nearest rival to this supremacy was Buffalo (Brown and Watson 1981). Another staunch competitor was the new town of Hamilton at the head of Lake Ontario, where George Hamilton purchased land in 1815, laid out a townsite, and promoted it in 1816 as the seat of administration for the Gore District. Aided by the Welland Canal, the opening of the Burlington Canal in 1826, and the fact that as ships became larger they could no longer navigate the Desjardins Canal to Dundas but docked at Hamilton, this town had a population of nearly 7000 when it was recognised officially as a city in 1847. The local and regional circumstances of settlement within the Niagara Peninsula and inland from the frontier had changed remarkably through the early days of the Welland Canal era.

The Decline of Some Established Settlements

The obverse side of the coin to the growth of canal communities was that several other nearby places declined absolutely or relatively because of the canal achievement. Not every location gained through William Hamilton Merritt's ambitious endeavours. Ironic are the changed circumstances of John DeCew (DeCou), who owned a sawmill and an oil mill south of the place later named DeCew Falls. He had supported Merritt on his original survey and was an early shareholder in the Welland Canal Company. The canal would have flowed past his mill property and its associated village with its school, church, blacksmith shop, and two mills had the original route prevailed. However, the new canal route bypassed the mill complex and also cut off its water supply.

Merchants transferred as early as 1826 from Niagara (now Niagara-on-the-Lake), at least one moving to St Catharines where a new store was opened opposite Merritt's home (*Farmer's Journal and Welland Canal Intelligencer*, 28 June 1826). Niagara had previously lost out to York (now Toronto) as the capital of Upper Canada, but it remained the district town of the Niagara District and with diminished territory the county town for the united counties of Lincoln and Welland. Smith (1851) described Niagara as 'once a place of considerable business, but since the formation of the Welland Canal, St. Catharines, being centrically [sic] situated, has absorbed its trade and thrown it completely in the shade. The town, however, is airily and healthily situated, and is a pleasant summer residence, and will remain a quiet country town.' An aspect of this relative decline is that the oft-mooted lateral canal cut from Thorold to the Lower Niagara River was never achieved because of the American presence on the opposite bank.

The portage communities of Queenston, Chippawa, and Fort Erie soon lost their erstwhile portage trade to the Welland Canal. Their attempted redemption by the horse-drawn Queenston-Chippawa Railroad was not successful as a realistic competitor. Chippawa enjoyed temporary relief as the canal terminal on the Niagara River, but this was only a short-term revival.

Brown's Bridge, which had carried stagecoaches over the Welland River, did not outlast the construction of a new bridge over this river at Welland. In the south of the peninsula, the village of Crowland (Cook's Mills) at the centre of the township became less significant, and growth at Port Maitland was jeopardized by the greater flow of traffic to Port Colborne.

Across the Niagara Escarpment, where most of the mills along the canal developed, St Johns in the Short Hills went into severe decline despite the established industrial base that existed. Smith (1851) stated, 'It is an old village, but the formation of the Welland Canal has prevented its growth.' In Thorold Township, the new village of Thorold took its name from the township and became the business centre, displacing the earlier centres of Beaver Dams (now Beaverdams) and DeCew.

The impact of the Welland Canal was positive and negative. It favoured many new centres, but debilitated others. Either way, a powerful new base of future possibilities had been created across the Niagara Peninsula by the mid-nineteenth century during the era of the First Canal, and into the period of the Second Canal before the railway era of expansion dawned.

PART 2

The Second and Third Canals and Their Communities from the 1850s to the 1910s

5

An Expanding Infrastructure of Development

The period from the early 1850s to the outbreak of the First World War in 1914 involved the active years of the Second Canal and the opening of the Third Canal in 1887. Other factors included the arrival of three railways during the 1850s and later routes during the 1870s. By 1914 an interurban streetcar system also interconnected the canal communities and linked them across the Niagara River into the American system.

After Canada became a nation in 1867, protective tariffs shielded industry at the time when a strong new transcontinental nation was being established on an east–west axis across British North America. The canal communities gained during this expanding period of the national economy because of their proximity to the American border, a location that helped to attract U.S. companies. Also, as the provision of power moved from dependence on coal to hydroelectricity, this new resource was obtained first from lock sites along the canal, then by diversions from the canal, and more abundantly at low cost from generating stations on the Niagara River.

A core theme in this period becomes the transition from a commercial to an industrial era in the city-building process of Canadian cities. The first phase, which lasted until the 1870s, witnessed 'a degree of autonomy from imperial domination ... an increase in interregional trade and the growth of small-scale manufacturing related to the commercial role of urban places.' This was followed by 'the "new industrialism" which created a national network of communications and transportation and tended to centralize metropolitan power in the major central Canadian cities' (Stelter and Artibise 1982). Not all urban places passed systematically through these two phases in some deterministic fashion, but most of the communities along the Welland Canal entered an industrial phase in their growth process during the period of the Third Canal in the latter part of the nineteenth century.

The Second Welland Canal Matures

The Second Canal carried increasing volumes of freight (Bouchette 1867). The total volume (table 2) increased from 767,210 tons (696,012.9 tonnes) in 1854 to over a million tons (907,200 tonnes) from 1861 to 1863 and in 1866. Steamers increased from an 1854 total of 1069 to 1552 by 1867, whereas sailing vessels declined from about 5000 in 1863 to some 4000 by 1866 and 1877. Even so the canal could handle neither all of the expanding tonnage nor the increasing size of vessels on the adjoining lakes. An expensive alternative was to transfer the cargo to smaller vessels for the canal passage and, if the hull dimensions of the lightened vessel could pass through the canal, then these smaller vessels could be used in conjunction with horse and wagon to transfer part of the cargo and reload at the opposite end of the canal.

As a partial answer to this inconvenience and the possible loss of trade to rival routes, the government in 1853 increased the lock depth to 10 feet (3 m) by bolting timbers to the lock walls and the gates, and the bottom of the channel was widened from 36 feet (11 m) to 50 feet (15.2 m). The terminal harbours at Port Dalhousie and Port Colborne were also improved by dredging, and their protective breakwaters were upgraded. These changes permitted some vessels of larger draught, but not those of greater width, to use the canal.

A more suitable solution was to use the Welland Railway (to be discussed below), which opened in 1859 between Port Dalhousie and Port Colborne, to supplement the canal's functions. It permitted ladened vessels to off-load at Port Colborne. All or part of the cargo would then be transshipped to Port Dalhousie, where it would be reloaded onto the same or another lake vessel.

The adverse circumstances of limited canal dimensions became a problem of increasing severity through the expanding number of larger-sized steam vessels operating on each lake that were not capable of transferring from lake to lake. A further severe handicap was that the smaller dimensions of the downstream St Lawrence Canals hindered the through passage of vessels that had passed through the Welland Canal. The Welland admitted steamers up to 185 feet (56.4 m) long and 44 feet (13.4 m) wide, while the maximum capacity of the St Lawrence Canals was restricted to 142 feet (43.3 m) by 26.5 feet (8.1 m).

A through system existed between the Upper Lakes and the Atlantic, but it was imperfect and deteriorated in the 1860s because of the increasing need to ship American and Canadian grain from the area west of the Great Lakes. Of the 6740 vessels that passed through the canal in 1871, 42.8 per cent were Canadian and 57.2 per cent were American. Steamers provided for 30.8 per cent of the movement, with 69.2 per cent in sailing vessels, but at least sixty propellers on the Upper Lakes could not use the Welland Canal because of their size. In 1855 there had been only some twenty vessels in this category, and in 1845 all the vessels engaged in the grain trade on the Upper Lakes could pass through the canal. The canal, despite its expanding volumes, was becoming obsolete. This necessity to carry larger boats, and the desire to pass them through the Welland Canal rather than along some American route, led to strong demands for the enlargement and improvement of the Welland waterway during the 1870s.

In the meantime the railway era had arrived as a new competitor to transport bulky products, western grain, other agricultural products, iron ore, and coal. The total volume of freight carried through the canal in 1863 was 933,000 tons (846,418 tonnes) (table 2, pp. 136–7), but the canal was changing from a system used primarily for the movement of through products between external destinations. It had become a major lifeline in an expanding North American economy, now carrying mostly through, rather than locally generated, traffic. Ironically, with the increasing production of wheat on the Prairies, the Niagara Peninsula's crop output changed to higher-value agricultural products that included fruit and grapes. The more the amount of wheat that flowed through the canal, the smaller was the quantity grown locally.

Railways Are Added to the Canal Scene

After the early 1850s, railways introduced a series of remarkable changes to the landscape along the canal (Jackson and Burtniak 1978). Those on east–west alignments crossed the canal and provided a competitive inter-regional mode of transport that rivalled and soon ousted the canal as prime mode of long-distance travel and movement. An immediate consequence was that the Grand River Navigation Company as an extension of the Welland Canal failed soon after the arrival of railway competition in the 1850s. The successful American Erie Canal had also passed its heyday and was in the process of severe decline by the 1860s.

In contrast with the general decay that railways wreaked along canal systems of navigation in Europe and North America, the Welland Canal

TABLE 2 Traffic through the Welland Canal, 1854–1867 (commodities and total freight in tons)

Year	Vessels*	Steamers, paddle, and propellers		Sailing and other vessels		Flour	Corn and meal	Oats and meal	Rye	Wheat	Fire-wood	Sawn logs	Merchandise and other freight	Total freight
		No.	Tonnage	No.	Tonnage									
1854	C	690	57,396	2,691	327,111	40,022	113,495	10,434	134	76,961	36,764	20,606	468,794	767,210
	A	379	114,777	2,103	448,454									
	Total	1,069	172,173	4,794	775,565									
1855	C	1,069	91,948	3,211	413,124	27,828	115,313	5,776	894	166,520	61,323	28,305	448,374	849,333
	A	355	105,931	2,144	440,464									
	Total	1,424	197,879	5,355	853,588									
1856	C	1,142	99,250	2,789	353,405	38,700	140,436	8,073	3,458	227,337	68,915	17,122	472,515	976,556
	A	417	107,462	2,418	619,129									
	Total	1,559	206,712	5,207	972,534									
1857	C	377	44,274	2,667	374,710	41,949	79,790	1,719	112	213,174	68,763	16,984	478,581	901,072
	A	1,006	180,484	2,189	548,966									
	Total	1,383	224,758	4,856	923,676									
1858	C	230	25,335	2,298	335,559	44,336	102,470	10,123	345	245,954	64,255	18,917	368,712	855,112
	A	823	145,277	2,349	642,600									
	Total	1,056	170,612	4,647	978,159									
1859	C	159	27,043	2,026	291,126	37,494	35,900	2,546	145	159,077	46,302	9,446	418,611	709,611
	A	857	136,892	1,487	401,857									
	Total	1,046	163,935	3,513	692,983									
1860	C	555	69,439	2,249	311,620	43,738	139,445	2,495	278	274,405	54,081	16,026	423,316	944,086
	A	930	151,248	3,604	706,202									
	Total	1,485	220,687	4,853	1,017,822									

Year		C1	C2	C3	C4	C5	C6	C7	C8	C9	C10	C11	C12	C13
1861	C	576	55,137	2,540	381,356	58,737	156,745	1,550	2,310	313,526	68,648	17,680	351,287	1,020,483
	A	874	179,456	2,718	711,723									
	Total	1,450	234,593	5,258	1,093,079									
1862	C	417	63,556	2,938	447,799	68,395	162,099	3,716	4,395	479,853	99,877	18,735	406,704	1,243,774
	A	1,053	192,962	2,871	772,525									
	Total	1,470	256,518	5,809	1,220,324									
1863	C	530	79,023	2,895	442,795	75,614	100,821	92	2,428	370,194	114,973	7,691	469,307	1,141,120
	A	1,335	195,261	2,139	613,028									
	Total	1,865	274,284	5,034	1,055,823									
1864	C	289	37,095	957	161,187	27,599	13,807	150	56	101,438	28,816	8,142	142,325	322,343
	A	372	63,564	651	184,260									
	Total	661	100,659	1,608	345,447									
1865	C	999	127,392	2,842	466,169	55,514	59,877	16,296	3,718	224,982	118,852	18,210	388,427	868,078
	A	1,101	183,580	1,405	358,665									
	Total	2,100	310,972	4,247	824,834									
1866	C	915	111,471	2,755	443,165	49,693	124,361	16,879	1,949	227,243	144,639	20,453	416,300	1,001,517
	A	937	181,632	1,542	341,046									
	Total	1,852	293,103	4,297	784,211									
1867	C	733	98,395	2,360	394,444	28,355	144,168	11,160	1,986	154,173	113,206	31,424	203,788	933,260
	A	819	181,492	1,393	319,607									
	Total	1,552	259,887	3,753	714,051									

Source: Bouchette 1867

*C = Canadian; A = American

continued to expand, adding new routes and larger dimensions and increasing flows of traffic during the latter half of the nineteenth and into the twentieth centuries, a rare phenomenon of the period. Within the Niagara Peninsula the railways added new transportation possibilities to the existing range of canal-oriented industrial sites. As the main lines and their many spur routes then helped considerably to attract new industries, subsequent industrial sites had both rail and canal modes of access or access by rail or canal only. In all instances, the canal wharves, railway stations, and freight yards continued to be serviced by local unsurfaced roads and the ubiquitous horse-drawn wagon for local, short-distance movements.

The new stations and the rail yards became important focal centres of activity within the existing pattern of settlement (see maps 3, 4, 6, 7, 8, 9, 11, and 12). The tendency was strong for services to be drawn to these sites rather than to the canal as before, a directional pull that could substantially change the community format.

Railways, often accompanied by the telegraph for long-distance communication, introduced year-round linkages between communities previously dependent on the seasonal vagaries of canal and river travel and the post office, and added a new foundation for urban growth in the fledgling urban system that existed. They interconnected local places and made the external world more accessible.

These changes to the settlement pattern emerged gradually as demands on the rail system grew. Track had to be relayed and bridges reconstructed as heavier and longer loads were carried. The early wood-burning locomotives were replaced by coal-fired engines, which added speed and different servicing requirements at the stations and yards. Passenger travel became more comfortable, sleeping cars assisted overnight journeys, and refrigerated cars for freight both assisted the movement of some commodities to urban markets and helped the agricultural transition from wheat to fruit in the Niagara Peninsula. Along the canal, in addition to these changes, a visitor-tourist function was added at the lakeshore beaches at both terminals, and St Catharines, with spa hotels, was promoted as a health resort. In the meantime, ships continued to carry freight and passengers through the canal, but their destinations were limited to the linear confines of the Great Lakes–St Lawrence route. The economy had been strengthened, not weakened by these changes. Railways supplemented, but they did not supplant, the Welland Canal.

The first railway in the Niagara Peninsula was the horse-drawn Erie and Ontario Railway on wooden rails with iron straps, which opened along the Niagara River between Queenston and Chippawa in 1839. It handled freight and passengers, and its purpose was to defeat the Welland Canal and return trade to the portage communities. Its successor, with a revised charter and as a steam operation, opened between Niagara (now Niagara-on-the-Lake) and Chippawa in 1854. Later, as the Erie and Niagara Railway, it crossed the eastern arm of the canal at Chippawa by a low wooden trestle and swing-bridge so as not to interfere with canal navigation, then ran to a dock at Fort Erie for a ferry crossing of the river. After the opening of the International Bridge in 1873, the route ran direct into Buffalo and connected this expanding city with Lake Ontario. The railway provided a potential competitor to both the Welland and the Erie Canals for inter-lake traffic, but it did not succeed in either regard as direct, through movements to the coast were preferred.

The Great Western was the first railway to cross the Welland Canal (at Merritton, 1853) and to provide it with long-distance competition. The first through train between the Detroit and Niagara Rivers followed next year. This line was extended from Hamilton to Toronto in 1855, and next year to Montreal along the north shore of Lake Ontario. Other branches followed in quick succession. The most important of these new links was to the Suspension Bridge constructed across the gorge of the Niagara River, completed in 1855. The Niagara Peninsula was now connected by a continuous land route by rail with the cities and manufacturing areas of the United States.

The Great Western connected at Hamilton with shipping on Lake Ontario. At St Catharines the municipal council expressed a desire for a railway bridge across the Second Canal along approximately the line of Welland Avenue that would be close to the town's commercial centre, but these hopes were not fulfilled. The railway crossed Twelve Mile Creek closer to the escarpment where docks could not be provided, and no direct interaction between the rail and canal modes of transport followed.

The economic gain of having rail and canal facilities in close conjunction was awarded to Merritton, where the Second Canal was crossed at a lock on the straight line of track. As stated by Keefer in 1852: 'It must be remembered that the place where the road passes the Canal, from its situation, and the extensive water supply at command, is likely to become the seat of a manufacturing Town of some importance.' Merritton with its shanty communities was about to be transformed by industrial developments that took advantage of the water power available at the canal locks and the now dual advantages of both rail and water transportation facilities.

In the south of the peninsula, the railway era began with the Buffalo, Brantford and Goderich Railway, which in 1856 became the Buffalo and

Lake Huron Railway. The intention behind this American route to the Upper Lakes was 'a desire, on the part of the populous city of Buffalo, to render tributary to herself the rich peninsula of Canada West, and also to divert the stream of eastern and western travel and freight away from the Suspension Bridge route to her own hotels and stations' (Keefer 1864).

This line crossed the southern entrance of the Second Canal at the head of its harbour at Port Colborne in 1854 . Extended to Paris the same year and to Goderich in 1858, it crossed the Niagara River by shuttle ferry to Fort Erie, and its straight line of track then ran inland from the shoreline of Lake Erie to a station located at Port Colborne close to the entrance lock of the Second Canal and the village centre. The harbour was serviced by rail along its eastern side, where a grain elevator soon provided a direct link between the water and rail modes of transportation.

The Feeder Canal was crossed at Dunnville, again providing the potential for interaction between the rail and the canal systems of transportation. The line of track that lay inland from the Grand River, following it to Brantford, sounded the death knell for the Grand River Navigation Company. This railway attracted much of the earlier freight and passenger traffic that had moved through the Feeder Canal.

The Welland Railway

The story of the Welland Railway, which helped so considerably to promote the canal, began when the Port Dalhousie and Thorold Railway was incorporated in 1853. Ancillary to the canal, it was projected to run between Merritton, the station for Thorold on the Great Western Railway, to Port Dalhousie on Lake Ontario, where steamboats to carry passengers and freight would be used. This provided the Great Western with an alternative to its home port of Hamilton, and Buffalo with a potential outlet on Lake Ontario.

A revised charter extended the route along the canal to Port Colborne in 1856, and the name was changed to the Welland Railway in 1857. Its first through lake-to-lake train ran in 1859. This railway was supported strongly by William Hamilton Merritt, who argued that the rail and canal modes of transport could be complementary. His argument was that the canals would carry the bulky, heavy, long-distance loads, whereas the railways would carry the lighter and more valuable freight and would also serve during the winter months when the canals were closed to navigation.

The relationships between the Welland Canal and the Welland Railway were clearly enunciated. As stated in the Prospectus for the Welland Railway of 1856: 'Immediately on the arrival of steamers at either end of the Welland Canal, light and valuable freight will be transferred by Railway from one Steamer to another in the same line, having previously passed through the canal, and ready to proceed to her port of destination on the opposite lake. By this arrangement, heavy freight will not be subject to trans-shipment, while light freight will secure speed and certainty.'

A later prospectus of the same year noted: 'Passengers, livestock, butter, cheese and valuable merchandise will be transferred to the Railway, while timber, lumber, grain, iron, salt, and cheap bulky articles will go through the Canal, and each will add to the business of the other.' Coal and wheat as basic commodities that would be carried by the railway were added next year.

The opening of the north–south inter-lake Welland Railway across the peninsula connected with the pre-existing Buffalo and Lake Huron route at Port Colborne and the Great Western line at Merritton. Both terminal ports were strengthened as stations, yards, and elevators added on the east side of the canal at Port Dalhousie and Port Colborne served these canal communities and the canal traffic, to add a further impetus to industrial development.

As the Welland Railway ran parallel to the canal from lake to lake, every canal community now had access to a nearby station and its yards on this line of track. Spur lines constructed to industrial sites broadened the railway impact and segmented the communities through which they passed. The railway also connected each canal community to the other centres along the canal. Merritton and Port Colborne were the main-line junctions with connections east–west across the frontier and to all places served by the expanding Canadian and American networks. The potential for spatial interaction over a greatly extended area had increased enormously, and was extended when new routes across the peninsula were added.

Planning the Third Canal

The Second Canal faced severe problems associated with the regular supply of water drawn from the Grand River, which was inadequate as a continuous source of supply during certain seasons of low rainfall because of increasing settlement, forest clearance, and a diminished run-off. For example, the canal's annual report for 1861 notes: 'The scant supply afforded by the Grand River has been much felt, during the past season, so much so, that the Upper Level [between Thorold and Port Colborne] was frequently, for Weeks together, from one to two feet [0.3 to 0.6 m] below the established height. This will no doubt be partially remedied by staunching the Dunnville dam, the work of which has been put in hand.'

Ten years later in 1871, when the water in the Feeder Canal was again low and tonnage on the canal was increasing, the mills above Allanburg had their water supply shut off to conserve water for navigation. It was then recommended that the lock walls, banks, and waste weirs be raised to obtain a permanent depth of 12 feet (3.7 m) between Allanburg and Port Dalhousie. Canal improvement was necessary for more reasons than to carry grain from the West. A guaranteed supply of water to the mills remained an important consideration.

Confederation in 1867 was a major factor in achieving the Third Canal, as the Second Canal now became the responsibility of the dominion government. Three schemes were favoured to carry the trade from the west (Langevin 1870). One group argued for the enlargement of the Welland Canal through increasing its bottom width, widening and lengthening the locks, and constructing an additional line from the summit level at Thorold to the outlet on the Niagara River. A second group urged that, in addition, the whole line of canals to the St Lawrence should be widened and deepened. A third group recommended a shorter overall route, some advocating a route across the peninsula of southwest Ontario between Toronto and Georgian Bay and others a line of canals through the Ottawa valley between Montreal and Lake Huron.

A critical fact was that 'the most important of the Canals of Canada were designed, not only with a view of affording an unobstructed passage for the staple products of its soil to the Ocean, but of attracting, through the same channel, a portion of the freight passing from the West to the Atlantic, and that, notwithstanding all the advantages offered by the St Lawrence, the bulk of the traffic referred to continues to find its way to the seaboard over the railways and Canals of the United States' (Canal Commission 1871). Pride in the new nation and the competitive desire for trade from the West to flow over its own territory were important factors in retaining and improving the Welland Canal.

The situation by 1871 was that about 24 per cent more flour and grain had moved through the Erie Canal via Buffalo and Oswego since 1861 than had moved on the New York Central and the Erie Railways. Nonetheless, although production in the grain-growing states had demonstrated large and steady increases, the means of communication were insufficient to handle this capacity. As the commissioners for the State of Illinois had in 1863 noted: 'For several years past, a lamentable waste of crops already harvested has occurred, in consequence of the inability of Railways and

Canals leading to the seaboard to take off the excess' (Canal Commission 1871).

On the Canadian side, this commission argued, no canal 'has been so successful as the Welland.' Its tonnage and the revenue from tolls had doubled since 1849, but the greater proportion was American tonnage using vessels of the largest size to pass through the canal. There were more Canadian than American vessels, but their size was smaller. 'There can be no doubt that, as in the case of the Erie Canal, it [the Welland Canal] has not equalled the requirements of the growing commerce dependent on the facilities it affords for speedy and cheap transport ... The business of the Canal has comparatively stood still, although it certainly affords the best and most convenient avenue of communication' (Canal Commission 1871).

Steamers, that is, screw vessels rather than paddle wheelers or sailing vessels, now prevailed on the lakes. They were increasing in number and size, and had 'all the advantages in respect to rates of Insurance, expedition, safety and competition with railways, all important elements in the transportation of the bulky produce of the West ... [Yet] three-fourths of the tonnage of the lakes cannot pass the Welland Canal – a fact in itself sufficient to show why its traffic does not increase' (Canal Commission 1871).

On the basis of such arguments, as an expression of nationhood, and in order to retain and attract more of the wheat trade, a new and improved Welland Canal was perceived as vital to the commerce of the country, including not only the wheat trade but also other merchandise including lumber and the copper and iron trade of Lake Superior, which 'has been crippled on account of the want of cheap transport' (Canal Commission 1871).

The commission stressed the need for a consistent lock size throughout the system, and recommended a uniform scale of navigation with locks 270 feet (82.3 m) long and 45 feet (13.7 m) wide, with 12 feet (3.7 m) of water over the sills. These dimensions would prevail for the Welland and the St Lawrence Canals, and for the proposed canal at Sault Ste Marie.

On the Welland Canal, a new source of water supply other than from the inadequate Grand River was required. The excavation to bring the canal down to the level of Lake Erie, started in 1846, had been delayed by the 3-mile [4.8-km] width of the Onondaga Escarpment, which sloped gently to the south, and by further slides at the Deep Cut as the works of excavation proceeded.

On the assumption of an assured supply from Lake Erie, the commission recommended that the harbours at Port Dalhousie and Port Colborne be deepened to 15 feet (4.6 m) to provide a safer entrance for vessels drawing

12 feet (3.7 m) of water, that a second lock be added at Port Colborne to allow more water into the canal, and that the aqueduct at Welland either be deepened or another aqueduct constructed. Further, it was argued not only that the Second Canal be retained for the use of smaller vessels during the period of canal construction, but also that it be deepened permanently by replacing the lock gates and the temporary timber along the lock walls and weirs with masonry.

In the same manner as for each of the two previous canals, several alternative routes to cross the Niagara Escarpment were considered (Page 1872a, b). Favoured by St Catharines, because it would retain the canal next to the business centre of that community, was the line that would take the canal from Marlatt's Pond above Thorold to Twelve Mile Creek at St Catharines, then use the Creek to Port Dalhousie. But the locks of the Second Canal were too close together and it was impossible to construct larger locks without stopping navigation. As the valley was also too confined to carry two locks, it was 'imperative to seek another line where ample basins can be established between the locks to admit of the passage of vessels, and capable of holding an abundant supply of water for working the lock without drawing down the levels' (Canal Commission 1871).

Other routes were considered from Thorold to the Lower Niagara River at Queenston and Niagara. As slope intervened, the Niagara route was more acceptable, but its disadvantages included the rapid current in the river, late ice from Lake Erie in the spring, shoals off the entrance to the river, and high banks not suited for a harbour. As Port Dalhousie did not have these disadvantages, and as it was also the shortest and the cheapest route with established harbour facilities, the commission concluded that 'the direct Line from Thorold to Port Dalhousie is the best route that could be selected for the enlargement.'

Alternatives then included various alignments with locks and ponds to the south of Twelve Mile Creek, intersecting with the existing canal at different points up to the Deep Cut. The high land traversed by the Deep Cut rose gradually east to the area of Fonthill Kame, and this high land in the centre of the peninsula also made impractical the thought of a new canal route across the peninsula that might terminate at Port Maitland on Lake Erie and at the mouth of Twenty Mile Creek on Lake Ontario. The substantial additional costs of creating a completely new route rather than making use of existing canal facilities wherever practicable made this alternative and ideas such as a canal to Queenston speculative rather than practical, though the tendency was for each local community to push for what were considered its own best interests.

The consideration of various alternative canal possibilities emphasizes that the achieved Third Canal was the outcome of engineering decisions. The impact on the different canal communities was not part of the decision-making process. The form of the canal communities had resulted from the First and Second Canals, and they now had to adjust their character to suit the Third Canal. The fixed ingredients in the design process for the Third Canal were the two terminal ports and their harbour facilities, the intermediate channel between Port Colborne and Allanburg, and the challenge of crossing the embarrassing height and slope of the Niagara Escarpment.

The New Canal Landscape

The commission's recommendations were accepted, and contracts awarded in 1873 to reconstruct the Third Canal (map 1) to the required depth of 12 feet (3.7 m) (Heisler 1973; Monro 1872). The task of bringing stone, timber, cement, and other materials to the construction sites often devolved on the Welland Railway. For the Second Canal the delivery of these materials had been via the First Canal, which ran next to its successor. On the Third Canal this approach was no longer feasible, as the canal's northern route now followed a new channel distant from the old.

Divided into two divisions for the supervision of the work, the northern length of 11.8 miles (19.0 km) between Lake Ontario and Allanburg included enlargement of the harbour at Port Dalhousie, twenty-five locks, twelve road bridges, two railway bridges, and a tunnel so that the Great Western railway could pass under the canal. The length between Allanburg and Lake Erie added a guard lock at Port Robinson and a new aqueduct and new lock to the Welland River at Welland. A syphon culvert carried Lyons Creek under the canal, and a new lock and expanded harbour were required at Port Colborne. Six road bridges and three rail bridges crossed the canal. In each division, the project was massive in concept and in reality. Vaster than the Second Canal but now with more mechanical equipment available, the works again brought in a considerable labour force.

Two years later, in 1875, under pressure from marine and trading interests and to keep abreast of the increasing size of vessels, the depth capacity for the Third Canal was increased to 14 feet (4.3 m) (Committee on Transportation 1875). Except at the aqueduct over the Welland River at Welland, the completed canal opened to its 12-foot (3.7-m) depth in 1883; the enlarged version opened throughout the line of the canal in 1887 (Douglas 1884). The cargo capacity of vessels using the canal had increased more than threefold up to some 2700 tons (2249 tonnes), but the downstream St Lawrence Canals were not available at this navigable depth until 1901. The Welland Canal system was again substantially in advance of its downstream partners, and cargo volumes passing through the Third Canal only zoomed after the downstream deficiencies were remedied.

The major ingredient of the Third Canal, the new channel constructed between Port Dalhousie and Thorold, required enlargement of the entrance piers at Port Dalhousie to improve the harbour entrance, and a new lock was constructed on the east side of the weir that contained Martindale Pond. The twenty-six locks along the canal, now consistent in size and again of stone construction, were 26.5 feet (8.1 m) wide, 150 feet (45.7 m) long between the gates, and 14 feet (4.3 m) deep.

The canal then cut across Martindale Pond to May's Ravine on its eastern bank, where Lock 2 was constructed. This was followed by a straight new cut south-east across the farmland of Grantham Township. The canal then crossed the Niagara Escarpment to Marlatt's Pond at Thorold, where it curved west before following the Second Canal until joining it at Allanburg.

From Allanburg to Port Colborne the new canal required the widening and deepening of the old channel. As the bed of the new channel in the south now lay below the level of Lake Erie, the former entry lock at Port Colborne became a guard lock and the lock of the Second Canal was used to supply and control the flow of water into the Third Canal. The lock at The Junction (Helmsport), where the Feeder Canal met the main line of the canal, became a lift lock to cope with the differences between the two levels of water. The Feeder Canal and the Grand River were displaced as the source of water supply for the main line of the canal, but the feeder was retained for a while to permit diminishing navigational use by smaller vessels, and a raceway (now Prince Charles Drive) was cut from the end of the channel to the Welland River.

All twenty-five lift locks along the canal and their water-power potential were now centred on the northern length of the canal between Allanburg and Port Dalhousie. This advantage for industrial development was strengthened by a basin constructed at Carlton Street in north St Catharines, where ships might receive and discharge freight and where day boats passing between the lakes might dock for supplies and passengers. New swing-bridges crossed the canal on the main roads near St Catharines at Ontario Street, Niagara Street, and Queenston Road, where the bridge masters were provided with small offices. At Facer Street off Niagara Street, where the main road to St Catharines and the bridge across the canal encouraged the subdivision of land isolated from this city by the canal, the area became a mixed ethnic quarter as immigrants from central, southern, and eastern Europe made their homes in the vicinity.

Below the crest of the escarpment, the Third Canal crossed the Great Western (later Grand Trunk) Railway where it climbed the slope at the grade of 38 feet (11.6 m) to the mile. The original plan was for a grade crossing, but the railway authorities insisted on a tunnel for safety reasons and to avoid delays at a swing-bridge. The railway was reconstructed in a curved cutting that was 713 feet (217.3 m) long with a central tunnel. In use from 1881 to the mid-1890s, this tunnel was replaced by a double-track, steam-operated bridge over the canal on the original line of track. A road tunnel under the canal was also constructed to carry St David's Road between Thorold and this village.

Where the Third Canal crossed the upper waters of Ten Mile Creek, this drainage was diverted into a culvert constructed under the canal, and the scenic waterfall and gorge where the creek crossed the escarpment fell victim to the demands of the canal engineers. The Welland Railway that had followed the Second Canal was diverted for about a mile, and graded approaches took it over the canal by two swing-bridges.

Each of the mountain locks had its own weir and a pondage area, which functioned as a basin so that the largest vessels could pass without difficulty. The locks and their reservoirs were comparable in principle, but together they were much larger in size and covered a greater area than their predecessors. Now with a greater fall at the locks and a greater volume of water available from Lake Erie, new sites for industrial development might reasonably be anticipated. However, they did not materialize because the days of water power were now limited and the Department of Railways did not permit the sale of water rights. The water was required to operate the locks; water rights remained available along the Second Canal. One estimate of this apparent anomaly is that 1000 horsepower was being lost at each of the Third Canal locks (*Daily Standard*, 30 November 1909).

For the route of the Third Canal to Lake Erie south of the escarpment, the established channel of the Second Canal was followed along its eastern bank to where it joined the Second Canal at Allanburg. From there the former channel, widened to 100 feet (30.5 m) at the bottom and deepened to below the level of Lake Erie, was followed to Port Colborne.

A new cut-stone aqueduct with six arches, 460 feet (140.2 m) long and 85 feet (25.9 m) wide, was constructed to the west of the Second Canal at Welland. As this at first leaked and there were cave-ins at the side, a second contractor had to be appointed to complete the project.

At Port Colborne, where the harbour was deepened to below the level of Erie, piers at its east and west ends were extended into the lake. The contractors built dams at the required places, then pumped out the water so that this work could be undertaken when the channel was dry. As elevators, wharves, and industrial undertakings soon took advantage of these piers, railway tracks were extended south and land behind the piers was reclaimed from the lake for industrial purposes.

The entry lock was replaced on the east side of the old lock, and new swing-bridges carried road and rail traffic across the canal. A raceway to control the supply of water into the canal constructed to the west of this complex involved tearing up West Street, then in part covering the raceway with a road and leaving its width next to the canal open.

The lock at Allanburg was no longer required because of the now lowered canal depth to the level of Lake Erie. The lock at Port Robinson was retained, but enlargement was not necessary as shipping on the Lower Welland River had declined. The canal outlet to the Niagara River, the Chippawa Cut, was no longer required. Eventually, the flow in the lower length of the Welland River was reversed and this canal channel became part of the water-supply system to the Sir Adam Beck generating station at Queenston Heights.

When the Third Canal was completed, its route was 26.8 miles (43.0 km) long, shorter by 0.75 miles (1.2 km) than the Second Canal. The winding route of the Second Canal along Twelve Mile Creek and the canal valley, where most of the locks had previously existed, was abandoned for through navigation. Port Dalhousie remained the port of entry on Lake Ontario, but the long-established and industrially important canal communities of St Catharines, Merritton, and Thorold were each bypassed by the new system. They lost the considerable advantages of through passenger and commodity flows, as well as the service trades that had previously been enjoyed from the canal. Even so, the water rights and the supply of water for industrial purposes in the hydraulic raceway and at the locks in the valley bottom remained.

Further, the upper and lower lengths of the Second Canal down to Thorold and up to St Catharines from Lake Ontario remained open and in active use by smaller-sized vessels to serve the mills in these centres. An example of this continuing but partial use of the old channel is provided by a vessel that made regular weekly trips between St Catharines and Montreal. It left Montreal every Tuesday at 1:00 p.m. On the return journey it carried barrels of flour, passengers, and called at Toronto, Kingston, Brockville, Prescott, and Cornwall. Despite railways and their growing importance for travel, the water mode remained important and a major link between the ports around Lake Ontario.

A towpath was constructed along the Third Canal as schooners still used the system, but sailing vessels were now a diminishing breed as steam tugs replaced horses and mules. A tug might now tow several barges at a time, and it and several barges could be accommodated together in the enlarged locks. The new canal was designed to accommodate the new size and type of vessel, but it had also to make provision for the old. Gas lighting was also installed at the locks and along the line of the canal.

The year 1887, when the Third Canal opened to its increased depth of 14 feet (4.3 m), was in many ways a turning point in the canal's history. Now with fewer locks, a shorter route, and a faster passage with larger vessels that could carry some 3000 tons (2721.6 tonnes) or 75,000 to 80,000 bushels of grain, the canal for the first time in its history was situated wholly in a man-made channel. It also now had the advantages of a more regular and reliable source of water from Lake Erie.

Even though the State of New York had reduced tolls by 50 per cent in the early 1870s, the Erie Canal through attrition by the railways was no longer a serious competitor to the Welland Canal. The largest eastward trade from Buffalo was in 1862. By contrast, traffic on the Welland Canal continued to grow, and the Third Canal played a significant role in the development of Canada's grain export trade and its steel industry.

In 1870 there were more sailing vessels, both American and Canadian, using the Welland Canal. Twenty years later, there was almost a reversal. By 1890, the average tonnage of vessels had more than doubled, steamers carrying most of the cargo were now to the fore, and it took fewer than half the number of vessels to move about the same freight tonnage. American vessels were fewer, but they tended to be larger than their Canadian counterparts and to carry a larger portion of the traffic through the canal. It however remained that only the smaller-sized canal freighter could pass through the locks of the St Lawrence Canals between Prescott and Montreal, which meant some transfer to the railways of Montreal-bound grain at Port Colborne or Prescott.

These events in close conjunction with the Third Canal were also associated with the period during which Canada changed from its colonial, pioneer commercial economy to an advanced industrial one. Along the Third Canal this transition in the canal communities was from a local manufacturing base that had accompanied the canal since its inception to new activities serving national and international markets and employing a greatly increased labour force in larger factory units.

The Railway Network Is Expanded

The second phase in the development of the local railway network took place after the American Civil War (Jackson and Burtniak 1978). The Canada Southern Railway opened in 1873 as an American route across Canadian terrain between the Niagara and the Detroit Rivers. The Great Western Railway responded with a corresponding Canada Air Line Railway that opened the same year and linked the same two places along a comparable route. As both routes crossed the canal in the vicinity of Welland, a considerable momentum to growth was added at this location.

A further strengthening factor was that three rail bridges now crossed the Niagara River. In addition to the earlier Suspension Bridge across the gorge, the International Bridge crossed the entrance to the Niagara River at Bridgeburg (Fort Erie) in 1873, and in 1883 the Niagara Cantilever Bridge opened across the gorge below the Falls to carry Canada Southern – Michigan Central railway. The communities along the Welland Canal, especially in the south, were now closely connected with the rapidly expanding American economy that was so strongly to shape their future destiny.

During the 1870s and the early 1880s, the lines of the Great Western, the Buffalo and Lake Huron, and the Welland Railways were absorbed steadily into the Grand Trunk network. This monopolistic situation introduced the St Catharines and Niagara Central Railway, which opened between Niagara Falls and Thorold in 1887 and was extended to St Catharines the next year. This railway crossed the Third Canal east of Thorold, where the railway intended to construct two wharves next to the canal for combined rail and water purposes and to follow part of the canal's heel path through the town to serve its industries, but the canal authority contended that the railway was not needed and that a bridge across the canal would present a serious hindrance to navigation.

Support was strong for the railway. In the lead columns of the *Thorold Post*, John Page, the canal engineer, was described as a 'known obstructionist,' a 'fossilized functionary' who personified 'officiousness and arrogance' and whose 'red-tapism, officiousness and flunkyism' stood seriously in the way of the rights of the people of Thorold, who stood to gain much from the intended railway. As will be evident here and in several subsequent instances, canal and community interests have not always been in harmony.

The economic battle between the Welland Canal and the railways for supremacy in the long-distance trade was matched by one between competing railways, with their affiliations, agreements, and amalgamations, to secure the same trade. The canal communities had to weather these changes, an example being that when the Great Western obtained running rights over a portion of the Welland Railway in 1873, its broad-gauge track was converted to standard gauge.

And when the Grand Trunk in turn took over the Great Western in 1884, the Welland became a branch line but its steamer links from Port Dalhousie were retained. The paddle steamer *Empress of India* was introduced on the Toronto run that year. With one run a day in each direction, this ship connected at Port Dalhousie with a fast train to Niagara Falls and Buffalo. This service lasted well into the 1890s, but with only one train and one boat it could not compete with the twice-daily trip via the Canadian Southern and the Erie and Niagara Railways that permitted one-day visits between Toronto, Buffalo, and their intermediate stations, and also had the advantage of following the more scenic route along the Niagara River.

A further railway, the Toronto, Hamilton and Buffalo, opened between Hamilton and Welland in 1895. A through service to Buffalo followed two years later. A branch line was constructed to Dunnville in 1914 and this was extended to Port Maitland in 1916. A regular ferry service connected across Lake Erie to Ashtabula, Ohio. Hamilton was now connected by rail to Lake Erie; a major import was coal destined for the steelworks at Hamilton, and mainly newsprint was shipped in the southbound direction. A new cross-peninsula competition to the Welland Canal had emerged, a service that lasted until the Fourth Canal opened.

The Third Canal in Action

The standardization of the lock sizes and the now prevailing use of steamers led to the extensive development of a particular type of vessel on the inland canal system. Known as the 'canaller' and designed to fit the locks, she was the equivalent of a self-propelled barge with machinery at the stern, a navigating bridge towards the bow, and a long, almost box-shaped cargo hold in between. The vessel had a maximum length of 262 feet (79.9 m) and a carrying capacity rather less than 3000 tons (2700 tonnes). Larger vessels were built to sail from the Lakehead (now Thunder Bay) to Port Colborne, where their 15,000 tons (13,500 tonnes) of cargo was transferred either to smaller canallers for passage through the Third Canal or to the Welland Railway.

By 1888 the canal system between Lake Superior and tidewater had fifty-three locks, of which the enlarged Third Canal carried twenty-six. The freight carried on the Welland averaged almost 880,000 tons (805,593 tonnes) over the three-year period from 1886 to 1888. Wheat and corn were the prime commodities, then coal and forest products. Between 1888

TABLE 3 Cargo through the Welland Canals, 1867–1914 (thousands of short tons)

Year	Wheat	Total agricultural products*	Manufactures and miscellaneous**	Forest products	Total coal	Total mine	Total all freight	Total up	Total down
Second Canal									
1867	–	–	–	–	–	–	933	245	689
1868	275	536	89	281	103	230	1162	332	850
1869	314	504	192	374	103	162	1232	386	857
1870	432	622	114	285	98	254	1312	365	947
1871	435	699	139	322	125	279	1478	396	1082
1872	240	538	138	371	187	286	1333	410	923
1873	356	580	80	382	339	458	1506	456	1050
1874	413	647	51	310	324	381	1389	405	984
1875	254	418	34	222	321	364	1038	340	698
1876	202	410	58	262	288	370	1100	329	771
1877	254	464	33	311	324	367	1175	312	863
1878	192	403	29	227	295	330	969	268	701
1879	275	439	34	181	193	212	866	216	650
1880	242	442	32	202	110	144	820	–	–
1881	128	269	43	228	128	147	687	220	466
1882	215	306	22	211	238	252	791	282	508
1883	153	373	26	267	307	339	1005	368	637
1884	145	306	5	200	274	327	838	354	484
1885	124	274	17	221	248	273	785	–	–
1886	154	415	18	249	271	298	980	327	653
Third Canal									
1887	222	395	14	210	145	159	778	242	536
1888	161	420	17	201	224	241	879	297	582
1889	127	542	23	250	268	270	1085	327	759
1890	118	519	30	237	202	230	1016	299	717

1891	199	367	13	367	225	228	975	—	—
1892	232	527	7	210	212	212	956	270	685
1893	258	805	10	247	233	233	1295	282	1013
1894	271	591	4	209	204	204	1008	249	759
1895	203	486	9	215	159	160	970	234	636
1896	321	789	10	257	223	224	1280	291	989
1897	325	817	25	256	176	176	1274	224	1050
1898	208	720	14	231	162	175	1140	—	—
1899	198	460	9	197	98	124	790	153	637
1900	138	376	14	224	47	105	719	118	604
1901	152	291	15	166	49	148	620	106	514
1902	225	351	22	206	64	86	665	85	581
1903	250	537	54	246	148	166	1003	270	733
1904	165	374	26	257	114	154	811	185	626
1905	254	577	18	302	173	195	1092	232	860
1906	327	650	28	370	148	154	1202	223	979
1907	489	895	25	402	267	292	1624	390	1224
1908	732	976	36	356	317	335	1703	411	1292
1909	590	898	87	630	378	411	2026	642	1384
1910	587	1035	58	622	577	611	2326	724	1601
1911	562	1090	540	250	619	657	2538	843	1695
1912	796	1206	626	228	709	792	2852	820	2026
1913	1005	1685	549	338	946	999	3571	1005	2566
1914	1500	2116	361	360	949	1023	3861	856	3005

Source: Urquhart 1965
*Excludes animal products that are small in volume.
**Includes animal products.

and 1899 over one million tons (907,200 tonnes) were carried in most years (table 3). To encourage this movement, the Canadian canals were exempted from tolls in 1903, an exemption declared 'permanent' in 1905. By 1910 to 1912 the freight carried had doubled, and in 1913 and 1914 it was over 3 million tons (2.7 million tonnes). By 1914 the total tonnage of 3.9 million tons (3.5 million tonnes) included 2.1 million tons (1.9 million tonnes) of agricultural products, with about half that amount being mine products. There were also some 360,000 tons (326,592 tonnes) of both manufactures and forest products. Constructed and expanded principally to carry wheat, and now carrying more and more grain, the Welland Canal had now broadened its trade to become an integral part of the Canadian and the Great Lakes economies of North America.

With this expanding traffic, the desire of the shipping companies to avoid the costs and time delays of trans-shipment, and for the larger vessels to be able to move from the Upper to the Lower Lakes, were powerful factors that led to demands for further canal enlargement. As before, each new Welland Canal had become obsolete fairly soon after its opening, and plans were soon made to enlarge the canal again. Demand and more advanced technologies were the masters that determined the size of the canals, along with the response of the canal communities to the changing marine scene.

The transition in vessel size and type on the Upper Lakes is quite remarkable. The larger-sized steam vessel that could move through the canal under its own engine power dominated the situation after 1872, but sailing vessels were still actively used. By 1914, when 3692 vessels passed through the canal, 10 per cent still made use of sail. Canal towpaths, towboys, their animal teams, and stables gradually became redundant. They were replaced by coal docks and coal depots, machine repair shops, the faster passage of fewer vessels, a reduced demand for ship's chandlers, and a greater reliance on the terminal ports rather than the larger number of intermediate points of call. The end of the sailing vessel was not sudden, but its gradual obsolescence slowly exerted many effects on the canal communities and their services that had arisen when sail was the dominant mode of transport for both passengers and freight.

A significant date is 1888, when the steamer *Rosedale* arrived at Chicago from London, 'which naturally excited considerable interest, as it not only proved to Americans the possibility of sending grain direct from Chicago elevators to London without trans-shipment, but also proved to Canadians a like possibility of sending the products of the Northwest direct from the elevators of Port Arthur [now Thunder Bay]. The passage occupied 35 days, and the steamer was the first to follow the direct route [via the

Welland Canal and the St Lawrence route] from London to Chicago' (Canada Statistical Abstract 1888).

As more vessels used the canal and as road traffic expanded, the swing-bridges on the major highways became a matter of increasing concern. Delays on the highway crossings became more serious, and when a bridge was under repair either through maintenance or because of damage by a passing vessel (a possibility that increased as the weight and speed of vessels grew), the only way to cross the canal was by boat or through lengthy detour. As stated in a letter of 1872 referring to Welland, 'Our bridges are a bad lot. The two over the canals are in need of being replaced. The swing over the new canal ... [and] great traffic over it of wood, timber, etc. demand that it should be kept good and strong. The fact is that we should have two bridges over the new canal at this place – a Main street and a Division street bridge. One bridge is insufficient for travel over it and many a jam has ensued. Besides there should be one bridge for use whilst the other is being repaired' (*Welland Tribune and People's Press,* 28 March 1872). These criticisms increased as the streetcar and later the automobile were used.

Vessels passed through the canal locks on Sundays until 1845, but the Sabbath Observance Laws then enforced that Sunday was traditionally the day of rest. The issue was whether Sunday should be a holy day or a holiday. Shipowners, railways, and the other providers of public services wanted work from their employees, and in a society of increasing complexity a work-free Sunday was challenged as services had to be provided on a regular basis throughout the year. Until 1876 the locks were closed to ships on Sundays from 6:00 a.m. to 8:00 p.m., but in 1897 the Sabbath Observance Laws were again enforced and the locks were closed for the full twenty-four-hour period every Sunday.

Water Power and Property Leases

In 1867, when the Dominion of Canada acquired the Welland Canal Company and with it the right to take and appropriate from the Niagara and Welland Rivers as much water as was necessary for the canal, these privileges were vested in the Department of Public Works and later the Department of Railways and Canals. Diversions for power were permitted, subject to the prior needs of navigation.

Before 1881 water supply for the canal came from the Grand River. From 1881 to 1886 some water was drawn from Lake Erie to supplement the original source of supply, and the total diverted had doubled to some 600 c.f.s (17.0 m³). Between 1887 and 1898 the diversion increased to some

1000 c.f.s. (28.3 m³), primarily for power purposes at DeCew Falls, where the last power unit was installed in 1913. By this date the total diversion had reached 2400 c.f.s. (68.0 m³), of which 1000 c.f.s. (28.3 m³) was allocated for power generation at DeCew.

By 1867, some sixty-four water leases had been granted along the canal (Woodruff 1867a, b). They emphasized St Catharines, where eighteen leases had accumulated at several lock sites to serve a range of activities (table 4). Next in importance came Dunnville and Merrittsville (Welland), each with seven leases, followed by Allanburg with five and Port Robinson with four. These centres each had flour mills with runs of stone, sawmills with upright and circular saws, and perhaps a shingle factory. Water was also used for power at floating docks in Port Dalhousie and Port Robinson, and at wharves in Welland (3), Wainfleet (2), Port Colborne (1), and Port Dalhousie (1). At Thorold there was a water lease for a cotton factory, but payment was in arrears and no water was used that year. A carding machine was operated at Allanburg and Dunnville had three carding machines within a fulling mill. Water rights, an important aspect of the canal scene, were an asset that might be shared with others or sold when a property was for sale.

The amount of available water increased considerably with the arrival of the Third Canal and the switch in source to Lake Erie, but water as an asset for power to drive mill wheels was then waning as coal-powered machinery came into vogue. The contrast is marked between the large number of mills and other activities that lined the Second Canal from Thorold downstream and the absence of mills next to the Third Canal along this length.

Even so, water supplies from the canal remained important. Reviewing the situation by 1911, Denis and White confirm that the largest user of canal water was DeCew Falls, where the amount of developed electrical energy at 50,000 horsepower nearly equalled that of the Toronto Power Company at The Falls. Port Dalhousie used 500 HP in rubber manufacturing and St Catharines 3955 HP to produce a range of products in nine different plants, with half being used at a carbide works. Merritton used 1775 HP and Thorold rather more at 2305 HP from nine plants. Two companies used the canal to produce electrical energy at Welland, and one at each of Port Robinson and Marshville (now Wainfleet).

Further major users were the municipalities as public piped systems of water supply replaced wells to serve residential, commercial, and industrial premises. At St Catharines the first water supply, drawn from the Hydraulic Raceway at Merritton, was carried through hollowed pine logs to a reservoir on Church Street, the highest point in the town. Later, a

TABLE 4 Water power and industrial leases, 1867

Location	Lessees	Machinery	Yearly rent ($)
Port Dalhousie	R.Laurie & Co.	Grist mill	187.30
Port Dalhousie	R.J. & W. Laurie	Flouring mill	240.00
Port Dalhousie	R. & J. Laurie	Lot 1/4 acre	20.00
Port Dalhousie	R. Morrison	Sawmill	121.00
Port Dalhousie	Alexander Muir	2 dock lots	176.00
Port Dalhousie	George A. Clark	2 Wharf lots	100.00
Port Dalhousie	Donaldson & Andrews	Dry dock	100.00
Port Dalhousie	James Moore	Lot	20.00
Lock 2	Michael Kearins	Lot	10.00
Lock 2	John L. Ranney	Flouring mill	260.00
Lock 3 to 11	St Catharines Water Power Company	Surplus water	500.00
St Catharines	Calvin Phelps	Flouring mill	150.00
Lock 4	Calvin Phelps	Wharf lot	40.00
Lock 5	R. Collier	Sawmill	167.66
Lock 10	Thomas Jones	Grist mill	140.00
Lock 11 to 22	Welland Canal Loan Co.	Surplus water	720.00
Lock 12, 13, 14	Gordon & McKay	Cotton factory	240.00
Lock 16	John Brown	Cement mill	160.00
Lock 20	Wm. B. Hendershot	Sawmill	181.00
Lock 21	Wm. Beatty	Sa mill	216.00
Lock 22	Wm. Beatty	Tannery	63.60
Lock 23	Wm. H. Ward	Machine shop	50.00
Lock 23	Wm. H. Ward	Sawmill	146.00
Lock 23	John Brown	Wharf	40.00
Lock 24	Jacob Keefer	Flouring mill	222.00
Lock 24	Brown & Ross	Flouring mill	130.00
Lock 24	Park & Cowan	Flouring mill	160.00
Lock 25	John Brown	Cement mill	80.00
Lock 25	Alex. Christie	Flouring mill	160.00
Thorold	Chitty & Woodward	Cotton factory	100.00
Allanburg	Wright & Duncan	Flouring mill	270.67
Allanburg	Wm. H. Merritt, Jr	Sawmill	87.10
Allanburg	J. & H. Bowman	Pail factory	66.00
Allanburg	Tucker & Rannie	Sawmill	300.00
Allanburg	Wm. Pennock	Shingle factory	66.00
Allanburg	P.S. Mussen	Lot 1/4 acre	20.00
Port Robinson	J. &. J. Abbey	Dry dock	150.00
Port Robinson	Abbey & McFarland	Dry dock	79.20
Port Robinson	Donaldson & McFarland	Grist mill	86.00
Port Robinson	R. Baird & Co.	Grist mill	206.00
Port Robinson	John Donaldson	Wharf lot	8.00
Welland	Killens & dockstader	Saw mill	156.00
Welland	Dunlop & Seeley	Grist mill	216.00

TABLE 4 (concluded)

Location	Lessees	Machinery	Yearly rent ($)
Welland	Dunlop & Seeley	Sawmill	214.00
Welland	E. Seeley	Old aqueduct	20.00
Welland	M. Cook	Grist mill	192.00
Welland	E. Mead	Wharf lot	25.00
Welland	A. Sherwood	Wharf lot	25.00
Junction	John A. Hellmes	Wharf lot	25.00
Marshville	John Graybiel	Grist & saw mill	160.00
Broad Creek	L. McCallum	Saw mill	143.00
Port Maitland	Imlack & Hickes	Grist mill	138.00
Dunnville	Jacob Turner	Grist mill	180.00
Dunnville	Samuel Darling	Grist mill	86.67
Dunnville	L.J. Weatherly	Carding works	53.34
Dunnville	H. Mittleberger	Sawmill	77.34
Dunnville	Chisholm & Muir	Sawmill	138.67
Dunnville	A.S. St John	Grist mill	120.00
Dunnville	Brown & Merritt	Plaster mill	113.00
Haldimand	Oldfield & Noxen	Sawmill	237.34
Haldimand	J. Clark & Brother	Sawmill	66.67
Haldimand	J.C. & R.H. Kirkpatrick	Grist mill	153.34
Haldimand	Beatty & Baird	Grist mill	149.20
Port Colborne	A.K. Schofield	Wharf lot	25.00
Port Colborne	John Gordon	Wood yard	25.00
Dunnville(?)	C.J. Brydges	Water pipes	20.00
Total			8759.10

Source: Woodruff 1867a, b

water-supply reservoir owned by the municipality and in operation from 1879 was supplied from the canal at Allanburg to two storage reservoirs behind earthen dams across Beaverdams Creek at the crest of the escarpment. This reservoir, stocked with carp and landscaped around with trees and shrubs, also became popular as a destination for local family outings.

At Merritton in the mid-1880s a water system fed down the slope of the escarpment from a filtration plant in Thorold Township that was supplied from the head lock of the old Second Canal. At Welland the first waterworks system, constructed in 1888, took water from the canal to a pumping station on the Welland River at the end of the Feeder Canal. Ten years later Port Colborne established its waterworks, again with a pumping station that drew lake water from the canal. The canal provided a convenient location for the supply of pure lake water to the municipalities along its length, furnished an income that supplemented the canal operations, and was a resource of considerable advantage to each canal centre.

Hydroelectricity, the Streetcar, and Industrial Location

The new resource of hydroelectricity at first relied on the canal. Small generating plants were located at some of the escarpment locks, hydraulic raceways served from the canal to DeCew Falls where the Cataract Power Company began operating in 1898, and a smaller enterprise at Welland also provided power by the turn of the century. When the reliance on power switched to the larger generating stations at The Falls, Welland only some 12 miles (31.1 km) distant was the closest canal community to this immense power reserve. The Canadian Niagara Power Company, at first Plant No. 3 of the Adams Plant on the U.S. side of the river, produced the first power in 1905. Ontario Power Company arrived the same year and Toronto Power in 1908, together providing a significant and expanding source of abundant and cheap power (Carr 1992).

Hydroelectricity introduced the streetcar. St Catharines had a horse-drawn system from 1879. When an electric service replaced the horses in 1887, power was generated at Lock 12 on the Second Canal in Merritton. In 1899, the Niagara, St Catharines, and Toronto (N.S.&T.) Railway was incorporated to acquire the assets of the St Catharines and Niagara Central Railway. The N.S.&T. then changed from a steam-operated to an electric railway in 1900 (Mills 1967).

In 1901 the N.S.&T. service was extended to Port Dalhousie, where new docking facilities and an amusement park were added on the west side of the harbour. Two steamships, the *Lakeside* and the *Garden City*, were used to tap the Toronto market to provide passengers for the amusement park and the railway to Niagara Falls. In 1911, when the streetcar line was extended to Port Colborne, the expectation was to recapture the passenger and freight traffic between the two lakes that was then controlled by the Welland–Grand Trunk sequence .

The arrival of the new electric streetcar provided a convenient and frequent mode of inter-communication between the canal communities, industrially for connections to all the major east–west railway lines and the regular interurban travel that was now available for the first time to carry workers on their journeys to and from work. No longer had employees to live close to their places of employment. They now had the freedom to live in one place and to work elsewhere, and the canal communities were thereby encouraged to spread and interlink through this residential convenience. A point of considerable interest is that the system connected across the international border into the United States, but not west to Grimsby and Hamilton, again expressing an American linkage with the canal zone.

Hydroelectricity, being cheaper, cleaner, and easier to use than coal, also introduced major changes to the pattern of industrial location. Together with the improved Third Canal, railways, and water from the canal, a powerful new inducement to urban growth had been created, a theme to be discussed in chapters 8 to 10 community by community.

The water diverted to hydro plants at The Falls from the Niagara River, and water taken from the canal when not returned to the Welland River but flowing down the valley of either the Second Canal or via DeCew Falls into the Lower Twelve Mile Creek, reduced the scenic grandeur at The Falls. As reported by the International Joint Commission in 1906, 'It would be a sacrilege to destroy the scenic effect of Niagara Falls ... More than 36,000 cubic feet [1019 m³] per second on the Canadian side of the Niagara River ... cannot be diverted without injury to Niagara Falls as a whole' (Denis and White 1911). Of this amount, the volume recommended for diversion by the Welland Canal was 1800 c.f.s. (51.0 m³).

The flow of water into the Third Canal had now become the subject matter for international regulation, which provides a reminder of the fact that the prime justification for the First Canal was to provide a bypass for shipping around The Falls, and that water used in the canal, whether from the Grand River or from Lake Erie, did not cascade over The Falls. This diversion at first caused no consternation. There was plenty of water in Lake Erie and its catchment area included all the Upper Lakes and their catchment areas. However, the wheel of fate had come full circle. Circumstances at The Falls were now dictating the volume that could be used for the canal and its ancillary purposes.

The Externalities of Regional Setting

The Reciprocity Treaty of 1854 abolished import duties on a range of natural products when moved either way between the United States and the British North American colonies. Of signal importance was the new freedom to move grain and flour, timber, coal, and iron ore. American vessels could now use the Welland Canal, the St Lawrence, and the other canals on the same terms as British and colonial vessels. To the old channels of commerce through the Welland Canal and the St Lawrence were added new bonds with the United States.

The Niagara Peninsula with its through canal for water traffic, and as a narrow land bridge that connected Canada and the United States across two of the Great Lakes, gained considerably. As the most important trans-border location for Canadian-American links it had only two distant land competitors, the crossing at the Detroit and the St Clair Rivers to the west and the St Lawrence River to the east beyond Kingston. Beyond the boundary of the Niagara River, and now closer in time distance because of the railway network, were the major new American manufacturing centres to the south of Lake Erie and a quadrant of urban-industrial development that stretched from Cleveland through Pittsburgh, Baltimore, and Philadelphia to New York.

After Confederation in 1867, Canada's 'national policy' became to establish and then protect nascent manufacturing industries, develop the national economy, and thereby become independent and avoid absorption into the United States (Easterbrook and Aitkin 1958; Easterbrook and Watkins 1967). This involved land and transportation policies, including the deepening of canals and the support of railways to develop the trans-continental economy. The new reality was that in Canada a dominion from sea to sea now expanded both westwards and to the north, and links were strong with the greatly expanding American market to the south.

The western Prairies were opened for settlement, and new-found mining and lumber towns emerged in northern Ontario and the other provinces. It was a growth situation from which the Welland Canal benefited enormously. New possibilities for trade to and from the new, expanding localities encouraged manufacturing in southern Ontario to meet the demands that were created elsewhere in the nation. These new extensive markets were reflected in the changing patterns of trade that have been indicated (table 3), and are reflected in the industries that grew along the canal banks.

Protective tariffs took on a particular importance after 1879. Before this date, raw materials and most semi-finished goods entered the country free of duty, manufactured goods were taxed on their value, and a higher rate was imposed on luxury goods. In 1879 the general level was raised from 17.5 to 20 per cent in order to protect goods that were made or could be made in Canada. The production of iron and steel was given substantial new protection, manufactured industrial equipment and machinery had duties of some 25 per cent, and finished consumer goods were higher, at about 30 per cent. Higher protection continued to 1887, and remained in force until the Depression period of the early 1930s.

The communities along the Welland Canal became part of this protected economy and changed from a commercial orientation to an industrial form of development. With the United States nearby on the opposite bank of the Niagara River and with the conveniences of direct road, rail, streetcar, and canal access to this growing market, many of the new industries involved either American financing and control or they were branch plants of established American enterprises. The term 'intervening location' has been used to describe this reality, that is, a location between the American parent plant and yet within Canada to serve its growing population and economy (Ray 1971). Canadian localities that were closest to the great American manufacturing cities and had the best accessibility by rail and water routes were the most likely to attract American subsidiaries.

American companies began to operate in places along the Welland Canal in the 1870s, and more profusely as the century advanced. They jumped over the tariff barrier, and were encouraged warmly by the municipalities and their business elite for the multiplier effect they were expected to bring to the urban economy. Industrialization became the key to growth. The initiatives and technological abilities of manufacturing industry, the size of their employment rolls, the wages earned, and the volumes of output became the keynotes of urban success and population size.

No one factor is of single importance here. It was more a combination of several favourable factors that provided the impetus to growth. An American-owned company might locate in a canal community, bring in raw materials including coal via the canal, use power from an American company at The Falls, export via an American railway into the United States or serve the Canadian market to save on tolls. It would be welcomed by the local communities because it spelled growth and vibrancy, new services, the attraction of American capital, employment, and urban expansion. It was not an era that closely examined either industrial pollution or the social costs of the new developments.

An additional factor was that Canada's imperial links favoured an expanding trade through the Welland Canal. The high rates of nineteenth-century urbanization and industrialization in Great Britain and western Europe created a strong and persistent demand for food, processed products, and raw materials. Canada was one strong supplier to meet these needs, whereas in the United States the same factors led to a greater domestic absorption of its agricultural, forestry, and mineral resources, a shrinkage of their resource availability for export, and the transfer of some overseas demands to the Canadian side of the border.

The stimuli of increasing demands, and the associated production then outward movement of goods and raw materials led to a strong national demand for a continuous passage to seaboard. The cost of movement per mile of ton load to the British market was gradually reduced, while at the same time the price paid for exports such as wheat on the London market rose. Through these factors, the period that extended up to the outbreak of the First World War was advantageous for the Welland Canal. An inflow of British and American capital promoted production in the canal's industrial communities and aided the development of the nation to a high level of prosperity. The context of the Third Canal was that of expansion.

As industry arrived and expanded, and as European immigrants were attracted to the working-class industrial labour force, a further technological advantage to industry was that the water-power privileges available at the locks and along the hydraulic raceways along the Second Canal were not abrogated. They remained intact and might be purchased despite the successor Third Canal. In addition, U.S. coal imported through the canal encouraged the steady transition from water to steam power at industrial sites along the canal. A further source of power was the discovery of gas under Lake Erie and Welland County in the south of the peninsula.

The broadening basis for urban and industrial expansion in the canal communities must also be associated with the advancing provision of public and private services as improved school, library, health, banking, insurance, recreational, and other cultural facilities were added steadily to the increasingly urban mode of life. Although the transportation network, the power ingredient, and the industrial base were critical ingredients that spurred growth, the associated service elements also expanded as the population was housed, fed, clothed, and supplied with an increasing array of goods and public utilities. It was not only that the canal created growth, but rather the canal was an essential basis of economic opportunity in conjunction with many other public and societal factors.

These external factors also included the expanding agricultural base and the intensifying settlement in the rural areas to the east and west of the canal. Former wheatlands and mixed farming areas in North Niagara made the transition to fruit orchards and vineyards that received national recognition as the Niagara Fruit Belt, and in the south of the Niagara Peninsula the demands from the expanding market of Greater Buffalo required a continuous flow of many agricultural products.

While these many local, regional, and national events took place, as the century advanced the Niagara Peninsula increasingly found itself off-centre to the main thrust of Canadian urban development centred on Hamilton, Toronto, and the other towns of southern Ontario (Smith 1982; Spelt 1972, 1973; Watson 1943a). It also had the substantial American competitor of Buffalo as a major port, rail, and industrial centre advantageously located on routes to and from the North American West. In effect sandwiched between Hamilton to the west and Buffalo–Niagara Falls to the east, and partially under the influence of both, the Welland Canal communities could grow in but a limited manner in contrast with these greater centres. Hamilton grew more strongly than the canal communities during the late-nineteenth-century period of rapid industrial expansion (Weaver 1982; Dear et al. 1987), and the same experience of ebullient growth applied to Buffalo and its metropolitan area (Brown and Watson 1981; Whittemore 1977).

The Niagara Peninsula, at the centre of both pioneer and canal progress, gradually became a fringe Canadian locality as the nation advanced to the west and north. It became tributary to both American interests and to those stemming from the persistent advance outwards of Toronto's metropolitan sphere of influence. Given the large number of American companies and branch plants, the canal zone may even be described as an American industrial outpost in Canada (Hansen 1970; Marshall et al. 1970; Watson 1945a).

To such regional detail, we might add the spate of technological advance that gradually changed local circumstances throughout the Western world. The Welland Canal communities were not unique in having railway developments, canal improvements, hydroelectricity, the streetcar, and mechanized factory production. Competition was becoming more universal than during the previous commercial era of development.

An important change within the context situation of the Welland Canal as a significant part of the Great Lakes system of waterways was the construction between 1889 and 1895 by the Canadian government of a ship canal at Sault Ste Marie, Ontario. This provided the last link in a through Canadian water transport system upstream from Montreal to the head of Lake Superior (Heisler 1973; Passfield 1989). The first contract for the Cornwall Canal had been completed by 1882, the Lachine Canal was enlarged in 1884, and the Welland Canal in 1887. Trunk railways and the existing but inadequate canal system were not sufficient to capture the developing trade from the American and the Canadian prairies, and the alternative of a through water system in turn eliminated trans-shipments on the Welland and the St Lawrence canals. This was a local disadvantage for economic activities and employment related to the canal, but the new and expanded canal facilities provided a greater gain for the national good, and the urban situation along the canal had again to respond to the stimuli of external circumstances.

6

The Changing Character of the Canal Communities

Several features of municipal life illustrate the changing character of the Welland Canal communities as the nineteenth century progressed into the twentieth century. Starting with administrative structure and population size, there are the novel industrial circumstances expressed by the rise of larger-sized companies and the growth of service trades and enterprises that minister to the increasing population. The latter has two aspects, the absolute growth in the size of the population and the growing per capita demands of that population for services as technology and the requirements of society advance.

Municipal Status

The incorporation of a place as an administrative village, town, or city provided an important recognition of growing status. It meant that each place now had its own council and a local decision-making process, the power to levy taxes on assessed property to pay for local improvements, and more local initiatives through a degree of local autonomy. The Municipal Corporations Act of 1849 established criteria of population size for each advance. An incorporated village had to have at least 1000, a town 3000, and a city 15,000 people. This legislation was liberalized in 1873, when a village was lowered to 750 and a town to 2000. Ambitious towns might then petition the province for the dignity of city status when a population of 9000 was reached (Bloomfield 1983).

Within this hierarchy, St Catharines was the first centre on the canal to obtain promotion. It achieved the status of a village in 1827 when the First Canal was under construction, and was followed progressively to this rank of importance by Thorold and Chippawa in 1850, then by Merrittsville

(Welland) in 1852. Dunnville followed in 1860, Port Dalhousie in 1862, and Port Colborne in 1869. Merritton in 1874 was not followed by any other canal centre until Humberstone was incorporated as a village in 1912.

Port Robinson presents the interesting example of a place not seeking an improved administrative status. Empowered as a police village by Welland County Council in 1856, it reached the requisite total number of inhabitants to become a village, but did not consider that worthwhile advantages would result if this status was granted. After the population declined, municipal status was requested but the population was then deemed insufficient and Port Robinson remained unincorporated as part of Thorold Township.

The advance to incorporation as an officially recognized town with civic officials and new functions was first achieved by St Catharines in 1845. Thorold in 1875, then Welland three years later, and Dunnville in 1900 obtained this new esteemed honour. No other canal community was successful in this regard before the outbreak of the First World War.

At the higher level of a city, only St Catharines received this greater degree of stature and its accompanying jurisdictional and administrative privileges before 1914. This important event, celebrated in 1876, firmly established the 'Garden City' as the major urban community and the prime regional focus both along the line of the canal and within the Niagara Peninsula.

The rapidity of growth along the canal and the emergence of substantial communities is reflected on a relative basis by the time period that existed between the emergence of a village and the creation of a town. This interval was 18 years at St Catharines and 20 years at Welland. It took 25 years at Thorold and 40 years at Dunnville. Other towns in Southern Ontario made a more rapid transition. It took Guelph four, Galt seven, and Kitchener 16 years. The advance from town to city at St Catharines took 31 years, but it was 13 years at Hamilton and 24 years at Guelph.

This municipal progress is not inconsistent with other measures of growth, as population growth and industrial expansion each heralded a growing degree of centrality within the limited hinterland areas of the canal communities. The first weekly newspaper began in St Catharines during 1826. Thorold then Dunnville and Welland followed during 1849, 1852, and 1854, respectively. Merritton in 1874 and Port Colborne in 1891 were much later in the provision of a regular local publication. The even greater provision of a daily newspaper was achieved only by St Catharines in 1859.

County Status

At the level of county government, Niagara District Council, centred on

Niagara (now Niagara-on-the-Lake), started its career in 1841. Haldimand, Welland, and Lincoln were then each part of its administrative extent. Agitation for separate municipal and judicial organizations began in the southern part of the district most distant from Niagara. When Haldimand was formed in 1850, Cayuga on the Grand River was named the county town rather than Dunnville despite the latter's greater importance as a canal-river port.

By an act of 1851 it became lawful for the County of Welland to name a place as its county town. Rivalry between competing centres included the older settlement of Cook's Mills at the centre of the county, and Fonthill for the beauty of its setting at the centre of the peninsula. Port Robinson, where the canal connected with the Lower Welland River, with centrality on the canal and water systems of navigation, was also considered. Tradition has it that the cornerstone of a county building was laid, but Welland (then Merrittsville) became the county site, presumably because of its even greater centrality on the current canal and river systems of navigation.

When Welland became the county town of Welland County in 1856, the question of the county town for the remaining extent of Lincoln County now became one for that unit to settle. The town of Niagara, located awkwardly in a lopsided position at the extreme north-eastern part of the county and disadvantaged by the canal, wished to retain its traditional esteemed position as the first capital of Upper Canada. A severe conflict arose because upstart St Catharines as a newcomer settlement on the canal also sought the county town role.

St Catharines was more central within the county. It had an expanding and more important industrial base through the First and Second Canals, an asset that was recognized when the provincial legislature made provision for the county town to change from Niagara to St Catharines. The municipal electors at Niagara voted against this bylaw, and those at St Catharines voted in its favour, an embattled position that continued until St Catharines became the county town in 1862.

The earlier emphasis of pioneer settlement on the frontier along the Niagara River had now been transferred inland to the new canal settlements. The frontier communities with their focus on the military and the merchants at Niagara had lost their traditional portage and its lucrative trade around The Falls, first to the rival venture of the Erie Canal and then to the Welland Canal. They were now further eclipsed through the transfer of county-town status inland to the new canal communities, initially to Welland and then to St Catharines.

Of note is that a comparable inland movement in the prestige of settlements had also taken place on the American side, where Lockport on the Erie Canal displaced Lewiston on the Niagara River as the county centre of Niagara County. The new canal environments had indeed displaced the old patterns of settlement in many respects. Niagara remained a town, Chippawa and Fort Erie became villages in 1850 and 1857 respectively, but neither achieved town status during the nineteenth century.

County status offered several advantages that were incentives to growth, including county buildings, the courts, and their offices for the sheriff, the judges, the clerk of the peace, the county attorney, and their staffs. These functions expanded over time to include the county jail, the registry office, the county clerk and treasurer, and, as new county functions were added, officers such as the inspector of schools and various licence commissioners. The police and fire departments might be located nearby. The incentives of political importance and government buildings aided economic growth through the establishment of these new activities, their associated high social status, and the contacts of the new senior officials with the decision-making process of government. There was also the important relative factor of differential growth in contrast with nearby communities that did not obtain a higher administrative status. The fact of a territorially defined area provided the county towns with an extensive hinterland area over which their administrative control was exerted, and from which there was a considerable movement towards the market, stores, offices, and services that were available. St Catharines and Welland gained, while the other canal communities did not obtain the same incentives to growth.

The two new county offices on the canal were both prestigious buildings of high architectural quality. At Welland the combined jail and courthouse, designed as a three-storey Palladian structure of Queenston limestone, presented its powerful image to East Main Street where the business centre of the community was developing. Its site, which overlooked the Second Canal, the Welland River, and the aqueduct that carried the canal over the river, symbolized the significant roles of the two waterways in the urban then county achievement of Welland.

At St Catharines, the town hall and market house that had been built in 1849 became the courthouse and the county offices. As worth as a county town received more architectural acclaim than status as a town, this change in function from a previous building helps to explain why the architecture of the building that became the county offices was more subdued than its companion building at Welland, even though Lincoln County was more populous than Welland County. At St Catharines the courthouse had an inland setting, unrelated to the canal.

Even so, the renovated courthouse, again of classical design surmounted by a cupola and with a limestone facade, introduces the interesting point of similarity yet divergence between the two canal centres. In both county towns, the buildings were designed by Kivas Tully, an Irish-born Toronto architect prominent in the design of Ontario's early institutional buildings that included Victoria Hall in Coburg and the original Trinity College, the Customs House, and the Bank of Montreal in Toronto.

An Advancing Hierarchy of Settlement

Municipal advance and promotion were encouraged when boards of trade with their enthusiasms for local growth were established. These boards, which through time became chambers of commerce, were concerned with the civic and industrial promotion of their respective towns, their need for improved services such as fire and police protection, and the provision of gas then water and electricity services, street railways, hospitals, and public libraries. In conjunction with the municipal and often the county councils, the boards sought strongly to encourage decisive growth strategies that enhanced the town's centrality over the surrounding hinterland area.

The process of municipal boosterism started in St Catharines in 1872, Welland added this theme to the municipal scene in 1889, and Thorold created its board in 1893. It was an era during which the ethics of expansion and growth prevailed (as it still does), with towns competing against each other to expand and offering inducements such as free sites and services for a stated period of time at the expense of its taxpayers to attract industrial commitments.

Banks demonstrate another aspect of local progress. They reflected the relative importance of a community and helped to advance its business activities through loans and other financial support. By the early 1880s St Catharines had three banks, Thorold and Dunnville two, and Port Colborne and Welland one each. As an indication of the influence of larger centres over the nearby smaller ones and the extent of their local catchment areas, St Catharines was described as the banking town where Port Dalhousie and Merritton conducted their business. Thorold provided this service for Allanburg. Humberstone banked in Port Colborne, Port Robinson and Marshville (Wainfleet) banked in Welland, and Dunnville was the banking town for Stromness and Port Maitland (Dun 1884). The Imperial Bank, with agencies in St Catharines, Welland, and Port Colborne, had the greatest representation in the canal communities. The Canadian Bank of Commerce served St Catharines, Thorold, and Dunnville. Only St Catharines had the Bank of Toronto. Thorold's second bank, the Quebec Bank, also served St Catharines.

Within their respective communities, the bank buildings occupied prestigious locations, typically at the heart of the business centre, on its main street, and at an important street corner. The buildings also had some element of architectural prestige such as a classical motif to indicate their importance. The bank managers were listed as people of some consequence together with the leading businessmen, industrialists, and the municipal officials and officers in directories of the period.

A regional hierarchy of substance had started to emerge, with St Catharines dominant along the northern length of the canal then Thorold, to a lesser extent, where the brow of the Niagara Escarpment was crossed. Port Colborne, soon in ascendancy over Humberstone, was the larger of the two terminal ports. Welland, with its county status, grew as the centre for the middle length of the canal, and Dunnville was pre-eminent at the western end of the Feeder Canal and for the lower valley of the Grand River. Allanburg and Port Robinson were heading towards eclipse under the shadow of the nearby canal centres at Thorold and Welland.

The break-up of Niagara District and the creation of counties divided the canal communities between different administrative units. Previously each had been part of the one Niagara jurisdiction, but after the division into counties the county boundary between Lincoln and Welland separated Merritton from Thorold, and that between Welland and Haldimand intervened between Marshville and Stromness on the Feeder Canal. The newly created counties of Lincoln, Welland, and Haldimand now administratively divided the string of related communities that had emerged along the Welland Canal and its tributary arms.

Inter-municipal, municipal-county, and inter-county rivalries became part of the canal scene. They were aggravated when St Catharines became a separate jurisdiction from Lincoln County, and the self-importance of egotism displayed through municipal boosterism, local newspapers, and assertive boards of trade placed one canal location before its neighbours in their avid search for growth. The economic unity created by the canal was now divided politically as competing realms of county and municipal government were formed along the canal banks.

The Expansion of Population

By 1871, the canal communities that had existed for forty or more years together held a total population of slightly over 16,500 (table 5). Port Mait-

TABLE 5 The population of the canal communities, 1871–1911

	1871	1881	1891	1903	1911
Chippawa	922[a]	664[a]	523[a]	460[a]	707[a]
Port Dalhousie	1,081[a]	1,129[a]	879[a]	1,125[a]	1,152[a]
St Catharines	7,864[a]	9,631[a]	9,170[a]	9,946[a]	12,484[a]
Merritton	1,000[b]	1,798[a]	1,813[a]	1,710[a]	1,670[a]
Thorold	1,635[a]	2,456[a]	2,273[a]	1,979[a]	2,273[a]
Allanburg	300[b]	150[c]	100[c]	250[e]	250
Port Robinson	600[b]	700[c]	660[d]	600[e]	200[f]
Welland	1,110[a]	1,870[a]	2,035[a]	1,863[a]	5,318[a]
Humberstone (Petersburg)	400[b]	1,100[c]	800[g]	600	600[f]
Port Colborne	988[a]	1,716[a]	1,154[a]	1,253[a]	1,624[a]
Marshville (Wainfleet)	200[b]	125[c]	150[d]	—[e]	200[f]
Stromness	100[b]	100[c]	150[d]	100[e]	100[f]
Port Maitland	80[b]	50[c]	70[d]	75[e]	75[f]
Dunnville	1,452[a]	1,808[a]	1,776[a]	2,105[a]	2,861[a]

Sources: [a]Census of Canada for the village, town or city unit; [b]Lovell, 1871; [c]Lovell 1882; [d]Ontario Gazetteer and Directory, 1842–3; [e]Ontario Gazetteer and Directory, (Ingersoll), 1902–3; [f]Union Publishing, 1910; [g]Might's Directory, 1892

land and Stromness on the Feeder Canal each held a hundred or fewer inhabitants. Marshville (Wainfleet), at the mid-point on the Feeder with mill privileges and at the centre of land reclamation for agricultural settlement on Wainfleet Marsh, had grown to only two hundred persons. The two settlements at the Deep Cut were also small, with Port Robinson at six hundred being twice the size of Allanburg to the north. Both had locks and mill privileges, but the former the additional advantages of a link to the Lower Welland River and also shipyard facilities.

The remaining centres, now or soon to be classed as municipalities, were composed of a group that contained a population of about 1000 or more. These included the entry ports of Chippawa, Port Colborne, and Port Dalhousie; of the inland centres Merritton and Welland had this population size by 1871. Thorold and Dunnville, each with a population of some 1500, were more important, but St Catharines was firmly in the lead with a population of almost 8000. This total was nearly half of that contained within all the canal communities and nearly five times the size of Thorold, its nearest competitor.

Although a canal of comparable merit through its flow of vessels along the whole line had been established, the emphasis of its emerging settlement pattern was neither equal nor continuous along the waterway. The greater strengths occurred in the north, across and below the Niagara Escarpment where the abrupt change in levels had encouraged mill enterprises to take advantage of the abundant water power that was available.

Change over the next forty years to 1911 witnessed a total growth in population throughout all the canal communities of nearly 11,800 persons, an increase of over 70 per cent. The largest volume of change, at St Catharines, brought the total population of this city up to almost 12,500. Its nearest competitor was now Welland, with a population of over 5300. Thorold had been displaced from second place, and the emphasis on development had to some extent been transferred to south of the escarpment.

Dunnville and Thorold, with populations of 2861 and 2273 respectively, now followed Welland in the hierarchy, then Merritton and Port Colborne in the 1600 range, Port Dalhousie with some 1150 people, and Chippawa with 700. Allanburg and Port Robinson next to the Deep Cut had now each shrunk to about 200 people. The smaller centres along the Feeder Canal had even lower populations and were relatively of little consequence in the regional pattern of settlement.

When settlements were incorporated, this designation extended over a defined administrative area. As population expanded and more land was needed to provide industrial sites, houses, and services, the administrative limits were extended into the adjacent rural townships to meet these urban demands. This process of annexation took place at St Catharines in 1850, 1854, and 1876, at Welland in 1858, 1878, and 1905, and at Port Colborne in 1869 and 1912.

The growth in population depicted in table 5 was neither consistent nor regular. Every canal community witnessed some form of population decline in the period between 1881 and 1901 and, even when the population was static or revealed but a slender increase, this still meant some outward migration because of the general excess of births over deaths. The communities along the Welland Canal, whether in the larger towns or the smaller villages, did not expand continuously or at either the provincial or the national rates of growth during this period.

The tendency was to boom during the periods of canal construction and to decline when 'normal' circumstances resumed. As noted for Thorold when the Second Canal was constructed, 'for a decade or more the town enjoyed the richest kind of boom, while the many million dollars of government money flowed freely around, raising values to an abnormal figure, and creating a spasmodic prosperity. Upon completion of these works, the temporary population floated away' (*Thorold Post,* 24 May 1889). As the article continued, 'The great tide of trade and travel, diverted into the new canal, did not nourish the town, could not be so easily tapped as under the old order of things, and people began to look surprised.' Again, with the Third Canal: 'This, however, was only a breeze compared with the grand boom of the many years preceding, and when the enlarged canal was completed ... Thorold found itself down to its natural level.'

The canal communities grew, but they also declined because of the canal. Periods of population decline occurred at St Catharines and Dunnville between 1881 and 1891, at Thorold over each decade from 1881 to 1901, at Merritton successively from 1891 to 1911 and at Welland from 1891 to 1901. The Third Canal on its new route that opened in 1887 had bypassed the settlements at the northern end of the First and Second Canals. In the south, Welland and Port Colborne had lower populations in 1901 than in 1881, and at Chippawa a population decline was recorded at each successive decennial census from 1871 to 1901.

Cultural Variants

In referring to the larger centres for which census data on religion is available, one notes that by 1871 the Church of England had the highest numbers of supporters in Port Dalhousie, Thorold, and Port Colborne. In St Catharines the Roman Catholics were at the head of this numerical sequence, and in Welland and Dunnville the Methodist groups were in the top category. Roman Catholicism and the Church of England were the two largest groups in each of the above centres, except for Welland, where Methodism was followed by the Church of England. As the various religious divisions also included Presbyterian and Baptist groups, the canal communities had remained essentially Protestant despite the fact that Catholic Irish labourers were used to construct the canal and some absorption of this group into the local communities had followed. Some religious divergence rather than precise similarity had started to take place between the canal communities.

At the same date, by place of origin, the first, second, and third cultural groups in Port Dalhousie, St Catharines, Thorold, and Port Colborne were from the British Isles. The sequence in each instance was the Irish, followed by the English, then the Scots. At Welland and Dunnville the English were first and the Irish second. The German group was third in Welland, a place taken by the Scots but followed by the Germans in Dunnville. A multi-ethnic social and cultural character had started to emerge and the canal communities were no longer either wholly British or Protestant as was their tradition.

By 1911 the data by religion and place of origin had each changed. At St Catharines and Port Dalhousie, Anglicans were now in the lead, followed by Roman Catholics at Port Dalhousie and Methodists in St Catharines, where the Catholic population had declined to third place. At Thorold, Merritton, and Dunnville, Methodism in the first place was followed in Thorold by Roman Catholicism and in Merritton and Dunnville by Anglicanism. In Welland the religious lead was held by the Roman Catholic and then by Anglican adherents. This sequence was reversed in Port Colborne, but in each instance the next three places were Methodist, Presbyterian, and Baptist. New groups included Salvation Army, Jews, Dissenters, Lutherans, and a small number of adherents to other faiths.

By place of origin, the English came first and the Irish second in St Catharines and Thorold. This sequence was reversed in Port Dalhousie and Merritton, but in each instance the Scots were in the third category. The situation was different in Welland, where the English were followed by people with a German place of origin, then closely by the Scots and the Irish. Recognizable other groups with over a hundred persons included Germans in Port Dalhousie and Thorold. St Catharines after its English and Irish citizens had French, Dutch, Black, and Jewish groups. Welland had the distinction of Austro-Hungarian then Jewish and French groups in its ethnic-cultural heritage.

Religion had become more splintered over the forty years since 1871. The reasons were in part doctrinal, but they also involved the arrival of new immigrant groups and their varied places of origin as large-scale industry expanded, and a working-class labour force augmented the resident population. The canal centres had become more varied culturally over the intervening years.

Business and Industrial Progress by the Mid-1870s

Dun and Wiman (1878) may be consulted to compare the industrial and service circumstances that had arisen by the mid-1870s. All industrial and business activities are valued in this source at their 'estimated pecuniary strength' or monetary value and assessed under thirteen ratings ranging from less than $1000 to $1,000,000 and over.

If the arbitrary value of $40,000 be taken as the line of divide between large and small companies, then the vast majority of the recorded enterprises in 1878 were small with limited capital and credit security. These included the many trades and professions of the day such as taverns, inns, and hotels; the general stores, grocers, and butchers; stores selling dry goods, clothing, shoes, millinery, and fancy goods; the local craftsmen such as blacksmiths, tinsmiths, cabinet and furniture makers, harness and carriage makers and coopers; and the merchants selling fuel supplies of coal or cordwood. Livery stables, tug owners, and towing operations were each recorded. Druggists, bookstores, jewellers, chandlers, doctors, dentists, stationers, printers, and booksellers are also each listed in various places along the canal.

As an indication of the range of local services that might exist in a small canal community, Marshville (Wainfleet) had two each of shoemaker, blacksmith, wagon-maker, and general store; and one each of hotel, cordwood supplier, carpenter, lumber supplier, tavern, and grocer. At a higher level in the urban hierarchy, in Port Robinson the additional services included billiards, tailor, butcher, builder, baker, cabinetmaker, and physician. Welland with over eighty establishments had many further functions that included auctioneers, liquor seller, bookseller, drugist, harness maker, printers, a milliner, jeweller, and dry-goods seller.

Taken together, the canal communities by the late 1870s had engendered slightly over five hundred establishments. By distribution they emphasized St Catharines with 60 per cent of these outlets, then Thorold with 20 per cent, and Dunnville and Welland with rather less. Port Colborne and Humberstone were at a lower scale of importance. Except for Marshville (Wainfleet) with its Lee Edwards general store, all large establishments lay along the northern length of the canal between Port Dalhousie and Thorold. St Catharines dominated with fifteen out of the twenty-three, Thorold had four, Merritton three, and Port Dalhousie one of these units. As neither Allanburg, Port Robinson, Welland, Humberstone, Port Colborne nor Dunnville had attracted any of the larger endeavours, the canal now strongly emphasized the north over the south, and within this division the vicinity of the Niagara Escarpment was of particular importance.

The major companies were Muir Brothers, shipbuilders at Port Dalhousie. Merritton recorded the Lybster Cotton Manufacturing Company and the John Riordan paper mills, both with a pecuniary strength of from $75,000 to $150,000. N.&O.J. Phelps, lumber, followed in the $40,000 to $75,000 range.

At St Catharines, its distinctive and superior business characteristics included Dennison, Belden & Company, contractors, the largest company recorded on the canal system, with a strength in the $500,000 to $750,000 category. Four manufacturing industries with a strength of from $150,000 to $300,000 followed: Sylvester Neelon, with mills and shipping interests; Norris & Company, millers; James Norris, miller and shipper; and John Riordan, papermaker. Two individuals in this category of wealth were A.L. St John, broker, and Samuel D. Woodruff, capitalist.

The next category of from $75,000 to $150,000 included Beatty, McCleary & Company, lumber, and the Taylor & Bate brewery. The lower grouping of from $40,000 to $75,000 included the two industries – Lewis Schickluna, shipbuilder, and R.H. Smith, saws – together with four stores: Andrew Jeffery, hardware; W.J. & J. McCalla, grocer; J.D. Tait, fancy dry goods, hats, and furs; and Richard Woodruff, dry goods.

Two regional contrasts might be noted. Hamilton, not a canal community but relying upon the Welland Canal for many of its imports and exports, now greatly surpassed St Catharines in importance. Here were over a thousand establishments, with some forty in the largest size categories. By contrast, Niagara and Chippawa as two of the Niagara River communities that had preceded the canal were now each relatively unimportant. Niagara had forty-eight small establishments but none of high pecuniary value, and Chippawa with forty establishments included only James F. Macklem, tanners, in the $75,000 to $100,000 range.

The Situation by 1911

Within thirty years, business circumstances along the canal changed markedly. Certainly the very large number of small service establishments remained. They included the numerous stores, the service outlets, and the trade, government, and professional services to be found in a hierarchical fashion along the canal and its feeder system. Family names often long established and an expression of social continuity, along with a central business area within each community as the focus of retailing and commercial activities for the town and its agricultural hinterland, could both be recognized.

The distinctions of change were most remarkable for the larger enterprises. Now mostly manufacturing, their numbers had increased greatly. The earlier pattern of an almost wholly northern emphasis had been modified as a number of new well-known names of considerable employment capacity and substantial financial backing were added to the canal communities. Bradstreet (1911) provides a suitable guide to these new achievements, but as more categories with different parameters and a different classification system are now involved, the division between large- and small-sized establishments will be interpreted against the yardstick of those with a pecuniary strength of over $50,000.

Fifty of these more important business and industrial concerns were now recorded. St Catharines had maintained its sway over manufacturing activities with half the larger establishments. Thorold had six, Merritton four, and Port Dalhousie two such establishments. A substantial increase had also taken place in the south of the peninsula, where the villages remained small, but Welland and Dunnville had six and Port Colborne two of the larger establishments.

In the three commanding communities of St Catharines, Thorold, and Welland, over the decade from 1881 to 1891 federal census data indicates a decrease in the number of manufacturing establishments (table 6). This decline, by almost one quarter, was from 225 to 171. At the same time capital investment, numbers employed, wages, and the value of factory products each increased. The same type of trend persisted over the period from 1891 to 1911. Using St Catharines as a case example, while its population increased from 9170 to 12,484, the number of manufacturing establishments halved, their capital increased more than threefold, the value of their products rose by almost the same proportion, and the average number of employees per establishment more than quadrupled from 12 at 1891 to 54 by 1911.

Repeating this story at Welland, the town's population more than doubled over these two decades from just over 2000 to 5300 persons. By contrast, the number of manufacturing establishments decreased from thirty-six to twenty, the employees in manufacturing tripled, capital per establishment increased over tenfold, and the value of the products rose nearly sixfold. The increase in employees per average establishment expanded from six to thirty-four, a proportion lower than in St Catharines but nevertheless representing a very considerable change in emphasis and attitude.

New large industrial establishments had become a major feature of the canal. With higher levels of and a greatly increased scale of activities from the mills of earlier years, this feature resulted from the arrival within the canal communities of large, externally financed and mostly American-controlled industrial enterprises. It was not primarily the consequence of internal growth within the community through local entrepreneurship, and the locational factors, no longer the local market, were regional if not national and international in their scope.

A few of the new industries took over the earlier canal-oriented sites of water-powered mills, but most had necessarily to occupy nearby rural-fringe sites on former farmland. They were unable to locate in the centres of

TABLE 6 The industrial expansion of the canal communities, 1891–1911

Town or village	Date	Population	Number of establishments	Capital in $	Number of employees	Salaries and wages in $	Cost of materials in $	Value of products in $
St Catharines	1891	9,170	108	1,721,661	1,310	442,588	1,420,976	2,444,680
	1901	9,946	40	1,841,423	1,900	603,584	1,075,119	2,070,513
	1911	12,484	58	5,919,728	3,139	1,390,966	2,606,248	6,024,217
Merritton	1891	1,813	16	1,087,475	634	211,318	358,727	719,287
	1901	1,710	8	1,768,875	600	270,733	481,772	1,036,350
	1911	1,670	6	911,838	347	154,841	364,265	840,775
Thorold	1891	2,273	33	488,700	311	94,045	288,463	495,946
	1901	1,979	10	324,410	248	75,366	163,022	317,946
	1911	2,273	14	680,610	381	162,231	728,492	1,233,133
Welland	1891	2,035	36	175,290	215	44,023	131,385	233,738
	1901	1,863	8	100,363	128	57,257	65,222	152,087
	1911	5,318	20	1,877,576	684	286,752	711,137	1,375,374
Port Colborne	1891	N/A	N/A	N/A	N/A	N/A	N/A	N/A
	1901	1,253	N/A	N/A	N/A	N/A	N/A	N/A
	1911	1,624	4	217,600	239	151,420	140,622	777,543
Dunnville	1891	1,776	41	98,090	112	34,757	63,910	141,255
	1901	2,105	16	208,353	318	62,190	175,890	268,090
	1911	2,861	19	1,321,495	728	185,787	649,712	1,290,260

Source: Canada Yearbook, 1913

the established communities because of their extensive space requirements for buildings to house machinery, the need to store materials on each site, and their transportation requirements by canal and rail for incoming raw materials and the outgoing flow of manufactured or processed products.

The large number of workers required to work in each plant were accommodated in their own small homes or in bunkhouses for single men that were provided by the industries, and municipal services and utilities were but slowly extended to meet the needs of higher resident populations. Industry and the urban milieu each expanded, the one in conjunction with the other, but with the focus on employment and industrial expansion more than the quality of either the environment or living conditions. Industrial towns had emerged, whereas previously commercial centres dominated the canal scene (Watson 1945a, b).

The differences between the new and the old modes of life were many. A large number of people were now employed in the new plants, against few previously. The plants were concentrated in a few selected places, often pursuing the same or related activities, so that industrial specialization and manufacturing differences between the centres became the norm. There was a growing incidence of pollution in the air from factory chimneys, in the water and soil from effluent disposal, and inside the factory buildings from smoke and dust as water power was replaced by coal. There were also the occupational hazards of working with machinery for long hours in depressing and repetitious factory surrounds that typified the industrial revolution across North America. The statistics of urban population growth and the census data of expanding manufacturing output tell only a part of the total story about industrial expansion.

The new manufactories generally involved imported labour, frequently associated with new waves of immigrants from locations in southern or eastern Europe. An increasing social distinction grew between the labouring hands in the factory and the established merchants, the more skilled industrial groups, the managerial element, and the owners who invented then improved the machines that were used. As the occupational structure became more varied, this became reflected both in the workplace and in the range, size, location, and elegance of the nearby homes. Housing, living conditions, and the quality of life had their sadder aspects in the new industrial factory towns that arose along the canal. As indicated in one survey, the immigrants' houses were often overcrowded, and rents were high for poor and often inadequate facilities that suffered from overcrowding and insufficient baths, toilets, yard space, and garbage removal (Methodist and Presbyterian Churches 1916).

The impact of periods of unemployment and recession was severe, partly because of the higher numbers that were involved, and also because of the often external control of the new plants by American companies taking advantage of local resources and avoiding duties through the tariff situation. Decisions were no longer made locally with nearby personal, business, and municipal interests primarily in mind, but in more distant locations across the border of the Niagara River.

The rise of a labour movement and trade unions to redress the sadder elements of this situation became a part of the canal's industrial history. Locals were organized to achieve higher wages, shorter working days, and better working conditions. The right to a day of rest on the Sabbath became an issue, as were regular payments and suffrage for the working man. Socialism argued for safety in the workplace and employer's liability. Immigration by Chinese and 'foreigners' raised severe forebodings and was frequently resisted by the established members of society (Forsey 1981). The structure of society had been broadened greatly by industrial advance, as had the nuances of employer-employee relationships.

St Catharines increased the number of international union locals from three in 1880 and 1890 to twenty by 1902. In Port Colborne the sailors struck for higher wages in 1872. The Sailors (or Lake Seamen's) International Union, formed in 1877, extended along the canal from Port Colborne to Port Dalhousie and had a local in St Catharines. In 1875 the Journeymen Stonecutters Union of Thorold struck for three months against wage reduction and demanded an eight-hour day.

The spiral of new growth bypassed some of the smaller village settlements along the canal but urbanized others, a process that created a greater differentiation between the larger urban settlements and the smaller villages that had previously been established across Niagara's rural landscape. A new pattern was being established, with Niagara not being so very different from other localities in southern Ontario, or indeed as prevailed across much of western Europe and the United States. The industrial revolution, though late by contrast with these other world areas, had indeed reached the banks of the now Third Canal and its tributary system of waterways by the early 1900s.

A Changed Pattern of Regional Settlement

The nature of the canal communities and the classification used by the 1850s had changed remarkably by the 1910s. Now with the abandonment of the Feeder Canal and the Lower Grand River for canal purposes, Wainfleet, Stromness, Port Maitland, and Dunnville no longer functioned as canal centres. The canal had been a vital part of their earlier existence and it continued to flavour the form and the functions of these villages and towns, but except for redevelopment, conservation, and some historical reflection, the canal no longer played a vital role in their urban future. The same type of story held at Chippawa where, as the century advanced, the Welland River was no longer an intrinsic part of the canal system.

The story at St Catharines, Merritton, and Thorold was somewhat different. Here the established business centres were bypassed for the continuous passage of vessels through the Third Canal, but major canal features such as sections at each end of the former channel and the hydraulic raceways were retained for their water-power capabilities. The industries that had located along the canal were therefore not necessarily adversely affected, and navigation continued to serve St Catharines and Thorold.

Through the extensive advance of large-scale manufacturing industry, Welland, with its higher rates of employment and its status as a county town, moved upwards in the hierarchy of settlement and soon became more than twice the size of Thorold to the south. Merritton, a slow starter, also became an active industrial centre. At Port Dalhousie and Port Colborne, the port functions remained important and active, but with the advance of manufacturing both centres might also be classified as industrial towns. Throughout these changes the industrial city of St Catharines – with its extensive marine hinterland for the import of raw materials and the export of finished products – remained the dominant regional centre of the Niagara Peninsula. Within its boundaries lived more people than in all the other centres along the canal combined.

The regional pattern of settlement along the canal had changed from a cruciform shape with arms to the Niagara and Grand Rivers to a more limited I-shaped form that extended between Port Dalhousie and Port Colborne. Unified by the canal and now with an economic base that included large-scale manufacturing activities, the corridor of canal development was by the 1910s divided between two counties (Lincoln and Welland), one city (St Catharines), two towns (Thorold and Welland), four villages (Port Dalhousie, Merritton, Humberstone, and Port Colborne) and two unincorporated settlements (Allanburg and Port Robinson).

As these settlements were each expanding with housing, services, and industries into the adjacent rural townships of Grantham, Thorold, Crowland, and Humberstone, fifteen different administrative units now helped to shape the canal scene and its communities. Their core localities may be grouped into four categories, based on their canal status and their varied experiences in relation to the system of waterways:

1 Chippawa, Wainfleet, Stromness, Port Maitland, and Dunnville on the atrophied and abandoned canal arms to the Niagara and the Grand Rivers
2 Port Dalhousie and Port Colborne as the long-established and continuing lake ports at the northern and southern points of entry to the canal system
3 Welland, Allanburg, and Port Robinson as persistent centres along the central length of the canal throughout the First, Second, and Third Canal periods, with Welland moving to urban dominance
4 Thorold, Merritton, and St Catharines, which were bypassed when the Third Canal was aligned along a new route to cross the Niagara Escarpment

7

Two Lake Ports, and the Eastern and Western Arms of the Canal

Port Dalhousie and Port Colborne each advanced as the canal changed from the Second to the Third Canal system, but considerable differences in growth and opportunity existed. By contrast, the ports at the eastern and western arms and the small centres along the Feeder Canal declined from their earlier apex of activity during the period of the Second Canal. They survived as a reflection of the past and retained their canal ethos through their earlier years of association with the waterway.

Port Dalhousie

The two lake ports were not identical twins. Each might have been abandoned as the Second Canal changed to the Third, but each was the anchor feature that greatly influenced the route for the Third Canal. Port Dalhousie advanced in population size and economic strength at first through its harbour services and facilities, including especially its shipyards and ship-repair facilities and the milling complex based on water power that developed on the island in the centre of the weir that created Martindale Pond. The village layout, which later extended under pressure from the Welland Railway to the west bank, was divided by and centred upon the canal, the harbour, and their mutual shipping activities.

The western side, the first to develop, grew with a T-shaped business and commercial centre that faced the harbour and spread along Lock Street at right angles to the first lock of the Second Canal (map 3). Here were the stores, the taverns, the post office, and ships' chandlers that supplied the various needs of the vessels in the harbour and those passing through the canal. The eastern or Michigan side grew later and focused on the Welland Railway, its wooden grain elevator, station, roundhouse, railway sidings, and coal wharves. The two sides, separated by the width of the outer harbour, were linked by the circuitous road across the weir and by a casual ferry.

When larger vessels passed through the Third Canal, the need for terminal grain facilities at each end of the canal gradually ended. As a consequence, when the grain elevator at Port Dalhousie burned down in 1910 it was not replaced. By this time the N.S.&T. had built its track on the west side of the harbour. The steam road on the east side could not compete and, after the lapse of a few years, this was electrified to provide a direct service to Niagara Falls. Because of the difficulties of pedestrian access between the two sides of the harbour, the vessels from Toronto now moored to load and unload passengers on both sides.

Page's Historical Atlas of 1876 noted a population of about 900. 'From the fact of Port Dalhousie being a port of entry, and that it has a fine farming country in its vicinity, it is growing very fast ... The canal gives good facilities to manufacturers, who utilize its power in various establishments. The place also has two ship yards and one or two grain elevators. The harbour of Port Dalhousie is safe, and vessels can find refuge here at all seasons. The place contains a number of fine residences and many comfortable homes ... During the summer months a steamer runs between this place and Toronto.'

The large number of hotels, taverns, and drinking establishments serving teamsters, tow boys, sailors, labourers, and residents became a marked and unruly feature of The Port. In the words of Cumberland (1913), 'The "Port" in those days of the horse [Second] canal ... was merely a turning place for the canal crews. Its one principal street facing the canal basin, had houses on one side only, mostly drink shops, with or without license, with a few junk and supply stores intervening. Its immediate inhabitants, a nomad collection of sailors and towing groups, waited for another job ... It was a little haven not far from the realm of Dante's imagination.'

Shipbuilding and repair provided a linchpin in the community's economic base after construction of the Second Canal. Alexander Muir, joined by four of his brothers, used a floating dry dock from about 1850 and then, from 1867, a permanent dry dock on the west side of Martindale Pond close to the first lock of the Second Canal and next to the business centre of the village. Mostly schooners were produced and also a few steamers, a unifying characteristic being that the ship's names generally began with an A as in *Ayr*, *Alexander*, or *Arctic*. The Muir brothers built, owned, and sailed these vessels on the canal and the lakes to carry grain and timber, and some Muir ships crossed the Atlantic and made the return journey from ports in the British Isles.

The Abbey brothers were also in Port Dalhousie, as were Donaldson and Andrews individually and in various partnerships with each other and additional persons, such as Donaldson, Andrews, and Ross. They constructed further hulls and completed vessels that included tugs and propellers (the term used for vessels driven by screw propellers). This yard, located on the eastern (Michigan) side of the harbour by 1860, was displaced when Lock 1 of the Third Canal was constructed. The close affinity of the port with shipping increased after the Third Canal opened and continued to be reflected in the many hotels, taverns, and inns of varying quality that provided accommodation, entertainment, and drinks and functioned as social centres.

Lake fishing, with a small community of fishermen living below the bluffs on the harbour side next to the village, was another port activity. The port also provided crew members for many of the vessels that sailed the canal and on the Great Lakes. Cutting blocks of ice from Martindale Pond was a winter activity. This product was exported by rail and sold throughout the year from an ice house to vessels to keep their food fresh.

With shipbuilding and servicing vessels then the prime economic activity, an important industrial advance took place in 1886, when the Maple Leaf Rubber Company (later Consolidated Rubber, now Lincoln Fabrics) commenced operation in the former Lawrie flour mill on the weir of the Second Canal. A head of 14 feet (4.3 m) was available to generate hydro-electricity for the plant, and this was then used to light the village streets and some homes. Within two years over 130 hands were employed and housing had to be provided. After fire demolished the premises in 1898, the company rebuilt on a grand scale with two large five-storey brick mills linked at an upper floor level over the intervening road (now Lakeshore Road) that followed the weir. One mill was the factory and the other the office building and warehouse.

By 1907, when the Consolidated Rubber Company took over this factory 320 workmen, many from Scotland, were employed and sixteen cottages had been provided (Nichols 1907). Rubber footwear, with all types of boots and shoes, was manufactured for domestic and foreign markets. The output included mining, hip, sporting, and lumbermen's boots, with some forty brand names and many varying styles.

Other marine businesses at this date included William Hutchinson, later managed by Edward Quackenbush, who supplied vessels passing through the canal with coal from trestles and chutes for some thirty years. Older by five years, a partnership between Ed Murphy and Frank Scott operated a general store (now Murphy's Restaurant) at the corner of Front and Lock

Streets next to the First Lock of the Second Canal. It served as a ship's chandlery, with farm products, groceries, hardware, meat, crockery, and glassware. The store carried a full line of ship's supplies, was the agent for steamboat, telegraph, telephone, and express companies, and did a large insurance business. The post office was in the store of James Stanton, which sold fancy and staple goods, canned and bottled items, and family supplies.

The development of Port Dalhousie as a summer tourist resort began in 1892 when the St Catharines, Grimsby and Toronto Navigation Company established a passenger service by steamship between Toronto and the pier at Port Dalhousie. The situation was favourable to resort expansion. The sandbar, in contrast with areas below the nearby bluffs with no beach, now had access by railway and by steamer. It was also an era of increasing wealth with more leisure time for recreation, and the Victorian restrictions on public bathing, bare bodies, and uncovered legs were ending.

This tourist trade was promoted more vigorously when the electric N.S.&T. streetcar line laid its tracks along Main Street and across the beach to the steamer pier. This trolley company leased the federally owned lands on the west side of the canal and started the development of a permanent amusement park. Excursion passenger traffic was brought in by steamer from Toronto, and the trolleys carried fun seekers from across the peninsula. Recreational facilities were added, including change houses for the bathers, concession stands, a baseball diamond, a merry-go-round, a dance hall, amusements, rides, and side-shows, and the sandbar across the entrance to Twelve Mile Creek was widened by importing sand from the Fonthill area by streetcar. Through the residents, visitors, and the seasonal cottagers who built on the bluffs outside Lakeside Park, Port Dalhousie had become a popular Ontario summer recreational resort and featured a midway of some prominence.

As an adjunct to the now prosperous Niagara Fruit Belt, the steamers on their regular scheduled runs, which averaged four or five a day at the height of the summer season, were also loaded with baskets of fresh fruit for delivery to the Toronto market on their return trip. This service provided an important reminder of the continuing impact on the environment stemming from the construction of Merritt's waterway. Neither peaches nor tourist passengers were then envisaged, only a commercial waterway, but the canal has frequently had to adapt to new circumstances throughout its history.

Another major event was that, with the arrival of the Third Canal, Martindale Pond was no longer used for through shipping. Under the initiative of G.W. Hodgetts, manager of the local branch of the Toronto Bank, the Canadian Association of Amateur Oarsmen, founded in 1880, agreed to stage their annual championships on this sheltered length of water. The Royal Canadian Henley Regatta had been born as an annual event and the pond rapidly became a permanent national rowing course of international acclaim and the venue for a major North American rowing event. A grandstand for nine hundred spectators was constructed to overlook the finishing line at the northern end of the course facing the Maple Leaf rubber factory, and in 1881 a shellhouse and clubhouse were built at the southern end of the course. Despite the incidence of pollution (McNabb 1969), a valuable additional seasonal feature had been added to the port's shipping and recreational attractions.

Port Colborne

Despite the lower availability of water power to attract manufacturing industries, by the late nineteenth century Port Colborne was the larger of the two entry ports in population size. Incorporated as a village in 1869, seven years after its northern counterpart, it became a town in 1917, some thirty years before Port Dalhousie.

The harbour entrance up to its guard lock provided the fulcrum around which Port Colborne developed. The association between the two banks was much closer than at Port Dalhousie, where the two sides of the harbour were more divorced from each other. In 1850 a general store opened next to the new inland lock of the Second Canal, and this replaced the earlier small grocery enterprise at the lakeshore lock that took the First Canal through the sand dunes along the shoreline of Lake Erie. The harbour's importance was that sailing vessels waited here, both to be locked and towed down through the canal. When upbound they might have to wait for favourable winds to ease their passage along or across Lake Erie. The harbour also accommodated vessels laid up through the winter months. In addition to its functions as a canal and lake port, Port Colborne functioned as a port for the productive Lake Erie fishing industry.

The significance of the harbour and its commercial basis of activity increased greatly with the arrival of the Buffalo, Brantford and Goderich branch of the Grand Trunk Railway in 1854, followed four years later by the Welland Railway. The port now had east–west connections as well as north–south links to Port Dalhousie, and was superior to the northern outlet, which had only the Welland Railway. Industry had not developed where the canal dropped into Lake Erie, and its first expression was the provision of the Grand Trunk elevator on the eastern bank of the canal at the terminus of the Welland Railway. The wooden elevator with its dominant brick chimney added the visual dimensions of height and bulk to the harbour scene. Built to transfer grain from vessels to rail cars (hence the title of 'portage railway' for the rail link north to Lake Ontario), the elevator could handle the transfer of 6000 bushels an hour so that the lightened vessel could then pass through the narrow dimensions of the canal. In 1885 over 500,000 bushels of grain passed through this elevator.

As described in Sawle (1955), 'grain boats with deeper draft than the canal allowance had a portion of their cargo drawn out at the elevator and dumped into special railway grain cars. These were small box cars with small interior doors and trap doors for loading at Port Colborne and unloading at Port Dalhousie. The grain carrying trains were made up of 28 loaded cars and travelled down to Port Dalhousie every day. A permanent staff of employees was stationed at each elevator ... The lightened boats were towed to Port Dalhousie where their grain was awaiting them at the elevator.' This business of transfer from lake vessel to railway remained active even after the deeper 14-foot (4.3-m) Third Canal came into operation, and as some steamships also took advantage of the lightening service, this elevator was not dismantled until 1908.

Whether for through or waiting vessels, many opportunities existed to supply vessels and their crews with clothing, food, supplies, and equipment. Livery stables located at the entrance to the canal supplied horses to tow vessels through the canal, and these horses had to be shod and fed. Here was also an advantageous outlet for the sale of local farm products; and ships' chandlers catering for the marine trade became an important part of the scene. Often also associated with the ownership or partnership of tugs, barges, towing horses, and schooners, this assemblage of activity led to fifteen mercantile establishments being recorded by the mid-1880s. There were in addition seven hotels, two each of drug shops, bookstores, and wagon shops, and one butcher, bakery, lumber yard, and sash and door factory, 'all of which have a fairly good trade' (Welland County Council 1886).

Jacob North established a brewery next to the lake in the mid-1850s, and its output helped to quench the thirst of residents, sailors, and visitors alike over several decades. The plant was purchased and became the Cronmiller and White Brewery in 1875. As described in 1907: 'Since then the capacity has been made more than double, a complete ice plant has been installed, a bottling works has been added, where the celebrated Cronmiller & White Lager is prepared for the market, and new boiler and engine rooms have been built. The output is sold throughout this entire County [Welland], and the demand for it keeps the brewery running at capacity all the time' (Nichols 1907).

Though Port Colborne was never as great a shipbuilding centre as either Port Robinson, St Catharines, or Port Dalhousie, tugs were built and launched at canalside locations and a shipyard operated by George Hardison on the east side of the canal is credited with building the hulls for two tugs, a schooner, and a brigantine. In 1870 the firm of F. Woods and Sons (later E.G. Marsh) was established. Its blacksmith shop was involved in shoeing the horses that pulled vessels through the canal, in the repair of canal and lake vessels, and in the machining and fabrication of equipment used by local industry, including the brewery and glass works, and of the bits and drills used for the gas and oil drilling operations.

Railways and the new businesses also changed the visual scene. The high sand hills along the shoreline provided sand for the road beds, and its trees were felled to provide fuel for the locomotives. A railway was built along the shore for these purposes. In the words of an earlier description, 'There was a heavy growth of large trees of many varieties on all the points [headlands] in view [from the hills out]. The hills were covered with a thick growth of large hard maple trees with an undergrowth of cedar' (Carter 1923).

The economic base in association with the canal was extended by tourism, which took advantage of Lake Erie's sandy beaches, along with its shallow and warm waters, and the remains of the forest cover. Besides the novelty of rail access to and from the nearby American city of Buffalo, the railway also provided access to the beaches from the communities along the canal for summer relaxation. By 1887 the *History of the County of Welland* recorded that Port Colborne enjoyed 'a well deserved popularity as a healthful lakeside resort for the people of neighbouring cities who wish, during the heated term, to escape from their hot, dusty, bustling streets and enjoy a few months' recreation ... Accommodations on quite a substantial scale have been provided, and each summer large numbers of people from Buffalo, St Catharines and other cities, pass a few weeks under canvas in the cool groves along the beach, enjoying the boating, bathing and fishing, for which the place is becoming famous.'

Erie Park was laid out east of the harbour by the Welland (Grand Trunk) Railway in the 1880s. The park was provided with bathing houses, picnic tables, a long pavilion for large parties, family shelters with tables and benches, and a covered raised platform for music, singing, and dancing. A 'ferry,' really a punt for ten passengers, connected with the stores in the business centre of Port Colborne. To the east was Lake View Grove, described in 1887 as 'a popular picnic ground for parties who drive from different parts of the county, and is also liberally patronized by city people who live in tents or board at the summer hotel on the premises, during the hot weather' (*History of the County of Welland*). And to the west of the village, at Sugar Loaf Hill, the description of 1887 is again of white tents in cool groves and the bracing Lake Erie breeze during the hot days of July and August.

Between the Sugar Loaf and the village in 1888 Peter McIntyre organized a company that purchased 40 acres (16.2 ha) of lakeshore land east of Port Colborne. 'Solid Comfort,' as the station was named, officially the Humberstone Club (now Tennessee Avenue), became an exclusive private resort for upper-income American southerners from the Memphis area. The estate was organized with spacious homes featuring pillared verandahs and bay windows. A dock then boat houses were built, together with a central dining room, a bowling alley, and an assembly hall used as a church, recreational centre, and meeting place. A casino was added in 1912. The re-routing of Sugar Loaf Street, the earlier Indian trail along the shore, to outside this estate, and closing of the gates at night, ensured privacy for the wealthy patrons of this foreign enclave on Canadian territory.

Industrial Expansion

The use of the natural-resource background is also manifest in the drilling for oil that began in 1866. Sulphur gas was discovered at a depth of 420 feet (128 m), and some was piped to a boiler, but this was discontinued after an explosion. Drilling was continued to twice the previous depth, but the well was abandoned when no oil was struck. In 1885 when the Port Colborne Gas Light and Fuel Company was formed, a well sunk to a depth of 1230 feet (374.9 m) yielded a flow sufficient to keep two hundred regular jets burning constantly.

A series of gas wells was developed by different companies, and gas lines were laid in the village. In 1892 natural gas was supplied to the Erie Glass Company on Welland Street east of the canal. This company was short-lived because of a disastrous fire and, though rebuilt, it closed down in 1893 because of financial difficulties. Using glass blowers from these defunct works, the Humberstone Glass Works was founded in 1891 and the Foster Glass Works in 1893. The role of the canal was to import sand from Ohio and Pennsylvania and chemicals from England. Products manufactured during the 1890s included glass canes, ornamental paper weights, jars, and black, green, and amber bottles.

Natural gas was also piped from the area to markets in St Catharines, Welland, and Niagara Falls and across the border to Buffalo. In 1891 gas brought in the McGlashan and Clark silver-plating factory to the village of Humberstone, but this enterprise was wrecked by an explosion and did not survive the decade.

The Erie Glass Company was the first industry induced to locate in Port Colborne in consequence of its newly discovered resource, natural gas. '"This is only one" said a hopeful citizen, ... "Others will follow when it is understood that we have plants of factory roofs shining in the sunlight all over town."' Many argued that gas should not be exported to Buffalo. 'If manufacturers know the whole product is to be kept in the country they will have more confidence and it would not be long till hundreds of manufacturers would be located between here and Fort Erie. You wouldn't catch the American government letting such a valuable product go out of the country' (*Welland Telegraph*, 28 October 1892).

This glass works, which opened in 1892, produced a full range of ale bottles and fruit and druggists' jars. It used soda ash imported from Liverpool, white sand from Akron, Ohio, local limestone, and natural gas (Bird and Corbe 1974). A marine aspect of this venture was the shipping of bottles and jars to Fort William for the Western Canada market. It is interesting that, as tourists arrived to view the glass-blowing operations, nicknacks such as glass canes and pipes were produced for the visitors to carry away as souvenirs.

Reflecting this ethic of growth, the village council in 1891 advertised that concessions would be available to factories establishing within its boundaries. Guaranteed were a free site, free gas for a glass factory for one furnace for one year, and exemption from taxation for ten years. In exchange the firm was not to employ fewer than fifty men and boys during the work season. As elsewhere, bonusing as a means of attracting industry was a suspect practice. Referring to Port Colborne, the *Thorold Post* (20 January 1905) noted: 'Last summer the Village voted a bonus to a brass company of $10,000, and after a portion of it had been paid, but not all, the company failed and the president was arrested for fraud. The whole history shows an abuse of the bonussing privilege, one witness declaring the Port people to be a lot of "easy chumps."'

A further local resource, limestone from the Onondaga Escarpment, was available. By 1887 Humberstone Township had three quarries, including one about a mile (1.6 km) to the east, another the same distance to the west of Port Colborne, and a third about 6 miles (7.4 km) to the east. White lime and building stone were produced, and often exported from the stone docks on the canal.

An addition mineral resource was sand from the dunes along the shore, which was shipped in canal boats to Buffalo, where a ready demand existed. By 1887, the 'shipping of sand promises to become an important industry' (*History of the County of Welland*). The Buffalo market was also important for the receipt of agricultural produce, ranging from wheat, barley, and maize to root vegetables, garden produce, meat, poultry, cheese, eggs, and fruit.

An important event was the arrival in 1906 of the Great Lakes Portland Cement Company, later Canada Cement. Bonusing provided the company with a fixed assessment of $6000 for a term of ten years, and the next year the ratepayers approved a fixed assessment for general rates of $10,000 for twenty years. Railway track was constructed between the company's dock on the Welland Canal, the plant, and its limestone deposits in Humberstone to the west of Port Colborne; the company also had limestone and clay pits in Wainfleet. Using natural gas as a fuel, the plant had a daily capacity of 1500 barrels. Production started in 1908, the original capacity was doubled in 1912 with the addition of new kilns and cement storage facilities, and coal-burning equipment was then installed as the reserves of natural gas had become exhausted. One gain for the local community was the construction by the company of a concrete road as a demonstration project along present Highway 3 to the Sugar Loaf, the first such road in Ontario.

Farmers from over an extensive surrounding area built the Humberstone Cheese Factory on Killaly Street East to dispose of their milk. This lasted from the late 1880s to the mid-1900s for the production of Cheddar cheese. The milk was brought in by wagon and, after separating the curds from the whey, the latter was returned to the farms to feed the pigs. This local activity reflected the urban role in rural development, with the town providing centrality, manufacturing space to process a rural product, a market, and a cash income for the rural community. The canal had created urban centres, and each continued to assist rural development, in part through the interactive operations of the chandlers and the ability to move rural products to markets via canal barges.

The milling industry with its single grain elevator also expanded. After the breakwater for the Third Canal was extended into the lake to protect the outer harbour, a million-bushel concrete Government Storage and Transfer Elevator was constructed in 1908 by the Department of Railways and Canals on the west pier at the southern end of the breakwater.

It could be reached by lake vessels drawing 20 feet (6.1 m) of water and was served from the now Canadian National (former Grand Trunk) Railway via its causeway across Gravelly Bay. Its purpose was 'to provide facilities for transferring water-borne grain cargoes from large lake steamships to canal-sized vessels or railway cars' (Department of Railways and Canals 1909). Described as a fireproof structure, constructed of steel and concrete, it had an initial storage capacity of 850,000 bushels with provision to increase to two million bushels when justified by sufficient business.

The earlier Grand Trunk elevator of the 1850s on the opposite side of the harbour, now redundant and obsolete, was sold for scrap and dismantled. The new elevator, enlarged in 1912, together with harbour improvements, made the port more attractive to vessels on the Upper Lakes and, in the highly competitive inter-urban and inter-national situation that prevailed for commerce, caused some switch of activity in the grain trade away from Buffalo to the Welland Canal.

Another feature of importance in this trading rivalry between competing routes was that in 1901 the Northwestern Steamship Company of Chicago began the operation of a fleet of four steamers between Chicago and western Europe. The impact of Canada's dominion status, its commission of inquiry into inland navigation, and the resultant achievement of the Third Canal were beginning to exert repercussions on urban development through the canal's new trading features.

The Hedley Shaw Milling Company, which became the Maple Leaf Milling Company, the largest in the British Commonwealth, located in Port Colborne on reclaimed land at the western side of the new harbour entrance east of the Government Elevator. This company had previously operated the Howland Mill in Thorold and the Imperial Mill at St Catharines, together with another at Kingston and a warehouse in Toronto (Maple Leaf Milling 1960).

Inducements to locate at Port Colborne were the granting of a fixed assessment of $10,000 for municipal purposes over a ten-year period and the national 'Milling-in-transit' agreement of the 1890s that made it possible to produce flour and enjoy the balance of the grain rate at points en route between the wheat fields and their large markets. The mill commenced operations in 1911. Wheat brought in by vessel from the Lakehead was milled into flour, and a cereal laboratory helped to bin the wheat according to its protein content. The daily output of 8000 bags of flour, each weighing 100 pounds (45.4 kg), soon doubled. The product was exported through Montreal and Quebec, and also through the Atlantic coastal ports in the United States with their more frequent sailings compared to Canadian ports.

This mill introduced the Ontario Bag Company, a subsidiary company with largely female operators, which worked in conjunction with the flour mill. The output included jute, cotton, and other bags for Maple Leaf and other firms that used bags to export their products. The barrel economy had now changed to other forms of packaging. In addition, the Port Colborne and St Lawrence River Navigation Company was formed as a subsidiary of Maple Leaf Milling to carry grain and flour east to Montreal.

Now with two competing public and private grain elevators at its canal point of entry, Port Colborne was provided with an impressive visual point of entry from Lake Erie and when the milling complex was viewed across Gravelly Bay from the town. An adverse aspect of this scene was that the canal breakwater and its piers critically changed the ecology of the beach profile from sand to mud and weeds. The life of Solid Comfort as an exclusive resort changed to that of a private residential area as the resort buildings were sold to local residents for demolition or conversion to new homes. As with Erie Park, industrial advance at Port Colborne was at the expense of its earlier tourist and recreational abilities.

The production of shoes began in 1905 when Charles Christian Knoll, who operated a shoe repair shop on Main Street in Humberstone and made custom shoes by hand, formed with partners the Comfort Shoe Company. This became the Humberstone Shoe Company in 1907, and later expanded on several occasions.

The major transition from a commercial port with ancillary industries to a manufacturing base of heavy industry arose with the opening in 1913 of the Canadian Furnace plant, a privately financed American corporation which later became the Canadian Furnace Division of the Algoma Steel Corporation. The plant located on the east side of the canal entrance, taking advantage of Port Colborne as a centre of railway and shipping facilities and of the orbit of Niagara Falls for its demanding supplies of electricity. Its displacment of the popular lakeshore picnic ground facility of Erie Park illustrates the fundamental clash between industrial development and resource-based recreational opportunities that has been the story of so many of the canal communities.

The daily output of 400 tons (362.9 tonnes) of pig iron required the input of 800 tons (725.8 tonnes) of haematic iron ore brought in by lake vessels from Minnesota and Michigan, 500 tons (453.6 tonnes) of coke, and 200 tons (181.4 tonnes) of limestone. To move and accommodate these raw materials, several lines of parallel track were laid by the Grand Trunk Railway. Docks were also built along the canal by the dominion government in conjunction with the expropriation of East Street for the widening of the Fourth Canal through the town centre. In addition, several millions of gallons of water per day for cooling purposes were obtained from the canal. The pig iron was exported to various automobile industries and their foundries in Canada and the United States, where it was used for engine blocks and pipe fittings.

The same year witnessed the founding of the Dwor Metal Company, which started with the buying and selling of scrap metal. Later, as Marine Salvage, it either purchased and sold Great Lakes vessels or had them broken up and sold as scrap, sometimes at Port Colborne and sometimes through towing to an overseas destination. Further iron industries were to follow, but the foundations of the steel industry at Port Colborne were established through the initiatives of Canadian Furnace and Dwor.

The Business Centre and Humberstone

Urban form and character continued to emphasize the harbour and its immediate surrounds (map 4). The harbour, extended and improved for both the Second and Third Canals, was faced by East and West Streets on opposite banks along the canal, which were linked by a road bridge across the head of the harbour. Retailing provision had developed on both sides of this bridge. The bridge lay between Clarence and Charlotte Streets, and not on a through cross-town route as was the function performed by Main Street at Humberstone. The view from the bridge focused onto business premises in both directions.

As interpreted from Lovell's Business Directory of 1882, West Street carried the greatest number of service establishments, forty-six against eighteen. Eleven hotels were the most prominent grouping, but retailing including the only bank provided considerable variety. 'East Street was typically a canal street with almost equal occupational groupings in hotel managers, canal clerks and ships chandlers ... with necessary retail outlets such as a blacksmith, general store and grocer. The canal offices were located on the east rather than the west bank, because of the towpath on the east' (Koabel 1970).

By the turn of the century, the business centre remained split by the canal into two parallel but restricted sectors. Each side was backed by residential development and the connecting link was by a convenient bridge across a relatively narrow canal. On the east side were a department store, the first movie theatre, the post office, the Grand Trunk (B.B.G.) station, a

Catholic church and school, and the Cronmiller Brewery. West Street had the greater variety of outlets as previously, but not the attractive power of the department store and cinema.

Together the two areas made for an intimate, closely grouped, H-shaped business district serving vessels in and passing through the harbour and also the local population. External links were by boat along the canal and to the ports around Lake Erie and the Upper Lakes. The railway on the eastern side of the harbour with its coal docks, grain elevator, and then other marine-associated industry connected via the Grand Trunk (former Welland Railway) to Port Dalhousie and via the Buffalo and Lake Huron (now also absorbed by the Grand Trunk) to the International Bridge and the United States.

The decision in the 1910s to build a new ship canal brought drastic change to this community and in particular to East Street. Land on the east side of the canal was expropriated to widen the harbour and construct a new channel from Lake Erie to Humberstone. This process of widening the canal process adversely affected the post office and customs building, the Imperial bank, a commercial block, a hotel, and two stores. North of the railway were two hotels, a cork factory, and several residences. Work on the canal stopped during the First World War, but the properties had then been acquired. Although the occupants of the expropriated lands retained possession during the war years, the die was cast for demolition in the immediate postwar years.

Some merchants relocated in the western retail area, some moved to Humberstone, and others closed their business, but no extensive outcry occurred as harbour improvement was perceived as bringing in additional business. Of greater concern was that the new lock, Lock 8 of the Fourth Canal, was moved north to a location near Humberstone. The intention was also to remove the bridge, critical for the economic success of the business centre, to the new location, but the merchants successfully petitioned the government to change this decision, and the bridge was moved only a short distance to the north. This made Clarence into a through street on the only direct east–west link in the business core, and effectively transferred the business centre to the west side of the canal.

The nearby but now subordinate centre of Humberstone, which carried the main east–west road in the south of the peninsula, had developed commercially and residentially along this main road on both sides of the swingbridge over the canal. As a regional and farming centre of some considerable local importance, Humberstone by 1886 contained 'three churches, one public school, one Lutheran school, a town-ship hall, a Temperance

Hall, five hotels, a machine shop, a foundry, a saw-mill, a planing-mill and sash and door factory, two cabinet shops, three wagon and carriage factories, eight shops, one organ-factory, two harness manufactories, four blacksmith shops, and being surrounded by good farming country, is in a very prosperous condition' (Welland County Council 1886). The same source noted that 'the Fort Erie and Stonebridge roads run along the top of this ridge [the Onondaga Escarpment]. The road is macadamized and gravelled nearly its entire length, and as it is the direct highway to the City of Buffalo it always presents an animated appearance.' Slightly to the east of the village, 'the Chippawa road [the link with the Niagara River] ... is good nearly the whole year round.'

The independent and separated villages of Humberstone and Port Colborne had linked steadily along the west side of the canal, and Elm Street later provided the route for the N.S.&T. interurban streetcar link between the two centres and their connection north into the regional system. The expansion of the now united urban unit had also been to the west along the sandy shoreline of Gravelly Bay towards and past its recreational resorts, and also inland from these centres of summer activity. The limitations to lateral expansion included severe drainage difficulties because of the marshy environment as well as the extensive limestone-quarrying operations.

Chippawa

At Chippawa (map 7), expansion continued after mid-century. Shipbuilding remained an important activity and many vessels were launched, including for the Erie and Ontario Railroad, which established a second dock on the north side of the Welland River as the terminal for its combined steamer services to Buffalo. These steam operations began in 1854. It is interesting that the steam side-paddle wheeler *Clifton* used on the Niagara River was built at Chippawa, whereas its companion the *Zimmerman* used on Lake Ontario to reach Toronto was built at Niagara. Both yards used the skills of Louis Shickluna, the St Catharines shipbuilder who provided an economic linkage of inter-association between the new canal and the older river communities.

The village reached its peak of importance during the 1860s, but population decline then set in to the end of the century. This decline and a diminished assessed value for buildings can be attributed to transportation features including the canal link along the Welland River, which, though surviving into the 1900s but carrying little traffic, fell gradually into disuse.

Through trade along the canal for Lake Erie increasingly went through Port Colborne, bypassing Chippawa and the Niagara River. Another factor was that the construction in 1855 of the railway suspension bridge downstream across the Niagara River effectively bypassed Chippawa and enabled more visitors to arrive at The Falls by rail on either the Canadian or the American sides. Later bridges across the gorge and at Fort Erie perpetuated the isolation, with Chippawa no longer being on routes for east–west traffic across the border.

By 1886 the incorporated village of Chippawa, with its station on the now Erie and Niagara branch of the Canadian Southern Railway between Buffalo and Niagara (now Niagara-on-the-Lake) and still a port of entry, was in a state of severe decline. The population had dwindled from 1450 in 1864 to 664 by the census of 1881. Six years later the foundry was deserted, the tannery was still in operation but produced only leather belting, and the distillery had been converted into a roller flour mill. The village services had declined to six stores, four hotels, two blacksmith shops, two shoe stores, and one tinshop. The total value of personal property exceeded $110,000, but it had been $185,000 (*History of the County of Welland* 1887).

When the N.S.&T. streetcar line reached Chippawa in 1893, this route was served from Buffalo by a dock on the river. When four years later this service was extended south along the Niagara River to Slaters Dock, the change was disadvantageous to Chippawa. The village lost its terminal facility and was passed through by most passengers en route to The Falls.

Increasingly bypassed by transportation systems of the day and always subordinate to The Falls in its visual appeal to the visitor, Chippawa was not assisted by the Fenian Raids, which indicated vulnerability to possible renewed aggression from the United States. The village also suffered from serious fires, as in 1881 when some twenty-six homes were destroyed. The population had sunk further to 460 by 1901.

A redeeming feature was the arrival of a printing press that grew into Gordon Kaumeyer Industries. A more substantial event was the arrival of hydroelectric production at The Falls, and with it the transition to an industrial community through the advent of the Norton Company in 1910. Attracted by the availability of the new but substantial power resource and the plentiful availability of water from the Niagara River for cooling purposes, the first plant produced 4 tons (3.6 tonnes) of silicon carbide a day. This crystalline compound, one of the hardest known substances, was used as an abrasive and heat-refractory material in high-temperature industrial situations. The company arrived from Worcester, Massachusetts, and the work force was immigrant. Services and houses expanded in response to this new employment opportunity, but Chippawa could no longer be described as an active canal community.

Dunnville

The possibility of extending the Grand River system of navigation from Brantford to London, Chatham, and Lake St Clair was unsuccessful, but had one of the schemes involving the Grand and the Thames Rivers been completed, an inland system of navigation would have extended across western Upper Canada (Guy 1993). But this was not to be. Railways took over from the canals and Dunnville lost first its position at the head of navigation along the Grand River and then its association with the Feeder Canal.

By 1877, the canal rather than the railway was valued in the achievement of growth at Dunnville. 'To the almost unlimited water-power available, Dunnville owes much of its early prosperity and subsequent growth. The shipping facilities furnished by the canal, and the excellent harbour at Port Maitland, have made it the market where nearly all the grain of the surrounding townships is handled; and the building of the Buffalo, Brantford and Goderich Railway has given Dunnville increased advantages as a business centre' (Page 1877). Even so, Dunnville had 'maintained a degree of prosperity, even in the last few years of depression, equalled by very few communities of its population in Ontario.' The population at about 2000 was increasing, the houses were mostly well built of brick, and some stores were described as 'large and substantial structures.' The two bank agencies were the only ones in Haldimand County. 'Dunnville has a large number of well stocked stores, and a few prosperous manufacturing establishments.'

Much advantage was taken of the town's scenic setting on the Grand River, and in particular the sharp contrast that existed upstream from the dam, where a lake had been created across the river's floodplain, and downstream, with its wide expanse of marsh, reeds, wild grass, and flags (Sorge 1977–89). The dam and the embankment upstream along the eastern side of the river provided an important extent of public open space for relaxation. Here the river and its banks were closely associated with boating, dip-net and rod fishing, visiting anglers, flocks of birds and ducks, and paintings and drawings by those of artistic temperament.

Less happily, spring flooding with the break-up of winter ice often caused severe damage in the town's business section and its low-lying residential areas. Disaster years included 1869, 1898, and 1913. The bridges, dams, and sluices known as the Long Bridge were frequent casualties of these inundations. Battered by ice and the fast currents of rising water levels, sections of the bridge had frequently to be repaired or replaced, and boats were then used to provide the necessary link between the two sides of the river. Flooding, a natural phenomenon of the slow-flowing river, had been aggravated through clearing timber from the upstream areas, and the dam did not have sufficient capacity at its waste weirs to cope with high river levels. Flood control, more so than in any other canal community, had become an important aspect of the urban scene and part of Dunnville's folklore.

Industrial development beyond the introductory pioneer mills and services included three short-lived soft-drink manufacturers during the 1870s and 1880s. Some barges were also constructed at Dunnville, but no extensive shipbuilding enterprise emerged. Later, a linear industrial sequence developed along both sides of the Buffalo and Lake Huron (former Buffalo, Brantford and Goderich) Railway at and close to its yards and the station, which bounded the town to the north.

An important seasonal activity was that the Grand River provided a source for winter blocks of ice to stock local ice boxes and to keep food fresh during the summer months. Ice was also exported. In 1898 five to ten railroad cars of ice a day were shippped via the Grand Trunk Railway to Buffalo. The 'mammoth ice house' of the Webster Citizens Ice Company of Buffalo, next to the Grand River and on a spur line of this railway, had a capacity of 45,000 tons (40,824 tonnes) (MacDonald 1992).

Hydroelectricity for street lighting and in the stores arrived in 1885 through the Dunnville Electric Light Company. Power was at first generated from the Feeder Canal through turbines at a sawmill and then a roller mill, but when a more substantial input was obtained from Niagara Falls, because of the greater distance and higher costs for electricity than along the main line of the canal, Dunnville did not have the same opportunities as locations closer to The Falls, and heavy industry did not arise.

In 1888 natural gas was struck on the east side of Dunnville. Other nearby natural resources included limestone quarries, and especially Dunnville Rock Products on the road to Port Maitland, and gypsum mines upstream along the Grand River. The agricultural hinterland and its products led to a cider mill for pressing apples in Byng. Other seasonal activities, again mostly for export and this time primarily in the rural areas as a cooperative effort by farmers, produced cream, butter, and Cheddar cheese.

Fall and spring agricultural fairs run by the local agricultural society existed by the 1880s. They were held on a 17-acre (6.9-ha) site after the harvest and when the spring planting was complete. The exhibition hall, a new bandstand, and an improved race track had been added by the 1890s. Fair days were important festive and social occasions, featuring band competitions, sideshows, entertainment, and bicycle races in conjunction with the displays of livestock, farm machinery, agricultural products, food, and handicrafts. Also important in Dunnville from at least the 1890s was the farmers' market held in the market square twice weekly on Tuesday and Saturday mornings. Commercial fishing was practised as well, with catches coming from both the Grand River and Lake Erie. Tugs docked on the lake side of the dam.

By 1893 booster advertisements in the Dunnville *Chronicle* stated that, with a population of about 2000, the town had eight natural-gas wells that were deemed sufficient to heat and light the town, in addition to an electric plant. Its mechanics' institute was endowed with 3000 volumes, and its industrial base included three roller flour mills, two stone flour mills, two sash and door factories, and one each of a foundry, woollen factory, machine shop, and sawmill. The town had six hotels, brick store blocks with glass fronts, and granolithic sidewalks. Dunnville was described as 'a good place for manufacturing enterprises, and cannot be excelled as a healthy, pleasant and cheap place of residence.' Its streets were 'lined with stately maple trees and many have the appearance of fine avenues.'

The transition to large-scale manufacturing began when Frank Lalor, a retail grocer who expanded into the dry goods business, established an evaporating works for drying apples. He approached the council for a loan of $6000 and a bonus of $4000. 'The canning factory I propose to build will be one of the largest and best in Canada. I will pay out to the farmers of this section not less than $50,000 per year for small fruits, vegetables and poultry. I will pay out not less than $50,000 per week for about eight months of the year for labor, and from $100.00 to $200.00 per week during the winter months to can makers and others employed about the factory' (Stretton 1950). With a bonus of $10,000 from the town, the bylaw passed with a substantial majority. The canning factory operated on natural gas. It processed small fruits, vegetables, and poultry, became perhaps the largest of its type in Canada at that time, and later operated as a branch of Canadian Canners.

The textile industry provided another area of industrial advance. Dominion Hammock Manufacturing (later Dominion Fabrics) was organized by James Camelford, a shawl weaver from Paisley, Scotland, in 1898 to manufacture a range of quality hammocks as well as carpets. The weaving was undertaken by hand through female employees, and the kettles used for the dyes were heated by natural gas. As the demand for hammocks languished, new looms were introduced for the manufacture of towels.

Monarch Knitting (later Monarch-Knit) followed in 1903 to manufacture hand-knitting yarn and finished goods such as outer wearing apparel. Frank Lalor was one of its three founders. This company specialized in knitted outerwear that included lumbermen's socks and women's sweaters. As the company expanded, the four-storey mill became Dunnville's largest employer of labour, with over five hundred employees on a 3-acre (1.2-ha) site. The head office remained in Dunnville, but branch plants were established in St Catharines, St Thomas, and Buffalo. With the addition of Imperial Knitting Mills and the Kitchen Carpet Factory, Dunnville now had four textile companies contributing to its distinctive and specialized economy.

In form, Dunnville grew initially along the wharf that extended down the canal from the swing-bridge, with mills located between the Feeder Canal and the Grand River (map 8). A large pond supplied from the Grand River acted as the head pond for the Feeder Canal, and it also fed into a cluster of smaller ponds that served the mills. One or more mills then located on each mill race. The town developed on the opposite or northern side of this water barrier and its milling complexes.

The main business centre expanded along Canal Street (now Queen), where the businesses backed directly onto the canal, with storehouses built on the canal banks or over the water. Other stores located along Lock and Chestnut Streets. Access from the Byng side across the Long Bridge over the Grand River here met the main through road that followed the northern bank of the Grand River, and Dunnville grew as a road and regional service centre around this node in conjunction with its marine then railway attributes.

Two streets were drawn parallel to the canal east of the mills and between the canal and the river. As a further and more extensive range of streets were added east of the canal inland from the main streets and again parallel to the canal, urban form very much reflected the town's water endowments. As this pattern did not conform with the township grid, a feature was the change in the direction of the streets that ran parallel to the canal at Chestnut Street. This street led to the Grand Trunk station grounds and acted as a sort of hinge in the urban structure.

Dunnville, which lost first its river and then its canal traffic to railway development and was no longer a trading port of substance, had sufficient resilience to survive and expand as a regional centre with industrial commitments that centred on the textile industry. But at this western end of the Feeder Canal and the canal system, the abandoned canal channel and its vicinity were perceived as unwanted open space and an impediment to urban growth. The channel, eventually purchased by the town council, was then filled with quarry overburden to provide new space for development.

Even Canal Street, which had been laid out parallel to the canal along its eastern bank, was renamed Queen Street, thus discarding from public memory the critical fact that it was the Feeder Canal that had created and promoted Dunnville over some seven decades of existence. For neither the first nor the last time in this narrative, the past had become unimportant. The future use of former assets based on the whims of developers rather than a public conscience was the most pressing concern.

Fortunately Lock Street, so named as it headed towards the guard lock at the head of the fore-bay that served the canal and the mill races, has survived as a community road name, and public open space with a historical plaque that commemorates the founding of Dunnville has been retained along the banks of the Grand River. The infilled former channel, now hardly recalled, is occupied by buildings, parking lots, and empty, often derelict open space, which is hardly a fitting memorial to so many decades of marine history in association with the canal and the Grand River.

Along the Feeder Canal

Feeder Canal became a misnomer with the arrival of the Second Canal and its use as the through route to Lake Erie, but the term remained in use because the channel also continued to supply the canal with water until its level was reduced to that of Lake Erie. The period of the Second Canal encouraged new development along the length of the feeder between Welland Junction and Port Maitland. In particular, a village grew around the canal turning basin at Broad Creek (Stromness) where the channels to Dunnville, Port Maitland, and Welland met (Paul 1983).

In 1855 Lachlan McCallum purchased a sawmill near this small community. He bought tugs, entered the towing and wrecking business on the lakes, and also developed a substantial general store that sold dry goods, hardware, and groceries. His business empire included farms, shipbuilding, a cheese factory, a blacksmith shop, and a hotel (Sorge 1977–89). His shipyard on the banks of the feeder produced tugs, barges, and a three-masted sailing vessel. His tug *Mary Ann* was the first to be registered by the new Dominion of Canada, and a river barge of 320 tons (290 tonnes) was the largest yet attempted. This shipbuilding activity declined through the advance of the railway, but a later period of boat construction arose next to the turning basin during the 1910s when Archie Mac-Clean produced commercial fishing vessels that sailed out of Port Maitland for use on Lake Erie. The village grew until this section of the Second Canal ceased to operate as a commercial waterway, but then declined as this route was superseded by the Third Canal. The feeder was used only occasionally afterwards.

At Port Maitland the changes brought by the impact of the entrance lock to the Second Canal are recalled in an account by a 'bush farmer' who became a miller. 'I have often seen three large vessels locking through it at one time. This made a water power possible out of the canal, which I leased from the government and built a flour mill on it … I rented my farm, eventually sold it, and moved over to the east [canal] side of the harbour, built a house, a store and saw mill, and commenced business, which was more congenial to me than this so-called farming, without a market for anything you produced' (Imlach 1900). This dealer also acted as a marine insurance agent. As wrecks along the shore were 'very frequent,' he bought cargoes of general merchandise, bales of cotton that he hung up over the four stories of his mill to dry, crates of crockery, hardware, and frozen groceries of all kinds that he then sold retail in his store – an unusual though doubtless lucrative way to stock a store!

A vivid later description depicts the scene at the entrance to the harbour at Port Maitland. 'Steamers here took in their supply of firewood, and great piles of it were to be seen in the docks. The harbour was filled with vessels all summer long, steamboats wooding up, schooners which had to put in for supplies or shelter, little fleets of Grand River scows and barges waiting till the lake was calm enough for the tugs to tow them across to Buffalo. Great rafts of timber often lay here for days. During the time of the American Civil War a good deal of round pine was bought to go through the canal, immense stacks some of them over one hundred feet [30.5 m] in length and three or four feet [0.9 to 1.2 m] in diameter. At that time the Southern ports were blockaded and shipyards had to get their masts and spars from Northern forests' (Tipton 1893).

Fishing, also an important activity, combined the advantages of a sheltered natural harbour at the entrance to the Grand River with the marine bounty of both lake and river fishing. Port Maitland became an important fishing centre. The fish caught included mainly herring, whitefish, pickerel, pike, sturgeon, perch, and catfish. This activity extended out from supplying the tugs, the fishing vessels, and their crews to encompass the making of nets, building of netting machines and ice houses to the storage, packing, and despatch of fish, including substantial exports to Buffalo.

The import of coal from the American side of Lake Erie was another activity. Either stored or trans-shipped to a barge for transit through the canal, this import was a major factor that brought the T.H.&B. Railway to the port to meet the regular arrival of the car ferry from Ashtabula, Ohio. The ferry dock was located on the east or canal side of the harbour, and the two sides were connected by a hand-operated ferry.

Towards the end of the century, tourism added its regular seasonal quota of summer visitors to the guest houses, farms, and available camp sites at Port Maitland and its shoreline vicinity. This eased the economy, but Port Maitland was not transformed. It remained a small community at the southern entrance to the canal. Port Colborne attracted a greater number of vessels to its port and for transit through the canal, and Port Maitland could not compete with these more extensive facilities.

The possibility that Port Maitland might become the southern port of entry to the pending Fourth Canal was seriously considered during the 1910s. The village had a reputation as the finest natural harbour on the Canadian side of Lake Erie and it was perceived as second on Lake Erie only to Erie, Pennsylvania, on the American side. Its harbour was also free from ice two to three weeks earlier than at Port Colborne and it had a good protected access in the worst of weather. However, this reasoning was not to prevail against the elevator and harbour developments that already existed at Port Colborne. Port Maitland, not to be located on the route of the Fourth Canal, in an isolated position, and with no road bridge across the broad Grand River, was to subside steadily into being a place of only limited importance.

At the middle of the long length of the Feeder Canal, the village of Marshville (later Wainfleet) expanded as a rural service community at the centre of Wainfleet Marsh and the township of that name (Raymond 1979). Wainfleet became the supply and administrative centre for the farming community and a central point of service for vessels on the canal (map 8). The village had a two-day, fall agricultural fair by the 1880s, with horse racing and exhibitions, that on occasion outclassed the Welland Fair.

Services typical of the era included wagon-making and blacksmith shops. The vicinity had numerous gas wells, and the cutting and baling of peat were carried out on the marsh. Industry included a local brickyard, a cheese factory, an ice factory that stored ice cut from the Feeder, and a saw-mill and grist mill on the canal. Their combined mill race excavated into the marsh drained to the north-east away from the village. Edward Lee owned the general store and mills. He purchased grain from the surrounding farmers, stored it through the winter months, then exported it by boat when the canal opened in spring. He also served as reeve of the township and warden of the county.

Communities such as Stromness and Wainfleet along the Feeder Canal, and the feeder itself as a transportation artery, stagnated as the waterborne trade from the Grand River along the canal declined through railway competition and as the majority of canal traffic increasingly used the main line rather than the eastern and western arms to the Niagara and Grand Rivers. The Feeder Canal ended its operation as a commercial waterway in 1881 when the Third Canal opened. With desultory traffic, the last known industrial use was in 1908, when a load of railway ties for use by the N.S.&T. streetcar service was sent through to Thorold. The irony of the Feeder Canal is that it carried water at a higher level than Lake Erie and became a barrier to drainage (McGeorge 1947). To offset this disadvantage, drains known as the North and the South Back Drains were constructed along both sides of the canal and connected by eight syphons constructed across the canal.

With the abandonment of the Feeder Canal for navigation, reeds grew in the canal bed, the banks became overgrown and neglected, and drainage problems became a topic of local farming and residential concern. Shrubbery encroached along the banks, branches extended out over the water, and fallen trees clogged the once active channel. The through flow of water was cut off as embankments were constructed across the canal to provide access to neighbouring farms and residential buildings. The water became stagnant. It filled with sediment and weeds, but occasional short sections might be used for punting, canoeing, hunting, and fishing. The few small centres that the canal had nurtured suffered accordingly, but the momentum that they had acquired as busy local centres from the canal carried Wainfleet, Stromness, and Port Maitland into the future as small villages but no longer integral features of the canal scene.

The federal Department of Transport retained responsibilities for the canal, and towpaths along the banks became public highways maintained by the township in which they were situated. Surplus lands beyond the roadway not required for canal purposes were sold to the adjacent landowners. In 1944 the Welland Canals Conservation Committee, formed by the Welland County Council, argued that trees should be planted along the Feeder Canal right-of-way. The superintending canal engineer did not object, but he opposed undertaking this project as a canal work.

At the same time, improved drainage for the marsh was required. Some

1750 acres (533.4 ha) were idle and worthless for agriculture as the Feeder Canal had placed an obstacle across the natural drainage of the land. It was agreed in 1949 that the Department of Transport 'should restore drainage conditions insofar as possible to what they would have been if the Feeder Canal had not been constructed.' This proviso required the enlargement of ditches on canal property, plus the construction of culverts across the canal, and funds were provided by the Treasury Board for this purpose. It is apparent that the canal and its controlling authorities remained of considerable importance for the landscape, even though the waterway itself was now abandoned.

8

The Inland Centres of Welland, Allanburg, and Port Robinson

Three communities developed along the central reaches of the canal. Welland, the largest, expanded where the canal crossed the Welland River. After it became the county town, its new administrative activities and status provided a further encouragement to growth. Even so, the transition to industrial development and large-scale foreign immigration did not occur until after the turn of the century. By contrast, Port Robinson and Allanburg remained small communities, the former dwindling from a greater importance, whereas the latter never advanced.

Welland

Because of its several water barriers Welland was a more segmented urban entity than any other location along the canal (see map 6). The village grew outwards from its core at the aqueduct in three separated sections: the northern section around North Main (now Niagara) Street, bounded by the river and the canal; a western section around West Main Street, circumscribed by the river and the canal; and the largest section to the south of East Main Street, bounded along one side by the canal. A fourth section to the north-east was a narrow strip of land (now Merritt Island) that extended between the canal and the river north to Port Robinson.

When the Third Canal arrived on the scene, its route was again constructed through the centre of Welland. A new canal route to the east of the town on a more direct route to Lake Erie was opposed from within the town as this was expected to lead to a severe loss of trade. The outcome was the widening of the Second Canal at the centre of Welland along the west side of the old channel. As its excavated materials were dumped into the now unused channel of the First Canal, the ponds, hollows, and islands that have been described were mostly eliminated.

One pond named after Dennis McCarthy, an Irish immigrant who had helped to dig the Second Canal and who had settled on the bank of this pond, had become an unsightly and unhealthy area where all types of unwanted materials were dumped. When the contract was awarded to straighten the canal to the north, some of the excavated materials were brought into the town and used to fill this pond. Maples and oaks were planted on this reclaimed land in 1911 and, seeded to grass, this obsolete and distasteful relict section of the earlier canal at the centre of the town was translated into the small green area of open space that became Merritt Park.

The new cut-stone aqueduct of the Third Canal with six arches 40 feet (12.2 m) wide was constructed next to its predecessor on the upstream or western side where the Fourth Canal is now situated. It opened in 1887 after a ten-year construction period, and was described in glowing terms as a work of art for the quality of its masonry. As the canal now carried more water and was deeper, the space above the river at the arches was reduced to the restricted but fluctuating height of perhaps 4 feet (1.2 m). The Second Aqueduct was left in service and survived until the 1920s. The larger canal boats used the Third Aqueduct but the two also operated in tandem, one for upbound and the other for downbound vessels. The lock connection upstream with the Welland River also remained for awhile, but this trading flow was now approaching its end.

The aqueduct of the Third Canal marked the end of the 9-foot (2.7-m) draft for the canal, but as this depth fluctuated with the wind direction and the level of Lake Erie, the canal had often to operate at a lesser depth. Barges might then get stuck on the aqueduct and tie up traffic, an incident that could last for a few days until the grounded vessel was removed. The opening of the new aqueduct also marked the passing of the old tow horses and the start of tug boats and the larger cargo boats known as propellers.

When the Third Canal was built, the two small bridges that connected East and West Main Streets were removed to permit construction of the new channel and the new aqueduct, and a replacement swing-bridge was constructed across the waterway at the west end of Division Street. As always, the old canal had to be kept in full operation during the construction of the new waterway, but this decision by the canal authorities made Main Street into a dead-end passage even though this was now the principal commercial street and the town had developed on both sides of the combined canal-river barrier. Strong civic and business objections arose as the trade on Main Street declined or transferred elsewhere.

A second bridge urged for Main Street began as a floating plank waterway, 4 feet (1.2 m) wide with a railing on both sides and steps at each end. It could be used only during the closed winter season of navigation, and it was then treacherous when covered with snow or ice. A new swing-bridge known as the Alexandra Bridge, with a steel span and walkways on both sides, did not open at Main Street until 1903. Access to the centre of Welland had suffered for some twenty years with the arrival of the Third Canal.

At Helmsport, where the Feeder Canal met the main line of the canal at a lock and provided its water supply, the connection from the Grand River was discontinued after 1881 as the canal's source of water was now derived from Port Colborne. However, the feeder remained in use as a supplementary source of supply for use when lake levels were low, and for the limited but declining trade from the Grand River.

At the end of the feeder, in 1887, the flow was diverted north through a raceway along the canal banks, then north to the Welland River, a route that was later filled and became Prince Charles Drive. The first waterworks were built at the end of this raceway in 1888. The fall of some 16 feet (4.9 m) at the bank of the river was used to drive the pumps that distributed the water through the town's water mains. The water itself was taken from the canal upstream of the feeder, then piped to this site. A new waterworks built upstream next to the canal came into operation in 1912.

The Welland Electric Light Company, established in 1887 on a site next to the waterworks, also used the fall at the end of the hydraulic raceway to generate power for two dynamos. In 1890 a system of arc and incandescent lighting using alternating current was introduced, followed five years later by the use of an engine for supplementary power. In 1906, the newly incorporated Welland Electrical Company brought in current transmitted from DeCew Falls, a more extensive source but still canal-based.

Some consideration was also given to diverting the Welland River to the edge of the Niagara Escarpment to provide a further power opportunity, but this idea was not pursued. Another aspect of water flow was that the metal industries continued to draw water from the canal for their cooling and industrial requirements. The volumes abstracted were greater than that used either by the city from the canal or for the flow required to work the locks along the canal.

The power resources used in the expanding industrial community were augmented by natural gas obtained from the Welland gas field that underlay the south of Welland County. Electricity, gas, coal, and natural gas, each with competing companies to provide their supply and with competition between each utility to provide heating and power, provided an industrial advantage for Welland and an edge over other communities that lacked both this choice and an abundance of supply in each category.

When Welland was described in the mid-1870s, its county and service facilities were to the fore. The industrial commitments received scant mention, and the future was perceived to be in commerce rather than manufacturing. 'The village has a grammar and two common schools, and the County Agricultural Society hold their annual meetings here, where they have a large fair ground with ample buildings upon it. A swing bridge across the canal connects the village, which is situated upon either side of the canal. The place has several churches belonging to different denominations, and a number of fine brick stores and some large mills and manufactories of various kinds. Welland is well situated to become a much larger and more of a commercial place' (Page 1876).

An indication of Welland's regional status and its expanding service catchment area is provided by the Welland *Tribune*. This newspaper started in Fonthill, but moved to a more central location in Welland with its county status and after it obtained superior railway connections. From its offices in Welland the paper merged with the Thorold *Mercury*, the Welland *Telegraph*, and the Port Colborne *Citizen* to serve an area that extended from Fort Erie to Dunnville and from Port Colborne to Thorold. Agricultural links across Welland County were stressed by the Welland County Fair, held in the fall on Denistown Street to the west of the canal from 1847 onwards.

The town had grown around its aqueduct and, with the addition of railways, extensive sites outside this earlier concentration became available with the advantages of access to canal and rail facilities. The industrial sites that arose were therefore either generally parallel to the line of the canal or on railways close to the canal. They might use water from the canal for forging and cooling purposes or directly for power, and enjoyed the transportation advantages of one or both the rail and canal modes with good effect for imports or exports. The town in its turn, while retaining the central area for most of its business expansion, grew up to and then beyond its industrial plants, which became enmeshed in the growing town.

The railways, on the same level as the flat landscape, were crossed by the city's roads at grade crossings. The lines of track divided the new developments into distinctive social areas, which might be inhabited by different ethnic groups. The railways therefore reinforced the canals and the river as divides. It was not an orderly pattern of growth, but a series of separated residential areas that grew as each new industry located in the town. The physical character of terrain virtually dictated that these developments were either to the east or south along the canal into Crowland Township, and led to persistent demands for these areas to be amalgamated with the town of Welland.

The Welland Railway, east of the town and with its station and yards on East Main Street, did not cross the canal, but for a while defined the eastern limits of Welland. The Canada Southern to the south of the town, again with yards and a station together with connecting track to the Welland Railway, crossed the canal. The urban layout now had a network of railways, freight yards, stations, and spur lines added to the earlier components of canal and river. Only a few through streets crossed these rail barriers, and even these roads were not necessarily continuous across the full extent of the urban environment. Welland, more severely than elsewhere among the canal communities, was divided into separated residential areas by the pattern of its several canal and rail transportation routes.

From Mills to Industry

The Welland Flouring Mills owned by Messrs. Seeley and Dunlop located about 1847 on the north bank of the Welland River. Served by a raceway from the north end of the first aqueduct, it was probably the first mill to be constructed in Welland. Next door, using the same raceway, were the Ellenwoods Woollen and Carding Mills and a sawmill founded by Mr Seeley. After the flour mill burned down in 1860, reconstruction took place under a new owner on an adjacent site. 'It was at that time run by water power. In 1877 Phelps Bros. ... put in steam in consequence of the water supply being cut off by the deepening of the canal. Since that date the mill has been run by steam power ... Recently the Messrs. Phelps have obtained the privilege, conditionally, of using the water power from the new [Third] canal, and at this time are contemplating the digging of a raceway in the near future' (*History of the County of Welland* 1887).

Another early water-powered grist mill was built about 1850 by Moses Cook at the south end of the aqueduct on the island between the First and Second Canals. As water drawn from the canal by this and the other small riverside grist, saw, planing, wool, and carding mills was returned not to the canal but to the Welland River, this flow was lost to the canal and not available for use in either downstream mills or for canal navigation. This loss into other drainage systems has been noted at Dunnville and Wainfleet. It differed from the return of water to the canal that existed along the Thorold–Merritton–St Catharines industrial strip.

The mills at Welland were harassed by low water conditions for over twenty years, and the supply of water for purposes other than canal navigation gave rise to a significant conflict situation as production and wages were each lost when the flow of water to mills was disrupted. Despite urgent appeals to the canal authorities and to the government by industrialists and labour groups to retain the diversions to mills, the government finally prohibited the use of the canal as a source of water power.

Ironically, this prohibition fostered technological progress at the mills. When the town obtained its first steam mill in 1877, 'the engine is rated at 35 horse power and the boiler at 40 ... Due to the contemplated withdrawal of water, the town would have been actually without gristing or flouring facilities unless for the substitution of steam' (Sayles 1963). When the former Cook's mill was acquired in 1878 to construct the Third Canal, the mill relocated downstream next to Main Street on the southern bank of the Welland River. Purchased by the Brown brothers and known by its location and machinery as the Aqueduct Roller Mills, the mill had necessarily to change to steam power and its burr stones were replaced with roller wheels that could grind a finer grade of flour. Thoroughly modernized on its new site, the mill exported fine flour from its riverside wharf to the Montreal market via the connecting lock to the canal.

These two mills provide excellent examples of the transition from water power to steam power, and testify to the late date for this technological advance. There are also the interesting points in each instance of earlier locations made suitable by the canal being disrupted by later canal works, and then of the disadvantage of enforced relocation being used to modernize and increase output. Land acquisition, though often regarded as worrying for its enforced disruption of the status quo, is not always as unfavourable as it might at first seem!

Other industrial activities included a grain buyer who purchased from local farmers and erected a warehouse at the station for the receipt, storage, and shipment of grain. Further, a wool and carding mill located next to the Seeley and Dunlop mill used the same raceway as its neighbour for water power.

For a time Welland's leading industry, the Welland Iron Works, founded in 1862 by Mathew Beatty close to the aqueduct next to the river on North Main (now Niagara) Street, moved to the canalside location where Welmet Industries later located. By 1887 Beatty and Sons had diversified from the production of agricultural implements into the manufacture of mill gearing, castings, and contractors' machinery such as road scrapers, hoists, steam dredges, scows, pumps, boilers and steam ditchers, derricks, and shovels. The company served an area from Quebec to the Western Territories, a firm indication of the extensive marketing area available to industry that located on the canal banks.

By 1907 their plant with concrete and steel buildings covered 14 acres (5.7 ha) on the canal banks. It had a total floor space of 80,000 square feet (7432 m^2) and employed 200 skilled workers (Nichols 1907b). The company manufactured steam machinery for the shipbuilding industry and marine construction, which included the first all-steel dredge to be constructed in Canada, the *City of Toronto*. Tugs were designed and built, and also dredges for use in the construction of the Third and Fourth Welland Canals. One record notes that 'the Welland Iron Works, already the most prominent industry of the town, and with the greatest number of employees, is rapidly extending its business' (*History of the County of Welland* 1887).

The first steam tug to ply the Welland River is reputed to have been built in Welland during the late 1850s. Other small vessels constructed at wharves along the west bank of the canal or in embayments on the river included the hulls of steam tugs and barges for use on the Welland River. Produced by small-scale local craftsmen, they indicate an accumulation of economic activity in and close to the centre of Welland. Timber rafted or poled down the Welland River provided the raw material for the shipbuilding enterprise that included steam barges.

A considerable degree of regional industrial linkage was involved. When the first steam tug to operate on the Welland River was constructed, the hull was built in Welland and the boiler and the engine were manufactured at Chippawa. In 1883, for a tug required by a Port Colborne businessman, the keel was laid at Welland, the contract to build her went to Port Robinson, the boiler was manufactured in Welland, and the tug was built on the river at Welland next to a planing works.

Metal fabrication was also undertaken by other companies at this time. Robertson Bros. began business as machinists and foundrymen at the aqueduct site west of the canal in 1879. Eight years later, they were involved in the manufacture and repair of agricultural machinery and contractors' plant. There were also the Welland Boiler Works, which supplied the above-mentioned steam tug, the Welland Pump Factory established in 1870 with a specialty of rubber chain buckets, and in the 1880s the Welland Iron and Brass foundry, formed to build steam winches for the canal vessels, scows, dredges, and barges.

The Frost-Wire Fence Company of Cleveland, Ohio, established in Welland during 1898 to produce fencing of coiled, high-carbon spring wire bound to horizontal wires by a patented wedge lock; with steel-framed wire gates it was used extensively to bound private properties including farms and railways. Output increased rapidly, reaching 750 miles (1207.0 km) of fence by 1901. At first located in an old church, the factory in 1901 moved into a two-storey building next to the Grand Trunk and Wabash tracks and station at the east end of Main Street.

After about three years the company proposed to move to Hamilton, claiming superior shipping facilities and lower shipping costs. To remain they sought tax concessions and a new building, but as 'bonusing' was not then pursued by Welland, this plant moved to Hamilton when that city offered greater inducements. Its buildings were taken over by Quality Beds, incorporated in 1904. The two-storey brick building employed 25 to 50 workmen by 1907, and about 3000 brass beds a year were produced.

A different type of activity included three steam planing, saw, and sash mills to meet local demands. One of them, the Welland Planing Mills, established in the 1880s, occupied a two-storey brick building between West Main Street and the bridge across the Welland River. In 1907 the ground floor was used 'for the manufacture of moulding, sheathing, man-tle, flooring and siding, and porch and stair work. The upper floor is devoted to the manufacture of sash, doors and blinds' (Nichols 1907).

With an abundance of clay in the vicinity, it is not surprising that a brick and tile works, with two brick kilns and a drain tile kiln, located on some 9 acres (3.6 ha) of land west of the canal. Established about 1855 by Thaldeus W. Hooker to manufacture bricks for the new courthouse, it provided the red stock building bricks used in most of the building blocks constructed in Welland's business centre and elsewhere in the nearby countryside. About five hundred cords of wood were required to produce an annual output of a million bricks and 100,000 drain tiles.

This use of local clay deposits provided Welland and its environs with a visual emphasis in its buildings that contrasted with the use of limestone in the vicinity of the Niagara Escarpment to the north at Thorold–St Catharines, and the Onondaga Escarpment to the south at Port Colborne. In a direct link with the canal system, the brick kilns fired by wood pro-vided one important market for the cordwood hauled along the Feeder Canal. The bricks might also conveniently be exported by canal, river, or feeder to their destination.

Despite these various developments, Welland by the turn of the century could still be described as a 'farm town, drawing much farm trade, and depending on that for the regular source of business' (Sears and Sawle 1902). A town hall at the junction of Division and King Streets facing the canal opened in 1901, and Welland's service centrality through its county status, its prime location on the navigation system, and its commercial facilities were supplemented by industries that, 'although not extensive, included a varied line. Apart from Messrs. Beatty's big iron works, the Frost Fence Works and three flouring mills, the manufacturing is principally for local and distant trade. The lesser concerns include, a boiler works, a foundry, two machine shops, a saw mill, and sash and door factory, four lumber merchants, a brickyard for the manufacture of brick and drain tile, three coal dealers, two large printing offices and an acetylene gas machine factory.' Six good hotels and between seventy and eighty stores and shops were also recorded.

The Transition to Mass-Production Industries

Major industrial expansion started with the arrival of Plymouth Cordage in 1905 to produce binder twine for the expanding western market, but production soon expanded into all types and sizes of rope (Morison 1950; Plymouth Cordage 1924; Unyi 1995b). The American parent plant in Plymouth, Massachusetts, was distant from this market, a 25 per cent duty was imposed on goods manufactured in the United States when imported to Canada, and sisal could be imported as cheaply to Welland through the canal as to Plymouth. Welland with its shipping and rail facilities to the West lay much closer to the Prairie market.

By 1907 this company owned 180 acres (72.8 ha) of land, bounded on three sides by railways and on the fourth by the canal. The mills occupied a series of brick buildings, over four hundred hands were employed, and the same welfare facilities as at the parent plant were established. Unique for Canada at that date, they were necessary because Welland was small, and its housing accommodation was inadequate to cope with the large influx of foreign workers including Italians transferred from the parent plant. Incentives were offered that included the powerful guarantee of both a job and housing.

The first industrial nurse in Canada was employed in the plant, and housing built on a tract of land next to the plant was provided at a low rental with maintenance-free living to workers in the plant. In 1907 the company had a boarding house and eighty-eight tenements. These semi-detached homes had plenty of space around each dwelling and, located on straight roads with the unimaginative names of First, Second, and Third

Streets, their setback from the road was substantial. The healthy exercise and cost savings of garden production were encouraged by a fall fair with prizes for the best flowers and vegetables. These houses remained company property until 1947, when the company adopted a policy of sale to the tenants. The company itself lasted in Welland until 1969, when the plant and company offices were demolished and the Welland General Hospital constructed.

A clubhouse, Plymouth Hall, built in 1908, provided a dance hall, billiards room, recreational programs, a kindergarten class, and courses in cooking and sewing. A skating rink was included and nearby was an athletic field with a grandstand and dressing room. There was an annual picnic, and Plymouth Cordage was the first company in Ontario to form a successful Credit Union (Unyi 1945). The equivalent of the English benevolent garden city, as at Bournville or Port Sunlight, had come to Canada through this prestigious development at Welland.

In 1906 Canada Forge from Titus, Pennsylvania, came to Welland and extended the industrial base into heavy industry with a range of new products. Frame buildings were constructed on a 4.5-acre (1.8-ha) site served by the Michigan Central Railway. Steel precision forgings were produced for mining, hydroelectric power developments, locomotives, cement mills, bridge companies, and the growing auto industry in St Catharines. In 1912 this company merged with a company from Guelph and Canadian Billings and Spencer of Hartford, Connecticut, which had started the production of steel castings in Welland during 1907. Using natural gas in the forging process, the plant on the lines of the Wabash and Grand Trunk Railways produced drop forgings in copper, bronze, iron, and steel for agricultural implements, railways, and mines. The combined operation turned out precision forgings of great variation in size and composition, ranging from flanges to precision tools for customers that included the developing auto industry.

Another company incorporated in 1906 started production in a frame building with 24,000 square feet (2296 m^2) of space near the grounds of Canada Forge. Known as the Supreme Heating Company, it produced stoves, furnaces, ranges, and boilers, patented for economy and to produce the maximum amount of heat from the use of wood and coal (Nichols 1907b).

The Electro-Metallurgical Company from Albany, New York, followed in 1907 to produce ferro-alloys and carbon electrodes for the steel and foundry industries and was Ontario's second largest hydroelectric contract. An American company started to refine cobalt and other ores in

1912; using the same method, Canada Zinc opened Canada's first zinc reduction plant in the former Quality Bed factory. This company had a dock in Crowland Township next to the canal south of the town. In 1914 Union Carbide erected a separate plant nearby on the same site to produce calcium carbide and metallurgical coke. (In 1922 this company acquired the assets of Electro-Metals.)

Dain Manufacturing from Iowa arrived on a canal site located even further south in 1909 to produce agricultural implements that included a new type of hay-handling equipment and spreaders for the Canadian market. An important locational factor was again the link by an American railway to the American parent plant. Dain, which gave its name to Dain City, was purchased by Deere and Company in 1911. (This name was changed to John Deere Manufacturing Company in 1918.) A broadening range of agricultural implements that included castings and lawn mowers were manufactured.

Agricultural links to the local area are expressed by the arrival of Vaughan Seeds in 1909. This company, which grew to become Canada's largest exporter of grass seed, was followed by Welland Seed, the Canadian branch of two companies from Dunkirk, New York, and then by the Goodwill and Sons canning factory.

Page-Hersey Iron Tube and Lead Company (Page-Hersey Tubes) arrived in 1909 as the branch of a plant at Guelph, but all the company's operations later located in Welland. Again located to the south of Welland, this mill (later Stelco) manufactured metal pipes up to the then distinctive diameter of 12 inches (30.5 cm). Water taken from the canal was used for cooling and pipe-testing purposes. Rail and water transportation made this location possible, as did hydroelectricity from The Falls. The success of this enterprise led to the addition of a second pipe mill in 1911. A competing works, Hamilton Tube of that city, was briefly successful.

The textile industry that had originated in 1851 with the water-powered factory of Ellenwoods Woollen and Carding Mills expanded when Bradshaw-Stradwick from Dover, New Jersey, set up its Welland plant in 1910 to make overalls and other work clothing to serve the Canadian market. National Textiles arrived next year to produce cotton overalls using raw materials shipped in from Quebec. Woods Manufacturing began operations on a 30-acre (12.1-ha) site as the Empire Cotton Mills of Montreal (Wabasso after 1959) in 1913. The largest employer of female labour in the community, its products included bag cloths, denims, towelling, and industrial sheeting as well as a variety of yarns, all processed from raw cotton. Although low paying, the textile industry with a female labour force added variation and diversification to the male labour force in the iron and steel industries.

Industrial Promotion

Industrial promotion was late to arrive but through its active encouragement and with the arrival of Plymouth Cordage, Welland moved into the modern industrial era. Between 1901 and 1911 the population of the town leaped almost threefold from 1863 to 5318. In addition to the Third Canal, the border tariff, hydroelectricity, and railways that so strikingly encouraged the expansion of American plants within Niagara's Canadian economy, emphasis must be laid on the industrial concessions applied by each municipality for its own supposed advantage and in competition for growth with its neighbours.

At Welland, boosterism was promoted after the arrival of Byron J. (Bide) McCormick from Port Huron, Michigan. This individual directed the industrial growth of the town from 1905 until the First World War, and together with W.M. German, the federal member of parliament for Welland County, provided the initiatives and the entrepreneurship that attracted Plymouth Cordage, Dain Manufacturing, Page-Hersey, and other companies. Boosterism also brought in other plants that either never produced or were short-lived, including motor manufacturing companies and a company to produce bags for the Canadian wheat and cement markets (Young 1976).

A bylaw of 1905 provided a fixed assessment from the town council for twenty years. All optioned lands were to be brought within the municipal boundary and provided with municipal services to their properties. In addition, the Department of Railways and Canals provided a turning basin, docks, and warehouse facilities next to the canal, and the Ontario Power Company constructed a power transmission line from Niagara Falls. Most industries sought and usually obtained some preferential consideration, either to locate or remain in the town. Electro-Metals obtained a fixed assessment of $10,000 for ten years; the loss to Hamilton of Frost Wire Fence has been noted previously. It was an era of gross interurban competition during which industries, through economic incentives and irrespective of their favourable or adverse impact on the character of the urban environment, received special recognition when they located in a specified community.

McCormick in 1913 stated that 'Welland will grant reasonable concessions in exemption from taxation, water, sewers, etc., to companies desiring to locate here ... No other manufacturing centre can offer inducements so alluring or commercial advantage so powerful.' These concessions generally included a free site, a cash bonus, or some combination of the two. There could be a fixed assessment for ten years, the provision of city water for manufacturing might reflect only the cost of pumping, or, if an enterprise was situated directly on the canal, a franchise for free water was available. In addition, sewers, sidewalks, roads, and other municipal services were extended to the new industries.

The obverse side of this municipal rush to expand under new industrial enterprises was high company profits versus low wages for the immigrant workers, the industrial pollution of air and water resources, the pressures of a sudden large-scale influx of immigrant workers and industrialization on the urban environment, and cultural antipathy between the largely English Protestant background of Welland society and the massive non-English, mostly Catholic, influx of 'foreigners.'

The boom of industry brought in an influx of men and their families to settle near the new plants. Mostly foreign-born, working-class labourers lacking in skills, with little or no capital, from alien cultures and often unable to speak English, the largely Hungarian, Italian, Polish, and Ukrainian workers nevertheless required the provision of homes, churches, schools, shops, clubs, and health and welfare services. While the factories were provided with free improvements out of taxes and their profits exported abroad, mostly to the United States, in the town and its residential areas there was a dearth of municipal revenue to provide even the minimum range of expected amenities such as parks and open space, clubs, and community halls.

As described by Sayles (1963), 'The industrial scene was distressing. Jobs had to be bought, wages were low, living conditions were bad. Poverty was everywhere. Many of the "non-Anglo-Saxon" residents had been in Canada only for a short time. Few of them had any contact with Anglo-Saxons except as victims of exploitation.' The new immigrant, as elsewhere in Canada and along the canal, 'was faced with the necessity of finding employment and lodging for himself and his family in a new land and an environment largely hostile to him. This assimilation for both the original residents and the stranger proved a formidable task' (Young 1976).

When in 1910 another 'car-load of foreigners' was brought into the town by the Grand Trunk Railway to work in the steel plants, the Welland *Telegraph* stated that 'Welland's foreign population seems to be keeping pace with the growth of the town.' Land promotion, in the southern and eastern sections of the town extending south into Crowland Township, where the new industries located, was widespread. 'Agents whipped up one of their typical real estate promotion campaigns. Industrial Commissioner McCormick made a pretty penny killing two birds with one stone. He brought factories to Welland and land-customers to his Welland Realty Company, selling $700,000 worth of real estate in five years' (Sayles 1963).

Farmland was purchased, subdivided, and sold. Real-estate advertisements appeared regularly under screaming headlines. Promotion reached its peak in 1912 when three real estate companies provided free rail excursions from Toronto to visit Welland. During another real-estate boom in 1913, rails and ties were even laid spuriously along Ontario Road to indicate that development was imminent. A platform was constructed and a large sign board presented a map of the supposed Roma Park subdivision. It showed imaginary locations for the canal, railways and factories. 'The docks will be here, the station there. The business section will be here, the hotel there and a theatre there' (Sayles 1963). The real estate companies, the auctioneers, and the land developers reaped their rewards as employment in the factories increased, wages multiplied, and the value of products shipped out by rail and canal increased. The other reality, however, was that 'For years ditches lined the streets and smoke blackened the air.'

As the immigrant workers had to struggle hard for reduced hours, better working conditions, improved wages, and some measure of job security, with industrial growth came trade unions and local associations, together with their demands for improved conditions. The labour movement had by 1913 formed the Welland Trade and Labour Council, and in early 1914 'foreign mobs' of unemployed demonstrated outside the town hall. Some men ejected from their boarding houses had been forced to sleep in the open, many families had not obtained work through the previous winter and families were existing with inadequate food, shelter, and clothing. These problems were found more with the unskilled labourer than the skilled workman, but agents continued to send immigrants to the city for work that did not exist. Work rather than charity was required, an issue resolved in 1914 when the First World War erupted.

As the town grew, its rural fringe in Crowland Township, which received most of the major industry, became known as 'urban Welland' or 'the factory section.' With accompanying working-class housing, it was created through industrial expansion and the availability of land for factories, but socially it involved the segregation of Welland's new immigrants. Earlier most of the mills, the services, and the accompanying population had been located close to the aqueduct, and the village centre had focused east of the canal along East Main Street. As the railways, the industries, and the extensive areas of open space along the canal lands exerted a new pull to the south and the west of the town, these localities became the prime areas of growth.

Their annexation, proposed by Welland, was opposed bitterly by Crowland Township, which desired incorporation as an industrial village. It had an industrial tax base and an internal social unity that included an integrated Polish-Slovakian Russian Lithuanian Society that provided sickness and death benefits and a Catholic parish. A Hungarian Club was started in 1913, an Italian-language newspaper was founded in 1914, and a Hungarian-language one in 1916. With the contentious conflict of attitude between the town and township over amalgamation that arose, the issue of amalgamation was postponed, and neither annexation nor a new municipality were then pursued.

The Industrial Advance of Welland

There can be no doubt about the sudden industrial transition at Welland. As stated by McCormick in 1913, 'In the past five years Welland has secured a new manufacturing industry on an average of every ninety days, making a total of twenty-five concerns to locate here during that time.' Of these companies, only two originated in Welland with local capital. Nine out of 14, including the largest companies, came from the United States. Two, Hamilton Tube and Chipman-Holton Knitting came from Hamilton; Canadian Automatic Transportation came from Toronto. More than being an integral part of the industrialization process in southern Ontario, Welland and the other canal communities were now part of the international, expanding north-eastern manufacturing belt that overrode the Canada–United States boundary at Niagara.

The industrial growth and the opportunities at Welland were impressive. McCormick in 1907 stated: 'In the past five years Welland has secured more Canadian branches of American manufacturing concerns than any other in the Dominion. The growth of industries and population is unparalleled.' And later in the same report, he noted: 'In the past four years WELLAND has shown a greater industrial increase than any other town or city in Canada.' Over the five-year period from 1904 to 1909, population had increased by over 300 per cent, the dutiable goods entered at the Port of Welland increased 3½ times and the free goods 10½ times. Welland's assessment increased about 350 per cent over the same period, its postal revenue more than doubled, and money orders increased more than fivefold.

After listing the industrial accessions to the town, McCormick (1913) noted: 'Each one of these concerns made a thorough examination of Canada and went through the same preliminary that you [as a potential new industrialist] are now experiencing. These above firms proved to themselves beyond all question of doubt that they could assemble their raw material here easier, manufacture their product cheaper and distribute their output quicker than at any other point in Canada. Many of the above-named firms refused cash bonuses from other towns and chose Welland for the commercial advantages it contained.' These advantages were then stated to be unequalled power facilities and cheap power, unrivalled rail facilities, lake shipping, cheap sites, and liberal concessions.

The power situation had changed rapidly from mill sites with water power, through coal-fired machinery, to the novel resource of hydroelectricity. By 1909 steam power was stated to be 'completely unknown' in Welland. The plants were run by electricity, but if coal was required, it came by rail from mines in Pennsylvania or Virginia and was sold at the price delivered at Black Rock, plus a cost that increased by distance: 40 cents to Welland, 50 cents to Hamilton, and 60 cents to Toronto. 'This rate ... puts Welland in a class by itself on rates for coal' (McCormick 1913). If a steamer was chartered, this would unload at Welland at a saving of 35 cents a ton on the above prices.

Welland had the great advantage of a location south of the Niagara Escarpment at a distance of only 12 miles (9.3 km) from the hydroelectric power companies at Niagara Falls, and an even shorter distance from 'the immense and phenomenally cheap' resource of the Dominion Power and Transmission Company at DeCew Falls. 'These companies quote rates in Welland lower than can be had in any other Canadian industrial centre. Welland's municipal lighting costs only $40 for each 2,000 candle power light all night, every night of the year. It is the cheapest lighting in Canada and power prices are on a par. The first electrical smelting works in Canada was located at Welland' (McCormick 1913). The cost of a 100 HP of electricity per year was stated in 1907 to be $15 in Welland and Niagara Falls, against $22.50 in Hamilton and Brantford, or $25 in Toronto. As business size and power consumption each increased, so too did the relative advantages of a location in Welland.

A further power advantage was the availability of natural gas for heating and lighting purposes. Two companies, Natural Gas and Provincial Natural Gas, were in competition for this business. In 1909 the cost at Welland was 12 cents for 15,000 feet (4572 m) per day, which increased to 20 cents in Hamilton and 26 cents in Guelph. As stated in the comparative evaluation of costs, 'Gas is the largest saving in Welland's favour, as our business requires much heat for forging, but for large users of power that would be the greater factor' (McCormick 1909).

The railway network was then described as 'a loadstone that draws many new industries.' Through operating rights on the lines of track owned by other companies, it provided direct access to the three Canadian transcontinental networks and into the even more substantial American system across the border bridges at the Niagara and the Detroit frontiers. Seven operating companies were now involved: Canadian Northern; Canadian Pacific; Grand Trunk; Michigan Central; Pere Marquette; Toronto, Hamilton and Buffalo; and Wabash.

This range, greater than at Hamilton or Toronto where only six companies were indicated, resulted in forty-six passenger trains stopping at one of the Welland stations every day. No switching charges between the lines of different companies were involved, whereas this would cost $2 a car in Niagara Falls and Hamilton, and $2.50 in Brantford, Guelph, and Toronto. Soft coal could be delivered in Welland at $2.50 a ton for a fifty-ton load, against $3 in Hamilton or $3.35 in Brantford and Toronto (McCormick 1909).

Railway rates fixed by the Board of Railway Commissioners for Canada were the same from Welland to the Pacific as from all other manufacturing points in Eastern Canada lying between the Detroit River and Montreal. Also, as elsewhere in Canada, switches or spurs were constructed by the manufacturing company. The railways then refunded a fixed sum for every car switched over the spur until the full amount was rebated. Welland's advantage was that it lay in the Buffalo switching area and enjoyed the Buffalo rate from the east and south.

Against these plentiful power and railway advantages, the ongoing importance of the canal was symbolized by Welland's municipal motto, 'Where Rails and Water Meet.' The southern length of the canal with its mooring facilities between Welland and Port Colborne was perceived by the town and presented in its publicity materials as Welland Harbour. Rates were cheaper than by rail for bulk cargoes inland to Lake Superior and downstream to tidewater and across the Atlantic. A daily freight service both ways is noted by McCormick in 1913, as is the fact that 'Welland holds the distinct position of being the only municipality in the entire Dominion of Canada owning and operating its own municipal wharf accessible by shippers and steamboat lines alike without charges of any kind.'

A further but independent aspect of the canal was its abundant supply of fresh water from Lake Erie to the factories and the town for industrial and domestic purposes. The rural vicinity was also extolled, including the two lakeshores and the rivers for relaxation, along with the agricultural setting on prime farmland that contributed to the services available in the town such as its banks, shops, and stores.

The proof of Welland's success lies in the number and size of the industries that soon arrived. By 1914 Greater Welland had developed a broad-ranging and specialized manufacturing structure along its railways and the canal, with interspersed ethnic residential areas between and close to the factories, and a core of commercial, county, municipal, and service activities along East Main Street. The town's specialized range of employment included the processing of primary metals, metal working in iron and steel, textiles, and activities associated with the building, equipping, repair, and servicing of ships, barges, and tugs on the adjacent waterways.

The Two Villages at the Deep Cut

Neither Port Robinson nor Allanburg could compete with the victorious industrial advance at Welland. Port Robinson, at its peak by mid-century when it was considered for selection as the county town, lost trade and centrality through not achieving this status. It mellowed through the period of the Second Canal as a local centre to meet the needs of its inhabitants and those living nearby in the rural areas, for the building of canal vessels in its shipyards, and as the link of declining importance between the canal and the Lower Welland River (map 5).

Wooden sailing ships, steamships, barges, and scows were constructed by the brothers John P. and James S. Abbey in the large ponded area that lay between the locks at the head of the canal cut to the Welland River and the main through line of the canal. A barquentine *E.S. Adams* constructed here traded to England in 1858 and returned safely. Other vessels served on the canal and plied in the trade on Lake Ontario, and yet others, including two paddle-wheel steamers for service on the Upper Lakes, were towed down the Welland River to Chippawa.

As tugs came into prominence along the canal, these and sundry other small vessels were produced around the 1870s by William Ross and his sons at a dry dock on the canal cut where it joined the Welland River. Described as the 'most prolific tug builder on the Welland Canal,' this firm constructed about twenty in its yard for service along the length of the waterway (Addis 1979). But the end of shipbuilding was close. The steamer *St Catharines*, built in 1874, was the last of the larger vessels to be constructed at Port Robinson (ibid.).

The Third Canal changed the situation. The lease for shipbuilding was terminated to permit the widening and deepening of the channel and less water was available to the shipbuilding areas. New locks were not constructed to connect with the Welland River, and the village was increasingly bypassed as larger vessels came into service on the canal. As described in 1887, the village 'at present contains four stores, three hotels, a post office, a telegraph office, a custom house, a dry dock, a flour mill and the various shops usually found in a village. The present appearance of Port Robinson indicates that it is a pleasant place doing a nice, quiet trade. There are some indications, however, that it has seen better days, and such is the case' (*History of the County of Welland* 1887).

The decline of the village and of its associations with the canal had started. The railway ended the steamboat services between Dunnville and Buffalo, and taverns and inns went out of service as tow boys and horses were replaced by tugs. This gradual process of attrition was hastened when a serious fire in 1888 destroyed a large portion of the business section, a hall, several shops, stores, groceries, and a bakery (Michael 1967). As later fires engulfed mills, a hotel, and other shops, the adverse circumstances of declining trade were intensified by such disasters, from which the village never recovered. In 1895 the office that collected canal tolls closed, and two years later the customs office moved to Welland. By this time the population had dwindled to about five hundred, and the business activities to a grist mill, two blacksmith shops, and two general stores.

Decline was not offset by industrial activity, which by the mid-1910s was limited to the Standard Steel Construction Company, Dominion Canners, and the Model Flour Mill. The former (later James United Steel then Ennis Steel) located in 1912 on 50 acres (20.2 ha) of land north of the village next to the canal and became the acknowledged pioneer in the peninsula for the fabrication and erection of structural-steel products that included steel buildings, bridges, and other welded construction works. This company from Pittsburg, Pennsylvania, may be perceived as the northern outpost of Welland's steel industry. Steel supplies were obtained first from Lackawanna, New York, by rail, and the canal provided a supply of water for cooling purposes.

Dominion Canners of Hamilton was smaller. Based on the products of local agriculture, this company occupied a 4-acre (1.6-ha) site close to Standard Steel Construction. Both industries were connected with railway sidings, both were located between the railway and the canal, and both could be served by either mode of transport.

Industry did not accrue to the north at Allanburg (map 10), which remained the smaller of the settlements located at the two ends of the Deep Cut. By 1887, 'Allanburgh is not the same thriving place it once was ... In 1850 the place contained about three hundred inhabitants, and boasted two grist mills, two saw mills, two woolen factories and a tannery. All these industries have ceased to exist. There are now in Allanburgh three stores, a post office, an office of the Great Northern Telegraph Company, and a number of shops' (*History of the County of Welland*). There were also two hotels, one in the village and the Black Horse Inn to the east on the Thorold Road.

Allanburg's association with industrial development was indirect. As the Pilkington Glass Works, Beaver Board, Ontario Paper Mills, and Coniagas Reduction occupied new sites south along the canal from Thorold, some houses were constructed in the village to accommodate their employees. The village centre lay closer than the centre of Thorold for local services such as shops and churches. Having lost its earlier independence but with this residential expansion, Allanburg may be viewed as a suburb on the fringe of Thorold.

An interesting point is that the banks of the Deep Cut between Allanburg and Port Robinson remained without industry. The canal had been initiated here and the area had the longest period of close association with all four canals, yet neither mills nor industry developed. This somewhat anomalous situation can be explained by the fact that the canal was here in a long, deep cutting. The banks, higher than the canal, had no access to water power. Nor did they provide the large, flat sites that were required by industry. The Deep Cut therefore became a divide on the canal system: to the north Thorold, with its advancing industrial base, and to south Port Robinson, with its industrial links to Welland. Regionally, Allanburg had links via Thorold, Merritton, and St Catharines across the Ontario Plain to Port Dalhousie and Lake Ontario, whereas the economic links of Port Robinson were south across the Erie Plain through Welland to Port Colborne and Lake Erie.

The peninsula was divided physically by the Niagara Escarpment, which separated the two plains, whereas the administrative division was between Lincoln County centred on St Catharines and Welland County centred on Welland. Economically, the canal communities overlapped the escarpment for purposes such as urban development and industrial achievement, and the Deep Cut now provided the dividing line between the northern and southern sections of the peninsula.

9

Across the Niagara Escarpment from Thorold to Merritton

At Thorold and Merritton, where the First and Second Canals crossed the Niagara Escarpment, the locks were most numerous and the greatest amount of water power of anywhere along the canal system was available. The urban landscape became one long series of mill races, mill ponds, and tail races. When through navigation followed the rerouted Third Canal after 1887, both centres lost their flow of traffic as this marine activity was transferred to the farmland on the eastern urban outskirts. Even so, the two communities retained their water-power capabilities from the Second Canal into the era of the Third Canal, and both communities continued to advance with new industrial undertakings.

Thorold's Industrial Strength

Thorold (see map 11) continued to grow from the versatile industrial base that has been introduced here (Kurchak and Lafferty 1973; Nichols 1907a; Town of Thorold 1950; Thorold Board of Trade 1953). One description states: 'From 1850 until 1888 were Thorold's brightest days. There was work for every laborer and mechanic; all the shops did a thriving business, and the factories found a ready market for all their wares' (Thompson 1897–8).

By the mid-1880s Thorold possessed 'four large flour mills, three of which are classed among the best in Canada. These are run on the Patent Roller process, and have an aggregate capacity of 800 barrels per day, and the product is exported chiefly to Europe and Newfoundland. The splendid woolen mills of the Thorold Woolen and Cotton Manufacturing Co. gives employment to a large number of hands in the manufacture of hosiery and woolen goods of all kinds. The Thorold Felt Goods Co., an incorporation for the manufacture of strong woolen felt for boots, horse blankets, paddings and linings of all kinds, where warmth and durability are required, also serves as a factor in the prosperity of the town. The Ontario Silver Co. ... manufacture ... silverware, nickel-plating, etc. In addition to these, there are foundries, machine shops, saw-mills, cabinet shops, and cement mills, all of which are doing a fair trade ... Thorold also possesses very fine and almost inexhaustible quarries, from which are obtained stone for building purposes, water-lime and cement' (Welland County Council 1886).

This industrial scene fluctuated considerably as the products, the companies, and their location each changed. Activities listed at 1880 included two grist mills and one each of a basket, bicycle, carriage, casket, cement, foundry, knitting, and pulp factory or mill. But defunct were a felt works, tannery, plough factory, sawmill, the silver plating of forks, spoons, and other silverware, two potteries, a cement mill, a chair and two furniture factories, a wagon factory, and a soap and candle factory.

Directories record this changing scene. In 1871 the record referred to the manufacture of flour, lumber, and cement. Ten years later these remain, but a large knitting factory was added. By 1887, in the *History of the County of Welland*, 'Thorold's principal industries are of the kind that benefit the rural population of Welland County as well as the town itself. Grain and wool, two of the staple products of the farm, are here made ready for the consumers of eastern countries, as well as of the distant parts of Canada.' The flour mills, the oldest in the town having been built in 1827 by George Keefer, and the Welland Mills, built by his son Jacob, had each been modernized. Rollers had now replaced stones, which permitted both a higher volume and a finer grade of flour.

Thorold Woolen and Cotton Manufacturing Company, described as an 'extensive factory ... well equipped ... [with] a capacity for completing about one hundred thousand dollars worth of goods annually, is one of the largest of its kind in Ontario.' In a different vein were the Ontario Silver Company, which had opened in 1884 and manufactured spoons, forks, and other types of silverware that were sold to the hardware and jewellery trade, and Thorold Foundry and Machine Shop established in 1859, which made mill gearing, shafting, general machinery, and contractors' plant. Two cement works, a flour mill, and a grist mill, together with a company that manufactured paints and shoe blacking, were also described.

By 1892 the record referred to flour, knitting, and sawmills, along with a cement works, foundry, and machine shops. By 1910 the *Ontario Gazetteer* indicated that Thorold still had a diversified economy, but the detail varied from the previous account as the directory noted the town as 'the seat of numerous manufactories including flour, paper, knitting and saw mills, foundries and machine shops, wood turning, asbestos steam pipe covering, cement works, 2 pulp mills ... Grain, flour, cement, woolen goods, stone, pulp etc. are shipped.'

The understanding of industrial change is complicated as many of the mills were multi-functional. A single upright shaft turned from the water wheel transmitted its consistent energy to the upper floors. Different sets of manufacturing activity, not necessarily under the same management, might operate at the different levels in their respective areas of floor space. For example, at Ward's four-storey mill by mid-century, turning and finishing were undertaken on the ground floor, tongue-and-groove joining followed on the second floor, carding was undertaken on the third floor, and a cloth dressing business was carried out on the upper floor.

The 1902 trade edition of the Thorold *Post* provides detail about the varied circumstances that then existed. The Williams Thorold Basket Factory, which had started in a small way in 1880, expanded in 1890 into a two-storey brick building at the corner of Ormond and Portland Streets, and then into nearby vacant lots and vacated buildings for the storage of timber and basket materials. Its production, closely related to the emergence of the Niagara Fruit Belt north of the Niagara Escarpment, approached two million one-quart berry baskets annually, plus an enormous production of larger-sized baskets as well as crates. The more traditional but also rural-based activity of the Fife Flour and Grist Mills, formerly run by James Norris and then by Monro and Roantree by the 1900s, had the capacity to manufacture a thousand barrels of flour a week.

Stuart's Machine Shop and Foundry (Manson Manufacturing after 1906) occupied a site that extended from the Second Canal to Ormond Street. In operation from 1865, in addition to the traditional blacksmith's forge it included a three-storey machine and pattern storage building, a moulding and casting factory, and a further two-storey building also devoted to the storage of patterns. Contractors' plant including steam drills, pulleys, derricks, and hoisting engines together with mill machinery of all kinds were produced. Industrial plants equipped with machines from this plant included the Riordon Paper Mills in Merritton and the Whitman Barnes manufacturing plant at St Catharines. As has been indicated for shipbuilding, a series of industrial enterprises relying on the products of others had emerged along the canal system.

Another activity was a carriage-maker and blacksmith at the corner of Clare and Ormond Streets. Production included a combination wagon that could be converted from a two- or three-seater wagon to a fruit-carrying vehicle. This enterprise also manufactured all types of sleighs and carriages, and had the ability to construct 'any kind of vehicle, from a wheelbarrow to a modern tally-ho coach, and this from the raw material to the finished product' (Nichols 1907b).

Another multi-purpose enterprise, J.M. Howell on Front Street next to Lock 23 on the Second Canal, had taken over the earlier business of Joseph Battle. The operations included the supply of hard and soft coal, sawed or split hard wood, and coal supplies to the blacksmiths. Winter ice cut from the canal and its ponds was stored for summer use in a building with a storage capacity of 800 tons (725.8 tonnes). Howell also owned the only skating rink in the town.

The textile industry was represented by the Penman Knitting Mills, a branch of a company from Paris, Ontario, that produced men's winter underwear. This four-storey mill on John Street next to the canal had a two-storey annex and a three-storey warehouse. In the main building, a stock and parcelling room occupied the upper floor, with spinning and knitting on the third floor, the second floor was used for carding, and the ground area for finishing. Scouring and dyeing were undertaken in the annex. The wool, about 100,000 pounds (45,360 kg) annually, was imported from Australia and South Africa.

The earlier cement works established in 1841 by John Brown, taken over by John Battle and now the Thorold Cement Works, covered 44 acres (17.8 ha) east of the Second Canal. After quarrying, the material was taken to the burning kilns, then transferred by rail to the mill next to the canal. The mill could store 20,000 barrels, had the capacity to produce 100,000 barrels annually, and employed some 60 workmen. The product was much used for the basements of homes and agricultural buildings, for road and rail works, and for building stone. The mill was served by a 1.5-mile (2.4-km) length of spur line from the N.S.&T. railway.

These companies enjoyed an advantageous transportation situation. In addition to the possibility of water transportation via the canal, the N.S.&T. had traffic agreements with the Michigan Central, Wabash, and Canadian Pacific Railways. As this line both crossed the Niagara River into the United States and was associated with steamers to Toronto, it was able to reach all the important points in Canada, the United States and Europe through direct bills of lading.

One might also use Thorold to illustrate how the leading manufacturers and storekeepers often also controlled the civic and cultural affairs of a class-conscious and class-responsive society. Thus, David Battle on different occasions was a member of the town council and a member and chairman of both the Separate and the High School Boards. A. McClenchy, who ran a flour and feed business, was a member of the town council, then mayor, a member of the Public School Board, and treasurer of the Protection Hose Company. J.H. Wilson of Maple Leaf Bakery was a member then mayor of the town council and a member then chairman of the Public School Board. As elsewhere along the canal, the owners of the major businesses held the reins of social and political authority within their communities.

The Advance to Large-Scale Manufacturing Industry

By 1902 the leading wood-pulp enterprise was the Davy Pulp Mill founded in 1852 between Front Street and the canal. Another plant for the manufacture of mechanically ground wood pulp was the Thorold Pulp Company, established at the turn of the century. Both industries occupied two-storey mills. In each instance their ground floors were devoted to the preparatory processes and the upper level to the finishing of the dry pulp for shipment. Both had a daily output of about 10 tons (9.1 tonnes); 10 cords of 4-foot (1.8-m) spruce were required to produce this amount of pulp.

The advent of hydroelectricity from The Falls in conjunction with the town's transportation advantages and its location close to the frontier brought in heavy large-scale manufacturing industry to the south and east of the town. An early arrival was Coniagas Reduction, established in 1907 next to the Third Canal. This industry processed ore from the mines at Cobalt in Northern Ontario into silver bars. The name is derived from the abbreviated title of its products: Co, cobalt; ni, nickel; ag, silver; as, arsenic.

Montrose Paper (later Provincial Paper) in 1903 purchased the McDonagh sawmill and started to produce newsprint at Lock 23 on a single papermaking machine of small size. The mill of stone and brick was named after the birthplace at Montrose, Scotland, of the mill's founder, William Findlay. Water rights originally secured in 1837 by William Beatty to draw water from the canal, and later defined as 850 HP, were purchased. Pulp brought in from Northern Ontario by rail was used to produce newsprint, including in 1903 a special commemorative edition of the *Thorold Post*. The steady expansion of demand for paper led to the installation of a second machine in 1913. Fifty workmen were employed in 1907, and the annual output was then about 1500 tons (1361 tonnes) of high-grade book, writing, lithograph, bond, and cover paper. Grenville Board & Pulp followed in 1910, but later moved.

Ontario Paper Mills (later Quebec and Ontario Paper Company, now Quno), owned by the Chicago *Tribune*, was incorporated in 1912 to build newsprint-making facilities at Thorold, and started production at its mill on the east bank of the canal south of the escarpment the next year. Low-cost hydroelectric power from the Ontario Power Company at The Falls, not direct water power as before, was used to turn the grinders. This enterprise was one of the area's first large-scale users of this new power source. The mill was also constructed as a combined pulp and paper mill, whereas most other paper mills purchased the pulp they needed from other plants. Some company housing was located close to the mill, which had two machines making 31,000 tons (28,123 tonnes) of newsprint a year for the parent company (Wiegman 1953). As timber limits were not then available in Ontario, this basic raw material was obtained from Baie Comeau on the north shore of the St Lawrence in Quebec, then transported to the mill in Thorold by ship.

The potential of the site at Thorold for both the transportation of incoming raw materials and the dispatch of the finished product via the Welland Canal was an important locational factor. Docks and storage areas for the pulpwood and the coal from Pennsylvania and enclosed areas for the export of newsprint rolls to Chicago were built at the side of the canal. Vessels were used both to bring pulpwood to the mill and to carry the processed newsprint to Chicago. The canal, in addition to its use for water transportation, also provided water for use in the mill's industrial processes.

The Thorold Mill of Beaver Wood Fibre, a branch of the Buffalo, New York, plant, was added to this pulp and paper concentration in 1913. Again using hydroelectricity from The Falls and the canal as a source of water supply, this company located south of the town with docks next to the canal to unload coal and pulpwood. Spruce from northern Ontario was also brought in by rail to the company's own siding. Newsprint was produced for its market in Buffalo, gypsum paper wallboard for use in walls and partitions in the Canadian market, and paper board for tubes, partitions, and cartons supplied a regional market in southern Ontario.

Two further large specialized companies added their products to the industrial scene during the 1910s period of growth. Pilkington Bros. Glass Works arrived in 1913 as a branch of the company from St Helens, in Lancashire, England. This glass factory was established south along the canal from the Ontario Paper Company in Thorold Township to make sheet, plate, and wired glass. An important locational factor was the need to safeguard the company's drawn-cylinder patent in Canada (Barber 1960). The housing built by the company for its employees became known as Windle Village. As service facilities were sparse, children from this subdivision travelled across the canal to a school at Beaver Dams until the company provided this facility.

In 1914 Exolon (the subsidiary of a plant at Tonawanda on the American side of the Niagara River, named after its fused aluminum-oxide abrasive) established an electric-furnace plant with power from the Toronto Power Company at The Falls on an open site west of the town served by the N.S.&T. (Exolon 1984). It produced silicon carbide or carborundum to produce artificial abrasives and soon added aluminum oxide to its output. The availability of cheap hydroelectricity was an important factor. Raw materials such as petroleum coke and bauxite were brought in through the canal. The raw products were sent to Tonawanda to be crushed and graded into a granular form for use in the abrasive paper, refractory and metal industries. This industry may be viewed as an extension of the abrasives industry that had developed at Niagara Falls, Ontario, of which Norton's at Chippawa was an important component.

The Service Component of Expansion

Through these industrial activities and other smaller ventures not included in this account, Thorold developed as the canal's major industrial centre outside St Catharines until Welland took over this role. The community developed in close association with, and in immediate response to, the canal as a waterway and through the copious availability of water that could be used directly as a power resource and in many of the manufacturing processes that were attracted to the canal banks. This flow of water also provided a multiple asset for other activities. For example, when a system of waterworks was installed in 1907, the water was taken from the canal about one mile (1.6 km) above the town. Thorold's Pumping and Electric Light Station was also supplied with water from the canal.

The continuing process of industrial achievement that heightened towards the end of the nineteenth century and into the twentieth century brought with it many service elements, with their employment, that grew in the local economy. This expansion applied to shops and stores, churches and schools, the public library, lodges and societies, and to the development of professional and municipal services (Sanders 1919; Thorold and Beaverdams Historical Society 1897–8, 1932).

Banking facilities may be noted as some indication of the town's growing status and regional importance. The Quebec Bank and the Bank of Commerce dominated the scene from 1877, but from 1897 to 1912 the Bank of Quebec was the only banking facility in Thorold, suggesting that Thorold depended upon the superior facilities in St Catharines for major transactions. With industrial success, the Imperial Bank and the Royal Bank opened agencies in Thorold in 1912, and in 1914 the Merchants Bank and the Bank of Montreal also opened their doors in the community. Thorold thereby had the grand total of five banks, but the Quebec and Royal soon merged to reduce this competition.

The business centre that accommodated the banks and other service activities grew next to the canal along Front Street on its eastern bank. A fire in 1877 destroyed many buildings on the west side of this street. An east–west transit would reveal the tow path on the western side of the canal, then the canal and its water system lined by various mill establishments, backed by the main street (Front Street) with commercial activities on both sides. The railway depot was to the north in Merritton and, after the arrival of the high-line N.S.&T. interurban streetcar system, its bridge dominated the townscape scene and bounded the extent of commercial development to the south. The close grouping of the centrally located shops and stores contrasted markedly with the extended linear spacing of the mill and then the factory enterprises along the canal to the north and south of the business centre.

By 1886 housing had developed about equally on the two sides of the canal to the east and west of the business centre. Social stratification was reinforced by a mortgage policy that viewed the west side as low class and favoured the east side. To the south in Thorold Township towards Allanburg lay the prime area for industrial advance in the later period after sites in the vicinity of the canal and its locks across the escarpment had been filled.

This urban extent was curtailed abruptly to the east when the Third Canal was constructed. Only a few blocks were left for expansion between the town centre and the bridge across the canal. This barrier impact of the canal was aggravated severely after the early 1910s by the purchase of land and then by the more extensive works for the channel of the Fourth Canal and its Flight Locks. Thorold's Lakeview Cemetery, founded in 1886 in a scenic location at the edge of the escarpment, was now isolated from the residential areas of the town. The Third and then the Fourth Canals took over land east of Chapel Street, where street patterns and access routes in the town of Thorold were changed and several houses relocated to new sites (Orr 1978, 1979).

Further restrictions to urban growth contained the growing town to the south. Here canal and water barriers increasingly intervened after 1875. Previously the First and Second Canals had curved towards the village of Beaver Dams (now Beaverdams), but severe limitations on growth began when St Catharines passed a bylaw to construct a waterworks. This facility diverted water from the Second Canal at Allanburg along a 7-mile (11.3-km) channel to two storage reservoirs behind earthen dams across the Beaver Dams Creek. This water system expanded when a lease of 1897 permitted further water diversions from the now Third Canal to the DeCew hydroelectric power project through a second channel from Allanburg. The earlier municipal water-supply reservoirs were then expanded considerably.

This water system was necessary to replace the antiquated supply system to a town pump in the Market Square at St Catharines and to provide electricity for homes and industry before this was available from The Falls. Nevertheless, the waterworks restricted expansion at Thorold. The village of Beaver Dams became an island and its extent curtailed through the backup from the dam and the storage reservoirs. It was physically confined to the south by these events.

Industry, locating along the east side of the canal south from Thorold and including the large new enterprises of Pilkington Glass, Beaver Wood Fibre, and Ontario Paper was not affected by these municipal and power water diversions. Their sites lay to the south of the water barrier that had been created, close to the sources of power production at The Falls and within the area of economical power distribution that increased in cost with distance, and also in a location where the sites could be served by rail spur lines from the N.S.&T. system.

On the other hand, the housing areas that developed in the vicinity of these industries suffered social and urban disadvantages. Thorold was boxed in to the east by the Third Canal and the works of the Fourth Canal, in part to the south by these two waterways and the earlier canal systems, and to the west by the Beaver Dams water complex. The low-income residential areas in and close to Windle Village that housed foreign immigrant labour with ethnic groups from southern and eastern Europe were separated spatially by the canal from the business centre of Thorold with its range of service and social facilities. When Thorold Township in 1914 passed bylaws to extend the N.S.& T. streetcar line to the Pilkington Glass Works, this line could not be constructed. It was possible to cross the Third Canal, but not the broad swathe of the works then under way for the Fourth Canal.

The residential localities south of Thorold in the area now known as

Thorold South became an ethnic and cultural outpost beyond the established urban canal community. Analogies include the Facer Street areas of St Catharines that lay beyond the Third Canal and the southward advance of Welland into Crowland Township.

Another boundary, this time administrative, lay to the north where the industrial continuity of the canal, its water power and mills, and the linkages and associations between Thorold and Merritton were divided by the combined municipal-county boundary across the centre of their mutual activities. To the north lay Lincoln County, Grantham Township, and Merritton, and to the south Welland County, Thorold, and Thorold Township. Thorold's urban associations were north along the canal, more than south towards its own county town of Welland. Thorold, physically and administratively had become an encased and confined community, more so than elsewhere along the canal.

The Mills of Merritton

The area centred on Merritton (Welland City until 1858) attracted a substantial array of manufacturing industry (Leeson 1974; Nichols 1907a; Steele 1970; Taylor 1989, 1990, 1992a). Against an earlier background of grist and saw mills at many of the locks, the later industrial commitment reflected cotton, paper, and chemical plants. Services concentrated in the town centre, which, like Thorold but smaller and with fewer functions, emerged in a compact form east of the canal. Merritton's process of urban growth really started in 1853 with the introduction of the main-line Great Western Railway. Its station and railway yards were sited east of the canal and north of the concession road (now Glendale Avenue) that crossed the canal's water system. Here, between the road crossing and the railway, rose the new village of Merritton (map 9).

Soon after the arrival of the railway in 1855, the private Welland Canal Loan Company that had purchased over 500 acres (202.4 ha) of land on both sides of the canal for industrial purposes advertised the new potential that existed. 'The Great Western Railroad crosses the canal at this point, and has a station on a portion of the above lands, thus affording uninterrupted communication from the seaboard to the Western States and Canada, by land or water, throughout the year, altogether forming a combination of advantages to the Manufacturer, Miller, Mechanic or Man-of-business, not exceeded, if equalled in any other locality.'

The same advertisement offered special inducements to attract industry and encourage certain types of manufacturing. 'Any company who will invest not less than $100,000 in erecting durable stone buildings, and machinery for the purpose of manufacturing Iron, Cotton or Woolen Goods – to be in operation by the 1st January, 1857 – may have the choice of the best site, with all the water power, and grounds for the erection of buildings required for the establishment, free of rent or any other charge.' The Second Canal was now being promoted as an industrial catalyst in conjunction with the new advantage of a major railway route that crossed its line of channel.

As with Thorold, the full story of these factories is incomplete. Even so, an astonishing assortment of industrial enterprises arose along the Second Canal to process forest and farm products for domestic and industrial uses. The mills included sawmills devoted to the manufacture of wooden products such as spokes and wheels, shingles, oak buckets, staves, barrels, window and door sashes, and other products cut and shaped by lathes. The milling industry and the processing of grain led to distilleries, the grinding of wheat into flour, and enterprises such as biscuit factories. Textile mills supplied with wool from sheep farms south of the escarpment and tanneries using the hides of locally slaughtered animals also made use of local resources.

Cement and plaster mills used products that came from local stone quarries along the edge of the escarpment. Metal working included iron and brass foundries producing a variety of moulded, cast, and shaped goods and the manufacture of hammers, knives, and other sharp instruments. There were coopers and blacksmiths. Bent goods such as wagon wheels, runners for sleighs, and buggy tops are each recorded in advertisements that appeared in local newspapers. If such a thing as an Eaton's catalogue had then existed, most of its featured items would have been made in Merritton or nearby in the mills and factories of the industrial corridor from Thorold to St Catharines.

Some companies operated their own merchant fleets. Noah and Orson Phelps, who operated a large sawmill from about 1855, turned out some 5 million feet (1.5 million m) of lumber annually as well as shingles and building laths. Their vessels plied regularly on the Upper and Lower Lakes. Even more substantial was the partnership between James Norris and Sylvester Neelon over some twenty years from 1850. Their activities extended across the different communities at the northern end of the canal. Neelon erected the Lincoln Paper Mills at Merritton and also operated a prosperous shipping business and flour mills, and had extensive lumber interests.

It was an interlocking economy. With each enterprise depending on mill races from the canal, local wheat was made into flour that went into barrels made from local timber by local craftsmen and onto ships built locally by other craftsmen who used timber, fittings, and marine engines that had been manufactured locally. The vessels manned by local crews then transported the flour, timber, whiskey, cotton, and the machine products of the canal communities to distant lake and overseas markets. The metal and machine work incorporated in these vessels was undertaken in brass and iron foundries, and the size of these vessels was determined by the dimensions of the canal locks that had given rise to the industrial transition in the first place. A distinctive, highly integrated, regional economy had soon emerged along the canal waterway, its locks, and the hydraulic raceways created from the canal.

The mills' products and ownership and the sites where a particular product was manufactured changed with bewildering frequency as new and different sets of machinery were added to the building. One spoke factory switched to oaken buckets, then became a flour mill before becoming a paper company. Another building combined a spoke factory with a shaft-turning factory, and it then developed into producing parts for the early auto industry.

Any one lock could have more than one industry served from the same or different mill races, one example being a brewery, flour and biscuit mill, oil refinery, hammer mill, and wheel works at Lock 10. Every successive directory indicates different names and different products. Moving forward through the period from 1882 to 1892 the change in description at Merritton over this decade is from '2 cotton mills, a paper mill, saw and grist mills' to 'large paper, cotton, saw and flour mills, knitting works, a foundry and other manufacturing industries.' To the latter is added the information that 'cotton goods, paper, knitted goods and other products of the factories are shipped.'

Important activities at 1902 included the McLeary and McLean's Lumber and Planing Mills. The site covered some 5 acres (2 ha) close to the Thorold town line. Active until well into the twentieth century, this canal-side mill could saw 20,000 feet (6096 m) of lumber per day. The planing mill manufactured sash, frames, doors, mouldings, and other requirements from hard and soft wood. In addition to building supplies and interior furnishings, timber was supplied for building construction and to meet the requirements of canal and rail construction and maintenance projects. 'A large business in imported shingles and lumber is done, the appearance of the site ... being that of some of the great lumbering establishments of the Ottawa valley' (Board of Trade edition, Thorold *Post*, 1902).

With a grocery and hardware outlet in Thorold, James Wilson had an iron foundry and machine store in Merritton. A series of connected operations included a machine and pattern shops, a blacksmith shop, and a wire-rope manufactory. The output for a market across Canada included the manufacture of mill and railway castings, contractors' plant, machinists' tools, and wire rope for power and hoisting purposes as well as machinery repair.

The record of major concerns (Nichols 1907a) included the Merritton Brass Foundry, established at the turn of the century on Merritt Street, which produced brass castings of all types for engines and machines from very small to a weight of a 1000 pounds (453.6 kg). Phosphor bronze was a specialty. The Merritton foundry established in 1870 provided further evidence for this type of activity. The production of castings and general iron work included the manufacture of wire rope. The Canada Wheel Works, first established in 1868 by Saunders and Phelps on 1.5 acres (0.6 ha) of land at Lock 10 of the Second Canal, employed some sixty workers by 1907 to supply carriage factories across Canada with shafts, poles, and buggy and wagon rims.

The Transition to Large-Scale Industry

Some companies grew from old foundations, an example being Hayes-Dana, a later producer of auto parts that started life about 1865 in Merritton as Patterson, Fotheringham and Phelps. This company produced bent goods such as wagon wheels, cutter and bobsled runners, and buggy tops. It became Canada Wheel Works and ultimately, in 1922, the Hayes Wheel Company of Canada.

The multifaceted array of early industry also led gradually into manufacturing specialization that by the end of the century emphasized first cotton, then paper and carbide as an adjunct of the electrochemical industry. The transition was to mass production with higher volumes of employment in larger premises for more extensive international and national markets, rather than the earlier small-scale almost handcraft production.

The textile industry was represented first by Beaver Cotton, which located at the foot of the escarpment about 1856; it became the Merritton Cotton Mill, and later until 1920 the site for Independent Rubber for the manufacture of rubber footwear. The Canadian Coloured Cotton Mills under Gordon McKay, later Lybster Cotton, followed in 1861. A gazetteer of 1869 mentioned an employment of 200 hands, the combination of steam and water power that drove its 200 looms, and an output of 10,000 yards

(9144 m) of cotton cloth and 1000 pounds (453.6 kg) of cotton warp for shirtings and sheetings. The smaller Beaver Cotton, literally across the road, then employed eighty hands, and used 50 horse power.

In 1863 John Riordon (Riordan) established the St Catharines Paper Mills to manufacture wrapping paper. This was followed in 1867 by a second mill upstream where the sulphite pulping process was introduced. This mill was serviced by a rail spur that ran to the property along Ker Street from the Great Western main line, and then by the N.S.&T. freight system. The earliest raw materials were rags, cotton waste, straw, and waste paper. By 1880 newsprint for the Chicago *Tribune* contained 10 per cent each of chemical wood pulp and straw and 40 per cent each of rags and ground wood pulp. Wood pulp started to predominate after the sulphite pulping process was introduced at the Riordon Mills in 1875. This invention enabled the softwood resources of the Canadian Shield to be used for paper production, and the Riordon Mills became the pioneer Canadian producer of cheap paper, supplying the Toronto *Mail*, which John Riordon later owned, and also making substantial exports to the American urban markets to the south.

Another advance in the paper industry occurred when the Lincoln Pulp and Paper Mill under Sylvestor Neelon began in the 1870s on the hydraulic raceway at the junction of the current Oakdale and Hartzel Roads. It at first employed fifty hands in the manufacture of manilla paper, white print paper, grocer's bags, and flour sacks. This mill was followed in 1890 by the first Garden City Paper Mill, which located on a downstream island in the canal, approximately opposite Lincoln Avenue. Then came the Lybster Paper Mill, where the cotton mill of that name had been in 1908, and Interlake Tissue (later Kimberly Clark), next to the Thorold boundary, in 1912. Its paper machine, which started in 1913, produced 10 to 12 tons (9.1 to 10.9 tonnes) per day of lightweight wrapping paper, carbon-copy paper, fruit wrap, and toilet tissue.

These companies, extending along the canal and its raceway from Thorold through Merritton and into St Catharines, produced a varied array of paper products, ranging through newsprint and fine papers to specialty products like glassine, a nearly transparent glazed paper that resisted the passage of air and grease, and other greaseproof papers. Over the short period between 1863 and 1912, the Thorold-Merritton canal corridor had become an important paper-making area of national and world repute.

The locational factors that influenced this growth had changed remarkably. Water power from the canal was now augmented by cheap hydroelectric power from The Falls. The canal as a means of transport brought in coal to produce steam for drying the rolls of paper and sometimes the raw material, as with Ontario Paper, but the ubiquitous rail access through spur lines from the main routes was also an essential ingredient for success in most instances. The transition in production sources was from straw obtained from local grain, rags and waste paper from the surrounding area and cotton waste from nearby mills, to timber imported from the forested environments of northern Ontario and Quebec. The Second and Third Canals had nourished a key area in the historical development of the Canadian pulp and paper industry.

Another major industrial development was that in 1862 Canada Carbide built a plant on Thorold Road (now Oakdale Avenue). A series of power houses were constructed at Locks 8, 9, and 10 of the Second Canal to smelt pulverized limestone with coal and coke in the furnaces to produce calcium carbide. The works became known as the Willson Carbide Company after Thomas L. (Carbide) Willson, dubbed the father of the electrochemical industry in Canada and described as 'a chronic inventor gifted in both recognizing the potential of his discoveries and funding their development' (Carter 1985). In 1892 he discovered the electrical process for the bulk production of calcium carbide and its important by-product, acetylene gas. He returned to Canada in 1896 and set up the manufacturing process in Merritton that provided the bright white light used extensively in railway coaches, light buoys in navigation channels, and lamps for the first auto vehicles. The gas was also used in greenhouses for promoting growth and in small towns across Canada that had neither gasworks nor electrical plants. The Merritton operations started in 1896 on a 5-acre (2-ha) complex (sold to Shawinigan Chemicals in 1928, and later the offices for the Lincoln Board of Education). By 1907 the output of about 1100 tons (998 tonnes) was sold across Canada.

Merritton Expands

Not unnaturally, with so many industrial achievements, Merritton was incorporated as a village in 1874. A school was built in 1871 followed by a library; the town hall was erected in 1879. In 1887 a piped supply of water was introduced from a waterworks and filtration plant to the south in Thorold Township. The Merritton Fire Department, formed next year, became noted for its active role in community affairs in the days before service clubs became the vogue. Demonstrating the close relationship between industrial and community advance, four churches were constructed over the decade from 1888 to 1898 to cater to the religious and social needs of Anglican, Methodist, Catholic, and Presbyterian congregations.

As elsewhere, a spiral of growth had been created through the new potentials introduced by the canal and its water system. Visual unity to some extent prevailed, providing a link between industry, the town, and its setting through the use of a reddish stone from the nearby Walker Quarries at the edge of the escarpment for some of the more prestigious buildings. This indigenous use of the same stone was found in the town hall, the Riordon Mill, the two cotton mills, the first Lincoln Mill, and some of the more affluent homes.

This unity continued when the streets of the town were named. By the 1870s these included Lock (now Bradley), Merritt (now Hartzel) as the main street, a second Merritt Street to the southwest of the town, and Canal (now Hastings) Street next to the canal. The rural backcloth was recalled by the consistent naming of streets after trees: Cherry, Cedar, Spruce, Oak, and Elm to the north of the Great Western Railway; Walnut, Chestnut, and Almond closer to the centre; and Hazel and Pine to the south. No other canal community had such consistency, or such a rich reminder of the rural vicinity in its urban make-up.

Merritton's business centre developed where the Great Western Railway crossed the canal and around the town hall located on Merritt Street, the main through road between St Catharines and Thorold. The size of the centre was less than might reasonably be expected from the array of industrial developments that have been introduced, the explanation being that Merritton's population and services were sandwiched between Thorold upstream and St Catharines downstream. These external centres had the considerable advantage of an earlier start. Both held a higher position in the urban hierarchy and had a greater array of service functions in their respective business centres. This sandwiching between two larger communities was an urban disadvantage that did not exist elsewhere along the canal. The residential areas surrounded this nucleus, but growth was restricted along Merritton's western side by the canal and its wide and continuous system of reservoirs. Beyond the canal, there was a marked transition to the rural areas of Grantham Township. To the north lay the railway tracks, the freight yards, and the station. With the escarpment slope to the south, growth was pushed mainly in an eastern direction.

The trend was also strong for Thorold and Merritton to link together as a linear economic and physical continuity north–south along the canal system. Merritton and St Catharines also gradually converged across the longer distance of their intervening canal channel, a process comparable to that which linked Port Colborne and Humberstone, but more extensive because of the greater array of economic activities. It involved both residential and industrial developments.

An important element in this process of urban convergence included the Hydraulic Raceway that fed water east then north in an open ditch to St Catharines. For the First Canal this raceway supplied water drawn from the Merritton Locks section of the canal; for the Second Canal supplies were drawn from just above where the Hayes Dana plant later located. In each instance the water then flowed along the top of the bank, where supplies might be diverted to certain of the paper and carbide works that have been introduced. This process of urban aggrandizement, assisted materially by the N.S.&T. streetcar system at the turn of the century and then by dispersal through the automobile, was eventually to interconnect Thorold with Port Dalhousie along the line of the canal from the escarpment across the Ontario Plain to Lake Ontario.

The change from the direct use of waterwheels by industry to power for public use in Merritton began in 1890. Electricity was generated from a waterwheel just south of the Lybster Mill, then used to light the town hall and supply a limited amount of lighting for the streets, the churches, and several homes. Here was the private harbinger for what eventually became the Hydro Commission of the Town of Merritton, but it again provides an illustration of how the canal contributed to urban development in a variety of different ways.

The east–west and north–south layout of the roads did not accord with the grid for Grantham Township within which the village, then town, grew (map 9). Where crossed by the road allowance for this grid, Glendale Avenue curved awkwardly to the south away from its true alignment to create the clumsy Glendale Avenue–Merritt Street interchange. The reason for the curve of Glendale was that the water system and its industries took precedence and determined the location of the road. Glendale was obliged to swing south to cross a lock of the Second Canal, and to pass around the Lybster Cotton Mill and close to Beaver Cotton. Here also was a basin on the canal, necessary for vessels to await their turn to climb up through the locks. As this basin also had wharfage to load and unload raw materials and finished goods, the canal fixed the route for the only east–west highway crossing of the canal.

Access to Merritton was by only four roads, now Hartzel Road to the north, Merritt Street to the south, and Glendale Avenue to the east and west. As these routes to and from the outside world did not meet at the town centre, Merritton was oriented more to the canal and its railways than to the highway network, and the village did not have a rural hinterland. Here are further explanations for the small size of the business centre relative to the massive industrial growth in the vicinity.

The Abandoned Second Canal

The Second Canal remained open for a period of time, downstream to Thorold and upstream from Lake Ontario to St Catharines, but the intermediate central section through Thorold and Merritton to St Catharines across the Niagara Escarpment was abandoned for through navigation. As this channel was gradually removed from active use as a waterway, its associated features including the business centres of Thorold and Merritton became separated from their nourishing waterway. The replacement Third Canal route to the east across the rural fringe provided a modern substitute with wharves in a new location. The canal impact thereby remained ongoing, but in a new location with two urban responses: to the old as it continued into the future and to the new with its modern set of circumstances.

As both ends of the Second Canal remained open for a while and the water rights that had been so instrumental in promoting industrial development remained intact, the canal's traditional functions as a shipping and trading artery and a source of water continued for a short period. Disruption existed through the routing of shipping to the Third Canal, but the canal centres were sufficiently well ensconced in the landscape and had developed so many roots that their lives as towns continued under the new economic circumstances.

10

St Catharines: Industrial Giant of the Canal Communities

St Catharines continued to expand as the major urban-industrial centre on the canal during the railway era of development (map 11). By 1907 it was recorded: 'The building of this great waterway [the Welland Canal] has been responsible for the erection of the beautiful, busy, and prosperous city of St Catharines ... As a manufacturing city St Catharines is one of the most important in the Dominion, a fact due to the magnificent water power she has in the Welland Canal, as well as the electric power derived from DeCew Falls. There are at present more than one hundred industrial establishments in the city. Another factor in the city's wealth is the excellent facilities she has for freight and transportation' (Nichols 1907a). The digest that follows uses newspaper, board of trade, directory, census, and other published accounts. (It is based on Jackson and Wilson 1992, and the title for this chapter is from chapter 12 of that book.)

The Railway Era of Development

Connections for passenger and freight services expanded to all parts of Canada and the United States through the city's location on the main-line Great Western (later Grand Trunk) route that crossed both the Niagara and the Detroit borders into the United States and served east to Hamilton and Toronto. In addition, the Welland Railway that became the Welland division of the Grand Trunk Railway linked south along the canal to Port Colborne to provide St Catharines with connections to all the east–west routes that crossed the peninsula.

Because the crossover point of the Great Western and Welland Railways was at Merritton and not St Catharines, St Catharines was neither a major junction nor a substantial node on the railway system as were Hamilton and Toronto, which both grew faster than St Catharines during the latter half of the nineteenth century. For St Catharines, a mere way station by contrast, the major feature of its future destiny remained the canal for trade and its continued and increasing potential for water power as hydroelectricity was added to the scene.

The Great Western route lay neither within nor through the town, but south of the canal valley where no possibility existed of combining spur lines with sites having canalside facilities and access. The main-line station with its freight yards was isolated to the west of the canal on Western Hill. Access from the business centre and the industrial and residential areas to these rail facilities, in a period when the railways served by horse-drawn wagons dominated the transportation scene, had to contend with the intervening obstacle of the deep and steep-sided valley of Twelve Mile Creek. Irksome delays at its swing-bridge as slow-moving vessels passed through the canal were numerous and considerable until the Third Canal permitted a fixed highway bridge to be located over the former canal channel.

St Catharines had in effect been bypassed by the Great Western Railway. Its in-town railway with station and yards at Niagara Street, with the addition of spur lines to serve industry as required, was the Welland Railway. This situation prevailed until the arrival of the St Catharines and Niagara Central Railway, and then its successor the Niagara, St Catharines and Toronto (N.S.&T.). Both were important for interurban passenger traffic, for their freight-carrying capacity, and for exchanges with other railway systems. Again with spur lines that served industry, their urban significance is indicated by the rail-oriented industrial sites that arose north-east from the business centre during the period up to 1914. These contrasted with the earlier canal-oriented industrial sites along the valley and banks of the Second Canal, and with the virtual absence of new industrial activity next to the Great Western station and along its line of track.

Railways influenced agriculture and encouraged the emergence of the Niagara Fruit Belt. Previously the mainstay of farming around St Catharines had been mixed farming for domestic and local consumption, with wheat and other grains providing an input to the mills and a surplus for export via the canal system. As the railways steadily opened and extended settlement across the Prairies in both Canada and the United States, more specialized products related to the favoured climate and soils of the Niagara Peninsula could be introduced in the vicinity of St Catharines. Refrigeration and fast, reliable deliveries to distant cities aided a rapid transition to fruit orchards, vineyards, and the production of vegetables and soft fruits.

As the Niagara Fruit Belt emerged as a distinctive feature of the national economy, creameries, canneries, wineries, and the manufacture of bottles, crates, and baskets developed in St Catharines and nearby in Thorold. Some of the fresh fruit and vegetables were exported through the canal and its fast packet services to lakeside destinations such as Toronto; some of the less perishable products such as barrels of apples and canned or bottled products were dispatched by boat to England; and wineries such as Barnes built wharves next to their canalside premises on Twelve Mile Creek for the export of their products by boat.

The board of trade report for 1900 noted that over the previous decade the fruit and vegetable-canning industry had grown from one to six factories. The output that year, described as 'a record in volume of output, was 250,000 cases, equivalent to 18,000,000 pounds [8.2 million kg].' Seven years later the board reported that 'the collective fruit industries of this city and the surrounding country are a great, permanent and constantly increasing source of wealth and prosperity to St Catharines.' The plants included the L.M. Schenck Canning Company, which had started by canning tomatoes in a small building on its farm, then moved to a larger factory and warehouse on Russell Avenue. Simcoe Canning and Garden City Canning had grown in a few years to be the largest in the business in the Niagara District. The industry itself was first in the number of factories, second in the amount of wages paid and in the value of the products, and third in the cost of the raw materials and capital invested.

By 1907 Canadian Canners, with head offices in Hamilton, had amalgamated three earlier large canneries. Simcoe Canning with 80 employees used about 1200 to 1400 barrels of imported sugar annually. Ontario Pure Food, formerly Erie Preserving, with 140 employees, had an output of some one-and-a-half million cans annually. Schenck & Co., with 125 hands, produced nearly two million cans a year. This work was seasonal, and it employed a largely female labour force, including recent immigrants from eastern and southern Europe from the Buffalo area. Fruit was brought to these canneries on a daily basis, then packed and exported to Dominion, American, and world markets.

Cold storage and warehousing facilities grew in conjunction with such activities. The St Catharines Cold Storage and Forwarding Company, a joint stock company, handled the storage of their products from farms and orchards, then shipped to the market from their warehouse on Welland Avenue next to the Grand Trunk (former Welland) Railway. This company also supplied farmers with their spraying materials and supplies such as baskets at prices close to cost.

The wine industry was represented by the Ontario Grape Growing and Wine Growing Company, founded in 1873 by George Barnes, which later became Barnes Wines. The annual output by 1894 was some 75,000 gallons (283.9 litres), shipped to destinations throughout the peninsula and, from its cellars next to the Second Canal on Martindale Road, by lake vessel to Toronto, Montreal, and other lakeside destinations.

The railways through their passenger services also encouraged recreation and tourism, most markedly at Niagara Falls and at each of the two lake ports. At St Catharines this was reflected in the development of the city as a select, upper-income spa resort that became known as Saratoga of the North. Tourists arrived mostly by train from the American South, but the canal also played a role as some guests arrived from across Lake Ontario by boat. After stopping at Port Dalhousie, the vessel moved up Twelve Mile Creek to drop off its passengers at a wharf on the former Second Canal at the Welland Vale Works.

Stephenson House, which opened in 1855 on Yates Street on the crest of the canal bank, initiated this development. It commanded vistas across the canal to the Niagara Fruit Belt, and its gardens were landscaped down the slope to the canal. The Welland House at the corner of King and Ontario Streets, which followed next year, combined the functions of a sanatorium with those of a first-class hotel. The third of these resplendent institutions, the Springbank Hotel and Bathing Establishment, founded in 1865 on Yates Street, again enjoyed a canal setting on the top of the bank and landscaped gardens down to the canal.

An interesting feature of these hotels was that, as guests from the American South during the American Civil War came as much for safety as for health reasons, St Catharines became a temporary centre for conspirators, spies, and intrigue. (One American leader was later quoted as remarking, 'If one were choosing the location from which to conduct clandestine operations against the North, it would be hard to find a better place than St Catharines' [Jackson and Wilson 1992].) These American links through tourism in St Catharines and at Port Colborne should not be viewed in isolation. The connections between the two societies may later have contributed something towards the arrival of American branch-plant industries. St Catharines and Port Colborne were both well known, respected by American visitors, and favourably publicized by their advertisements and by the upper-class clientele that they attracted. Why then not locate a branch plant in one or other of these favoured locales with their advantageous living conditions?

Mills, Shipbuilding, and Ship Repair

By 1871 four large flour mills existed. Each operated by water power, worked for eight months of the year, and together employed over fifty hands. The decline in number from earlier circumstances can be attributed to roller operations rather than grind stones, to the change in preference to harder wheat grown on the Prairies, and to the mills at St Catharines being bypassed by vessels using the Third Canal.

It was a situation of mill redundancy for the specified purpose of flour milling, but then the reuse of the available site and often also the mill building and its water rights. This constructive reuse for some alternative purpose was not atypical. It took place at Port Dalhousie where the Lawrie flour mill became a rubber factory, and at Merritton where a cotton mill became a paper mill. It meant the continuation of industry in the same locale. This trend provided St Catharines with an advantage for industrial location, for the city had a greater number of earlier mills than any other canal community. Along with the associated infrastructure of services, houses, and transportation arrangements, together with a skilled labour force, this permitted industrial change and employment in the new activities.

As the disadvantages of a given site were often its outmoded, multi-storey buildings, heavily built with stone, inflexible to internal renovations, and limited in outdoor storage space, some mills were demolished and the site used for some new industrial activity. The reuse and adaptation of either earlier industrial buildings or their sites became a staunch feature of the city's industrial history. It was both the end of an earlier industrial era and the start of a new industrial experience. Examples of this industrial rehabilitation included a mill complex on Twelve Mile Creek taken over by the Welland Vale Company to manufacture agricultural implements. A mill on the raceway at the south end of Geneva Street was used for the production of incandescent lamps by the Packard Electric Company. Another mill at the confluence of the canal valley and Twelve Mile Creek became the home of the Kinleith Paper Company.

Sometimes the traditional multi-storey, stone-built mill and the modern factory with operations on one floor level existed next to each other on the same site. A case in point was the REO factory, which produced handcrafted automobiles and used a modern one-storey building built about 1905. Its large windows for light and ventilation contrasted with the now outmoded but still functionally useful four-storey mill building next door. 'The Red Mill's belts, pulleys, gears and shafts linked to a water wheel or turbine require, for efficiency, a tall, compact structure, whereas at the nearby REO factory, electrically powered machinery can be operated all on one level' (Taylor 1992b). One building typified industrial heritage, the other modernity.

Largely through progress at the Shickluna yards, shipbuilding and the repair of vessels remained a predominant activity, but the technological transitions to be weathered included the fact that steam tugs, propellors, barges, and scows were now in regular use on the canal. The changes were from sail to steam, from wooden to metal hulls, and from small to larger-sized vessels as the period of the Second Canal advanced to that of the Third Canal. By 1856, it was estimated that Shickluna had produced twenty-four vessels over a fourteen-year period, including four that were propelled by steam and four three-masted schooners (Warwick 1978). This figure was out of some 268 steamers, propellors, and schooners then sailing on the lakes, and it excluded those that had been overhauled or rebuilt. Shickluna's output was greater than at places like Niagara, Kingston, or Montreal with their vaunted facilities.

In the 1860s Shickluna added a dry dock to meet the rising demand for repair facilities. He could now repair vessels without disturbing the work of shipbuilding in the adjacent yard. Although the figure is variable and uncertain, by 1880 when he died, perhaps 250 hulls had been built for canal and lake service. An interesting feature of industrial linkage within the community is that the engines and steam machinery for the vessels he constructed came from local machine shops such as Towers, Oille, and Yale. At the height of the yard's importance in the early 1870s, the enterprise stretched for something like 0.6 miles (1 km) along the banks of the canal close to where Burgoyne Bridge now crosses the waterway. Included within the area were blacksmith shops, a planing mill, a sawmill, two dry docks, and spar and lumber yards. The employment of about 250 men included Blacks who had escaped from slavery in the United States. The shipyard was operated into the 1890s after Shickluna's death by his son Joseph, but this was then mostly for repair work.

A smaller but nevertheless important and successful yard was operated by Melancthon Simpson at Lock 5 on the Second Canal from the 1860s. 'For over 15 years he built every type of craft conceivable, from barges, tugs and sailing ships to larger steamboats and propellers. The most famous of his St Catharines hulls were the *Persia* and the *Asia*. The former was known for its record of long service; the latter for the abrupt end it met when it sank with more than 100 lives lost in Georgian Bay in 1882' (Addis 1979).

The Transition from Traditional to Modern Industries

Manufacturing based on the rural output included the family enterprise of the St Catharines Tannery on Chestnut Street, founded in 1865 by W.H. McCordick then taken over by his son F.C. McCordick in 1900. The tanning vats were on the ground floor of a four-storey building, and drying was on the upper floor. The annual output averaged 15,000 hides, with specialties that included tanned and rawhide lace leather. A further 10,000 hides a year were produced at the Wood brothers' tannery on Thorold Road (now Oakdale Avenue) next to the canal; again the vats were on the ground floor. This company and other leather manufacturers produced harnesses, whip lashes, laces, horse collars, saddles, valises, and trunks.

Kegs of amber ale, lager, and porter were exported by boat from the Taylor and Bate brewery with its extensive areas for storage and brewing and stables on the Second Canal just below its confluence with Twelve Mile Creek. Many lighter soft drinks such as carbonated, aerated, and fruit-flavoured waters that included ginger ale, sodas, and drinks in syphon bottles were produced by other smaller establishments.

One large bakery, the Nasmith Company of Toronto, in 1905 purchased the Shelly Bakery on Queen Street. The city alone required 2500 loaves a day to meet its demands and, with the additional specialty of wedding and anniversary cakes, by 1907 this bakery's employment included twenty-four bakers in addition to other employees and managerial staff. Interesting for its close association with Italian immigration and the resultant gradual change in local eating habits are the name and product of the A. Puccini company on St Paul Street. Macaroni and the lighter vermicelli were produced from wheat paste. The equipment included mixers, dough kneaders, and presses, the ingredients first being mixed then kneaded and shaped by machinery.

There was no underlying field of natural gas, as to the south in Welland County. The St Catharines and Welland Canal Gas Company, initiated in 1853, had an extensive manufacturing plant next to the canal south of St Paul Street. By 1907 the area from Port Dalhousie to Thorold was lighted from this plant. Ammonia, tar, and coke were its by-products. The company as a store sold gas stoves, mantles, and fixtures, it had over 50 miles (80.5 km) of gas mains, and manufactured some 22 million feet (6.7 million m) of gas every year. Electricity, then in the ascendancy, was used in the lamps at the canal locks and bridges.

Urban growth, accompanied by the expansion of the building trades and the now general use of brick for the structure of houses and most public, office, and industrial buildings, was materially satisfied mostly through the large Paxton and Bray brickyard on Queenston Street in the east end of the city. Here brick kilns on about 18 acres (7.3 ha) of clay land manufactured the common stock building brick used extensively by builders and contractors in the area. Limestone from the Niagara Escarpment south of the town was also quarried and used in certain of the more important local buildings, for road construction, as agricultural lime, and as an export product.

The most important local resource, however, remained that of water power. Electric power for industrial purposes and lighting the city was first provided directly from the locks of the Second Canal to two companies, the Cooke Electric Company on the former Shickluna site and the St Catharines Electric Light and Power Company, which had been incorporated in 1884 to generate electricity upstream at Lock 5. These two initiatives helped to perpetuate St Catharines's industrial lead during the intermediate period when the new power resource of electricity was coming into vogue, a point underlined by the fact that by 1900 St Catharines had seventy industrial establishments that were powered either directly from waterwheels or indirectly through the intervention of electricity. The two power companies merged in 1901 as the Lincoln Electric Light and Power Company. Taken over some four years later by the Cataract Power Company, and then in 1906 by the Dominion Power and Transmission Company, its penstocks, turbines, and power house were at DeCew Falls. The fall over the edge of the Niagara Escarpment provided a head of 269 feet (82 m), greater than that which existed at Niagara Falls, where power was generated by diverting water from above The Falls to deep pits along the bank.

The territory served from the power station at DeCew Falls extended south through Merritton and Thorold to Welland, along the line of the canal for lighting purposes, and to the N.S.&T. interurban streetcar system for power. One hydro line served Hamilton and another to St Catharines crossed the Second Canal close to the business centre by reinforced concrete hydro poles. Tapering to a height of 150 feet (45.7 m) to avoid the masts of ships in the canal below, they were the tallest in the world. Designed and patented locally by J.L. Weller, superintendent of the Welland Canal and president of the Concrete Pole Company of St Catharines, the poles were used extensively by power and trolley companies. This was the only company of its type in Canada.

A further industry related to the novelty of electricity was the Flexlume Sign Company, which arrived in the city in 1907 to manufacture illuminated electric signs for day and night use. The signs with raised glass letters, lighted by an inside bulb, were enclosed from the weather, and varied in size from a few inches to several feet. The plant on the hydraulic raceway also made pressed lead glass and was reputedly the only manufacturer in Canada for both products.

With these spinoffs from hydroelectricity, it is as well to remind ourselves about the importance of hydroelectric generation from DeCew as perceived in the circumstances of the day. 'The results achieved by this company [Dominion Power and Transmission] have been of the greatest value to manufacturers in this field.' They had reduced the power cost from steam by from 50 to 60 per cent and increased plant mechanical efficiency by from 35 to 40 per cent, a saving of from 25 to 30 per cent in the investment that would be required for the installation of individual plant (Nichols 1907a).

A new, powerful source of industrial advancement stemming from the canal had been introduced to the urban scene. Not to be forgotten are also that the water for DeCew was diverted from the canal at Allanburg – an early date for this hydroelectric power achievement. Further, after the water passing through the penstocks at DeCew reached the canal, it there added its volume to the canal's industrial potential and assisted towards the attraction of Kinleith Paper and McKinnon Industries (now General Motors) to the banks of the canal. Both companies had high demands for water. From Twelve Mile Creek this was clean water, not adulterated as from the Second Canal. The situation was that even more water passed through the hydro channel than was used at the locks for navigation by ships. The canal had added the function of a hydro conduit to its now substantial and growing list of achievements.

Of the two sawmills and two planing factories that existed by 1871, one still operated at a canal lock using the traditional mode of water power, two had made the transition to steam power and were therefore no longer tied to a canal location, and one used a combination of both steam and water power. Woodworking included the fitting out of vessels in the Shickluna and Simpson shipyards, together with the usual activities of producing building supplies and as well as doors, sashes, laths, window blinds, and other internal fittings and furnishings. The largest enterprise was probably George Wilson's planing mill and box factory, which moved to a large 8-acre (3.2-ha) site at the junction of Church and Niagara Streets after fire destroyed its earlier building.

Several companies manufactured farming implements and edge tools. In 1871 the largest site was at the Welland Vale Works where William Hamilton Merritt had initiated the canal. Here, at Lock 2 of the Second Canal,

Tuttle, Date & Company had been succeeded by Welland Vale Manufacturing, with over three hundred workers by 1907. Axes, edge tools of all types, forks, hoes, and rakes were each produced, along with bicycles by the turn of the century.

Other metal-working factories at this time included a foundry, boiler, and machine shop that employed fifty men, another that employed half this work force, and an agricultural implements factory with twenty employees. The machine shops of the Welland Railway added a further fourteen hands to these numbers. Several other companies, all smaller, ranged through sheet-metal works, tin and copper smiths, tinware manufacturers, and pump manufacturers. An urban mainstay was the blacksmith, with six being listed in 1871, together with four carriage- and wagon-makers. One of the latter employed sixty hands, and in 1892 the St Catharines *Standard* noted that one carriage factory had just exported two sulkies to Moscow.

A direct link with the canal is provided by the City Machine Shop and Foundry, a company established in 1896 by A. St John and Robert J. Black at the corner of Welland Avenue and Catherine Street. 'He does a large export trade, and does all kinds of repair work on machinery of all kinds, and manufactures sheet iron pipes. A specialty is made of repairs on boats passing through the Welland Canal' (Nichols 1907). Other metal-working companies of note included the Dominion Pump Works operated by R. Stephenson, described as 'the oldest continuous manufacturer in St Catharines' (Nichols 1907a). This factory, which occupied a two-and-a-half storey building on Queenston Street next to the hydraulic raceway, supplied farmers of the peninsula with suction pumps.

James Cunningham, established about 1887, manufactured machinery and castings in a foundry on St Paul Street. Its specialty, which employed a workforce of sixteen by 1907, was cylinder castings for gas engines. A small metal-working operation on the same street, again indicative of the vast range of output from the factories in St Catharines, was the manufacture of metal weather strips by the brothers W.C. and R.H. Legg. Some 20,000 feet (6096 m) of weather strip was produced annually, together with a large amount of cabinet work.

The R.H. Smith Company had a large saw works on St Paul Street close to the central offices and waiting rooms of the N.S.&T. where saws of all types were produced. National Spring and Wire arrived in 1906 as the subsidiary of a parent plant in Albion, Michigan. Its factory on Vine Street was equipped with machinery to coil, knot, and japan wire for markets across Canada.

The New Industries

The division between the established and new industrial developments is not always clear, as the old frequently moved into the new. An example is Patterson and Corbin at the west end of Queenston Street, which produced horse-drawn vehicles and then, when the streetcar arrived, the standard vestibuled car used on streetcar systems across Canada. Its three buildings employed seventy people, but despite its national acclaim this company closed suddenly for uncertain reasons in 1897.

By 1900 the largest manufacturing concern in the city was the Whitman & Barnes Manufacturing Company. Located at Locks 5 and 6 on the Second Canal on Thorold Road (now Oakdale Avenue) at Westchester, it employed 160 men in the manufacture of mowers, reapers, binders, straw and paper cutters, planers, and special knives of all kinds. Formerly Collinson, Burch & Company, established in 1870 with a workforce of twenty men, this company was consolidated with Whitman and Barnes of Akron, Ohio, ten years later. As stated by Nichols (1907a), 'Of all the manufacturing industries of which St Catharines is the possessor, and of which she is justly proud, there is none that has more of a national reputation, or whose products are used to a greater extent all over the world.'

Closely associated with the development of hydroelectricity, a site on the former St Catharines fairground and race track on Westchester Avenue in East St Catharines in 1905 became the home of Jenckes Machine of Quebec. Here a boiler shop and foundry employed some 135 workmen by 1907 to manufacture boilers, tanks, hoisting, crushing, and mining machinery, steam engines, and turbine waterwheels. This output included the penstocks required for the current hydroelectric schemes at DeCew Falls and Niagara Falls. (In 1926, this site became the home of Foster Wheeler.)

New electrical, automobile, and automobile-parts industries were added steadily to the industrial scene. Incorporated in 1894, Packard Electric moved next year to Neelon's grist mill on the upper hydraulic raceway at the corner of Race and Geneva Streets, and converted these premises to produce incandescent lamps. When the assembly of electrical motors was added, some were used to operate the locks on the Third Canal, which makes the nice point of industry operating on an industrial site created by the canal then serving its creator in a new capacity.

Transformers and electricity meters were added to the growing range of products, and even greater diversity when the grist mill next door was acquired in 1905 to add a new motor-car department. The Olds car, designed and manufactured on a production-line basis, was tested along local streets. When this department was sold to the Reo Motor Car Company in 1909 (Reo being an acronym for R.E. Olds), Packard stopped making lamps and electric motors to concentrate on transformers. In the meantime, Reo cars rolled off the production lines at the rate of about a hundred a month. A very successful promotional venture was the first trans-Canada journey by motor car, undertaken in 1912 from Halifax to Victoria. The automobile age had arrived, with St Catharines briefly to the fore with this new mode of transport as its capital moved to Detroit.

Automobile parts, with St Catharines later to become known as a General Motors town, were however to prevail rather than the assembly of motor vehicles. This saga of industrial advance began when a store on St Paul Street with four employees manufactured buggy hardware in a small building at the rear on the hydraulic raceway. As with other carriage and wagon manufacturers, the forests of the region supplied hickory for wheel spokes and rims, oak for carriage frames, and walnut and butterwood for the coach and bodywork. Power was obtained from the hydraulic raceway, which provided the energy to drive the saws, lathes, and other woodworking machinery.

A new company named McKinnon and Mitchell was formed in 1878 to produce buggy and wagon gears and a patented carriage dashboard. This item had an iron frame that could be straightened if it became bent. With leather on both sides, it protected the occupants from water and mud splashed up by the horses' hooves. Also with whip sockets and a place for the reins, it became an important status symbol. As this company advanced, it became in 1890 the McKinnon Dash and Metal Works. In 1901 the decision was made to transfer the expanding operations of its malleable iron foundry and ninety-five employees to a 43-acre (17.4-ha) site on Ontario Street north of the town next to the Second Canal. Water from the DeCew water and hydroelectric works, added to the flow in the canal, was used for cooling purposes in the foundry. A year later McKinnon Chain with plants in Tonawanda and St Catharines was formed. By 1913 McKinnon's in St Catharines had 650 employees. The company had become the largest employer of labour in any canal community and the Niagara Peninsula, and exerted strong dominance over the regional pattern of employment that increased as subsidiary items were produced by smaller firms.

A related auto-parts activity was the manufacture of vehicle wheels by the Woodburn Sarven Wheel Company, which started about 1870 on the east bank of the Second Canal. By 1900, when about half the wheels for the Canadian trade were manufactured here, steel and rubber tiring departments had been added. Its significance lay in the fact that, together with other enterprises, in 1922 it became Hayes Dana, the auto-parts manufacturer.

The lowest paper mill on the canal and an extension of the pulp-and-paper industry that had developed along the Second Canal between Thorold and Merritton, the Kinleith Paper Company, incorporated in 1899, had a plant that covered two acres (0.8 ha) where the Second Canal met Twelve Mile Creek. This company, with three large steam boilers in operation by 1907, used water from the hydraulic canal and electric power. The specialty was high-grade papers for books, writing, and lithographing and bond papers. The output of 15 tons (13.6 tonnes) a day was achieved by 100 to 110 employees.

Another conspicuous contributor to the industrial might of St Catharines, the textile industry, employed a largely female labour force in its production processes. George and Thomas Warren established Warren Knitwear on St Paul Street about 1877 to manufacture cardigans, gloves, and hosiery. The output by 1907 is described as 'Jersey knickers, sweaters, lacrosse shirts, and all kinds of fancy knitted goods' (Nichols 1907a). The factory then had twenty looms and some sixteen operators. The machines were located on the ground floor; the upper floor of the two-storey building was devoted to the shipping department and storerooms.

Canada Hair Cloth, run by James and Hugh McSloy, followed in 1884 on the raceway next to St Paul Street, manufacturing a range of hair cloth for use by tailors that included lining and padding for men's and women's suits and other garments, and also horse-hair sofa materials for railway carriages. The textile was woven with lengthwise threads of cotton and widthwise threads of horse then goat hair. By the turn of the century the number of mills in St Catharines producing this product had increased to three, which together turned out about 2500 yards (2286 m) of cloth every week.

Another successful venture was to use looms for the manufacture of carpets. The Empire Carpet Company, which arrived from Paris, Ontario, in 1893, had a dye house and employed about fifty people by the next year. Thirty looms produced union carpets of all grades, including wool carpets as a specialty product. This company soon relocated next to the canal, where it had three separate buildings. One devoted to scouring and bleaching contained sixteen vats; another had eighteen power looms. As the woven designs were often obtained from Philadelphia, the centre of carpet manufacturing in the United States, a further indication is provided of the American industrial connections that existed along the canal corridor. The wool used was imported mainly from England.

In 1899 at another mill, the Knitting Works of W.E. Channell, wool combed from fleece in the establishment was used to produce woollen socks. The female labour force was described as among the best paid 'girls' in the city. Another mill known as the St Catharines Clothing Manufacturing Company combined manufacturing with retailing, thereby reducing costs by eliminating the profits of the middlemen. Perfection Knitting, owned by B.J. Leubsdorf from about 1897 produced high-grade sweaters in a factory on King Street. A different type of textile activity was the manufacture and repair by Frank J. Sutton on William Street of tents, awnings, flags, horse and wagon coverings, and other closely woven cotton and linen duck goods.

Now certainly a major Canadian industrial city, St Catharines's population increased by 24.5 per cent between 1901 and 1911 to almost 12,500 people. As over the same period of growth the proportion of manufacturing employees had increased to 65.2 per cent of all employees, the title of industrial giant used for this chapter is certainly appropriate. Perspective on this growth is provided by the fact that this population increase in St Catharines was more than twice as large as the provincial average of 15.6 per cent. The rates at Welland were seven times as high but from a much lower base, and Port Colborne also exceeded the provincial rate. Thorold was lower at 14.9 per cent, Port Dalhousie lower again at 2.4 per cent, and Merritton at −2.3 per cent declined.

As described in the St Catharines Board of Trade Yearbook for 1913, 'Industrially St Catharines is very busy. Within a five-mile radius of the City Hall, which will include Port Dalhousie, Merritt, Thorold and DeCew Falls ... are located about eighty industries. Among these number are several of the largest paper, pulp, rubber, woolen, glass, iron and steelware manufactories of Canada ... Any article from a tiny lock to a monster marine boiler can be manufactured in St Catharines.' The 'tiny lock' referred to the recent arrival of Canadian Yale and Towne, 'one of the latest and larger acquisitions to the industries of our city ... The Canadian branch of the immense works at Stamford, Connecticut.' This new plant manufactured the range of Yale products from cylinder locks to padlocks and chain locks for the Canadian and export trade. 'Buildings have been erected and plant installed, designed to accommodate five-hundred workmen and already three-hundred men and women are employed' (St Catharines Board of Trade 1913). The reference to marine boilers is probably to Jenckes Machine Company.

As at Welland, municipal boosterism was again an important aspect of industrial advance. In 1872 a motion by the town council proposed 'the offer of bonuses as premiums for the encouragement of new manufactures within the town To Petition the Legislature for the passing of an Act to enable this corporation to offer bonuses.' In the same year a board of trade was formed to advertise the city and its advantages as a manufacturing centre extensively. A free site, free water for ten years, and a further financial bonus might each be added. A policy was formulated that any manufacturer established in the town employing at least twenty men and with at least $10,000 capital would be exempt from taxes for ten years. In keeping with this policy, exemptions were given to St Catharines Saw Works, the Dolphin Manufacturing Company, the St Catharines Stove Company, the St Catharines Wheel Company, Dominion Mills, Welland Vale, and the McKinnon Company inter alia. Again, anomalies arose when companies that had purchased their own land located next to those provided with free sites, or when companies that had received a bonus moved all or part of their operations to new centres.

The Expansion of Service Activities

As the city grew so too did its several service industries, and this growth in turn added substantial further employment to the occupational structure of the city. As one illustration, Bissonnette and Case, gentlemen's outfitters on St Paul Street, in 1907 occupied the three floors and the basement of a building with a 40-foot (12.2-m) frontage and a depth of 100 feet (30.5 m). The store employed six salesmen, the tailoring department had two cutters and eighteen other operatives were kept busy on work to order. One cigar store employed seven workmen; another in the same trade with over sixty years of tradition had an annual output of almost a quarter-million cigars.

After the county seat was transferred from Niagara (now Niagara-on-the-Lake) in 1862, the former town hall became the courthouse and the county offices for Lincoln County. The town then grew as the centre of administration for each of the county, the town, and Grantham Township. Advances in areas of cultural and social esteem included the Grand Opera House for theatre and music lovers and the production of newspapers.

The farmers' market in the basement of the courthouse soon extended into the adjacent square. Held on three days a week (Tuesdays, Thursdays, and Saturdays), here was the bustling centre of retailing activity. One turn-of-the-century report counted 756 vehicles in and around the square. Of these, 146 were selling firewood, 60 were loaded with corn and hay, and the remainder sold farm produce. The number of stall holders was estimated at 236.

A full-time police force existed from 1856, when three constables were hired. By 1872 there were ten men under two sergeants and a chief constable. The first central hall for the fire department was constructed in 1867 on the main street (St Paul). This was supplemented by substations, and fire boxes were added to the street scene across the city.

A federal post office, announced in 1882, was constructed on Queen Street at the corner of King. The post office was on the ground floor, with customs and excise on the floor above. A free public library was approved by a city bylaw in 1888 and granted an annual sum for its operation. The armouries on Lake Street opened in 1906 and became the home of the Lincoln and Welland Regiment. New churches to a total of about twenty by 1914 were established as the expanding population moved gradually outwards from the centre where the first churches were established to meet the demands of new immigrants and new religions.

The two small banks of 1858 had increased to five agencies by 1869. With mergers and growth, three major banks had established by the mid-1870s and were still the main monetary institutions by 1900: the Canadian Bank of Commerce, the Imperial Bank of Canada, and the Bank of Toronto. Supplemented at this date by loan and savings institutions, insurance and assurance companies and agents, five further chartered banks contributed to this financial cluster by 1913 (St Catharines Board of Trade 1913a).

Schools and education each expanded. Grantham Academy became a collegiate institute, an elite type of secondary or grammar school, in 1872. A business college was founded ten years later, then in 1889 Bishop Ridley College as a private residential school for boys. Each supplemented the public school system of ward schools and Roman Catholic separate schools.

A hospital to meet the requirements of the sick poor opened in 1865 in a house on Cherry Street. It tripled its number of beds, moved to a larger building, and then purchased a property on Queenston Street in 1870 where further expansion again took place. A direct canal association is reflected in the name St Catharines General and Marine Hospital. A nursing school, the first in Canada, opened at this hospital in 1874. New wings were added in 1874 and 1895, and the number of patients increased from 254 in 1895 to 953 by 1914. In addition, a sanatorium for the treatment of consumptives was approved by the city council in 1905. Through concern for the health of the populace, a garbage collection and disposal service was initiated four years later.

An increasing range of retail outlets, professional services, and offices of all types were added to the business centre. They expanded east along St

Paul Street, to a lesser extent at right angles along Ontario and James Streets, and then along King parallel to St Paul. The McLaren department store on St Paul Street, with roots in a general store founded in 1848 by R. and J. Woodruff, was by 1900 organized into some twenty departments according to the type of goods sold. Men, women, and children each had their clothing departments, and stock ranged from furniture through dry goods to shoes, glass, hosiery, and toys. The largest such enterprise in the Niagara district, its acre (0.4 ha) of floor space was packed into a three-storey building, over a hundred sales people were employed, elevators linked the floors, and stock warehouses at the rear of the building were connected to the main store by covered bridgeways. With the interurban streetcar outside the front doors, this was truly a regional service activity of substantial magnitude.

Important institutions included the YMCA, the Orphan's Home, and Montebello Park with its regimental band concerts as an active centre for outdoor leisure and recreational pursuits. As these service facilities enjoyed equal or greater esteem than elsewhere along the canal, and were accessible from the smaller centres by streetcar, St Catharines was in the lead role among the canal communities industrially, commercially, and through many cultural, social, and recreational dimensions of society. The role of service giant of the canal communities was coupled to that of industrial giant as a direct response to the strengths that manufacturing had created.

Urban Form and Structure

The retail outlets along St Paul Street, the main thoroughfare, ranged across the full spectrum of retailing, from bakers to butchers to bicycle dealers, and from grocers to glassware vendors. They included goods produced locally and imports from the United States and Europe. Land use was mixed. Buildings and building blocks might have different activities under the same roof, such that ground-floor shops might coexist with upstairs offices or residential units. On a par with all but the largest of Ontario's centres and larger than most, the city's nearest urban rivals were Buffalo to the east across the boundary in the United States, and Hamilton to the west at the head of Lake Ontario.

The population of the expanding city, 11,496 by 1911, remained essentially British. Those with a birthplace in Ontario formed 77.7 per cent of the total population, and a further 19.7 per cent had been born in the British Isles: mainly England, some in Scotland, and even fewer in Ireland. Among the remaining 3.6 per cent were a few persons from Russia, Italy,

Germany, and China. Except for the substantial number of 636 persons born in the United States, no other foreign-born group exceeded a dozen in number.

Much of the new growth of population, industries, housing, and services was either north towards Port Dalhousie or north-east to where the Welland Railway, the Third Canal, and then the Niagara Central Railway helped to open up new areas of flat land. The urban profile from south to north across the channel of the Second Canal remained farmland, then industry along the valley of the former Second Canal and along its crest following the hydraulic raceway at the rear of properties along St Paul–Queenston Streets. Then came commercial and institutional activities, interspersed with middle- and upper-income residential areas. More industry lay north and north-east in association with the lines of railway track and lower-income housing.

It was a pattern growing and dispersing from the pioneer settlement that had formed at the Ontario–St Paul Streets junction, but which had then expanded along the canal industrially and St Paul Street commercially. This pattern had a marked asymmetrical appearance in relation to the point of origin and as a response to the natural S-shaped system of river valleys that were followed by the First and Second Canals. Within this restricted extent, specialized land-use areas had started to emerge: water-powered industry along the canal and its banks; retailing along St Paul; entertainment and resort hotels along Ontario and Yates; offices along James; churches, schools, and institutional uses along Church; and the newer industries with their associated lines of track to the north-east. Ethnic segregation included the Armenian community close to the McKinnon and Welland Vale factories, the Black community in the vicinity of North Street, and immigrants from southern and eastern Europe in the Facer Street area.

At the wider urban scale, some preliminary thought was given to one combined and interlinked administrative community. As stated in the Thorold *Post*, 'There is no reason why our Corporation [St Catharines] should not control the whole, and do away with the multiplication of offices, and the rivalry that naturally exists between the several parts. It would be of advantage to the whole to be united in one place, and no disadvantage could ensue. The town of Thorold would make a thrifty and flourishing east end; the village of Merritton with its factories would form a substantial centre; and the city of St Catharines, with its wealth of industry and concentration of interests, would play an important part as a west end.'

Two High-Level Bridges

When the Second Canal from Thorold to St Catharines was closed to navigation at the end of the 1881 season, an opportunity existed to remedy the inadequate highway circumstances next to the city's business centre that had arisen because of the canal and its slow passage of vessels. Only two narrow bridges crossed the canal, one at Lock 2 at the western end of St Paul Street and the other at Lock 4 at approximately the extension of Geneva Street. Development to the west of Twelve Mile Creek in the area that included the main-line railway station and to the south towards the Niagara Escarpment were each inhibited by this restricted access.

A memorial of 1912 from the city to the minister of railways and canals requested permission to construct a high-level bridge, later the Burgoyne Bridge, over the Welland Canal, as well as a grant in aid of its construction. This city 'has always been deprived of convenient access to or communication with that section of the said city; for that reason the said portions of the City have never been populated and are practically in the same condition as they were forty or fifty years ago and although ... really nearer to the business part thereof the growth of the said City has extended in a Northerly and Northeasterly direction.'

Access to the Grand Trunk (former Great Western) station and its freight facilities, some 0.75 miles (1.2 km) from the main business section of the city, had to overcome a steep descent of over 80 feet (24.4 m) into the valley of Twelve Mile Creek to cross the Second Canal. The station was separated from the city by the canal and the streetcar was unable to reach this station. When the Burgoyne Bridge opened in late 1914, the span of 1236 feet (376.7 m) replaced the lower swing-bridge that previously had to suffice to cross the canal and provided the only access into the city from the west. The astounding point is that the city had had no direct access to its main-line railway station for some sixty years – a highly unusual situation, but one that emphasizes that the canal was to the fore in the decision-making process despite its sometimes severe adverse consequences for the city.

Earlier that year the Glen Ridge (later Glenridge) Bridge as an extension of Ontario Street had been opened to traffic. Its purpose was to cross the valley of the Second Canal and open up south St Catharines for residential development. The unidirectional growth to the north-east would be offset and balanced by new opportunities towards the Niagara Escarpment.

Both bridges were necessary for the development of the city, previously hemmed in to the south and to the west by the Second Canal. Even so, the construction of each bridge and its approaches also destroyed important commercial properties at the west end of St Paul Street, including banks, offices, and retail stores in the central commercial district. This destruction helped to focus commercial activity towards the market and city hall on James and King Streets, which reduced the asymmetrical form of the central business district and changed the character of the town that had emerged over ninety years of canal history.

PART 3

The Third and Fourth Canals
Reflect Community Advance from
1914 to the 1960s

11

The Achievement and Character of the Fourth Canal

The year 1932 witnessed the much-delayed opening of the Fourth Canal. The First World War and then government priorities for the works of Ontario Hydro at Queenston had intervened. Until the 1930s many small coal-burning vessels steamed through the canal, leaving raw materials and receiving finished goods from the industries that lined the canal banks. The canal was an integral part of the expanding industrial and municipal economies, and its vessels were supplied and serviced from onshore as was required to meet their marine needs.

After 1932, much was to change, neither immediately nor suddenly but gradually and imperceptibly. The new and larger vessels passed through without stopping and required less service attention. The channel was wider and more of a barrier than previously. On land, the period of railway dominance in the movement of people and goods was about to end under pressure from the motor vehicle and the construction of modern surfaced highways.

The elements for examination in this chapter include the canal alternatives that were considered before selecting a route for the Fourth Canal across the Niagara Peninsula, and then the character of its construction including the detail of its locks, bridges, the northern and southern canal reaches, and the planting schemes that added the new qualities of beautification and ornamentation to the canal works. The canal is then viewed in its essential function as a national transportation artery, with the through movement of its cargo tonnages increasing steadily until 1959 when the St Lawrence Seaway Authority took over from the Department of Railways and Canals as the operating agency.

Possible Routes for a Fourth Canal

Many alternative possibilities both within the Niagara Peninsula and beyond its confines were considered before a new route that used sections of the Third Canal and its facilities was approved for what is now officially the Fourth Canal. Known during its construction period as the Welland Ship Canal because of the extensive new dimensions that were incorporated into the design, the canal has subsequently taken on its correct numbered sequence in the progressive evolution of the Welland Canal system of waterways.

A prime issue in the location of the canal was how and where to accommodate the enormous pressures on transportation facilities of the need to move grain from the Prairies to the British Isles and western Europe. A possible American route was to Buffalo, the same distance from the West as Port Colborne, and then by rail to Boston or New York. Another American possibility was to use the Erie Canal, which by the 1900s was in the course of reconstruction and deepening to a depth of 14 feet (4.3 m). A Canadian possibility was to a port on Georgian Bay, then by rail to Montreal. There were many alternatives, by rail and water transport, through Canada or the United States, in various combinations of these price-competitive facilities by the different modes of transport and with each having different national implications.

The great advantage of an improved Welland Canal route was the perception that it would halve the cost of moving grain. 'If we do nothing towards improving our water-ways, in a few years hence the grain trade of the North West will be diverted from this country [Canada] to American channels. If on the other hand, we deepen and enlarge the Welland Canal, cheapening transportation from the West to the Ocean ... we will not only hold our own in the North West grain trade, but we will bring large quantities of American grain to the Port of Montreal, and the situation will be exactly reversed from what it is at present' (Richardson 1907). The First Canal of 1829 was Canada's answer to the Erie Canal. Eighty years later this trading rivalry between the two neighbouring nations persisted, and provided a strong and persistent argument for an improved Welland Canal.

Many surveys were undertaken and projects developed during the period from 1890 to 1910 for a ship canal between the Upper Great Lakes and the Atlantic. These appraisals involved serious consideration of a direct route from Georgian Bay along the French and Ottawa Rivers to Montreal, which would about halve the circuitous distance south from Lake Huron through Lake Erie to the Welland Canal and then north-east to Montreal.

This Georgian Bay route involved several lengths of river and lake navigation, several short lengths of canal around the rapids on the two river systems and locks for the rise and then the fall from the summit level that lay close to Georgian Bay. Difficulties included the costs of excavating through the hard rocks of the Canadian Shield, the uncertainty as to whether a canal could compete with the Grand Trunk Railway and its movement of grain from Midland through Belleville to Montreal, and the additional elevators and port facilities that would be required along this northern route. The Georgian Bay route was shorter than via the Great Lakes, but more lockage would be required and the lengths of confined movement in a winding river course were much longer than along the favoured and well-established St Lawrence route.

In 1901 estimates were prepared for the cost of a 14-foot (4.3-m) navigation channel via Georgian Bay with 16 feet (4.9 m) in the open reaches. To achieve a 20-foot (6.1-m) depth, the cost escalated threefold. A later study in 1904 required higher costs again to achieve a depth of 22 feet (6.7 m). A major advantage of the Georgian Bay Canal was that its route lay inland from the boundary with the United States at Niagara and along the St Lawrence River, but 'the tremendous cost of the undertaking and the time that would be consumed by vessels in passing through the locks were serious drawbacks to the accomplishment of this project' (Heisler 1973). Even so, a Georgian Bay Ship Canal remained a distinct possibility. It became an election promise that was made by the defeated Liberal party in the 1911 federal election, and it was advocated in a British defence report of 1912.

The eastward movement of grain was not the only fact to be considered. Other bulk cargoes such as iron ore and coal were also important, as were the many smaller loads that were shipped to and from the ports around the Great Lakes, which if bypassed would lead to the decline or the oblivion of their marine trade. A solely Canadian route along a new Ottawa alignment would have less merit than an international route that served both the American and the Canadian trade, and that also interconnected the separated water systems of the Upper and Lower Great Lakes.

Given these several considerations, by 1907 the question of how best to enlarge the Welland Canal became an important centre of attention. The important factor was now the size of the canal that had to be constructed. Following the opening of the downstream St Lawrence route in 1901 to vessels drawing 14 feet (4.3 m) of water, improvements had been initiated at the Port Colborne entrance to the Welland Canal. The harbour had been deepened to 22 feet (6.7 m), the government elevator had opened in 1908, and large breakwaters were added to improve entry to the harbour.

This modernization of the canal's infrastructure provided both a strong argument for rejecting a Georgian Bay route and powerful justification for retaining a Welland Canal through Port Colborne rather than on some new alignment across the Niagara Peninsula. Geographical inertia, as has been indicated for the canal communities, was again a powerful force. Throughout the history of the Welland Canal, it has always been easier and cheaper to retain certain of the existing facilities and sections of the old route than to start anew.

The routes considered in the Niagara area included ones from Lake Erie starting at Selkirk to Hamilton Harbour and from Port Maitland to Jordan Harbour, but neither could compete in cost with retaining Port Colborne with its harbour improvements and grain-handling facilities. A second consideration was that the largest type of vessel using the system had to be accommodated, a factor that required locks to a depth of at least 25 feet (7.6 m) and a length of 600 feet (182.9 m), together with a much wider and deeper channel.

Another possible route seriously considered was on the American side of the Niagara River following a LaSalle-Lewiston route. Costs were estimated in 1919 by Colonel J.G. Warren of the United States Army Engineers for canals with bottom widths of 200 and 300 feet (61 and 91.4 m). The Warren Report also considered, but did not recommend, a longer and more costly canal from Tonawanda via Lockport to Ollcott on Lake Ontario. As this would pass through an area that had been built up over recent years, the purchase costs for the rights-of-way and property would be prohibitive. This detail is of particular interest because it seems to end the demand expressed since the 1820s for an alternative, competing, and parallel canal across U.S. terrain (Clinton 1918; Hill 1918), a scheme further jeopardized by the later construction of the Robert Moses Niagara Power Project across the LaSalle-Lewiston route and leading to the comment by Robert Moses, Chairman of the Power Authority: 'It seems obvious that there will not be an All-American Canal since most of the St Lawrence River is located entirely in Canada. It also appears that if shipping tonnage exceeds the present capacity of the Welland Canal, additional locks will be provided rather than a new parallel canal or a separate canal on the American side' (Moses 1958).

The need for a speedier passage also meant that the number of locks had to be reduced. As argued in 1911 by J.L. Weller (Muntz 1997), the superintending canal engineer, 'The question of the lift of each lock is a most important one. Masters of large vessels have a great dislike to locks, as they are afraid of injury to either the lock or their vessel every time they enter one. Consequently, to make the canal a success it will be necessary to reduce the number of locks to a minimum, and reducing the number means increasing the lift of each lock' (Weller 1910). Although no locks of great lift had yet been constructed anywhere, the science of canal engineering had progressed. As Weller observed, 'There seems to be no reason why very much higher lifts than have been formerly used cannot be used with perfect safety.' His thoughts had advanced from locks with a lift of 50 feet (15.2 m) 'to locks with a lift of 165 ft [50.3 m] each. Two such locks would overcome the total rise between Lake Ontario and Lake Erie, about 326 feet [99.4 m].' A large quantity of water, four times that required by a canal with eight locks, would be required, but this diversion from the Niagara River was dismissed by the canal engineer to be of 'small moment ... as the interests of navigation are paramount and the total required is relatively small.'

This gargantuan project, not then accepted, re-emerged when superlocks were considered for the Fourth Canal in the 1960s. On the Fourth Canal, the number of locks was reduced from twenty-seven on the Third Canal to seven with a greater depth of water over the sills. This change would reduce the time of passage through the canal by half.

Route Selection and the Construction of the Fourth Canal

Weller deemed that several routes for the canal were practical: retaining the present canal from Port Colborne to about 3 miles (4.8 km) below Welland, then branching off towards Fonthill and following Fifteen Mile Creek to Lake Ontario; following the present canal to Allanburg then west of Thorold to Twelve Mile Creek and Port Dalhousie; and along a new route from Morgan's Point on Lake Erie 4 miles (6.4 km) west of Port Colborne to Twenty Mile Creek near Jordan (Weller 1910).

With regard to the present Third Canal, Weller stated, 'The lock portion from Port Dalhousie to Thorold could be closed entirely, and I may say that this portion is now showing signs of age and by the time the proposed canal is completed the cost of keeping it in repair will be very considerable and it will really be ready for the scrap heap.' A possible use for a power canal had been suggested, 'but I do not consider that this is at all practicable or advisable, as any water available can be used to better advantage and at smaller cost down the old [Second] canal, or by the Hamilton Cataract Company's plant west of Thorold [DeCew Falls].' The length of canal from Thorold to Port Colborne could be kept open at 'very small expense for local traffic.' By any of these routes, the time for completion was estimated at five years. By 1912, the route via Fifteen Mile Creek, the shortest line across the peninsula, was found by borings to be 'entirely unfitted for the location of a ship canal' because of quicksand encountered in several of the lock foundations (Weller 1912). The two practical routes were to Jordan Harbour or Ten Mile Creek, with the latter being recommended by Weller and eventually becoming the northern port of entry to the canal system.

Against such engineering assessments, the proposed new canal involved a widened version of the Third Canal along the southern reach from Lake Erie to the crest of the Niagara Escarpment at Thorold, and then a new route for the canal along Ten Mile Creek for the northern reach of the canal between the Niagara Escarpment and Lake Ontario (Cowan 1935; Grant 1932). As throughout the history of the canal, the route was determined on the basis of engineering calculations. The community implications of the different routes and the urban-regional advantages and disadvantages of one route over the other received no consideration.

The work on the new Fourth Canal (map 1) commenced in 1913, the expectation being that the waterway would be open for navigation in 1918. However, when the First World War intervened, the project was suspended in late 1916 because of labour and material shortages and greater priorities elsewhere. Work recommenced on a small scale in 1919, but largely as a relief scheme to help towards the employment of demobilized troops. Deemed more important because of the urgent need for power was the Queenston-Chippawa hydroelectric project on the Niagara River. As this also required a large manual labour force, needed similar types of materials such as cement, and made extensive use of machine equipment, the results were higher costs on the canal project and a general strike for equivalent pay that caused the canal works to be closed down for a year from 1919 to late 1920. (For construction progress and engineering detail see Cameron 1929; Department of Railways and Canals 1920; Engineer-in-Charge 1931; Sterns 1928.)

As Cowan (1935) noted: 'The Canal work was thus for a time practically regulated by what was done on the neighbouring Queenston work, with the result that for several years conditions were extremely difficult, costs exceedingly high, and the work was being carried out at an expense which was greatly in excess of what was really necessary, in view of the fact that there was no real urgency about the matter, and that the Dominion was only appropriating yearly a modest sum for the undertaking.'

Postwar, an important change of design initiated by Alex Grant, the engineer in charge from 1919, was the change from single-leaf to mitre gates at the locks, which resulted in a gain of 20 feet (6.1 m) from the original 800 feet (243.8 m) as the usable length for each lock. This modification required alteration to the concrete work already undertaken at Locks 1 and 6, where the breast walls and the gate recesses for the mitre gates were each recast in the early 1920s.

In contrast to the pick-and-shovel operations of the First and Second Canals and the limited machinery that was available to dig the Third Canal, large-scale mechanical methods could now be used for the excavation of earth and rock. Even so, a labour force of nearly 4000 men worked on the Welland Ship Canal project at the height of construction in 1927. As previously, the importance of the canal works for local employment and the regional economy were considerable, but three serious accidents occurred during the construction of the canal. In 1925 a steel structure collapsed, killing three men and injuring others. In 1927 at Main Street, Welland, a coffer dam collapsed, which drowned two men; another died later. And in 1928 the most serious incident of all took place when a locomotive crane carrying an end post for a steel lock gate overbalanced and fell into Lock 6 to cause the death of ten men.

Two types of canal construction were involved: dry work was needed along the northern reach, where the channel, locks, and bridges were located on former agricultural land and, in the south, the Third Canal had to be maintained in full operation while sections of the old were incorporated into the new. The most complicated overall task was at the escarpment, where the old and new routes crossed at two locations. This required an intricate progression of the various dredging and construction works in order to retain the Third Canal as an operating waterway until the Fourth Canal was available to accommodate vessels (Cowan 1935). The schedule was also complicated at the centre of Welland, where canal traffic and the flow of the Welland River had each to be maintained during the course of the construction works.

With construction being easier, but still slow and intermittent during the Great Depression and after the first generators of the Queenston-Chippawa Development came into service in 1921, the Fourth Canal eventually opened officially in August 1932, but vessels had passed through the full length of the canal before that date. The opening ceremony was deferred so that dignitaries attending the British Empire Economic Conference might attend. The official delegates who gave addresses included the Canadian prime minister and the minister of railways and canals, and senior representatives from the United Kingdom, Australia, New Zealand, South Africa, the Irish Free State, Newfoundland, India, and

Southern Rhodesia. It was a proud international rather than either a local or a regional occasion.

At the ceremony itself the S.S. *Lemoyne*, the largest freighter on the Great Lakes, sailed into the upper end of the Flight Locks, and the canal was then opened formally by the governor-general of Canada, the Rt. Hon. the Earl of Bessborough. His address concluded: 'It is a privilege to dedicate this Canal to the trade of the World. I hereby declare the Welland Ship Canal open to the commerce of the World' (Cowan 1935). The vessel carried within her holds some 530,000 bushels (19,000 m³) of wheat. Her length, to almost fill Lock 6, was 633 feet (192.9 m), the beam was 70 feet (21.3 m), and her depth was 29 feet (8.8 m). An important new era of canal endeavour had started. To mark this event a commemorative publication covered the construction feat and many historic aspects of the earlier canals (Duff 1932).

Water supply to the Fourth Canal, now wholly from Lake Erie, remained at about 2500 c.f.s. (70.8 m³) during the 1930s. A major change was the addition of water into the north of Lake Superior from the Long Lac and Ogoki diversions. These waters might be used by Canada at either Niagara Falls or through the Welland Canal, where diversions were made to a second power plant constructed at DeCew Falls between 1943 and 1947. The total diversion from Lake Erie fluctuated from 5900 c.f.s. (167.1 m³) to 8200 c.f.s. (232.2 m³). Related primarily to the water levels in the lake, higher diversions through the canal at approximately 7800 c.f.s. (220.9 m³) were possible during periods in the early 1950s and the early 1970s, whereas lower flows at an average of about 7100 c.f.s. (201.1 m³) were experienced from the late 1950s to the mid-1960s.

The canal prism was 200 feet (61 m) wide at the bottom and 310 feet (94.5 m) at the water line. The locks were constructed to provide 30 feet (9.1 m) over their sills, but the canal reaches between the locks were excavated to only 25 feet (7.6 m), which allowed a permissible draught of 23.5 feet (7.2 m) and was capable of improvement when required (Department of Transport 1953).

The Northern Length of the New Waterway

The radical new alignment for the northern length of the canal was almost due south from the harbour constructed in Lake Ontario, and was named Port Weller after James L. Weller, who had designed the route and most of the detail at the locks. This recognition was thoughtful because no previous project had recognized its senior canal engineer. Alexander J. Grant, who had taken over as supervising engineer in charge in 1919 after the wartime closure of the construction project, might also have been commemorated in an appropriate manner.

Port Weller, wholly an artificial harbour, lay about 3 miles (4.8 km) east of Port Dalhousie where Ten Mile Creek had previously entered the lake. The outer entrance piers, that extended into Lake Ontario for about 1.5 miles (2.4 km) to where the lake had a depth of 30 feet (9.1 m), incorporated two parallel embankments that functioned as protective breakwaters and enclosed a harbour 800 feet (243.8 m) wide. The entrance converged like a pair of pincers to half that width, and to the south the approach narrowed to Lock 1, which was placed inland from the original shoreline.

Townsite plans were prepared for Port Weller east of the canal in 1913, and both flanks of the harbour were subdivided into lots in anticipation of industrial development. This expectation did not materialize. The Fourth Canal was now more a waterway for the through passage of vessels, and ships in transit generated less land-based development than during earlier canal epochs. Wharves, however, were provided at the side of the canal in various locations to serve the riparian communities and their industries.

Port Weller added to the string of settlement created by the canal. Located at the Lakeshore (former N.S.&T.) Road crossing of the canal system, a new industry was born when a gate yard in the basin above Lock 1 became the home of Port Weller Dry Dock upon its transfer from Port Dalhousie. The piers and their adjacent areas of reclaimed land later accommodated service developments that included a municipal beach, a pollution-control plant, marina activities, recreational space, together with piles of industrial sand and nearby residential developments.

From Port Weller the canal south to Thorold followed the valley of Ten Mile Creek, displacing this stream and the former scenic waterfall where the Niagara Escarpment was crossed (Department of Railways and Canal 1935). Locks 1, 2, and 3 constructed along this length were individual locks separated by short lengths of canal, like gigantic water steps across the gently sloping Ontario Plain. All were constructed in the dry, with deep foundations down to bedrock to carry their weight. At Lock 1, in order to ease the task of construction, a temporary dam excluded lake water from the works.

To the south the canal channel, mostly in a cut, required a raised embankment for a short distance around Carlton Street, St Catharines. The environs north of the Niagara Escarpment were then agricultural land, mostly under fruit and vine production as part of the Niagara Fruit Belt in Grantham Township. As the city of St Catharines grew, the canal provided a ready physical boundary along the eastern flank of this residential and industrial development, and more or less separated the urban from the rural domain. East–west connecting links across the canal were provided by lift bridges on the main highways at Lakeshore Road, Carlton Street, Queenston Street, and Glendale Avenue, and for the main-line Canadian National Railway.

At Lock 3 of the new canal some 6 miles (9.7 km) from the outer edge of the harbour, the Fourth Canal crossed the line of the Third Canal; it again crossed this earlier canal to the south of Thorold. Between these two points the Third Canal had taken a wide sweep to the east in order to cross the escarpment diagonally and to have space for the locks and their intervening basins. Fourteen closely spaced locks on the former canal had here offset the height difference of 186 feet (56.7 m).

On the Fourth Canal along the same length, the escarpment was now approached directly, the locks were cut into its face at right angles, and the abrupt slope now required only four locks – Locks 4, 5, 6, and 7. The first three of these units were double locks in flight known variously as the Twin or Flight Locks; that is, the upper gate of Lock 4 was the lower gate of Lock 5 and so on. As each lock was paired, this provision allowed for the double passage of vessels, normally upbound on the west side and downbound on the east side to conform with the marine code for passing vessels. Both flights could be used in either direction if required through heavy traffic or when the need arose for operational repairs on one side. Lock 7, again an individual lock, carried the canal to the crest of the escarpment.

The Flight Locks introduced confusion into the lock numbering system that had been adopted. The locks were numbered with engineering precision successively from north to south along the canal. Locks 1, 2, 3, and 7 each have a single chamber, whereas Locks 4, 5, and 6 (the Flight Locks) each have two chambers. The Fourth Canal is always described in literature as having seven lift locks, but in reality the four individual locks plus the three that are duplicated provide a total of ten individual lift locks along the length of the canal between Lake Ontario and its summit level at the crest of the Niagara Escarpment. This may appear to be a minor point, but it is a confusing situation when visitors enquire about the number of locks.

As a precaution against severe flooding should a lock gate be carried away, and to prevent all water in the upper reach of the canal from draining across the edge of the escarpment should this disaster occur, a guard gate and a safety weir were constructed above Lock 7 and double gates were added at the head of Locks 6 and 7. These devices also permitted the lower length of the canal to be dewatered during the closed winter season for lock repairs, including the refacing of the concrete lock walls as they weathered, were eroded by flowing water, or damaged by passing vessels.

The materials excavated from the locks and their intermediate channels were transported by a standard-gauge, double-track construction railway to Port Weller, where they were deposited and became the breakwater walls on both sides of the harbour. The track, 7.5 miles (12.1 km) long, followed the west side of the canal, and had its own operating rules (Harris 1914). This railway also carried needed supplies to the various contractors' sites as well as excavated materials to Port Weller. It reached a traffic flow at the height of construction in 1927 of 384 trains a day (Department of Railways and Canals 1935). Removed at the end of the construction works, it should have been retained and made part of the N.S.&T. streetcar system to provide a scenic loop route along the canal between St Catharines and Thorold for visitors.

As might be appreciated, the many elements of mechanical construction and excavation added much noise and dust to the route of the canal and their communities. At Thorold several buildings had to be removed from the eastern fringe of the town and the town's reservoir had to be relocated. Here, as elsewhere, highways and railways that crossed the canal had to be diverted temporarily while the new bridges were constructed on the original alignment. Every community along the canal was inconvenienced to some extent by the canal works and by the flow of traffic to the work sites. These annoyances applied especially to Thorold on the northern reach because of the close proximity of the canal to the developed extent of the town, and also at Allanburg, Port Robinson, Welland, and Port Colborne along the southern reaches.

The Flight Locks

The Flight Locks, the largest single structure on the new canal and unique at the world level as a featured attraction, deserve special attention. They provided a total lift of 139.5 feet (42.5 m), against 85 feet (25.9 m) for the three locks at Gatun on the Panama Canal. The maximum lift at each lock of 47.5 feet (14.5 m) was also much greater that at Gatun. In terms of world comparisons between these two ship canals, the Welland Canal was the greater achievement and no precedent existed anywhere for either the high lift that was involved at each lock or for the combination of six separated locks to cross the Niagara Escarpment. The twin locks in flight are similar in design to those on the Panama Canal, but there is a greater individual and aggregate lift in the Welland, and its locks are narrower and shorter and have less draught.

The construction details at the Flight Locks are more than noteworthy. They contained 1,170,000 cubic yards (894,530.5 m³) of concrete. The length of the continuous central wall was about 3350 feet (1021.1 m), and it had a width of 60 feet (18.3 m) throughout this length. The wall between Locks 4 and 5, in effect a dam, at over 130 feet (39.6 m) high exceeded the height of a ten-storey building.

Most of this concrete came from limestone excavated from along the line of the canal between Lock 7 and Allanburg, where a rock-crushing plant was erected by the contractors. Sand for this concrete was obtained by dredging at the mouth of the Niagara River. Brought in by scow to Port Weller, the sand was then carried by the construction railway to points where it was required. In addition to the locks with their concrete construction (in contrast with the stone locks of the Third Canal), the buildings erected on the locks were also of reinforced concrete. They housed the gate and valve machinery and the control rooms from which the locks were operated.

Impressive statistics may be used to emphasize the imposing character and the enormous size and cost of the mega-undertaking that was created (Cowan 1935; Department of Railways and Canals 1935). The locks, whether single or double, were each 829 feet (252.7 m) long between the inner gates, and 80 feet wide (24.4 m). The massive gates were each made of steel, five feet (1.5 m) thick (Sterns 1928). The lower gates, 82 feet (25 m) high, weighed 490 tons (444.5 tonnes) and the smaller upper gates, 35 feet (10.7 m) high, weighed 190 tons (172.4 tonnes).

The new waterway had a depth over the lock sills of 25 feet (7.6 m), but the lock structures were designed for a 30-foot (9.1-m) depth of water, and much of the channel was soon excavated to a depth of 27.5 feet (8.4 m). The minimum width of the canal prism was 200 feet (61 m) at the bottom, and 310 feet (94.5 m) at the water line. The volume of water required at each lockage was 21,000,000 imperial gallons (95.5 million litres), the equivalent of covering an area of 77.5 acres (31.4 ha) to the depth of one foot (0.3 m). In practical terms, this meant that a large pondage area had to be provided above each lock so that no rapid current would be created and the canal upstream would not be denuded of water when each lock emptied.

Locks 1, 2, and 3 required a total of 500 acres (202.4 ha) for pondage, taking advantage of the valley configuration on the east side of the canal to provide this new expanse of water. To the south, dams were constructed east of the canal to impound various creeks that were tributary to the Beaver Dams Creek. This created extensive ponds to the east of Merritton and to the east and south of Thorold. The pond for the Flight Locks covered 84 acres (34 ha), and that for Lock 7 covered 65 acres (26.3 ha). In addition, the Third Canal above Lock 3 of the new canal was retained to supply the lower reaches of the canal with water. Each of the lower lift locks had a regulating weir on its eastern side.

As these water barriers in addition to the canal channel were introduced to the landscape, they bounded Merritton and more especially Thorold along their eastern borders. They also introduced new water features into the environment as former slender river valleys became lakes, thereby creating a new potential for recreational, park, and landscaping opportunities. Not unnaturally, the many engineering features and in particular the Flight Locks became a major sightseeing factor during their course of construction, and upon completion they were steadily recognized for their tourist value and appeal as visitor traffic increased with the advent of the automobile. Niagara Falls and the boundary with the United States were but 10 miles (16.1 km) away. After 1940 the Queen Elizabeth Way, which used the Homer Bridge until the Garden City Skyway was constructed across the canal, provided a visual introduction to the canal from the major highway in the peninsula and direct access to the canal from southern Ontario and the nearby American states.

As a measure of the strong appeal generated by the canal as an important tourist destination, when Michelin in 1982 produced its first Canadian Tourist Guide, the Welland Canal and the drive between Port Weller and Thorold past the sequence of lift locks and the formidable attraction of the Flight Locks received a two-star rating, the same degree of merit as awarded to the northern length of the Niagara River Parkway between The Falls and Niagara-on-the-Lake.

This new tourist appeal of the Fourth Canal, which was not a feature of the earlier canals, included the combination of vessels in the canal, their passage through a lock, the character of the locks and other canal structures, and their agricultural setting against the backcloth of the Niagara Escarpment and within the scenic Niagara Fruit Belt. There was also increasing recognition that the remnant features of the old canals had significant heritage values. Locally, tourism was recognized when a building at Thorold next to the Flight Locks, used as an engineers' residence during the construction period, became a restaurant with an attached motel with room balconies facing the locks; and a building that faced the canal at Homer Bridge on this major through street became a restaurant. Some recognition for the Fourth Canal as a visual feature of tourist importance had started, but these elements were not part of the canal's design.

The Southern Length of the Canal

Between Lock 7 at the crest of the Niagara Escarpment and Lake Erie, the Fourth Canal – widened, deepened, and shortened by the elimination of bends – followed mainly the route of the Third Canal. At Allanburg, the two wooden swing-bridges that had crossed the Second Canal at Holland Road and modern Highway 20 had been replaced by a single steel swing-bridge in 1902 at the Highway 20 location; this again was replaced in 1930 by a vertical lift-bridge a few feet to the north during the course of the canal construction works. Traffic from the west at first crossed this bridge and followed the canal to Holland Road, but a bypass completed in the late 1960s at the eastern end of the bridge routed Highway 20 around the village and its streets became culs-de-sac.

When the Fourth Canal was completed, the channel was now divorced from Allanburg, and vessels plying the canal passed by with no cause to stop. The dry-dock facilities had been closed and their basin filled with spoil as part of the widening process. The water power that had supplied power to former mills had been diverted to DeCew Falls, where it was used for power-production and water-supply purposes. The community had been downgraded to a roadside settlement. The new canal bridge became an incident on Highway 20 and an irritable barrier for frustrated motorists when the bridge was raised. Engineering requirements in the early days of the automobile and the through flow of vehicular traffic on Highway 20 had taken over as major formative factors in village character.

At the Deep Cut, where the canal was incised into the highest point of relief along the line of the canal, the cut was now to a depth of 80 feet (24.4 m) below the surface, with a prism width of 200 feet (61 m) at the bottom. The canal had also been shortened and its bends eliminated by new cuts between Allanburg and Port Robinson. Modern engineering technology had overcome the difficulties that had beset the canal over its earlier years.

At Port Robinson, in a similar manner to the situation at Allanburg, the new highway bridge across the Fourth Canal was located a few feet from its predecessor. Work started on this project with the need to stabilize the ground on the eastern side of the bridge next to the Third Canal. This was partially complete by 1919 and resumed in 1922, but the bridge itself was not completed until the start of 1931 navigation season – a long period of disruption to village life by the canal works.

As the Fourth Canal cut across the basin and lock connecting with the Welland River, a rock wall placed along the eastern bank of the canal disconnected the two water systems. The enclosed basin was then filled from a hydraulic dredger that worked in the canal but discharged its load into the disposal area of the basin until a height was reached even to the adjacent land. The surplus water drained out through the surviving lock of the Second Canal into the Lower Welland River; the infilled area, left derelict for a time, eventually became a sports field. The village centre that had grown next to the First and Second Canals, their locks, and the canal basin was thereby deprived of its canal setting and Chippawa Creek (the Welland River) had also to be diverted into a new channel. Like Allanburg, Port Robinson now lay more on a highway route that led to the bridge across the canal, and was left with buildings, an old lock, and memories to recall its past history of association with the earlier canals and navigation along the Welland River.

Between Port Robinson and Welland, the Fourth Canal followed a more direct course next to the Welland River. The canal and the river were kept independent of each other, but owing to its winding nature the river had to be straightened and realigned on a new course to the east of the canal channel. These works created a long narrow island between the two waterways that eventually became park space, named Merritt Island after the canal founder. On the west bank of the canal, new landforms were created as this length along the canal was used as a spoil dump, a process that had also occurred between Thorold and Allanburg.

The canal engineers had wanted to provide a canal bypass around Welland, but as this proposal was opposed by the city and its business interests because of the anticipated loss in trade, a longer but less effective winding route was constructed along the line of the Third Canal. As the canal was now too deep to be carried *over* the Welland River, an inverted syphon culvert with six tubes carried the river *under* the canal to replace the stone aqueduct of the Third Canal (Grant 1928).

The works of canal construction from Port Robinson to Welland were late contracts. The idea had been to dam the river at Port Robinson and raise it to the level of Lake Erie as the summit level of the canal. This proposal would have flooded some 1600 acres (647.5 ha) of land and eliminated the aqueduct. As water in the Welland River would now be at the same level as in the canal in an extended reservoir, an interesting proposition was to divert some 5000 c.f.s. (141.6 m^3) to flow west from the canal then north to Twenty Mile Creek near Ball's Falls. There a powerhouse would have been constructed with a head of some 300 feet (91.4 m), about twice the height then available at Niagara Falls. 'The construction necessary would not disfigure Niagara Falls and as the power would be generated at a point 18 miles

[29.0 km] nearer to the Canadian market, this advantage would accrue to the consumer by lessening the cost of transmission' (Gardiner 1913).

The idea of a reservoir at Welland was aborted and, as the Third Canal was used to supply domestic water to the downstream municipalities, a pipeline from Lake Erie to reservoirs at Welland, Thorold, Merritton, and St Catharines was approved (Department of Railways and Canals 1920). Meantime, objections to flooding, the potential of pollution, and the threat of litigation caused the idea of damming the river to be dropped. Instead, the canal ran parallel to the river at a higher level and, as the route for the canal crossed the river bed on four occasions, the Welland River was straightened and watertight embankments were placed along the east side of the new channel. This canal, rather than the envisaged pipeline, has remained the source of water supply to DeCew Falls.

The contract for the syphon to carry the Welland River under the canal, rather than flowing freely under the arches of the canal as before, was not awarded until 1925. The syphon was then constructed in the dry on the north side of the aqueduct for the Third Canal, and the river was carried through a new channel to this intake. The syphon is an optical illusion. Its visible features are parallel to the canal banks, but the low-level tubes are set at an angle of 59 degrees to this axis in order to accord with the direction of flow in the river. The syphon, a highly complicated mechanism at this scale of construction, was second only to the Flight Locks as an engineering structure of world acclaim. Eleven million cubic yards (8.4 million m³) were excavated from the open cut at the syphon. This material was then used to fill in parts of the old canal bed, which became an extension of Welland's central business district used for car-parking purposes. A roadway and a path crossed the river along the west side of the syphon. The striking stone aqueduct of the Third Canal was demolished because it lay within the intended channel for the Fourth Canal.

The stone aqueduct of the Second Canal was retained, but as the land in its vicinity was filled to the top of the arches, the grand aesthetic quality of these arches crossing the river was lost. After some temporary use as the bed of a railway used to dump material into the former river bed and then a period of uncertainty as to its future, this aqueduct became an open-air swimming pool in 1940. Through the rerouting of the river and moving of the canal to the west, the county offices and the courthouse lost their distinctive setting next to the river and the canal. As the original channels of each were filled, the visual and symbolic meaning of downtown Welland at a central location on the water network of the peninsula was eliminated by the construction of the Fourth Canal. As visual compensation, Main Street focused onto the lattice-work of Bridge 13 of the Fourth Canal as a striking feature of the town's urban character.

The syphon impeded the former free flow of the Welland River under the arches of the Second and Third Canal aqueducts; during construction of the canal channel a tap drain took surplus canal water into the Welland River. These events caused severe deterioration in the quality of water in this river. 'The water ... one mile west of the present aqueduct was clear in 1920. The bottom was visible for quite a distance in the river channel. After the construction of the Fourth Canal, the water was always turbid and swimmers standing on the bottom would sink in up their calves. The river was never clean again' (Gorman 1966).

Another change was that, until the late 1960s, the river received untreated raw sewage from homes and industry. No sewage filtration plant existed, a situation subsequently remedied but assisted by the presence of the canal when in 1963 the Seaway Authority diverted drinking water to the Welland water-filtration plant. The unused water then flowed into the river to dilute sewage and reduce the turbidity of water below Welland.

The canal works at Welland and the new cut for the southern length of the canal required a home to house engineers working on the canal project. Built east of King Street next to the canal by the Department of Railways and Canals, this building survived the canal works and became the Welland Club. Less fortunate, the lock at The Junction, where the Feeder Canal met the Third Canal, no longer required, was infilled to the top of its walls and became a small park.

South of Welland to Ramey's Bend, the Third Canal was again widened and deepened. To eliminate the sharp curve of this bend, the Fourth Canal was constructed from the north of Ramey's Bend on a straight line to the inner harbour at Port Colborne. This left a confined pear-shaped island known as The Island between the new and old canals at Humberstone. Its only access was from Main Street West.

The southern length of the Third Canal was retained to provide the water supply for the Fourth Canal, and its control mechanism was placed under the now fixed bridge on Main Street West. Interestingly enough, even though the channel of the Fourth Canal now lay some distance to the east, Main Street remained divided into Main Street East and Main Street West. The changed circumstances introduced by the Fourth Canal were not recognized by renaming this street in the former village of Humberstone.

At the centre of Port Colborne, as the new and old channels ran side by side, the Fourth Canal at the head of the harbour provided the only location on the canal where the sequence of locks from the Second, Third, and

Fourth Canals survived side by side. The old guard locks of the Second and Third Canals were retained and, using the route of the now abandoned Third Canal, supplied water to the new Fourth Canal at Ramey's Bend. Also at the centre of Port Colborne, the greater width of the two former guard locks in combination with the new channel required the removal of many commercial buildings. This included the loss of the Canadian National (former Buffalo and Lake Huron) Railway station on the east side of the Third Canal and its rebuilding on the west side of the new canal.

The summit level of the Fourth Canal south from Lock 7 was 16 miles (25.7 km) long. Maintained at one foot (0.3 m) below the standard low-water level of Lake Erie, this channel had a length of 25 miles (40.2 km), or 27.7 miles (44.6 km) between the outermost ends of the two lakeshore harbours. No lift locks were required, but Lock 8 as a guard lock protected the canal against the fluctuating water levels in Lake Erie. Variations up to 11 feet (3.4 m) resulted from the force and direction of the wind. Lock 8 had a length of 1380 feet (420.6 m), the longest in the world, and a maximum lift of 12 feet (3.7 m).

Spoil excavated from the southern length was deposited in various locations, including at Port Colborne along the breakwaters and at Ramey's Bend to infill old lengths of abandoned channel. Between Welland and Port Colborne, spoil was dumped within dykes on land west of the canal, creating hummocky terrain with intervening ponds in an otherwise flat and marshy area that eventually evolved into an area that attracted ducks and birds and became the Mud Lake Conservation Area.

A spur breakwater out into Lake Erie at Port Colborne was added to the existing arrangement on the western side to improve entry to the harbour for the larger vessels expected on the new canal. The earlier breakwater had contributed to the silting of Gravelly Bay, which adversely affected the cottage and recreational developments that had taken place along this length of shoreline. It also encouraged the early freezing of the now enclosed bay, which promoted it as a locale for winter ice fishing. Winter ice now also prevailed at Ramey's Bend, where the abandoned length of the Third Canal no longer carried a through flow of water but instead became a skating and fishing pond for winter recreational pursuits, and its southern end a place where old lake vessels might be berthed then broken up for their scrap value.

Crossing the Canal

Twenty movable bridges spanned the channel at irregular intervals to carry the east–west railway and highway arteries that traversed the peninsula. The first bridge to be completed, the Michigan Central Railway bridge at Welland in 1911, was followed by a streetcar bridge at Thorold in 1916; the remaining bridges were constructed from 1927 to 1931. These bridges were numbered with engineering precision in their intended sequence from north to south, Bridge 1 next to Lake Ontario and Bridge 21 next to Lake Erie. Even though Bridge 2 on Parnell Road in St Catharines was not constructed, the numbered sequence was not changed. The bridges had two names: the canal authorities used their official number, but the public preferred to name the bridge after the street or railway that crossed it.

At Welland in 1927, the Alexandra Bridge was moved temporarily south to Division Street to provide for the flow of highway traffic at the business centre, and construction then began on a new vertical lift-bridge to span the canal at Main Street. Completed in 1930, its twin towers became an immediate local landmark, but when the bridge was raised, serious local traffic congestion arose. The bridge carried a street railway and almost all the highway traffic that passed through the centre of the city. The only bridge not to cross the line of the canal at right angles, it was askew at an angle of some 22 degrees to link the alignment of East and West Main Streets across the former township boundary.

All bridges were lift-bridges, except for two swing-bridges at Thorold and Welland that restricted the width of the channel. Seven were of the rolling lift variety, but the most favoured were the eleven vertical lift-bridges with spans that lifted to 120 feet (36.6 m) above the water level in the canal so as to clear the masts of passing ships (Atkinson 1928). As the vertical lift-bridges predominated south of the escarpment, a strong new vertical dimension in the flat landscape marked the line of the Fourth Canal across the flat Erie Plain.

When, because of the direction and intensity of the wind, the water level in Port Colborne harbour varied, the clearance under the last two bridges could be affected. The important point to note is the dual set of limitations on the size and design of vessels: their height, by the overhead clearance of the lift-bridges, and their horizontal clearance, by the width of the locks (Department of Transport 1953). This role of the fixed structures has always prevailed, and has contributed to the obsolescence of the canal as ever-larger vessels have come into operation.

By function, eleven of the bridge crossings were solely highway bridges. Of these, three added provision for the interurban streetcar network, at Lakeshore Road and Queenston Street in St Catharines and Main Street in Welland. These highway bridges reflected the main road network of the period. They confirmed then reinforced the significance of the important crossing points of the canal within the urban structure of each canal community and reinforced their importance. Other minor roads of lesser importance that had previously crossed the line of the canal channel were closed, which caused local inconvenience where routes had previously crossed the canal. These curtailed roads included Linwell, Scott, and Eastchester in St Catharines, Beaverdams in Thorold, and Killaly in Welland. Even so, the same street name remained on both sides of the new canal as a reminder of past circumstances.

The N.S.&T. streetcar link between St Catharines and Niagara Falls crossed the canal at Thorold, and five bridges carried railways only: the Canadian National main line at the Flight Locks; the CN Welland Division line near Allanburg; the Michigan Central main line and the CN Wabash Division at Welland; and the CN Buffalo-Goderich Division at Port Colborne.

The weight assumed for the design of the highway crossings was 100 pounds per square foot (4788.3 kPa). All bridges carried one or two pedestrian sidewalks that were 5 feet (1.5 m) wide. The narrowest roadway was 17.4 feet (5.3 m) at Port Weller, where the bridge carried the streetcar tracks between St Catharines and Niagara-on-the-Lake along one side. The roadway width was generally 20 feet (6.1 m), but this was increased to 24 feet (7.3 m) on the Thorold-Allanburg Road and to 30 feet (9.1 m) on Queenston Street in St Catharines, at Allanburg, on Main Street at Welland, and on Clarence Street at Port Colborne. This provision identifies the more important roads of the day and the locales expected to have the heaviest traffic flows when the canal was designed.

Canal regulations gave precedence at the movable railway bridges to canal traffic, with the proviso that 'no unreasonable delay shall be caused by any vessel to railway traffic ... If the signal for the bridge is given by any approaching train while a vessel is between a quarter of a mile [0.4 km] and a half mile [0.8 km] distant from the bridge, the vessel shall slow down, stop if necessary, and await the passage of the train' (Department of Transport 1947). If the train was in the signal block area, it had priority and no power was available to operate the bridge.

At the highway bridges, the same regulations restricted speeds to 15 miles (24.2 km) per hour for automobiles and 10 miles (16.1 km) per hour for motor trucks and buses. These regulations slowed highway traffic; when the bridges were raised all highway traffic was brought to a halt. The bridges became points of frustrating delays to highway traffic on the main routes as the volumes of flow increased, as vehicle ownership expanded, and as bridges designed in the 1910s and completed in the 1920s had to operate within the unanticipated high-level traffic conditions of the 1960s.

Nine other crossings carried high-voltage wires that radiated out from the hydroelectric power projects at Niagara Falls and Queenston. As the height of these wires had to be safely above the masts of ships, the tall pylons that carried them across the canal were visually prominent. In addition, some twenty submarine cables were laid under the bed of the canal, and at the locks ducts were built into the walls and floors, for use by public utilities. The canal presented a visual and a physical barrier to urban development, but the provision made for utilities permitted urban or industrial advance across the canal.

The Northern Length of the Abandoned Third Canal at St Catharines

When Port Weller displaced Port Dalhousie as the northern port of entry to the canal system, Port Dalhousie's 132-year reign as a vital shipping centre came to an abrupt end. Lock 1 of the Third Canal was decommissioned by the St Lawrence Seaway Authority in 1968 and its southern end filled, a decision that disconnected Martindale Pond from Lake Ontario for navigation by even the smallest of boats. However, as the water level in Martindale Pond was retained, this saved the Henley Regatta course for rowing and as a scenic-recreational resource.

This retention of Martindale Pond also later permitted the small Heywood hydroelectric power station operated by the St Catharines Hydro-Electric Commission to open in 1990 within the weir next to the above-mentioned lock. The adjacent lock tender's office, the only such structure on the canal, also survived, as did the lower harbour with its two long piers dating back to the period of the Second Canal. On the east side of the harbour, the former railway lands were purchase by the City of St Catharines to encourage marina and recreational developments. Port Dalhousie entered a new era as a residential suburb of St Catharines, with a pride in its long years of canal history and with recreational, historic, and open-space attributes that were unequalled elsewhere in the city.

To the south-east, the line of the Third Canal between Martindale Pond and the Fourth Canal was de-watered, filled to the top of its banks, and, where the canal was above the level of the surrounding country, the banks, locks, and weirs were generally bulldozed down to ground level. These projects were undertaken by the Department of Railways and Canals during 1934. The narrow swing-bridges were also removed, and their locations became part of the radial road system extending out from the growing city of St Catharines. This removal of the canal and improved road access ended the social and ethnic isolation of the Facer Street area, and the city advanced piecemeal by several subdivisions to the north-east across its rural fringe in Grantham Township.

The area closest to Martindale Pond was purchased by the Ontario Construction Company, which used the height of the locks as a convenient platform for its plant operations. Later filled and purchased by the city of St Catharines, it became part of a public park named Ontario Gardens. Lock 3 was buried and Lock 2 on the east side of Martindale Pond, though gateless and without flowing water, was retained as a feature in an open-space setting named Martindale Park, a continuation of Ontario Gardens. Along the former line of channel, land uses gradually attracted to the infilled land varied from areas of public open space such as Lancaster Park and Fairview Golf Course to residential purposes, including wartime housing during the Second World War, and light industry.

The former canal could now be identified by roads such as Secord and Nihan Drives, Wood and Sandown Streets constructed parallel to the former canal alignment and not conforming with the original survey grid for Grantham Township. Except for the non-conforming pattern of these roads and the remnants of Lock 2, the evolving townscape of North St Catharines now retained few memories of the Third Canal. As previously at Dunnville and Thorold, the city had effaced a half-century of identification with the Third Canal.

As along the escarpment length of the Third Canal, now abandoned for transportation but retained to provide a controlled flow of water to supply the northern length of the Fourth Canal, the lock gates, weirs, and reservoirs were each removed. The former canal channel retained free-flowing water that cascaded by a series of small waterfalls through the lock structures. Upstream were the ponds entrapped by the canal, some partly submerged locks, and the tunnel that carried the former Great Western–Grand Trunk Railway under the Third Canal. The site of the once proud Flight Locks became abandoned open space on the urban fringe of St Catharines.

Historical Respect and Amenity Considerations

In 1924 a resolution by the national Historical Sites and Monuments Board recognized for the first time that the old canals and their modern version under construction nearby were events of 'national importance, and that action be taken this autumn to commemorate the turning of the first sod by the erection of a cairn or boulder and tablet.' A cairn with a commemorative plaque was then erected at Allanburg at the northern entrance to the Deep Cut, close to where each of the earlier canals had passed and where the works of the Fourth Canal were under way. One hundred years earlier, on 30 November 1824, 'an interesting gathering of about 200 persons, took place at a flat near the head of one of the branches of the "Twelve," for the purpose of witnessing the important ceremony of "Turning the first sod of the Welland Canal"' (Merritt 1875). Then after William Hamilton Merritt had delivered the inaugural address, the spade was handed to George Keefer who, as president of the board of directors, removed the first shovelful of earth.

Later, when the Fourth Canal was nearing completion, the land in the vicinity had to be reinstated. Spoil had been dumped in marshy areas and at various dump sites and the channel was a wide trench gouged through a farming landscape. After the concrete work was finished, much remained by way of reinstatement. The environs had to be graded, drained, ploughed, and sodded. Features such as the construction railway, the offices and buildings used during construction, and the areas where equipment and supplies were stored had to be removed or restored to some form of beneficial use.

Forestation was necessary to provide windbreaks for vessels using the canal, to minimize the effects of wave action, and to resist erosion, as the roots helped to bind the embankments. There would be the potential of an income and valuable timber supplies from well-managed forestry areas, and the concepts of ornamentation and beauty then being applied along the Niagara River by the Niagara Parks Commission and along the Queen Elizabeth Way by the provincial Department of Highways might suitably be applied to the canal banks. Further, as the Panama Canal, which had opened in 1914, had soon become an international tourist attraction, it was felt that the Welland Canal should receive comparable acclaim. As stated by the canal's divisional engineer, 'Public sentiment expects and demands higher standards of maintenance of public and semi-public property than formerly. For example note the attention now given to their premises by railroad companies: the increasing number of provincial, county and forestation schemes [etc.] ... After all the public sees the watchcase rather than the works – the Welland Ship Canal Setting rather than the mechanism – which is taken absolutely for granted' (Jewett 1927).

When a Forestry and Management Branch was formed by the Department of Railways and Canals, its director reported that the landscaping project 'would extend along the canal zone from lake to lake, gradually developing the right-of-way by different stages into a national park and driveway. The pondage areas and the back waters would be planted with wild rice and celery, and receive government protection for land birds. Many seed-bearing and fruiting trees, shrubs and bushes would be planted to provide feed for the land birds, as well as being decorative and ornamental.

'Focal points of interest are laid out to formal and informal flower beds and shrubbery plots, following in a measure botanical garden methods. Many hardy exotic and indigenous deciduous and evergreen trees would be planted in groups and plant sites, and finally develop into individual arboretums along the canal-right-of-way.'

After construction work was finished, 'the canal will be a blue waterway and same could readily be stocked with suitable fish by the Provincial Games and Fisheries Department, providing for future angling.' Further, 'When the main roads and the side roads are built and connected up between Port Weller and Port Colbourne [sic], all outlying back areas of the canal right-of way would be made accessible for inspection of reforested areas, plantations, nurseries, etc. and such that may be developed into tourist camps.

'Safe bathing beaches on the east and west sides of Port Weller harbour could be developed by building groins or sand traps such as started last season. Suitable shelters and resting spots, seats, etc., could be provided for tourists and others along the canal right-of-way.' And then, with suitable pride in the potential, the report concluded: 'In the United States the Arnold Arboretum is spoken of as America's greatest garden. In England, Kew Gardens is the pride of the nation. In Canada the Welland Ship Canal will be Canada's National park and show place' (Waddell 1930).

Much then followed. In 1927 six nurseries and one seed bed were established along the northern length of the canal near Lock 2. Hardwood transplants and seedlings were obtained from the provincial forestry branch, and shrubs were obtained from the Wellington and Davidson nursery at Fonthill. Five more nurseries were opened in 1928, and two years later some 321,100 trees of various kinds had been planted (Cowan 1935). As reported in the 1934 Annual Report of the Department of Railways and Canals, 'What might have been a guant [sic] commercial waterway is being transformed into a zone of natural scenic beauty.'

For the first time in the canal's history, the vision of the future now combined transportation with the development of riparian land to achieve a canal within the amenity setting of a national park. 'It is the intention, if the full programme is carried out to make the canal limits something of a park, with woods and tended gardens at the locks, definite park areas, the pond-age and other water areas turned into lakes and sown with wild rice, and the whole turned into a bird sanctuary' (ibid.).

These expectations for the Fourth Canal were also expressed in the desire to preserve some of the old canal locks. As A.J. Grant, the engineer-in-charge, reported in 1930, 'Would this not be a good opportunity for the Department [of Railways and Canals] to spend a little money in the preser-vation of one of these old wooden locks as being to the canals of this coun-try what Stephenson's "Rocket" is to the railways of England ... The one near Lock No. 25 of the Second Canal would be the one to use, as it is cen-trally located in the Town of Thorold and would be easy of access to Tour-ists.' The Rocket is now housed comfortably in London's Science Museum as an important relic of the industrial revolution, whereas Lock 25 has decayed into neglect and was not restored for public display.

The concept of applying amenity considerations to the Welland Canal received a boost in 1936 when the idea of a Round-River-Canal Tour was examined by the Niagara Parks Commission. 'Commissioner the Honour-able George S. Henry conceived the idea of extending the present boule-vard system along Lake Ontario from Niagara-on-the-Lake to Port Weller, then following the Welland Canal to Port Colborne on Lake Erie and from this latter point to the southern terminus of the Parkway at Fort Erie – making a complete Round-River-Canal Tour of some ninety miles' (Niagara Parks Commission 1938; Way 1960). This scheme was deferred until completion of the then current work projects, which included the res-toration of Fort George on the Niagara River Parkway, and was then aborted with the outbreak of the Second World War.

The Completed Canal System

The canal itself opened steadily for navigation in sections that depended on the work schedule at the locks. The first length north from Port Colborne through to Thorold opened in 1929, followed in 1930 by Locks 1, 2, and 3. Vessels now ascended the canal through its new port of entry at Port Weller. They then followed the lower reaches of the Fourth Canal and crossed the escarpment by using the locks of the Third Canal. Rejoining the Fourth Canal above Lock 7 to the south of Thorold, they then fol-lowed the Fourth Canal from its summit level to Port Colborne. With this temporary route in operation, the section of the Third Canal between Lock 3 on the Fourth Canal and Port Dalhousie could be de-watered. There remained the need to divert the flow of water from the upper length of the

Third Canal into Lock 7 and through the Flight Locks to bring the new navigation system into operation.

As sailing vessels had now been displaced by steamers, a towpath was no longer a design requirement. Instead bollards embedded in the lock walls served as mooring posts. They also followed the length of the canal to secure vessels in case of some mishap or if they required attention such as winter repairs. Service roads also followed the length of the canal on both sides to provide access to the banks for maintenance and repair of the canal and its installations. These private roads were maintained under the jurisdiction of the Department of Railways and Canals and, although public access was discouraged, they tended to be used informally for access, to view the canal scene and as through routes along the canal.

Tolls to pass through the Third Canal were temporarily abolished by the Dominion government in 1903. In 1905 the exemption was made permanent. Thus, when the Fourth Canal opened no tolls were charged. It was permissible to levy tolls through the Boundary Waters Treaty of 1910, but any such levy had to be applied alike to vessels of Canadian and American registry. The reasoning was that, by not levying tolls, trade would increase and this traffic would be beneficial to the country at large.

On the Third Canal, linesmen who accompanied the ships as they passed through the canal were paid for handling the mooring lines. On the Fourth Canal linesmen were placed at each lock to safeguard the larger vessels and to prevent damage to the lock gates and walls. With day and night service, this involved three shifts of eight linesmen at each lock, a not inconsiderable addition to local employment. Fees were charged for a vessel to pass through a lock to cover the costs now borne by government. Described as fees not tolls, they were nevertheless the renewal of a cost applied to shipping for passing a vessel through the canal. Shipowners objected to what they perceived as an additional tariff to pass a vessel through the canal.

The smaller number of locks, their more rapid speed of operation, the shorter length of the Fourth Canal, and the higher speed of vessels in the canal reaches each contributed materially to a more effective and efficient canal operation. Vessels could now pass through the system in about eight hours, not including entry into the canal, where delays might occur. As this shortened time period and the use of larger vessels also meant that each transit required fewer services from the land, the canal communities now played a diminished role as integral components of marine activity and the canal scene.

To provide the power that was required to operate the machinery at the locks and bridges and for heating and lighting, a hydroelectric power plant was built at the foot of the Flight Locks. This came into operation in 1932, and provided the strong vertical element of its tall surge tank as the complement to the fall of water through the locks. Water taken from the canal at the head of Lock 7 passed through penstocks that were constructed in the west wall of the locks. Transmission lines at 22,000 volts followed the canal from Port Weller to Port Colborne, and substations with transformers were located at each lock and bridge. The canal was also lighted for its whole length and a self-contained automatic telephone service linked all the locks, bridges, substations, the powerhouse, and administrative buildings along the canal. The Fourth Canal was constructed as an independent, self-reliant unit with direct contact with its riparian communities along the banks, around the harbours, and where the bridges crossed the canal.

Dock and port facilities were provided at several points along the canal. Their detail and the record of progress is recorded in Green's Great Lakes and Seaway Directory and in the annual reports of the federal Department of Transport. By 1953 the latter recorded from north to south two concrete wharfs on both sides of the harbour at Port Weller, the canal administration building on the west side at Lock 1, and the dry dock administered by Port Weller Dry Docks upstream from Lock 1. At Homer, close to the QEW, a wharf with a freight shed and a turning basin served St Catharines.

Thorold with a more substantial array of facilities had a turning basin above Lock 7, which meant that vessels arriving from Lake Erie had not to use the Flight Locks and pay its tolls, and then wharves around this basin. These wharves included the public Thorold Wharf next to Provincial Highway 58 (later the route through the Thorold Tunnel) and a rail connection with Canadian National Railways; here was located the warehouse of the Niagara District Warehouse and Forwarding Company. There were then two wharves operated by the Ontario Paper Company for loading newsprint, and for unloading pulpwood and coal; another wharf suitable for unloading bulk materials and leased to private industries; and the Beaver Board Wharf, in part leased to other companies and available for public use.

South from Thorold a turning basin had been provided at Port Robinson but, as no wharfage was yet provided in this area, the potential of possible industrial development had not materialized. At Welland, the Welland Centre Wharf was leased to various building-supply and coal companies, with a small area available for public use; and the Welland South Wharf, with connections to the Canadian National and Canadian Pacific Railway, was leased to two local industries, Stelco and Union Carbide.

At Port Colborne, around a turning basin located north of Lock 8, were

Ramey's Bend Wharf next to Provincial Highway 58, with connections to the two railways noted above, and leased to a coal and a flour-milling company, and West Docking, leased in part to a milling company, part undeveloped, but with highway and rail access both being available. Lock 8 had the Port Colborne Marine Post Office to serve vessels in the canal with a 24-hour facilities during the navigation season. And then around the harbour about 3¾ miles (6 km) of wharf wall were available for winter mooring, and hence for the seasonal repair of vessels. These facilities were supplemented by an anchorage area of some 37 acres (15 ha) within the west breakwater.

The two harbour piers jutting out into Lake Erie were occupied by the Government Elevator and the privately owned Maple Leaf Elevator, accessed by track of the Canadian National Railway laid on rock fill to the west of the harbour wall. They served the mill piers, coaling stations, and an oil station as a marine-oriented industrial area. Along the east wall of the canal channel were the yards of the Canadian National Railway, two metallurgical companies, and a coaling station. Port Colborne harbour, in existence since the early 1830s, had a much more active marine atmosphere than prevailed at either Port Weller or Port Dalhousie at the canal system's northern end.

An unanticipated impact of the Third and then the Fourth Canal was that marine life could now move or be moved upstream from Lake Ontario to Lake Erie. It might attach itself to the hulls of vessels, or be carried in their ballast tanks and then be discharged into the new ecological environment. This process applied to the Atlantic lamprey eel that devastated fish stocks in the Upper Lakes. The sea lamprey that had existed in Lake Ontario was first found in Lake Erie in 1921. Established by 1940, its destruction of lake fish through attaching itself to the host fish then sucking its blood and flesh led to the severe decline of lake trout and other freshwater fish.

Zebra mussels were probably introduced into the Great Lakes in 1985 or 1986 when one or more trans-oceanic ships, after passing through the Welland Canal, discharged ballast water into Lake St Clair (Snyder 1992). By 1989 they had colonized Lake Erie, were reported in all the Great Lakes, and were expected to spread through most lakes and rivers in Canada and the United States. As the mussels fed on phytoplankton, the food chain might be adversely affected, which in turn would exert an impact on the Great Lakes fisheries. Docks, breakwalls, and the bottom of vessels were affected and, as the cooling water inlets of boat engines became clogged, engines overheated. Also serious was the attachment of mussels to the intake systems of power and water-treatment plants, leading to reduced pumping capabilities and occasional shutdowns. An interesting point is that canals constructed in the late 1700s allowed mussels to spread from the Caspian region through eastern Europe, and rapid canal expansion during the early 1800s permitted the zebra mussels to penetrate much of the continent, as well as Britain.

The Trading Scene

Tonnage passing through the Third Canal, at 3.9 million tons (3.5 million tonnes) by 1914, had reduced to 2.2 million tons (2 million tonnes) by the end of the First World War. By 1930, the last year when the Third Canal was in full commission, it had risen to 6.1 million tons (5.5 million tonnes). The impact of the Fourth Canal for trading expansion was then immediate. By 1932 the tonnage had increased to 8.5 million tons (7.7 million tonnes), and 5712 vessel transits were recorded (table 7).

The circumstances of the Second World War caused a relapse to 10.1 million tons (9.3 million tonnes) by 1943. Each postwar year from 1946 to 1951 then recorded an increase, leading to 16.2 million tons (14 million tonnes) by 1951, that is, more than double the tonnage that passed through the Third Canal in 1926 or 1927 (Urquhart 1965). The cargo passing through the canal system then rose steadily to over 20 million tons (18 million tonnes) and 8072 vessel transits by from 1955 to 1958 (St Lawrence Seaway Authority 1959).

Cargoes passing through the canal also changed considerably, as a reflection of Canada's industrial growth and the development of cities around the Great Lakes. In 1915 the sequence was wheat then coal, each at over 930,000 tons (904,000 tonnes), followed by lumber, boards, and timber, and then by pulp, wood pulp, and paper. Wheat and timber were traditionally the main cargoes, but coal for domestic use and for iron and steel manufacturing had increased greatly in importance. By 1940, wheat had doubled to 1.8 million tons (1.6 million tonnes), but this grain was now far behind coal at 4.6 million tons (4.1 million tonnes). Oil for domestic and commercial heating and for manufacturing purposes followed, then pulp, wood pulp and paper, and iron ore.

By the 1950s manufactured goods were higher or about equal to agricultural products, and by 1959, when the Department of Transport turned the canal's operation over to the St Lawrence Seaway Authority, the sequence of commodities carried through the Welland Canal was led by mine products at 13.3 million tons (12.1 million tonnes), or 48.6 per cent of the total

TABLE 7 Cargo tonnage through the Welland Canal, 1915–1958 (thousands of short tons)

Year	Wheat	Total agricultural products[1]	Manufactures and miscellaneous[2]	Forest products	Total coal	Total mine	Total all freight	Total up	Total down
Third Canal									
1915	955	1,397	329	399	936	1,125	3,061	756	2,305
1916	336	694	185	266	1,180	1,401	2,545	500	2,045
1917	435	563	184	244	1,301	1,500	2,491	313	2,177
1918	162	288	230	124	1,403	1,533	2,174	182	1,992
1919	547	779	239	92	986	1,061	2,171	188	1,982
1920	357	438	205	170	1,383	1,462	2,276	201	2,075
1921	859	1,874	205	155	773	843	3,075	276	2,800
1922	1,752	2,572	280	157	292	382	3,391	329	3,062
1923	1,915	2,562	254	170	596	770	3,756	339	3,417
1924	2,994	3,645	421	213	580	759	5,037	395	4,642
1925	2,261	3,500	519	288	1,059	1,274	5,640	594	5,046
1926	2,349	3,344	616	293	664	964	5,215	680	4,535
1927	3,913	5,047	787	356	567	1,058	7,247	915	6,333
1928	3,946	5,294	912	334	482	899	7,440	970	6,470
1929	1,685	2,439	944	332	747	1,055	4,770	868	3,902
1930	2,546	3,095	1,105	367	1,328	1,521	6,088	951	5,137
1931	2,146	3,161	1,194	350	2,008	2,570	7,274	987	6,288
Fourth Canal									
1932	2,927	4,123	1,722	313	1,990	2,379	8,573	1,354	7,184
1933	2,833	3,683	2,085	340	2,500	3,087	9,194	1,818	7,377
1934	1,787	2,769	2,032	501	2,890	3,979	9,280	2,204	7,076
1935	1,880	2,715	2,292	367	2,708	3,576	8,951	2,322	6,629

Year									
1936	2,327	3,183	1,859	538	3,179	4,857	10,437	2,706	7,731
1937	1,872	3,583	1,901	497	3,880	5,767	11,748	3,103	8,645
1938	2,653	5,987	1,754	433	3,127	4,459	12,633	1,956	10,677
1939	2,697	3,849	2,745	449	3,687	4,685	11,728	2,252	9,475
1940	1,824	2,789	3,675	446	4,584	5,997	12,906	1,803	11,104
1941	1,577	2,274	3,724	512	5,002	6,719	13,230	1,511	11,719
1942	428	717	3,264	422	4,828	6,705	11,108	928	10,180
1943	643	810	2,855	287	4,512	6,164	10,116	811	9,305
1944	1,130	1,478	2,966	280	4,951	6,593	11,316	982	10,334
1945	2,726	3,475	2,897	453	4,014	6,138	12,962	1,127	11,835
1946	1,263	2,084	3,038	376	3,584	5,082	10,580	1,416	9,164
1947	1,528	2,405	3,228	501	3,877	5,671	11,806	1,945	9,861
1948	1,536	2,470	3,233	523	4,724	7,148	13,373	2,135	11,239
1949	2,890	4,476	3,501	504	3,391	5,211	13,692	2,141	11,552
1950	2,025	3,672	3,588	532	4,687	6,949	14,741	2,732	12,009
1951	2,808	4,118	4,076	614	4,842	7,390	16,198	2,752	13,445
1952	2,698	4,960	4,316	580	5,364	8,054	17,911	2,289	15,622
1953	2,795	5,607	4,138	500	5,966	9,297	19,542	2,582	16,960
1954	2,858	5,339	3,172	516	4,988	7,618	17,514	2,396	15,118
1955	2,733	5,336	3,470	510	5,422	9,515	20,894	4,260	16,634
1956	3,195	6,085	3,880	530	5,626	12,572	23,060	5,069	17,997
1957	2,764	5,054	4,296	561	5,503	12,462	22,373	5,141	17,232
1958	3,630	6,653	4,183	524	4,411	9,915	21,274	5,006	16,269

Source: Urquhart 1965

tonnage, with iron ore then coal dominating this movement. Agricultural products came second at 8.8 million tons (8.0 million tonnes), or 32.0 per cent of the total tonnage. The dominant grain, Canadian wheat, was followed at a much lower level by American corn, Canadian barley, and then by American corn, American barley, and American wheat.

The total for Canadian grains was only 1.4 times that for American grains. Manufactured goods were third in importance at 4.4 million tons (4.0 million tons). The greatest single impact on the canal was that these cargoes now generally passed through the canal without stopping. Before 1932, the larger vessels on the Great Lakes, unable to pass through the canal because of their size, had either to unload and use the elevators at Port Colborne or transfer their cargo to smaller boats or the railway. With the larger dimensions of the Fourth Canal they could now pass through to Kingston and need no longer stop at Port Colborne. Port Dalhousie had been dead-ended, and the vessels had no reason to stop at Port Weller.

The canal communities had increasingly to rely for their existence and growth on industrial and service pursuits that in many instances had been introduced as an outcome of the canal. These proved sufficient to maintain and often to expand urban virility, despite the declining relevance of marine traffic for urban development. Inertia from the past was again exerting its powerful impact on the future.

12

An Interlude in the Process of Urban Growth

The chapter heading here refers to the transition period in the growth of settlement along the canal between the outbreak of the First World War in 1914, when transport by rail was dominant and the motor vehicle had just started to appear on the urban horizon, and 1959 after the Second World War, when the railway had started to decline and the motor vehicle to predominate. Regionally, the development and growing use of hydroelectricity, the diminished use of the Second Canal as a source of power for industry, the increasing capacity but not the expansion of the railway network, and public transit by the streetcar contributed a set of parallel events that helped to change the regional environment. On the social side, the resumption of large-scale immigration from Europe helped to staff the new industrial enterprises and contributed materially to urban growth. At the same time, the adverse factor of pollution sullied the reputation of the canal environment along the valley of the Second Canal north from Thorold to Lake Ontario.

Hydroelectric Power

By 1931 when the Fourth Canal came into operation, water was taken from the canal system at some thirty points. The largest volume was abstracted at Port Robinson to serve the power and water plants at DeCew Falls, but water was also withdrawn at Thorold in ten locations, St Catharines (8), Merritton, Wainfleet (2), and Welland (1).

As the generation of hydroelectric power continued to expand tremendously, three important changes of emphasis were of direct consequence for the Welland Canal and its communities. The first change was that more water had to be drawn from the Fourth Canal at Allanburg than previously from the Third Canal at the same location in order to provide for increased power generation at DeCew Falls. Second, at The Falls the emphasis on power generation moved north to the greater head and power potential available at Queenston Heights.

The third change relates to power distribution. As the power grid expanded across Ontario, this policy of expansion at first reduced then eliminated the power advantages that had existed within the canal communities because of their cheaper rates through proximity to DeCew and The Falls. Rather than concentrating on the provision of low-cost power in close proximity to the Niagara River as on the American side, where an intense area of manufacturing industry developed from Niagara Falls to Buffalo, Ontario Hydro dispersed this power and by 1920 supplied fourteen local electrical utility systems, over 275 municipalities, and some 340,000 horsepower from The Falls. Then in 1966 Ontario Hydro equalized power rates across Ontario, which nullified any cost advantages that may previously have been held by the canal communities.

As the urban and industrial demands for power grew, the Hydro-Electric Power Commission of the Province of Ontario (later Ontario Hydro), built or obtained its own generating stations. At The Falls the American Ontario Power Station Company, purchased in 1917, was expanded in 1918. The Toronto Power Company fought the issue of public power, but its Niagara Plant was acquired in 1922. Operated by Ontario Hydro until 1974, its water allocation was then transferred to the more economic and efficient Queenston generating stations. Ontario Hydro in 1929 also purchased from its private owners the Hamilton Cataract Power, Light and Traction Company (formerly Cataract Power) at DeCew Falls along with its transmission systems.

In 1917, Ontario Hydro moved to take advantage of a fall of about 290 feet (88.4 m) against some 180 feet (54.9 m) at The Falls if water were diverted from above the rapids to the brow of the Niagara Escarpment at Queenston Heights. The Queenston-Chippawa power development was approved by the Ontario legislature that year, the underlying factor being international control over the amount of water required to flow over The Falls by the Boundary Waters Treaty of 1910. The same flow of water now had to be used with greater efficiency and the treaty allowance, equivalent to some 408,000 horsepower at The Falls, would produce closer to one million horsepower by diversion to Queenston Heights. The first unit of the Queenston-Chippawa Power Development came into service in 1921. Its head of 294 feet (89.6 m) compared with 180 feet (56.7 m) at Ontario Power at The Falls and 262 feet (86 m) at DeCew Falls.

Later named the Sir Adam Beck–Niagara Generating Station after the commission's first chairman, the Queenston project was related closely to the canal system. Not only had it competed successfully during the post-war construction phases of the Fourth Canal for labour and supplies, it also took advantage of the Chippawa Cut of the First Canal. To supply the generating plant with its head of water, the Lower Welland River was deepened to 40 feet (12.2 m) and its flow reversed for 4 miles (6.4 km) in order to draw water to Queenston Heights from the Upper Niagara River. The former canal channel of the Chippawa Cut and the canalized Lower Welland River had become a feeder to the new power project.

As the demand for power continued to escalate, the Sir Adam Beck and DeCew plants received major expansion after the Second World War. The former obtained a more than threefold increase in capacity when Sir Adam Beck No. 2 opened in 1954. At DeCew new penstocks, expanded storage, and a new generating plant were added during the 1940s. As the flow of international water from Lake Erie through the Welland Canal was involved in this additional diversion, to satisfy treaty requirements water was diverted from Arctic-flowing rivers to Lake Superior and the Great Lakes system. This augmented flow added some 5000 c.f.s. (141.6 m³), which could be used successively by a number of power stations along the Great Lakes–St Lawrence system. The Welland Canal, in addition to the production of power from the diversion of its waters, now also provided a conduit for the transfer of water to downstream locations for hydroelectric production. The canal's versatility had extended far beyond ships, trade, and its industrial and community developments.

Water Power and the Second Canal by 1929

Water power from the canal, of major importance throughout the canal's history, remained of considerable consequence despite the advancing use of hydroelectricity generated from the Niagara River. In 1929 the Chief Engineer's Branch of the Department of Railways and Canals reported on the physical condition of the various industries that had or were operating under lease or attornment along the Second Canal and its Hydraulic Raceway between Locks 1 and 25. As two major categories existed, where water was still being taken from the canal and where former water-powered industries had changed to the use of hydro-electricity, this study is of industrial transition from the traditional source of power to the new, expanding, and now more typical resource of hydroelectricity.

Typically, when industries still used canal water, the electricity generated at the plant was used for motors and lighting. Outside power was not generally purchased to meet normal requirements, although some might be obtained to meet high load requirements, and power was no longer supplied to outside users. City water was supplied to the plant for toilet and fire-protection purposes, and the waste water and sewage were discharged via the tail race into the canal. These arrangements operated for ten hours a day for six days a week, and for twenty-four hours a day where the machinery had to be kept in operation on a continuous basis.

At Lock 1, Port Dalhousie, Canadian Consolidated Rubber had three turbines installed to manufacture various types of rubber footwear from crude rubber. Upstream at Lock 2, Welland Vale Manufacturing with their output of pitchforks, garden rakes, hoes, axes, and hammers used turbine operations to run machinery in the forge, electric-welding, and machine shops. In this plant some additional power was provided by Lincoln Electric for load and lighting requirements.

Garden City Paper situated above Lock 7 had two turbines with process water taken from the canal to manufacture tissue paper from blanket pulp stock. In Thorold were the Provincial Paper Mills above Lock 21, the Thorold Pulp Company at Lock 22, and a pumping station for the town of Thorold between Locks 23 and 24. Its water supply obtained from the canal above Lock 25 was taken by a main to a settling reservoir, then pumped to the town's system with about 400 gallons (1514 litres) per minute being supplied to the Provincial Paper Mills.

More frequent than the above were those industries that had closed or made the transition to another source of supply. For instance, canal water was no longer used at Lock 3 where the plant of Lincoln Electric Light and Power, which had generated about 500 HP, had closed in 1911. By 1929 this site was used principally as a voltage regulator on the main transmission lines of the Hamilton Cataract Power, Light and Traction Company.

Kinleith Paper below this lock had operated from about 1900 until 1924. When in service it had drawn about 10 million gallons (37.9 million litres) of water a day from the Hydraulic Raceway to produce some 25 to 30 tons (22.7 to 27.2 tonnes) of bond and high-grade papers per day. At Hedley Shaw Milling on the southern end of the Hydraulic Raceway, the spillway and the foundation of the mill were still evident by 1929, but the structure had burned down about 1921. Another property on the south side of the raceway, described as the Jamieson Black Estate had been used as a flouring mill. This had ceased operation in 1926 and the wooden head race and flume were stated to be in poor shape.

Canada Hair Cloth, to the south of St Paul Street in St Catharines and described as 'a well preserved brick structure with modern equipment for the manufacture of hair cloth and linings,' had two turbines that operated from the mid-1890s until 1928. The volume of process water drawn from the raceway had averaged about 102 c.f.s. (2.9 m³), but this supply was now obtained from the City of St Catharines. Packard Electric to the east, supplied from the same raceway, manufactured transformers, switchboard's and house motors. Its water lease lapsed in 1917, when power was furnished by Ontario Hydro and the Lincoln Electric Power Company.

At Lock 5 were the ruins of the Richard Collier plant. The lease that had covered power sufficient to drive one upright saw and its equivalent machinery had recently been acquired by General Motors from McKinnon Dash. Also disused and owned by McKinnon Dash, the former Whitman and Barnes plant with wheels set within wooden flumes installed at both ends of the building turned line shafts that were approximately 350 feet (106.7 m) long. Hammers, wrenches, pliers, screw drivers, and various types of forging were each produced until 1926.

Canada Carbide owned power units and generated electricity at Locks 8, 9, and 10. At Lock 8 the power was transmitted to Lock 10, where it was used by the company in the furnace room until this unit closed in 1924. At Lock 9, where power had been transmitted to Lock 10, duplicating the detail at Lock 8, the plant had not operated since 1922. The principal plant for the manufacture of carbide from charcoal and lime was adjacent to Lock 10. Sold, then renovated by a Toronto company to manufacture nitric acid, this plant operated for a only a short period before it came back into the hands of the Carbide Company (Shawinigan Chemicals), but operations ceased in 1924.

Hayes Wheel operated at Lock 11 to manufacture forgings, axles, and housings for various Canadian motor companies. The plant, a modern brick structure, usually worked for twenty-four hours a day, six days a week. One turbine was installed at Lock 11, and power was also obtained from Lock 12 and from Lincoln Power and Lincoln Electric. The powerhouse at Lock 12, erected by the St Catharines and Niagara Street Railway Company, had been taken over by the Lincoln Paper Company before sale to Hayes Wheel. Its water lease also included Locks 12, 13, and 14. Hayes Wheel stated in 1929 that less power was available than the estimated output. This loss in power was attributed to 'the presence of a coating of wood fibre stock on the inlet of the runners.' With regard to the proposed substitution of electricity in place of water power, the changeover was deemed feasible and more reliable, but the cost would greatly exceed the rental paid under the existing lease.

The Alliance Paper Mills operated two plants on the east bank of the canal. The Lincoln Mill, which manufactured blanket wood pulp and paper between Locks 11 and 12, closed in 1928 through the cancellation of the attornment for the removal of water, only seven years after its modern power station had opened in 1921. The entire plant was described as being in 'excellent condition and today would have a probable value of $35,000 were water available for its operation.'

The Lybster Mill between Locks 14 and 15 manufactured high-grade paper. Three steel penstocks fed to the wheelroom from a concrete forebay, but low efficiency at some of the hydraulic installations led management to state that 'the substitution of electricity for water power developed would be favoured.' This proviso came with the expectation that 'the probable withdrawal of the water from the Old Canal would no doubt lead to the Government making a substantial subsidy in the matter of assisting the various industries to procure an adequate supply of water satisfactory for manufacturing purposes' (Chief Engineer's Branch 1929).

Independent Rubber at Lock 15, a stone building constructed as a cotton factory then converted to manufacture hair cloth, then taken over in 1910 by the rubber company, used two separate hydraulic installations to manufacture rubber boots and overshoes. The east flume served three vertical turbines that provided power to a single line shaft that served fourteen motors. The west flume had one turbine that supplied the power to operate four machines. These operations lasted until 1922 when the plant closed down, and much of the machinery was then removed and installed in the plant of Consolidated Rubber at Port Dalhousie, which held a lien on Independent Rubber's property.

The Riordon (Riordan) lease covered surplus water from Locks 16 to 20 for use in development at four sites. As described in 1929, 'The entire plant has been closed down since about February 5th, 1921, and is rapidly falling to ruin. There are various buildings of frame, stone, brick and reinforced concrete construction in which doors and windows are open or broken and in many cases large sections of the walls have fallen. The bulk of the machinery has been removed while the greater part of what is still in place has been partially dismantled or demolished.'

At Lock 23 a mill duplicating the name of Thorold's Provincial Paper Mills was a modern paper manufacturing plant. This mill had formerly drawn water from the canal, but the use of this water was discontinued in 1927 through pollution by wood fibres from the mills above. Two paper-making machines operated using 600,000 gallons (2,271,240 litres) a day, but this demand was now supplied from the town of Thorold's pumping station between Locks 23 and 24. At the former Monroe flouring mill, a three-storey frame structure also located between Locks 23 and 24, the plant remained in use, but it was now used entirely for feed grinding and was in operation on average for only two days a week. Maple Leaf Milling below Lock 24 had operated as a flour mill until 1925, but the stone building was now used as a storehouse for flour and meal being distributed to neighbouring centres by truck.

At Lock 25 the Davey Pulp and Paper plant had originally operated as a ground-wood blanket factory, but the wet machines had been removed and the grinder was used occasionally when a shutdown occurred at the Foley-Rieger plant, which was also controlled by Davey Pulp. This plant, also another operation that worked twenty-four hours a day for six days a week, took water from above Lock 25 of the Second Canal to produce blanket ground wood. The output varied from 7 to 8 tons (6.4 to 7.3 tonnes) of ground wood per day. Another Davey plant below Lock 25 had a pulp stock machine for the manufacture of building board. This plant had originally been used for the manufacture of hydraulic cement, but after conversion to meet the requirements of Rocmac Cement, it lay empty for seven years until purchased for experimental purposes.

These accounts about the conditions along the abandoned escarpment length of the canal system have portrayed the advance, decline, and also the reinvigoration of industry in association with the canal. Some further evidence has been provided for the remarkable and sudden changes in the urban industrial structure. In part 2 of this volume the theme was expansion, encouraged and promoted by industrialists, the city fathers, and the public alike. Now, a few decades later, the reverse feature of industrial instability and an uncertain industrial future is starting to appear. Some large and established industries were prospering and expanding. In other instances decline and oblivion were more to the fore. Industry created, extolled, and built up in association with the First and Second Canals had now become a transitory feature in the evolving urban landscape. The canal environment had changed to an unsure and perhaps capricious instrument of urban growth and achievement.

There was considerable controversy over the matter of whether compensation should be paid for the cancellation of water privileges. The view of riparian owners was that the land, no longer required for raceway purposes, should be returned to the original owners with compensation for the investment by industry in the installation of turbines and equipment fed from this raceway. The alternative argument by government was that the provision of hydroelectricity was more efficient and reliable, albeit more costly than the cheaper use of canal water, and that the industries had received the advantages of a cheap supply of power over the years.

The Railway System

Railways reached their zenith of importance during the period from 1914 to the late 1950s. The railway network did not expand, but the number, size, and capacity of trains increased, as did the number of passengers and the amount of freight that they carried. During this railway period of development par excellence, all of the larger industrial sites along the Fourth Canal either had railway access or were approached by operating spur lines. Many sites had both canal and railway facilities, and those along the abandoned northern length of the Second Canal used the freight services of the interurban streetcar service.

By the early 1950s this situation of railway pre-eminence started to change. The flows of passenger traffic diminished as customers transferred from the streetcar to automobiles and buses for shorter journeys, and to the airplane for long-distance travel. The railways remained important for the through movement of freight, but as traffic transferred to trucks on the highways, freight yards and industrial spur lines serving factory sites and docks became disused and abandoned. The number of trains per day was reduced, stations were closed for passenger services, and the less important lines such as the Erie and Ontario route along the Niagara River were abandoned. Indicative of changing circumstances, the last steam locomotive on regular passenger service passed through St Catharines in 1959. The next year all freight operations in the Niagara area used diesel rather than steam.

The relative importance of the canal and railways in the location of industry in the canal communities was measured by Martine, who in 1961 concluded: 'The Welland Canal was the most important factor in attracting industries to this area. All the Canal towns, with the exception of St Catharines, appear to have benefited greatly from the presence of the Canal. About half of the industrial establishments interviewed [20 out of 44] list it as a major locational factor. The Welland Canal was most important as a transportation factor, although it was also significant as a source of water power and industrial water ... In St Catharines, rail or road facilities and proximity to market become most significant.' When Martine examined the relative importance of rail or road facilities, she stated: 'This factor becomes most important in Welland where it was cited as a major factor by almost half of the industries interviewed. This was to be expected, for ... it was the railways that were mainly responsible for the city's rapid growth after 1905. It is interesting to note that rail or road facilities played little or no part in attracting industry to Thorold, Merritton, Port Dalhousie, and Port Weller.'

Cheap hydroelectric power ranked as the third locational factor, 'being most important in Welland and Thorold. However, ... it was never expressed alone – that is, cheap power attracted the industry to this general area but another factor actually determined the particular industrial site.' When the major and the minor factors influencing the locational decision were combined, then the significance of the canal as an agent for industrial development increased substantially. It rose to 70 per cent of the responses, against less than 40 per cent for rail or road facilities.

The Streetcar Network

Centred on St Paul Street in St Catharines, all divisions of the interurban N.S.&T. (Niagara, St Catharines and Toronto) streetcar network operated during the 1920s. These operations included routes south to Thorold then east to Niagara Falls, north to Port Dalhousie, south via Fonthill along the west side of the canal to Welland and Port Colborne, and east to Port Weller and Niagara-on-the-Lake. Only Allanburg and Port Robinson to the east of the canal had no direct access to the regional streetcar system.

The frequency of service was hourly to Port Dalhousie and to Port Colborne, but generally at twenty-minute intervals in urban areas, such as across St Catharines between McKinnon's (later General Motors) and Victoria Lawn Cemetery. In addition to its heavy passenger traffic, the N.S.&T. carried considerable volumes of freight on flat or in box cars, often at night to avoid conflicts with passenger travel. This cargo included cement going to work projects on the Fourth Canal, and freight traffic from the spur lines that served industrial sidings at places such as Welland Vale and the Garden City Paper Mill in St Catharines, Interlake Tissue in Merritton, and Walker's Quarry in Thorold. Freight was carried to the steamship services at Port Dalhousie and to the railways in Welland until 1922 to compete with the Grand Trunk monopoly, a movement that finished when the Grand Trunk became part of the Canadian National system. Placed under Canadian National management in 1925, the length of the N.S.&T. expanded to its peak extent of 104.3 miles (138 km) in 1928 (Mills 1967).

The peak passenger years had been in the early 1920s, but passenger traffic then declined through the spread of paved highways, the advance of the automobile, and competitive bus services along the same or comparable routes. The contraction of the routes began in 1931, when the tracks between Port Weller and Niagara-on-the-Lake were abandoned and a bus service was introduced. After some wartime revival and new traffic peaks, services between Thorold and Niagara Falls ceased in 1947; the main-line passenger service between St Catharines and Thorold ended two years later.

At Port Dalhousie two events preceded the end of regular steamer services across the lake to Toronto. The *Northumberland* burned at her moorings in 1949, and next year *Dalhousie City* was sold. The streetcar passenger operations between St Catharines and Port Dalhousie ended and buses took over this route. In 1959 the last scheduled electric interurban passenger movement in Canada, the linear link between Thorold and Port Colborne that interconnected the canal communities, ended. Freight services fared somewhat better, and lasted longer. Electric freight operations continued between St Catharines and Port Colborne until mid-1960, when all freight services were changed to diesel.

After their abandonment for passenger and freight services, the lines of track were generally either uprooted or covered with asphalt, and the overhead wires and poles were removed. Only certain routes remained. Used by diesel freight only, they included General Motors on Ontario Street in St Catharines served via Welland Avenue and Louisa Street, but with no passenger traffic and now operated by Canadian National as part of its railway service.

The Advance of the Automobile

The great predator of the streetcar, and then of bus and railway traffic, was the motor vehicle. No longer an oddity by 1914, it had grown in popularity since the 1890s, first as a rich man's toy, but then mass production soon made widespread ownership possible. In 1908 Henry Ford launched his Model T assembly line, and over the next twenty years some fifteen million of these vehicles were produced. Specialized vehicles serving a vast array of purposes were steadily added – fire engines, ambulances, buses, and a vast array of trucks and commercial vehicles for local and then more distant deliveries.

This growth slowed during the two world wars and the Depression, but by the 1950s the motor vehicle had become the most widespread mode of transportation. Its essential features were the freedom and ease of personal, family, and freight movements for short and long journeys over an expanding highway network that penetrated virtually everywhere. It was a major revolution in transportation and the life of society, more so than even the First Canal or the first railway, important as they were in their day.

The provision of space to accommodate the motor vehicle had necessarily to follow. The vehicle, at first described as a horseless carriage because of its appearance, and then as a bone-shaker because of its uncomfortable travel on rutted roads that were either muddy or dusty through weather conditions, demanded improved conditions along the highways and within urban areas. The first task, a provincial responsibility, was the need to rebuild to much-improved standards the pre-existing pattern of roads that linked the major towns and villages. These main routes had to be widened, paved with hard surfaces, crowned, and provided with side drainage. More gentle slopes and curves were introduced along the paved new intercity routes, and bridges or embankments were constructed to cross obstacles such as the canal, drainage channels, river valleys, and steeper surface depressions.

This new world of transportation was recognized when the Fourth Canal was designed in the 1910s. When it opened, four of the new canal bridges carried the main provincial roads that traversed the peninsula inland from the American boundary and its bridge crossings of the Niagara River. Highway 8 (now Highway 81) from the Queenston-Lewiston Bridge crossed the Fourth Canal at Homer as Queenston Road at Bridge 4, and also the abandoned Second Canal at Burgoyne Bridge next to the centre of St Catharines. The two highway bridges that crossed the Niagara River at The Falls led to Highway 20, which crossed the canal at Bridge 11 in Allanburg. And Highway 3 inland from the Peace Bridge at Fort Erie branched via Main Street and Bridge 19 in Humberstone and to Bridge 21 on Clarence Street in Port Colborne. The other canal bridges carried the county and the more important municipal and township roads that together formed the regional network.

Meantime, in addition to the provincial and canal developments, every municipality, business, and owner of a vehicle had necessarily to make their own considered improvements as the motor vehicle expanded in numbers. Automobiles, vans, and trucks required an abundance of new surfaced roads, and then a host of devices to direct and control the growing volumes of traffic. Also, in addition to these necessary provisions for the moving vehicle, parking space for the new mode of travel had to be provided at every destination, including houses in residential areas, in shopping localities, and at places of work, indeed at every type of building and land-use activity.

As the canal communities had each grown on the basis of pedestrian, horse, wagon, carriage, canal, then railway, and often streetcar provision for public movement, the task (still continuing by the 1990s) was gigantic. The form of each canal community was now transformed. It could reasonably be described as having both an inner and an outer area, the inner area with close historical associations with the canal and the outer area more distant from the canal where the layout and urban form were more closely associated with the automobile.

The motor vehicle reshaped every community, the centres along the canal being no exception to the response that prevailed generally across North America. Some buildings in the business centres were purchased and cleared to provide the necessary space. If a larger older home was used for a business or professional activity, the trees, bushes, and grass were cleared as the open space became a parking lot. Mature trees that lined roads were felled to widen the width of passage for the motor vehicle. The paved streets became painted with lines and signs were added to provide directions, control speeds, regulate parking, and advertise the services available in nearby premises. The business centre and their qualities of townscape became dominated by machines rather than, as had been their primary use over the decades, by pedestrians and buildings. Recreational areas such as Port Dalhousie that had depended on mass transit by railway, streetcar, and steamer for their recreational success now had to face the automobile and the choice in destinations that it introduced.

Vehicles were given priority over perambulating pedestrians. St Catharines had a campaign against jaywalking in 1923. The city's first traffic light was installed in 1929 and meters to control parking on some streets in 1947. With through traffic along the city's main street and an active business centre compounding the traffic situation, a report of 1950 urged the elimination of all parking space along the main streets and the extensive provision of parking areas in the business centre. In 1954 a one-way street system was introduced to resolve the problems of traffic congestion in the town centre. The traditional two-way artery of St Paul Street at the top of the canal bank became one-way to the east, and King Street carried westbound traffic. Welland soon followed with a similar type of one-way traffic control in its town centre.

As journeys could now be made in the family automobile and as more and more families gradually had this facility every canal centre grew outwards with residential suburbs on the urban fringe, strip development along the main highways, and all forms and types of urban development in isolated patches on both the urban fringe and in rural areas. Light manufacturing and service industries were now more free to locate where they pleased, and reliance on railways for the movement of goods and the street-car for the movement of people each diminished. The journey to work by car now enabled all employees to live where they chose, rather than close to their place of employment as previously.

In so far as the canal was concerned in this new urban process, the most radical new feature was that canal communities with some hundred years of achievement and change, relatively compact and cohesive in form, clus-tered along the canal and each with distinctive attributes of character, roads, services, and buildings, now spreadeagled outwards. Adjacent cen-tres gradually merged and coalesced, so that the urban structure became larger, more amorphous, and less clearly defined as a series of separated centres. The canal was seen as generating a ribbon of development, and the canal environment became an extended yet interlinked urban agglomera-tion between Lake Erie and Lake Ontario within which only the old, now historic centres retained certain waterfront associations inherited from the past as canal communities. And as the towns grew on their urban fringe, shops and stores moved outwards to serve this population. This trade was lost to the urban centres and their decline as active retailing nodes started. Urban renewal schemes became necessary in an attempt to retain former qualities at Port Dalhousie, St Catharines, Thorold, Welland, and Port Colborne.

During this period of urban transformation, an increasing flow of goods continued to move through the Fourth Canal as the Canadian and Ameri-can economies each expanded, and as the heartland of North America around the Lower and the Upper Great Lakes added population and new developments to its industrial strengths. Within the context of interna-tional transportation change, the canal communities rather than the canal and its maritime trade were now more closely affected as the emphasis in transportation moved to the motor vehicle.

Multi-Lane and Scenic Highways

Two new types of regional highway, the scenic Niagara River Parkway and the multi-lane Queen Elizabeth Way (QEW), exerted their impact on the canal scene. At the Niagara River the vision in 1904 was that work should commence without delay on a 'well-built boulevard, properly ornamented with shade trees, constructed and maintained over the whole distance' from Fort Erie to Niagara-on-the-Lake (Way 1960). The first section from Fort Erie to Victoria Park at The Falls was completed by 1915. Delayed by the

First World War, the parkway eventually reached Niagara-on-the-Lake by 1931. This was the period when the Fourth Canal was under construction and approaching completion and, as has been noted, an extension of the parkway to the Welland Canal as a circular tour was envisaged.

It remains a matter for some puzzlement why the attractive idea of a comparable scenic parkway along the Fourth Canal was not seriously pursued by the senior levels of governments. A parkway would have linked these canal works of world-class fame. A parkway might easily have been constructed in conjunction with the service roads that followed the canal. The land was owned by the dominion government, it was a period of economic depression during which relief projects eased the severe unemployment, and the vision of a lake-to-lake boulevard had been well received by the public and in the press.

The absence of a Welland Canals Parkway may with hindsight be perceived as a severe omission from the plans that were prepared for the Fourth Canal. These designs were based on the engineering needs to permit the passage of vessels and even the provision of tourist viewing areas did not occur until after the Second World War. By contrast, the ideals and potential of a scenic parkway were extended to the construction works for the route that became the Queen Elizabeth Way by Thomas B. McQuesten, a member then chairman of the Niagara Parks Commission, who had considerable experience with the achievement of the Niagara River Parkway. Later, in 1930, as the Ontario minister of highways he emphasized that, although the QEW had been designed primarily to improve traffic flow, amenities and landscaping should be added so that this new highway route would also serve as a parkway (Barnsley 1987) – an attitude similar to that expressed by the canal engineers in the late 1920s.

The QEW, which started in 1916 as an idea for a new highway to be constructed between the Lakeshore Road (Highway 2) and the Dundas Highway (Highway 5) to relieve traffic congestion between Toronto and Hamilton, was initiated as a labour-relief program during the Depression. By the early 1930s the concept had been extended from Burlington to the American border at the Peace Bridge in Fort Erie, redesigned so that the opposing lanes would be divided with access restricted to controlled points at cloverleaf interchanges. The route was to be landscaped along its length as a 'park' road (Ministry of Transportation and Communications 1975; Stamp 1987).

The Middle Road / Niagara Highway was opened officially by King George V and Queen Elizabeth at St Catharines, and was then renamed the Queen Elizabeth Way (QEW). This route was extended from Niagara Falls to Fort Erie as a two-lane, gravel-surfaced road in 1941, the west-bound lanes were completed in 1946, and both sections were paved and opened for traffic in 1946.

Like the canal and railway previously, the QEW had only a limited number of access points. The highway's purpose was the long-distance, fast movement of vehicular traffic but its repercussions on settlement were drastic. As the multi-lane concept spread, a much closer interaction was achieved between different regions and cities than had ever formerly existed. The improved highway network of the 1920s and the 1930s had loosened the bonds that had previously contained the physical extent of cities, but the QEW and its like were to sever these completely. The new points of emphasis in the landscape became the points of interchange between the expressway and the regional highway network and their vicinities.

Specifically in relation to the canal, the Henley Bridge, located where the QEW crossed the abandoned route of the Second Canal at Twelve Mile Creek in St Catharines, was designed as a Viking ship. Pride in the new highway venture was expressed in the form of a concrete prow in the median lanes that seemed to be breasting the oncoming waves of traffic. The design of the bridge in the form of a vessel also symbolized the former mode of marine transportation in the valley of Twelve Mile Creek below. As a contributor to urban development, this bridge also helped to offset the long-term isolation of the rural area that lay to the west of St Catharines because of the barrier presented by the Second Canal to movement.

Where the QEW crossed the modern Fourth Canal, traffic was at first diverted to Bridge 4 on Queenston Road in St Catharines. Here hectic delays to road traffic arose as slow-moving freighters passed under the raised spans of this bridge. The answer lay in the provision of a high-level bridge, which opened as the Garden City Skyway in 1963. Situated next to the earlier bascule bridge, which was retained to carry local traffic, two bridges of very different design now crossed the Fourth Canal on the eastern fringe of St Catharines.

The Skyway carried six lanes of highway traffic. It soared majestically to a vertical clearance of 123 feet (37.5 m) over the waterway below to allow the free passage of ships at all times underneath. The bridge itself had nine central spans totalling 2096 feet (638.9 m) in length and the approach spans totalled a further 5112 feet (1558.1 m). The canal had obtained an exceptional new landscape feature and the most impressive of all the bridges to cross the Fourth Canal. Ironically, although its piers were set within the canal, this bridge was not added to the numbered sequence of canal bridges.

TABLE 8 The population of the canal communities, 1911–1951

	1911	1921	1931	1941	1951
Port Dalhousie	1,152	1,492	1,547	1,723	2,616
St Catharines	12,484	19,881	24,753	30,275	37,984
Merritton	1,670	2,544	2,523	2,993	4,714
Grantham Township	2,439	4,412	5,364	7,052	15,411
Thorold	2,273	4,825	5,092	5,305	6,397
Thorold Township (including Allanburg and Port Robinson)	2,085	4,739	4,451	4,967	6,522
Welland	5,318	8,654	10,709	12,506	15,382
Crowland Township	1,667	3,826	4,999	6,638	12,086
Humberstone	–	1,524	2,490	2,963	3,895
Port Colborne	1,624	3,415	6,503	6,993	8,275
Humberstone Township	3,417	2,304	2,407	2,945	3,923
Total	34,129	57,616	60,838	84,360	117,205

Source: Census of Canada, 1911–51

For the first time in the history of the canal, however, land and marine traffic now each had an unimpeded route.

The Garden City Skyway also reflected the considerable increase in highway traffic flows, well above what had been contemplated for the narrow lift-bridges that crossed the Fourth Canal when it was designed in the 1910s. A new era of transportation had dawned. Its pressures on the streetcar and railway modes of movement have been noted, but the canal could still compete successfully with the highway truck for the long-distance movement of heavy, bulk commodities.

Population Change and Urban Expansion

Despite the two world wars and the Depression years, over the forty-year period between 1911 and 1951 the population in the canal communities and their adjacent townships increased more than threefold, from 34,129 to 117,205 (table 8). The largest absolute increase was from 12,484 to 37,984 at St Catharines, but Port Colborne recorded a higher fivefold increase in population size. The highest township increases stemmed from the movement outwards from St Catharines into Grantham and from Welland into Crowland, which demonstrated suburban expansion and the new freedoms of residential-workplace location introduced by the automobile.

By 1951, St Catharines remained the dominant centre on the canal, its population being two and a half times that of Welland at 15,382, its nearest successor. Port Colborne, third in importance, followed with a population of 8275, then Thorold (6397), Merritton (4714), and Port Dalhousie (2616). In national ranking, St Catharines was in twenty-sixth place, whereas in 1871 it had been eleventh. The city had grown, but its relative status had diminished through the western growth of the Canadian economy. Welland by 1951 had dropped to 79th place, with Port Colborne being 130th and Thorold, at 175th, even lower. The urbanized canal corridor had become relatively less important as an urban entity both nationally and provincially as other urban areas advanced to prominence.

With respect to the employed labour force, St Catharines at 1951 still dominated the canal corridor with 16,235 employees, of whom a third were female. At Welland, the second in employment importance of the canal communities, 6534 workers were employed but rather fewer, at 30.7 per cent, were included in the female labour force. Manufacturing, the most important occupation in both centres, employed 5262 persons (32.2 per cent of the labour force) in St Catharines. A further 2076 were active in manufacturing in Welland, with this proportion being 31.8 per cent of the labour force. Together, these two cities had 167 manufacturing plants (Canada Year Book 1955). The emphasis in both places was on metal products: 72.4 per cent of the employed labour force in manufacturing at St Catharines and 59.3 per cent in Welland.

This situation with regard to the growth of population and its manufacturing base is reflected in the associated administrative changes that improved the status of the canal communities and recognized the nature of their growing urban importance. Merritton became a town in 1918 after wartime expansion, and Port Dalhousie followed after the Second World War in 1948. In 1917, during the First World War, Port Colborne became a town and Welland a city. The urban sequence by 1951 remained that of an assertive lead by St Catharines, followed in sequence by Welland, Port Colborne, Thorold, Merritton, and Port Dalhousie.

There was also considerable expansion of the municipal boundaries as the various canal centres spilled outwards across their restricted administrative bounds into the rural areas of their adjacent townships (Krushelnicki 1994). Port Dalhousie expanded to the west along its peninsula between Martindale Pond and Lake Ontario when it became a town. St Catharines in 1946, and again in 1948, expanded its area south towards the Niagara Escarpment as a reflection of the new access created by Glenridge Bridge across the valley of the Second Canal, to the north up to and across

the QEW and in the east towards the Fourth Canal. Merritton expanded to
the north in 1934, and west and east in 1946. Port Dalhousie, St Catharines
and Merritton, each centred on earlier canals, had now in essence become
one interwoven but multicentred community through these surging suburban developments.

Along the southern reach of the canal, Welland expanded to the south in
1917 and then to the north in 1950. Humberstone grew to the east of the
canal in 1935. Port Colborne, having taken over the new harbour area in
Lake Erie beyond the shoreline in 1912, advanced to the east in 1950, then
amalgamated with the village of Humberstone in 1952. Port Colborne and
Welland on their outer fringes were each advancing towards the other.

This outward dispersal from the established canal centres reflects their
expanding industrial activities and the requirements for homes and services
generated by this expansion. The old and modern canals remained at or
close to the core of the southern communities in Port Colborne and
Welland. In the north, the Fourth Canal was situated more on the rural
fringe, but with older cores associated with earlier canals in their historic
cores of business activity at St Catharines, Merritton, and Thorold. Port
Dalhousie, the anomaly, retained its canal setting, but pleasure activities
had taken over from through lake vessels, though some continued to be
laid up to undergo repairs.

The Abandoned Second Canal and Its Communities

After first navigation and then water rights were eliminated at Thorold, in a
similar fashion to Dunnville an extensive section of the old canal, from near
the Thorold Tunnel through the old town north to the township line, was
drained and the channel filled to provide space for a range of different
activities. In the south, some lengths were used as public open space and a
park area close to the business centre provided a linear strip of open space
as a setting for the adjacent housing developments. Other sections of the
old canal next to the business centre were used for public buildings, such as
the Thorold Fire Department, and to provide parking space at the rear of
businesses on Front Street. To the north, former canal land was taken over
for business and industrial expansion, including an extensive provision for
car and truck parking and outdoor industrial storage.

Many remnant features of the earlier canals survived on or next to the
former canal bank (Orr 1979). They include the Welland Mills built in 1846
of local limestone, an outstanding example of nineteenth-century industrial
architecture and originally one of Upper Canada's largest flour-milling

enterprises. Close by, Maplehurst, a three-storey mansion built in 1864 by John Keefer, president of the Welland Canal Company and a mill owner in Thorold, provides a glimpse of earlier residential opulence. The restored fire hall of 1877 on Albert Street sits next to where this road crossed the canal at a lock. Most of the two- and three-storey brick buildings that housed stores and services also survived as a continuous development along both sides of Front Street, to a large extent in their original form as an important heritage endowment. Later they became the focus of an urban renewal project after a nearby shopping centre and other malls slowly devitalized the once active business centre.

To the north across the slope of the escarpment between Thorold and St Catharines, when the Hydraulic Raceway, the reservoirs, and the mill channels were de-watered, their sites either secured new activities or decayed into derelict land. A few mill buildings remained and the pulp and paper industry has remained important, but many of the industries at the lock sites became memories. The Merritton business centre survived as a small but distinctive retailing node, with buildings reminiscent of the canal era, but their trade potential was jeopardized severely by the nearby regional shopping mall of the Pen Centre and by strip commercial development along Hartzel Road to the north (Merritton Residents' Neighbourhood Improvement Committee 1979).

In St Catharines, a steady expansion took place in the number and type of retailing establishments. The three outlets classified as department stores at 1921 increased to six by 1938. Most stores were the traditional small, family-owned enterprises with a specialized concentration on a limited range of goods, but the change to multiple external ownership was expressed when the locally owned McLaren's Department Store, in existence from the turn of the century, was taken over by T. Eaton Company from Toronto in the 1920s.

The long-time tradition of central area shopping, promoted by the streetcar but then discouraged in favour of suburban shopping as travel by the motor vehicle increased, began to change in the late 1950s when city council approved its first outlying shopping centre. The fledgling Pen Centre, a regional shopping mall, opened near Merritton and long strips of commercial development spread along the main traffic arteries. As suburban shopping expanded, gradually eliminated from the city centre were its grocery fruit and vegetable, meat and poultry outlets, as well as the supermarkets that had developed there. The challenges of change for the traditional character of a long-established canal community were considerable, as St Catharines, with deep roots in the canal, became but one focus in a sea of retail outlets, and was even subordinate to the regional Pen Centre by the 1970s.

Port Dalhousie fared somewhat differently, declining as an active shipping centre after the opening of the Fourth Canal when through vessels no longer used the port, but regaining vitality as a tourist and recreational environment, a process spurred by the purchase of an old warehouse, ship's chandlery, and ice house and their renovation into a restaurant and series of boutique stores by the 1970s. Old Front Street facing the first lock of the Second Canal, itself partially restored and placed in a park setting, had taken on a new future, this time firmly based on the reactivation of the canal and its harbour for marine and recreational activities.

In St Catharines, the upper raceway between the Canada Hair Cloth mill and the rear of premises along St Paul Street was translated into a pedestrian walkway, but elsewhere the space of the former raceway was leased to the city and used to provide space for sewers and roadways. Its route became etched into the modern landscape through the appropriately named Race Street and the curves of Gale Crescent and Oakdale Avenue along the crest of the former canal bank.

Of the streets that ran parallel to the canal, former Lock Street next to the Mountain Locks as the climax of the Second Canal system has been renamed Bradley Street, but Street of the Seven Locks would have been more relevant. Two buildings of load-bearing freestone walls built by the government in the 1850s to house lock-tenders and their families have survived, but most of the industry associated with these locks has been lost, and neither a museum nor displays have been established that even start to depict the canal's impressive industrial history. Unlike Great Britain, where many remnants of the industrial revolution are cherished in open-air parks and museums, neither Ontario nor Canada have yet to any great extent recognized their industrial heritage, within which the Thorold-Merritton corridor is to the fore.

The now extinct Second Canal has been retained as a body of water, albeit severely polluted and often confined within chain-link fencing, as across the escarpment where it flows as a turgid stream that cascades through gateless locks in a series of waterfalls. The abandoned canal has survived as an open but polluted body of water from near the Thorold townline to Merritton, but it carried such an objectionable flow of effluent that it had to be hidden from public sight and culverted as a sewer to pass under Glendale Avenue and again next to the business centre of St Catharines. Out of sight, out of mind would appear to have been the public policy, but concealment did not resolve the problem. It only hid the issue from public view.

The Severe Incidence of Water Pollution

After the channel of the Second Canal had been abandoned by shipping and with sedimentation of the channel no longer presenting a problem for passing vessels, the length of canal from Allanburg through Thorold and Merritton became a convenient conduit for the disposal of all sorts of industrial waste. With water from the Fourth Canal south of Thorold, the flow by the mid-1930s was some 700 to 800 cubic feet (19.8–22.7 m³), to which were added the discharges from Shriner's Creek, which carried paper-mill wastes, and Dick's Creek with sulphite wastes. In St Catharines, Twelve Mile Creek emptied into the canal with a flow of some 1200 cubic feet (34 m³) per second from DeCew Falls, which, before reaching the power station, had passed through Lake Gibson, which acted as a settling reservoir. A further flow from the Fourth Canal at Allanburg was obtained at the St Catharines water-filtration plant at DeCew Falls, and these two flows were augmented by Beaver Dams Creek, with effluent from Certain Treed Products (Beaver Board).

By the early 1930s, the raw sewage discharged with no effort at treatment by the five municipalities (the city of St Catharines, the towns of Thorold and Merritton, the village of Port Dalhousie, and the township of Grantham) was from a contributing population of some 36,300 through 28 outlets. The water consumed was about 5.8 million gallons (26.4 million litres) per day. Certain Treed Products via DeCew Falls added 350 million gallons (1591.1 million litres) per day to this total. This waste was estimated to be from an equivalent of 40,000 people, making the total for sewage and waste combined approximately 450,000 people.

The flow of water in the old Second Canal was high. Its substantial velocity provided a favourable oxygen balance, but also mostly prevented the deposition of polluting material. The major receptacle for the deposit of solids became Martindale Pond, where 'in 1930 it was found that the appearance of the water was unclean, and at times a disagreeable odor was given off. When the bottom was stirred, bubbles were released for some time. The surface of the water carried masses of gray-green material in a rotting condition. Fibrous material was found in this ... [When the weather turned warm] the whole surface, from the top of the Henley course to the outlet were literally boiling and large masses of fermenting material came to the surface ... The odor at this time was disagreeable and the appearance was very unsightly' (Bell and Berry 1935). The sludge varied in depth from 3 to 8 feet (0.9 to 2.4 m) with maximum deposition along the upper edges of the watercourse.

Along the reach of the canal through Welland and Port Robinson, this issue of polluted canal waters did not apply. At Welland effluent was diverted away from the canal into the Lower Welland River, where 'each of iron, copper, mercury and phosphorus exceeded provincial water quality objectives and degraded sediment quality was found in several locations' (Pope 1993). This opportunity to transfer pollution downstream to other than the canal communities did not apply north of Allanburg, where the canal channel lay at almost the lowest point in the landscape. After its abandonment for the generation of local industrial water power in the 1920s, the channel provided an obvious, convenient, though deleterious, source for the receipt of waste materials.

At Thorold in 1936, then with a population of nearly 5000, the yearly consumption of domestic water at some 328 million gallons (1491.1 million litres) was exceeded by the use of almost 520 million gallons (2363.9 million litres) of unfiltered water by the Provincial Paper and Interlake Tissue paper mills. There were forty-eight sewers and drains, of which twenty belonged to the municipality. 'A large proportion of the area lying between the two canals drains to a dump located on Carleton Street ... The system of sanitary sewers was constructed about 1912 ... All sewage is discharged untreated into the second canal' (Gore and Storrie 1939).

At Merritton a large proportion of the water supply also went to industry, and after use this was discharged directly into the canal. Interlake Tissue took about one-third of the amount supplied for domestic purposes and to the other paper industries, a steel and a chemical company, and an abattoir. As ninety-one sewers and drains received all wastes, including those of industrial sewers from Alliance Paper Mills, Hayes Steel, Interlake Tissue Mills, and Canada Vegetable Parchment Company, there was further comment about the unwholesome consequences: 'The town is sewered for the most part with sanitary sewers although some are combined sewers and others storm sewers. All sewage is discharged directly into the second canal without treatment' (Gore and Storrie 1939).

When the Hydraulic Raceway is added to this unhealthy situation, no respite exists from the dismal circumstances that had been created. 'One large combined sewer from the town together with all the wastes from the Lincoln and Sulphite Mills of the Alliance Paper Mills Limited and also the wastes from the Canada Vegetable Parchment Company's plants have their outlets into the hydraulic race. As at present operated there is no dilution water in this race except that originating from the stormwater run-off. In its lower reaches the hydraulic race has been abandoned and filled in and the flow is intercepted and returned to the canal below Lock 5 in St Catharines' (Gore and Storrie 1939).

At St Catharines the average daily water consumption during 1934 was 4.45 million gallons (20.2 million litres). Of this volume about one-third was used for industrial consumption in the city's metal working, fruit canning, sawmills, electrical and automobile industries, automobile-parts, textiles, pulp and paper industries, and the remainder by residential, commercial, and institutional users. 'Practically the whole of the city is adequately served with a system of combined sewers ... All sewage from the city is discharged into Twelve Mile Creek at 17 different points. There is no treatment of this sewage except that of dilution' (Gore and Storrie 1939).

When the effluvia from Thorold and Merritton is added to the scene, we reach the grand but ominous figure of 156 untreated sewage outlets along the Second Canal. Its water system – formerly so instrumental in promoting industrial and urban growth, a trading artery of national and international importance, an object of great engineering esteem, and the central theme of expansive urban boosterism along its length – had in reality become a degraded sewer. This effluvia, now diluted to some extent by water flowing along Twelve Mile Creek, through the city's water-works, and from the hydroelectric plant at DeCew Falls, passed downstream to the settling basin of Martindale Pond.

Here, in addition to the village of Port Dalhousie and its summer tourist trade, a sheltered section of the Welland Canal contained within high banks had introduced the annual Royal Canadian Henley Rowing Regatta in 1903. Rowing was a most acceptable use for this course, but dunking the successful coxswains in the water was never too healthy a practice! With all the local waste from the shipyards, the ships in port, and the homes in the village also being dumped without treatment into the canal, it is no small wonder that the locale's sandy beach and recreational swimming, fishing in the lake, and the village as a summer resort were each placed in jeopardy.

From Pollution to Sewage Treatment Plants

Given this aggravated situation of upstream pollution, it is no small wonder that at St Catharines, as the environment became more polluted, the wells more contaminated, and effluvia in the canal more evident, the cherished title of Saratoga of the North as a resort spa became blemished then outdated. The decades of social opulence and glamour came to an end, and two of the spa resort hotels, upon decline, became schools. At the third, the baths had fallen into disuse by 1910, and a part of the building became a maternity hospital. Despite refurbishing, the remainder of the hotel failed to attract many visitors. The treatment of the abandoned Second Canal, now an ignominious sewer, had brought disgrace to the city.

The minutes of the city's board of health indicate various adverse aspects of the situation. By 1892, ice supplied to the city had to be cut from the Third Canal, not from the Second Canal. When further sewers and drains were constructed, permission was granted by the federal Department of Railways and Canals for their discharge into the Welland Canal property. In 1904 the council and the board of health indeed favoured the provincial board of health classifying the old canal as a sewer outlet. After the city started a garbage-collection and disposal system in 1909, when the first service proved unsatisfactory because the incinerator lay too close to the business centre where smoke was a constant nuisance, the dump site was relocated next to Lock 4 of the Second Canal. In the same year it was requested that the old canal at the foot of Carlton Street be used as the outlet for the city's proposed trunk sewer.

Meantime the search for industry using water from the canal continued. In 1912 a memorial to the Department of Railways and Canals recommended that the Goodrich Rubber Company be given a grant of land and the use of water from the Welland Canal for manufacturing purposes. In 1919 blood from Moyer's abattoir was reported flowing into the raceway. In 1923 it was stated to be unwise from the standpoint of public health to drain the Second Canal because its dilution water was required to sluice out the system. Also, as the canal bottom was uneven, the channel could easily become a series of stagnant pools.

In 1929 petitions received from residents on Thorold Road (now Oakdale Avenue) objected to the bad smell of water in the Second Canal, and the windows of the St Catharines General and Marine Hospital on its bank had frequently to be closed to offset the smell. In 1931 and other years, petitions urged that the raceway and certain locks be filled because of their unhealthy condition as disposal sites for all types of garbage. In 1933 the city engineer was authorized to place a screen above the Henley rowing course at Port Dalhousie to stop refuse floating onto the course during the regatta. In the same year Welland Vale complained that the storm sewer from Welland Avenue discharged into Lock 2 directly opposite the intake into its penstocks. The canal had taken on an evil and obnoxious presence.

Complaints about sewage disposal into the canal culminated in 1934 when representatives of the four concerned authorities (Port Dalhousie, St Catharines, Merritton, and Thorold) met and forwarded their concerns to the Ministry of Health. One solution, to drain the canal, had been suggested by the Department of Railways and Canals in 1929, and a later proposal suggested that the canal might be filled to remove the eyesore, then made into a scenic highway along its historic route.

In 1934 the residents of Port Dalhousie, worried that their regatta, tourist trade, and beach amenities were in jeopardy through the deposition of sludge in Martindale Pond and its offensive odours, petitioned the federal government against the discharge of sewage into the old Welland Canal. In 1935 Twelve Mile Creek and the old canal were described in the St Catharines *Standard* as 'bad for swimming because of the strong current and from a health standpoint because of the amount of sewage which passes down the canal.' Negotiations were persistent from at least this date onwards for assistance from the federal government towards improving conditions along the Second Canal from Thorold to Port Dalhousie.

These conditions of environmental disgrace, typical of industrial cities across Canada as elsewhere, required extensive and positive action by industrial concerns to reclaim valuable and usable materials from the wastes that were previously deposited heedlessly into the canal waters. For example, paper fibre in the canal was an economic loss, and the paper companies tried to prevent this waste by installing 'save-all' machinery. Domestic and industrial sewage had also to be treated effectively in order to avoid the severe health hazards that were posed (Steele 1979). These important tasks for costly resolution over the decades ahead also placed the nineteenth-century ethos of boosterism, development, and expansion in a new light as conservation and environmental protection now had to be brought consistently into the equation of industrial and urban expansion.

The Gore and Storrie report of 1939 recommended that 'a sewage treatment plant be constructed by the city of St Catharines in co-operation with the towns of Thorold and Merritton.' An intercepting sewer was required through these towns 'to which all municipal sewers and other sewers carrying sewage be connected.' Meantime, 'no special provision [should] be made ... for dealing with sulphite waste liquor.' The reason for this statement was that 'at the present time there appears no practical means of dealing with this troublesome waste except by adequate dilution.' It therefore followed that waste effluent from the various paper companies would be discharged into Beaver Dams Creek, Shriner's Creek, and Lake Gibson. In the latter instance, dilution water to the minimum volume of 1200 c.f.s. (111.5 m³) would be provided from the Welland Canal. The Hydraulic Raceway and the abandoned Second Canal would also be used 'for the accommodation of clarified effluents from mills, drainage water and dilution water,' a sad change indeed from the former active use of this raceway for mill development from the 1830s on.

With respect to the canal channel, certain ponded areas would be filled in accordance with a well-ordered plan to provide a continuous open channel, with culverts and conduits where necessary. The waterfalls and cascades at the locks would be maintained 'as the aeration resulting therefrom is most beneficial in helping the natural oxidation of organic matter.' The whole 'will not produce offensive odours or prove a nuisance,' or so it was argued. The costs for this remedial treatment were estimated at $588,000 for the intercepting sewers and its branches, with a further $410,000 for the sewage-treatment plant to be built at Port Dalhousie in the vicinity of Martindale Pond.

This provision of sewage-treatment plants was not achieved until several years after the Second World War. The first unit at Port Weller did not open until 1951, and this was for primary treatment only. Until then, all domestic and industrial wastes continued to be dumped into the various waterways, including the canal heading for Lake Ontario, and the plant at Port Weller soon became grossly overloaded as Merritton and Thorold hooked into the system. A more suitable secondary treatment plant did not open until 1971 (Proctor and Redfern 1971). The canal, its industries, and its towns had left behind a very unhappy legacy for future generations.

13

The Emergence of a Linked Urban-Industrial Complex

The overall situation between 1914 and 1959 was that of industrial growth from the foundations that had been laid over the previous century. Hydroelectricity from DeCew Falls and Niagara Falls was available on an abundant and cheaper basis than elsewhere. The regional N.S.&T. streetcar network interconnected the canal towns. The Third and then the Fourth Canal functioned strongly within the marine trading patterns on the Great Lakes. Both fostered development within the canal zone and provided an unrestricted source of power to municipalities and for industrial purposes.

The railway networks that crossed the canal connected directly with the expanding urban economies of southern Ontario and the growing communities in the West. Their crossings of the Niagara River at Niagara Falls and Fort Erie also provided important points of entry into the American economy to the south. Meantime, the motor vehicle in its various forms became of increasing significance for the movement of passenger and goods traffic. The Peace Bridge between Fort Erie and Buffalo had opened in 1927, other highway bridges crossed the international boundary, a pattern of paved roads served all parts of the Niagara Peninsula, and the multi-lane Queen Elizabeth Way provided improved highway access into the economies of both southern Ontario and the nearby areas of the United States. The setting was in an agricultural area extolled for its productivity and scenic qualities, and the towns offered housing, schools, and other services of merit (Brock 1930; Gayler 1994).

Port Dalhousie

The harbour and port activities here provided the mainstay of economic activities until 1932 when the Fourth Canal opened officially through Port

Weller. For some twenty prior years there had been the looming cloud on the horizon of a new canal that would forsake The Port and terminate Port Dalhousie's reign as a marine centre. It was also during this late period of the Third Canal before the Fourth Canal that the canal's northern port of entry with its changing ambience reached its height of activity. 'Port Dalhousie was in a period of peak growth and reaching maximum development at this time. Business establishments, both grand and small, were thriving and there was an abundance of food, security in employment, and considerable decadence' (Aloian 1978).

Served by the railway and the streetcar, the port's principal features were the Muir Dry Dock, its ships' chandlers, and many service establishments. Much had necessarily to change when the Fourth Canal and its northern port at Port Weller took over the canal trade. The milling industry moved to Port Colborne and the flour industry at Port Dalhousie failed. It was replaced by Maple Leaf Rubber, but this business also failed. Ships left as did the ship-repair industry after a short period of repairing or dismantling the diminishing number of Third Canal vessels, but marinas for sailing vessels and later for power boats then arrived as a replacement for the declining marine business.

Managed by Canadian National Railways, Lakeside Park expanded under the influx of arrivals from across the lake by steamboat and with further arrivals from across the peninsula by streetcar. In 1921 the park area was doubled. Low and swampy ground was filled with loads of sand brought in by streetcar from the Fonthill area, light standards were erected to promote the evening use of the area, and tents were replaced by wooden buildings painted to a consistent colour of green and white. The extensive renovations, completed in 1925, included a covered picnic pavilion with tables for over 3000 picnickers at one time, and the sand walkways were paved with bricks and wooden blocks to make walking easier (Turcotte 1986). Amusements, games, rides, band concerts, and firework displays were extensive. Lakeside Inn next to the pier provided the same quality of dining experience that was available on the CNR dining cars. A bathing pavilion provided change rooms, a large wooden water slide in the lake was a popular attraction, rowboats were available for rental, bicycle pedal boats and children's paddle boats could be hired, and the merry-go-round with its organ provided background music.

Success meant that the steamers *Northumberland* and *Dalhousie City* made two round trips from Toronto on weekdays, three or perhaps more on summer weekends. In 1929 they brought in 290,000 passengers and 112 group picnics were booked. From 1924 to 1951 the park became the scene of the annual Emancipation Day picnic to commemorate the freeing of Black slaves throughout the British Empire, an event that attracted up to 8000 people from church groups and social organizations across an area that extended from Toronto to Rochester, New York. Peak attendance, helped by gasoline rationing and large picnic parties sponsored by big-name companies for their employees, was reached during the early years of the Second World War. Decline was then rapid as streetcar traffic ended, the automobile presented a greater freedom of choice in recreational destinations, and severe pollution in the Second Canal detracted from the quality of Martindale Pond and the beach area. When the boat service from across the lake ended, Canadian National sold out, and Lakeside Park after a period of decline and decay eventually became a city park that barely reflects its former recreational glory.

The motor car may be blamed for the demise of Lakeside Park as recreational habits and patterns changed, but the growth of St Catharines and personal mobility to its business centre and factories because of the motor car also led to the revival of The Port as a desirable and attractive residential neighbourhood and ward of the larger city. The transition from an independent port to a residential suburb of character within the modern city involved many interrelated strands. On the marine scene, as recreational boating increased in popularity, pleasure vessels created a new activity along the walls of the outer harbour. Marina facilities gradually replaced the steamer berths of yesteryear, and the former commercial frontage that lined the west side of the harbour made the transition through decline and empty buildings to recreational and tourist-oriented service activities.

The Royal Canadian Henley Rowing Regatta on Martindale Pond steadily expanded (Brooks 1988). It enjoyed long-term success, remained an important summer feature, attracted more rowing events and spectators, and grew in international stature with the world's rowing fraternity. When it was decided in 1930 that Martindale would become the permanent home for Canadian rowing, the course was dredged, widened, and deepened by the Canada Dredge and Dock Company under contract from the Department of Railways and Canals. The area dredged was approximately 1500 feet (457.2 m) long, with the upper section enlarged to a width of 250 feet (76.2 m), and the lower length to a width of 125 feet (38.1 m). The specification conveyed a reminder of the polluted nature of Twelve Mile Creek: 'the material to be dredged consists almost entirely of silt, sand, decayed vegetable matter, ground pulpwood, and waste matter carried into the Old Canal through tributary streams and ditches and through municipal and industrial sewers.' A hydraulic dredge was used, and the dredged material was spread through pipelines from the dredge to other areas of the pond selected by the canal engineer on the basis of depth and currents.

On land, a new boathouse and a concrete grandstand to accommodate 3000 spectators were constructed. The fascinating transition in the environment at Martindale Pond had been from a meandering river to a man-made pond for canal purposes, then a century of usage by sailing, then steam vessels to rowing skiffs, and from the often hectic scenes of a commercial waterway to a seasonal sporting activity. An element of continuity is the controlling impact of the physical background. As stated by Coombs in 1950, 'The high banks on either side protect the course from sudden squalls of wind and the oarsmen are assured that they will not encounter rough water. We think that it is no exaggeration to say that there is no finer regatta course in the world.'

When the canal transferred to Port Weller, the Muir Brothers Shipyard and Dry Docks steadily declined. They were no longer on the through route of canal navigation, and as neither Lock 1 of the Third Canal nor the dry docks could accommodate larger-sized vessels using the Fourth Canal, the major operations in the yard transferred to Port Weller Dry Docks next to Lock 1 of the Fourth Canal. The old docks on Martindale Pond remained active until the 1960s, their major tasks being repair work on the diminishing number of the older canal vessels that remained in service and the breakup of outmoded vessels when they became redundant for use either on the canal or in lake service.

Of the manufacturing industries, Maple Leaf Rubber at first expanded. This company was bought out in 1915 by Canadian Consolidated Rubber, which in 1920 built an apartment block in the village for female employees. The plant continued to operate a large business until 1929, when the factory that employed some 360 female workers manufacturing rubber overshoes closed.

NuBone Corset in 1940 purchased Dalhousie Hall, where the women who had worked in the rubber factory were boarded, and used the building as a factory until a fire in 1960 led to its conversion as an apartment building. Meantime, after the Second World War the former rubber plant became the home of some five small companies that manufactured products such as reflective insulation, starch, and tape, and later was the home of Lincoln Fabrics for the production of broad woven industrial fabrics, an interesting example of a still operative canalside mill. Nearby, the Canadian Legion Hall occupied part of the complex used as maintenance buildings for the Third Canal after this was no longer required for canal operations.

As the original part dates back to the 1880s, here is another important example of the constructive reuse of former canal buildings.

Port Dalhousie, with its central location in the Niagara Fruit Belt and its export capability to Toronto by boat, was also a busy centre for the fruit and vegetable canning industry during the summer months. Boese Foods at times employed up to 500 women on each shift, while the Port Dalhousie Canning Company might employ 150 women packing cherries, strawberries, tomatoes, and pumpkins. Both companies were eventually taken over by Canadian Canners, then closed through the rationalization of production in larger plants and because of the cheaper importation of fruit from abroad. Small scattered plants that were difficult to maintain were gradually discontinued in favour of a major fruit-processing plant at St David's.

An increasing feature at Port Dalhousie was its closer dependence on nearby St Catharines. At first by the N.S.&T. network, then by bus and later by automobile, residents were now free to travel out to the greater employment opportunities of the larger city. In the reverse direction, some families moved out to homes in Port Dalhousie for the attractions of its physical setting between Lake Ontario and Martindale Pond, the many older homes of quality that had been constructed in this setting, and the favoured seasonal climates because of location next to Lake Ontario.

As these changes gradually took place, Port Dalhousie's urban character changed slowly from that of a self-contained port community with the administrative designation of a police village to that of a distinctive residential-recreational suburb of the larger city. The climax to this increasing inter-association occurred in 1961 when the town of Port Dalhousie and the city of St Catharines amalgamated into one unit, reflecting what had earlier happened between Port Colborne and Humberstone at the southern end of the canal.

St Catharines

The parent city of the canal enterprise, although no longer on the canal, continued to grow significantly. An immediate task with the declaration of the First World War in 1914 was to secure the railway bridges and locks of the Third Canal against possible sabotage. These precautions, though proved unnecessary, were taken because of the many German and Austrian workers on the Fourth Canal. Another perceived danger, as the United States was not then engaged in the war, was the large number of German and Irish settlers in Buffalo.

New activities during the First World War included the acquisition by

Frank Kromer and H.C. Specht of a property on Carlton Street to weave silk and manufacture underwear. Another American Company, Welch Grape Juice (later Powell Foods, then Cadbury Schweppes) constructed a factory next to the Grand Trunk (former Great Western) station on Western Hill. The canal was used for their import of some concentrated fruit and juices. Sanitary Dairy also started operations. War contracts were extended to Canadian Crocker-Wheeler (later English Electric) for shrapnel shells and electrical equipment, to Packard Electric and the Reo motor company for the manufacture of shrapnel shells, to Foster Wheeler to produce boilers for naval vessels, and to Canada Hair Cloth for uniforms.

McKinnon's, which changed from the production of saddlery and hardware for the horse and buggy era to parts for the automobile industry, was another major benefactor of the war effort. The plant on Ontario Street at first supplied saddlery and wagon hardware to the Canadian, British, and French governments for use by the cavalry and on horse-drawn vehicles. In 1916 a three-storey building was added to manufacture 100-pound shells and fuses, and in 1918 they began to manufacture automotive radiators and various parts for trucks and military vehicles.

The city gained permanently as the new and the old industrial premises used for wartime purposes added new machines and modern buildings to cope with the war effort. These augmented facilities were then available for peacetime use during the ensuing interwar years when the Fourth Canal opened to the east of the city, the Third Canal closed across the north-eastern outskirts of the city, new docking facilities on the Fourth Canal replaced those that were lost on the Third Canal, and the city's manufacturing output became more diversified. Empire Rugs was established in the early 1920s, using a female labour force to manufacture tufted and woven cotton and nylon rugs. With the addition of a further silk company, St Catharines obtained a deserved reputation as the Lyons of Canada.

McKinnon's was taken over by General Motors in 1929. When GM decided to remain in the city, the Delco building was added on the east side of Ontario Street to manufacture starting motors and generators. Large-scale reconstruction was undertaken at the plant by GM in 1936 to produce axles, and a new forge and foundry were built.

A company from Cleveland, Ohio, arrived in 1930 to manufacture woven wire cloth and screening materials. In addition, some twenty metal plants then manufactured agricultural implements, greenhouses, carriage and harness hardware, chain, tools, and electrical apparatus of various types, boilers, radiators, newsprint, and paper products. Canneries, wineries, the making of grape juice, jams, and meat products, and the processing of fruit and vegetables were each associated directly with the rich agricultural environs. As a part of this versatile output, the Taylor and Bate Brewery on the bank of the Second Canal added a new bottling plant in 1925 and further machinery two years later, and by 1929 produced nearly 1.3 million gallons (4.9 million litres) per annum. Even so it succumbed during the Depression.

Over the ten-year period from 1926 to 1936 because of the Depression and their inability to compete as markets shrank or changed, several plants closed. These included a paper mill, two planing mills, a silk mill, a cannery, and several small metal industries. But with new arrivals, the number of manufacturing plants remained steady at about a hundred. Employment, at about the 3000 level from 1921 to 1925, increased until 1930. Followed by a Depression low of 3139 in 1933, it then expanded to reach about 5500 by 1938/39.

The new industries that more than counterbalanced decline in some of the older activities included the production of the zipper, or metal slide fastener, by Canadian Lightning Fastener from 1925. Other companies added woollen garments and rugs. Thompson Products arrived from Cleveland, Ohio, to produce valves and pistons, and by 1938 had contracts to supply General Motors and Chrysler with automobile parts. Other big names included Foster Wheeler to manufacture steam power, oil-refinery, paper-mill and other equipment and heavy machinery, and English Electric, which manufactured motors, generators, transformers, and electric locomotives.

The Second World War introduced a renewed period of growth. Building permits increased more than four times between 1938 and 1941, and remained higher than at 1939 throughout the war years. The population increased annually during the war, rising from about 28,625 in 1940 to some 35,096 by 1946, and the number of employees in factories doubled to 11,109 by 1943. McKinnon's expanded to a peak employment of 4200 in 1942. As male employees left to join the armed services, employment nearly tripled as women played an industrial role of increasing importance. The same type of story took place at Thompson Products, Hayes-Dana, and other companies.

The demand for all-out production during the war years gave the economy a substantial boost that continued as peace resumed, when some established industries moved to fringe locations. A prime example was the transfer of the now obsolescent Muir Dry Dock from Port Dalhousie to the east side of the Fourth Canal at Port Weller. The new site, then owned by the federal government, was a little-used dry-dock facility for storing gates, lock valves, and gate-lifting vessels. The dry dock had the advantage of being flooded and emptied by gravity rather than by pumping, and was served by the N.S.&T. streetcar line. The yard expanded with ship repair, rebuilding, and reconstruction work and also the dismantling of vessels. Employment reached five hundred employees by 1950.

Purchased by Upper Lakes Shipping in 1956, Port Weller Dry Docks then produced a great variety of vessels that have included self-unloading bulk carriers, tankers, tugs, dump scows, self-propelled barges, car ferries, vessels for the coast guard, ice breakers, and ice-strengthened ships (Gillham 1992). It became one of the most important industrial enterprises in St Catharines and on the Great Lakes, employing a highly skilled but fluctuating labour force.

In a second major suburban move, again to the banks of the Fourth Canal, McKinnon's (now General Motors) purchased 145 acres (58.7 ha) in 1952 close to the Flight Locks. A malleable and grey-iron foundry and an engine plant opened there in 1964, and an administrative building next year. With 4500 employees by the 1960s, this was the largest single concentration of labour at any location on the canal. The canal was used as a source of water supply for cooling purposes, to stockpile sand at Port Weller, and for the import of coal. Also intended was the export of vehicle parts by water across Lake Ontario to the GM plant at Oshawa. However, as this meant that engines would have to be stockpiled at the plant in order to obtain full loads for a vessel, with further storage at the Oshawa end to meet demand, a regular daily trucking service was initiated instead. (The switch to such land transport prevailed after the multi-lane highway system came into being.)

At Thompson Products, the existing plant had reached its limits and ground was broken in 1952 for a new large plant linked under the road to the old one on the opposite side of Louth Street to produce blades for jet engines. As demand slackened, the company increased production of engine parts, front suspensions, and steerage-link components for the automobile manufacturing industry.

St Catharines was growing by external expansion into its fringe rural areas. Also, with large manufacturing enterprises such as the Port Weller Dry Docks and General Motors now on the east bank of the canal and with the service offices and yards of the St Lawrence Seaway Authority on the west bank facing General Motors, the new northern length of the Fourth Canal was starting to present as diversified an economic landscape as had previously followed the earlier canal systems.

The area on both sides of the QEW, previously crossed by the Third

Canal and with an industrial dock at Carlton Street on the Third Canal and then the Homer Dock on the Fourth Canal, also provided a suitable locality for industrial growth. Packard Electric moved from its confined mill site on the Hydraulic Raceway where the Reo motor car had been produced to a 35-acre (14.2-ha) rural fringe site in this new location. Drawn to the QEW for its anticipated transportation advantages when neither residential nor commercial development existed in the vicinity, the company merged with Ferranti, an English corporation, in 1958 and became Ferranti-Packard, manufacturing transformers and meters.

To encourage the location of smaller industries in this area with its site advantages of proximity to the QEW, an innovation by the city was the provision of industrial sites nearby in the Bunting Road area. It was a large open area with rail access, the city was close for its labour supply and commercial services, and the Homer Dock on the Fourth Canal was used by Ferranti-Packard and companies such as Niagara Structural Steel and Foster Wheeler for their export abroad of bulky products.

The city was growing in many ways, including population and industrial advance, the provision of new and improved services, and the expansion of suburban housing in adjacent jurisdictions. Between 1945 and 1955, the city's population grew by a modest 15 per cent, but over the same period the growth was 52 per cent in Merritton, 60 per cent in Port Dalhousie, and 260 per cent in Grantham Township. The city, now restricted within its obsolete municipal boundaries, was overflowing into adjacent territories and, as housing, industrial, and commercial facilities were steadily and progressively added to the fringe, the city crossed its confining valley canal barriers of the Second Canal to the south, the canalized Twelve Mile Creek to the west, and the Third Canal to the north-east. This advance also steadily absorbed into its municipal embrace the earlier canal communities of Port Dalhousie and Merritton.

With the takeover of so much intervening land in Grantham Township, the canal became an administrative boundary for the first time in its history. The municipal boundary followed the Fourth Canal between Carlton Street and the Flight Locks, but where McKinnon's and the Port Weller Dry Docks had moved from the city to the far side of the canal, the municipal boundary was extended to include these areas within the urban curtilage for their taxation advantages.

St Catharines had changed from a compact urban community that had grown next to the First and Second Canals to a multi-nodal centre that now physically embraced three of the earlier canal communities: Port Dalhousie, Merritton, and the core area of St Catharines, that is, the extent now represented by the business centre and its residential fringe between Welland Avenue and the valley of the Second Canal. The urban embrace now extended across the full width of Lincoln County from Lake Ontario over the Ontario Plain to the Niagara Escarpment.

Merritton

Merritton grew in close relationship with Thorold to the south through the water supply available from the canal and in close association with St Catharines to the north through industrial development at each lock downstream along the canal. Because of its intermediate location and the economic linkages involved, Merritton may be described as a sandwiched canal community. Elements of inter-connection between the three urban units included the N.S.&T. streetcar system and a series of interassociated industrial enterprises along the Second Canal. Shared affinities with Thorold included a combined high school, a considerable home-work interchange, and service facilities such as the town's waterworks in Thorold Township.

When annexation was under discussion in 1955, Eric Hardy suggested that Thorold, Merritton, and parts of Thorold Township should be included within an enlarged St Catharines, but this rationale was opposed in Merritton with its prideful independence, and was not pursued because the intermediate town boundary between Thorold and Merritton was also the county line between Lincoln and Welland Counties. The reorganization of municipal affairs remained within the confines of Lincoln County, Merritton was brought kicking and screaming into the ambit of St Catharines, and Thorold remained separated from both its sister town of Merritton and the enlarged city of St Catharines. The annexation process with its focus on antiquated administrative boundaries going back to the 1780s had ignored the crucial role of the Welland Canal in transforming the pioneer pattern of settlement and its introduction of new economic and social circumstances to the Niagara Region.

Industrially, by the early 1930s Merritton employed about 1200 persons in the production of paper, sulphite, flour sacks, macaroni, and forgings. The paper mills included Alliance Paper, Garden City Paper, Interlake Tissue, and Vegetable Parchment. Lincoln Paper produced 25 tons (22.7 tonnes) daily of manila fibre, greaseproof paper, and glassine, a nearly transparent glazed paper resistant to the passage of grease; nine tons (8.2 tonnes) of rope, manila, and wrapping papers; and 40 tons (36.3 tonnes) of sulphite. At Interlake Tissue the annual output in 1919 was 3000 tons

(2721.6 tonnes). A second paper machine in 1921 and a third in 1954 greatly increased the production of fine tissue and toilet papers. In 1961 the company was purchased by Kimberly-Clark.

Other important industrial names included Shawinigan Chemicals, Hayes Wheel and Forgings, the Haynes and Boardman Foundry, Machine and Wire Cable, Muratori Bros., and Prest-O-Lite. Merritton now had no cotton mills, but it remained an important paper-making centre even though the important Riordan Mill ceased operations in 1923. By 1932 part of the former extensive plant was an abattoir operated by Oscar Jacobson, and this was later taken over as a meat-processing plant by Essex Packers. The former extensive site of Riordan Paper had been subdivided between several smaller industrial concerns and commercial activities.

In 1916, on the site where Canada Wheel Works became Hayes-Dana after a complicated series of takeovers, the production of wheels and spokes expanded, then diversified during the 1920s to include components including brake drums, axles, and drive shafts. Specialization involved the transfer of operations from the company's Chatham plant to Merritton in 1932, and production increases during the Second World War to manufacture truck components and aircraft parts. Employment had increased to 1500 by 1945. Seeking to expand in Thorold, Hayes-Dana that year purchased the former Pilkington Glass Works to manufacture drive trains. A new office building was constructed in 1953 on Oakdale Avenue (Hayes-Dana 1972).

In conjunction especially with General Motors and Thompson Products in St Catharines, the St Catharines–Thorold industrial area had become an important centre for the production of automobile parts, with an important spin-off into smaller manufacturing companies. When companies were interviewed for a study undertaken in 1971, twelve out of eighty in the St Catharines–Thorold area were found to supply components to the auto industry. The series of industrial linkages extended through custom tool-and-die shops, electroplating firms, machinery fabricating and repair shops, the bonding of metal to rubber, and the supplying of brake drums (Jackson and White 1971). These firms in turn had their own suppliers.

While it would be unwise to attribute all of these activities to the Welland Canals for either their water-power or transportation advantages, certainly the canal was one of the major component factors that had initiated the area's industrial evolution and created this novel grouping of inter-related industries. This facility of inter-municipal and inter-industrial strengths has been noted previously for the relationships between wheat production and milling activities, for shipbuilding and ship repair, and for the food industries that arose in conjunction with the Niagara Fruit Belt. The phenomenon of inter-linkage is not new, but its incidence through the motor industry was now more extensive than before.

Thorold

It was stated at the end of the First World War: 'Despite the fact that we have no war industry, our expansion during the war period has been greater than ever, which argues well for our continued prosperity, without any break for reorganization' (Sanders 1919). Customs activity increased 2.4 times, and railroad freight almost four times. The Third Canal now bypassed the town centre, but the Second Canal remained open from the south to serve Thorold's mills, and there was the line of pre-war mills that had been attracted over the years immediately before 1914: Coniagas, Ontario Paper, Beaver Board, and Pilkington, plus the older Provincial, Thorold, and Davey Pulp companies and Brantford Felt and Roofing. The canal was in full use as a source of supply for coal, water, and imported raw materials. It was used in conjunction with the railway for the export of products. At the Provincial Paper Mills, 'The demand for their goods has so increased that plans are being formulated to considerably increase the size of this plant, and installing two extra 147-inch [3.7-m] machines' (Sanders 1919).

Exolon started its operations in 1914. In a further development, Domtar Paper, founded in 1918 to produce building paper and roofing felt from old papers, exported much of this product to Montreal and Quebec, and water from the canal was used in the production process. Industrial change included the departure of Coniagas to be closer to the mines, Penmans' Wool Mill became the Muratori Macaroni Plant, and in 1945 the abandoned Pilkington Glass Works was taken over by Hayes Steel Products (Hayes-Dana).

Spun Rock Wool arrived in 1934 to manufacture rock wool for the thermal insulation of homes and industry. Its raw product, a dolomitic shale discovered below the surface limestone during the construction of the Flight Locks, provides a unique link between the canal and industrial development. The shale turned into a liquid at high temperatures, and when cooled threw out long fibres that could be made into blankets of varying thickness and type to insulate against heat. Also a soundproofing material, it was used by the British Air Ministry during the Second World War to deaden the noise from testing aircraft engines and to reduce the noise from wind tunnels. Retailing was undertaken by a company named Superior Insulation, which advertised 'BLOWN ROCK WOOL INSULATION for Schools, Churches, Homes and Industrial Buildings.'

By 1950 the Walker Brothers Quarry served the district with crushed stone, shaped building stone, and agricultural lime. The Exolon Company produced aluminum oxide, silicon carbide, and other electric-furnace products. Operations at a smaller scale included an ice-manufacturing company, a foundry and a machinery company, bottling works, a lumber and planing mill, and a winery. Freight was handled at two docks on the Welland Canal, one operated by the Ontario Paper Company and the other by the Niagara District Warehouse and Forwarding Company. Designated the Port of Thorold, there was a customs house for the clearance of goods and both dock facilities had rail and truck access.

By 1954 foreign trade at the docks centred on importing a general range of items, inter alia jute from India; china, chemicals, sporting goods, and car parts from England; maraschino cherries from Italy; and iron sand for use in the local foundries from Scotland. Exports included goods from across the American side of the boundary, wine and canned goods, and rolls of newsprint from Beaver Wood Fibre through the dock of Ontario Paper headed to overseas destinations in foreign-owned vessels.

Of the paper companies, Beaver Board, set up to manufacture board and paper, added additional machinery in 1918 and again in 1925 when the production of newsprint was started. The Canadian head office moved to Thorold from Ottawa in 1920, and in 1928 the plant was acquired by an American roofing then gypsum-board manufacturer, Certain-teed Products, the first of several name and ownership changes. By 1932 the company was stated to have an 'immense output of beaverboard, the laminated interior finish noted the world over.'

The Ontario Paper Company with its office and mills in Thorold dominated the industrial scene and the pulp-and-paper industry with a daily production of 220 tons (200 tonnes) of newsprint. The Thorold Pulp Company manufactured 8 tons (7.3 tonnes) of mechanically ground wood pulp daily. Newsprint was added in 1933, using pine shipped to the plant from Savannah, Georgia. At Provincial Paper, the capacity of its earlier Montrose mill, which had acquired and consolidated other mills, had been doubled in 1920. By 1930 the daily production was 10 tons (9.1 tonnes) of coated paper and boards, and 70 tons (63.5 tonnes) of books and writing and catalogue papers. An interesting statistic is that over the forty-seven years between the start of production and 1950, the Provincial mill at Thorold had produced 7 million miles (11.3 million km) of paper, 2 feet

(0.6 m) wide, enough to reach from the earth to the moon twenty-one times (St Catharines *Standard*, 1950)!

Obviously, the short length of the old and the modern canals through Thorold to Merritton had become a major centre of the Canadian and international papermaking industry. The products ranged through pulp board, newsprint, book and writing paper, roofing and sheathing felt, mechanical ground wood, and sulphite wood. Within this context the Ontario Paper Company (later Quebec and Ontario, then QUNO, and most recently Donohue - QUNO) had a distinctive history (Plosz 1988). Newsprint has always been produced, but the relationships between operations in the mill and the canal for the inward and outward transportation of materials have changed remarkably.

The mill at first used wholesale suppliers. For those seeking an independent pulpwood supply, extensive timber concessions were not then available on the crown lands of northern Ontario. The vital raw material was obtained by leasing timber limits after explorations along the north shore of the Lower St Lawrence in Quebec during 1915. The company then established its own shipping company to bring pulpwood to the mill and to export the processed newsprint to the Chicago *Tribune*. The New York State Barge Canal was also used for the export of newsprint to the *Tribune*'s subsidiary, the New York *Daily News*.

Incorporated as the Ontario Transportation and Pulp Company of Thorold, this company first chartered vessels, but then established its own fleet for this work (Sykes and Gillham 1988). Given this inflow of raw material by boat, the increasing consumption of newsprint, and the important absence of a U.S. tariff on newsprint, a third paper machine was added at the mill in 1917, plus two more in 1919 and 1921. The first two machines produced 32,000 tons (29,030 tonnes) during their first year of operations, but by 1930 the output had increased to 102,000 tons (92,534 tonnes).

As much of the internal equipment was then obsolete, an extensive modernization program undertaken during 1930–1 transformed the mill in so far as the cost and the quality of the product were concerned. In addition, the Depression years witnessed the building of a new company paper mill at Baie-Comeau, and timber limits were finally acquired in northern Ontario where, after plant, roads, and wharves were constructed, harvesting began in 1938. Next year the first wood was shipped to Thorold through the Upper Lakes and gradually the mill's wood supply came from this source rather than from Quebec. The Welland Canal was still playing a significant role in the company's advance, but with different routes and sources of supply.

The Second World War was not the period of growth that has been depicted for other industries. The export of newsprint to the U.S. was viewed as unpatriotic to Canada and anti-British. When electrical power from Niagara Falls was rationed for essential war production, the Thorold mill received sufficient only to keep two of its five paper machines running. Timber was in short supply as woodsmen were called up for the armed forces, five of the company's eight ships were requisitioned at one time or another for war work, and substitute products including recycled newspaper were brought in from Chicago.

To further change the situation, new products also came into use. Alcohol was produced from the wood sugars in a plant opened in 1943 to make synthetic rubber. Research that started into the production of vanillin next year led to the operation of a full-scale commercial plant by 1952. The paper plant that employed over a thousand workers by 1950 had now diversified into the chemical industry, and Ontario Paper then played a lead role in pollution abatement, helping to relieve the disastrous situation caused by the paper industry and other activities that had followed the closing of the Second Canal.

Pollution abatement at Ontario Paper involved filtering out the solids, mostly residual wood wastes, removing the dissolved solid from the sulphite pulping process, and separating the industrial waste generated by the mill from its sewage. Meantime, as the postwar economic boom became a reality, the speed of the existing paper machines was increased and a sixth machine powered by steam and pump lines from Ontario Paper plant came into operation at nearby Beaver Wood Fibre. This operation provided a distinctive use for the Fourth Canal in that, before a new boiler could be added to the steam plant, a surplus naval corvette was berthed at the company's wharf and used to supply steam to the leased newsprint machine. Output was increased by 25,000 tons (22,680 tonnes) a year through this arrangement.

By the 1960s pulpwood was no longer shipped in to the plant from Quebec, and supplies from northern Ontario were brought in by rail. Meantime, the outflow to Chicago changed to rail only. As the company's shipping company carried less than 5 per cent of its cargoes on behalf of the paper mills, the fleet took on the role of a bulk carrier, but these operations were closed in 1983 through decline in the Canadian marine industry.

Transportation, at the root of so many decisions in industrial location and growth throughout this account, had a changeable and inconstant quality as new circumstances came into play. At the Ontario Paper Company almost everything had changed, from the sources of the raw materials

to their mode of transportation and to the nature of the processing operations within the plant. At one time over 80 per cent of all the raw materials
and finished products were moved to and from the plant by ships.
Undoubtedly the Welland Canal provided an imperative base for the
industrial successes that were achieved, but the rail system eventually took
over these marine operations.

All companies must face the many comparable dilemmas of market,
technological, social, and economic changes over time. Some weather these
unexpected storms of change. Others fail or move elsewhere as the basis of
their operations changes but, as indicated by the canal's pulp-and-paper
industry, the several diversified companies along the Welland Canal in the
same industrial category have had very different performances ranging
from closure to long-term success over the years.

The Villages at the Deep Cut

Neither Allanburg nor Port Robinson expanded over the wartime and
inter-war periods to any considerable extent, and both retained more of a
village than an urban character. At the first location, a commemorative
cairn unveiled in 1924 recognized where the sod-turning ceremony for the
First Canal had been held. This event provided some distinction to the village, but it was not sufficient to save the Black Horse Tavern, the nearby
inn to which William Hamilton Merritt and his party had repaired after the
sod-turning ceremony of 1824. This tavern succumbed in the late 1960s to
the appropriation of part of its lot for highway purposes and the building
was demolished.

Now more a village on the through highway route (Highway 20) to The
Falls, and situated on the fringe of the expanding St Catharines–Thorold
industrial area and in particular in close proximity to industrial expansion
taking place south from Thorold, Allanburg's future now lay more in residential expansion as a locale for commuting to external places of work than
as a canal centre. The relevance of the canal was the historic legacy of associations and buildings at the centre of the village and their significant basis
for conservation and development.

Shipbuilding and repair continued at Port Robinson. The Canadian
Dredging Company constructed at least ten dump scows for canal work
during the First World War, and in the early 1920s work began on reopen-
ing and enlarging a dry dock that was used to repair tugs, scows, and
dredges until this work was terminated by the works for the Fourth Canal.
These canal works again brought new activities to the village. Some men

boarded and bunkhouses were constructed. Socially, 'No local girl needed to be without a date. Some even found husbands' (Bourton 1930). The influx of labour for canal works, as always throughout the history of the canal, involved more than engineering.

At the northern end of the village, the wartime growth of James United Steel was assisted by the manufacture of structural steel for munition plants and included expansion to produce shells. Postwar developments were based on the increased use of steel in highway and railway bridges, buildings, storage tanks, and radio towers. The company was taken over by T.J. Dillon in the early 1920s, and a decade later its various units were combined as part of the United Steel Corporation. After the Second World War the fabrication of plate steel and its installation in bins, hoppers, and storage tanks was introduced. Trade was also boosted by the advent of television towers for the network then advancing across Ontario.

The remaining industrial activity, National Refractories, was introduced in 1926 to provide firebricks for the boilers of Canadian National steam locomotives. Using fireclay imported from Pennsylvania, the plant took over a wartime munitions plant that then produced washing machines. Diversification from firebricks was necessary with the switch of railways to diesel locomotives, and this led to a range of firebricks to line furnaces, stoves, and wood-burning fireplaces. Cooling water was obtained from the canal, which brought in coal and was also used for the occasional export of bricks, though transport by rail and by truck was more versatile.

Welland Grows

The period of the First World War was one of great industrial prosperity for Welland. The population expanded to almost 10,000, industrial production increased by 50 per cent, and imports through the recently created customs port tripled between 1916 and 1918. The climax to so much wartime change was the achievement of city status in 1917. In a town notorious for its muddy streets, the main roads were paved with brick, wood blocks, or tar pavements. Concrete walks were added. Water mains and sewers were extended, with much of this water being used in the factories.

The three public schools were enlarged and modernized, and a high school and then a vocational school were constructed. The market outgrew its small square and relocated in a larger area of space, there was increased business for merchants, and more work for builders and labourers. And conflicts rose over whether or not 'Urban Crowland' to the south, that is, the part of the township receiving the outward growth from Welland with its immigrant industrial population, should be absorbed into the more Protestant, traditional, and Loyalist society of Welland.

The expanding industrial strengths are attributed by Craick (1917) primarily to the Welland Canal, hydroelectricity, and the railway system. The canal, though now used less extensively by local companies for transportation service, nevertheless provided manufacturers in Welland with the advantage of competitive lake and rail rates and also its abundant supply of water for industrial purposes.

Output at the town's Hydro-Electric Power Commission, established in 1913, had increased almost fourfold by 1917. The municipal system was supplied from Dominion Power at DeCew Falls, and four high-tension lines brought in power at a cheap rate from the Ontario Power Company at The Falls. Until 1917 the town was served by 12,000-volt lines but, as these proved inadequate, a new transformer station capable of handling power from 46,600-volt lines was built.

Eight railway companies owned or had agreements to use lines that competed for business and radiated out from the town to the frontier, the rest of Canada, and the United States. These lines included the Grand Trunk used by Wabash; the Michigan Central used by Pere Marquette; the Toronto, Hamilton and Buffalo; the Grand Trunk; and the Niagara, St Catharines and Toronto electric railway, allied with Canadian Northern.

The established industries that expanded their operations under wartime demands included Plymouth Cordage for binder twine. M. Beatty produced hoists, dredges, scows, drills, and lifting equipment. British-American Shipbuilding in adjoining premises built the steel hulls of five large cargo vessels for the Imperial Munitions Board and the dominion government, and Canadian Steel Foundries produced many different types of castings. Dain Manufacturing added agricultural implements. Electro Metals produced ferro-silicon and electrodes. The new company of Volta Manufacturing was formed to produce electric regulators and winches, and Welland Winery was added. Page-Hersey made wrought, black, and galvanized pipe and tubular products, and Welland Iron and Brass in 1918 established a foundry to produce grey-iron and non-ferrous castings.

The spindles and looms of the Empire Cotton Mills (later Wabasso), with a capacity of about 7 million yards (6.4 million metres) produced clothing, fabrics, and bags. H.S. Peters produced overalls, and Chipman-Hoolton Knitting hosiery. Metals-Chemical manufactured chemicals, the large plant of Union Carbide acetylene and calcium carbide, and John Goodwillie canned goods. The companies that worked primarily on munitions included the two plants of Canadian Foundries, which also produced marine engines, and Electric Steel and Engineering (formerly Electric Steel and Metals).

Two immediate consequences of industrial growth were that, as a labour shortage developed, workers were attracted to employment in the Empire Cotton Mills from the Dominion Textile Mills at St-Grégoire de Montmorency in Quebec (Unyi 1995a). Twenty or so families came in 1915, and by 1919 some forty French-Canadian families, or about 250 people, resided in the town. With more immigration and through natural increase, by 1921 this number had nearly doubled; by 1941 the population had risen to 1345 persons. They lived on the east side of the city, east of the railway tracks in the area that became known as French Town (Frenchtown), between the two Quebec companies, Empire Cotton and Atlas Steel (Brochu 1989). The first Sacred Heart Church was created in 1919, French education started the next year, and eventually the establishment of a separate-school board led to separate educational provision at the junior and higher levels.

Another consequence of the labour demand was that, as Plymouth Cordage sought to meet high wage rates and the shortage of manpower, a Chinese labour force was imported and housed on the company's grounds in an inelegant bunkhouse complex reminiscent of a prison compound (Young 1975). The barracks were surrounded by a high chain-link fence, only a few Chinese at a time were allowed outside to visit the town centre, and their employer was held responsible for their actions. This group later dispersed to other jobs and places, and was only temporarily a part of Welland's ethnic structure.

Postwar depression arose as wartime contracts ended and plants closed, but then came both straitened industrial circumstances and the arrival of new plants. Among the former, Plymouth Cordage lost its advantages after the combined harvester-thresher was introduced on the Canadian Prairies, and as the demand for binder twine declined, the company's operations were reduced, workers laid off, the company's houses sold, and working conditions in the plant deteriorated from their earlier high standards.

In 1922 Union Carbide acquired the assets of Electro-Metals. In this now combined operation, a peak of over 2000 employees was reached in 1943, but this total then dropped through a reduced demand for ferro-alloy products after the war. Most of the raw material used in the plant, such as quartzite and manganese ore, coal and oil, was unloaded at the company's dock on what soon thereafter became the Fourth Canal; carbon was exported from there to various areas of the world.

Page-Hersey continued to expand. A third mill was built in 1923, followed by a seamless-pipe mill in 1930. A continuous-weld process was added in 1939, and in 1942 a plant followed to produce small-diameter tubing. Wartime production included shell forgings, naval boiler tubes, pipe for aircraft frames, bazooka guns, and mortars. After the Second World War three more tube and pipe mills were built.

John Deere carried out plant improvements in 1922 that included an office building, and in 1928 the capacity of its factory was increased. In 1952 a new plant that provided a fivefold increase in space was constructed around the old facilities, and further space was added in 1965.

The list of new companies is substantial, but not all survived more than a few years. Employment and the number of companies each fluctuated markedly (Young 1975). The new companies included Mead-Morrison in 1920, which carried on the production by Matthew Beatty of heavy industrial equipment, and Mansfield-Denman the same year as custom moulders of soft and hard rubbers and thermosetting plastics for the mostly Ontario market of the auto and appliance industries. In 1924 Welland Steel Castings (later Welland Electric Steel Foundry) introduced mild steel castings. Further new names included Joseph Stokes Rubber of Trenton, New Jersey, Welland Wires to manufacture light wires, and Scottish Canadian Fertilizer in 1921 on a site next to the railway as a chemical plant for the manufacture of several products to increase soil and plant productivity (Scottish Fertilizers 1931).

Dominion Yarns bought an empty factory in 1926 to spin cotton yarn blended with synthetic fibre for the company's plant at Dunnville, where it was dyed and used for towels and bedspreads. Of interest are the factory's location next to Wabasso in French Town, which provided a source for the female labour force required, and the renewal of economic links with Dunnville through the Feeder Canal.

Of particular importance to the town was the arrival of Atlas Steels in 1928 on the moribund site of former Frost Wire Fence, which had been used for a short period as a bedstead factory and then by a crucible company. This site reflected the fourfold advantages of railway access, water from the canal, hydroelectricity from The Falls, and the availability of a generous manpower supply that could be trained (Atlas Steels 1948, 1952). Although this enterprise was not sited next to the canal, the waterway was used to import raw materials such as coal and oil, and as a source of water. The plant was improved for the production of specialty steels including stainless-steel strip and tool and alloy steels, and the coal-fired furnaces were replaced by electric furnaces. Massive expansion during the Second World War led to an employment peak of some 3000 workers in 1942, when gun barrels, ordnance steels, and specialized steels were each produced. The focus postwar with reduced production centred on developing stainless steel for the Canadian market. The city, through this and other steel companies, became known as the stainless-steel capital of Canada.

Continuing this apparently limitless process of growth, E.S. Fox arrived in 1930 to undertake sheet-metal work and welding including for the marine industry, and Stokes Rubber arrived the same year. Commonwealth Electric arrived in 1936. The postwar period witnessed the arrival of General Drop Forge in 1946 for the production of dies, tools, and forgings, 'El-Mech' Tools the same year, then Switson Industries in 1947, Christie Bread in 1948 to produce bakery products, and Welland Tubes in 1956. When R. & I. Ramtite arrived in 1965, chemicals were imported from Europe and South Africa and finished products that included cement bricks and adhesives for various metals were exported, in both instances taking advantage of the canal.

Although the list of new industrial products is incomplete, stainless steel had come to dominate the industrial scene. As recorded by the Welland Development Commission in 1969, the three big employers were Atlas Steels (2516), Stelco (1572), and Union Carbide (979). Secondary steel industries included a number of smaller foundries that turned out a wide range of castings and forgings, and this impact extended outwards to machine tool and die shops. In addition, Wabasso, John Deere, and Mansfield-Denman each had over five hundred employees. A particular educational offshoot is that in 1968 the Niagara College of Applied Arts and Technology opened its campus on an open site north of the city.

That a considerable amount of industrial transition and change had taken place is conveniently summarized by comparing the 1930 and 1950 editions of Coombs's *History of the Niagara Peninsula*. At 1930 Welland had a population of 8942 and an assessment of $11.7 million. 'Its chief industries are the Cordage plant, Empire Cotton Mills, Welland Canal plant, and factories for the manufacture of furniture, chemicals, stoves, boilers and beds. In normal times these had over 2,500 on their payrolls.' Twenty years later Welland had 'over thirty industrial plants some of them such as Canada Foundries and Forgings Ltd., Atlas Steels Ltd., Empire Cotton Mills Ltd., Plymouth Cordage Co., Page-Hersey Tubes Ltd., Dominion Fabrics Ltd., being among the finest manufacturing plants in Canada. The employees number over seven thousand ... the total assessment is over fifteen million dollars.'

A specialized industrial structure had emerged. By 1970 this was dominated by primary metals, with 51.2 per cent of the total employment in manufacturing industries, a figure that included 35.0 per cent in iron and steel mills, 15.2 per cent in steel-pipe and tube mills, and 1.1 per cent in iron foundries. The next most important category, non-metallic metals, occupied 10.5 per cent of the total manufacturing employment, and included mainly the ferro-alloys, carbon electrodes, and the artificial graphite produced by Union Carbide. Next came the metal fabrication industries, at 8.7 per cent of the total manufacturing employment, which included structural-steel firms, tool and die shops, metal casting, and forging. Textile factories added cotton yarn and cloth mills.

The distinctive local resource of the natural gas industry remained important for use in many of these plants for heating purposes. As the flow from wells diminished, supplies were imported from the southern states in 1954 and then from Alberta through the Trans-Canada Pipe Line a few years later. Starting in 1950, and in order to maintain supplies, underground storage in areas to the east of Port Colborne augmented the volume of this resource that was available on a regular basis for heating purposes.

The impression must not be left that everything was always smooth and happy on the labour front at Welland. Accidents were common in the factories; for example, in 1942, when a reinforced-concrete floor collapsed at Atlas Steels, five workers were killed and a further twenty-nine hospitalized. As in industrial cities elsewhere, the labour and union demands were for improved working conditions, better wages, shorter hours, and some degree of worker security (Sayles 1963). There were strikes by foreign workers at Electro-Metals in 1917, at Canadian Steel Foundries in 1918, at British-American Shipbuilding in 1919 causing the plant to close, at Billings and Spencer and Page-Hersey in 1920, at Crowland for more relief work for a month in 1935, requiring police intervention, and in 1936 at Page-Hersey. From 1936 into 1937 a bitter strike among the female workers at the Empire Cotton Mill closed the mill for seven weeks.

In 1946 Electro-Metals offered increased wages per hour in exchange for a reduced work week, but as the total wages would be less, 1300 employees went on strike. The plant closed down and ninety-five days elapsed before a settlement was reached. In the same year a further strike occurred at the Union Cotton Mill. When, two years later, seamen working on the canal struck, vessels were not allowed to unload at the Electro-Metals dock in Welland. After twenty-five seamen in Welland jail began a hunger strike, 1500 citizens gathered in Merritt Park to support them. As recorded by Sayles in 1963, 'It was the first time in the history of the Niagara Peninsula that workers had left their jobs in support of united action.' Welland had advanced industrially because of the canal, and this in turn led to labour-management tensions and to workers' solidarity that cut across the many ethnic divisions that existed.

Ethnic and Cultural Diversity

Welland, including the northern part of Crowland Township into which the town had grown as industries extended south along the Third and then the Fourth Canals, became a socially divided and ethnically structured community. As people tended to group near the factories where they worked and close to friends, relatives, and other people who spoke the same language and had the same cultural background, distinctive areas arose: the established cultural group, largely middle class and British, but including United Empire Loyalists and later English, American, and Scottish immigrants, in the north and to the west of the canal; French Canadians to the east of the railway tracks close to the textile mills; Italians around Plymouth Cordage; and Polish, Croatian, Ukrainian, Yugoslav, and Hungarian groups close to the Page-Hersey and Union Carbide plants.

As Patrias (1990, 1994) has explained, peasant newcomers were attracted to the factories in Welland and Crowland through the promise of good wages and steady employment, but these prospects were then limited by language, religion, and education to unskilled or at best semi-skilled labour, with generally a surplus of applicants for any available job. To the host society, the immigrants were the 'foreigners' or the 'ethnics.' Degrading terms such as 'wops,' 'dagos' and 'polacks' were used. The new immigrants, set apart from the 'English,' the 'Scots,' or the 'Whites,' were obliged to take the worst jobs in the factories. In light industry 'they worked in establishments, such as Welland's Imperial Cotton Mill, that were notorious for underpaying or in chemical plants ... where they were exposed to noxious fumes. In the metal industry, they got the hottest, dirtiest jobs. Although they desperately wanted work in industry, former agriculturalists complained about the quality of air, the ventilation, the noise, and the pace' (Patrias 1990).

The broad range of its ethnic groups provided Welland with variety and diversity, more so than elsewhere along the canal. By 1931 Crowland could be described as 'one of the most ethnically heterogeneous communities in southern Ontario. More than two-thirds of the 5000 inhabitants were continental European immigrants, belonging to no less than eighteen ethnic groups (Patrias 1990; see also Burns 1994, 1995).

Ethnic differences arose, and members of the larger groups such as the Polish, French-Canadian, and Ukrainian communities established their own churches, benefit societies, shops, restaurants, schools, and social organizations. Even so, 'Despite this ethnic diversity and the existence of intragroup tensions, a common identity emerged among Crowland's foreign-born inhabitants and their children, based on shared experiences and lifestyles and reinforced by daily contacts in the neighbourhood and workplace. Perhaps the strongest integrative force among them was the discrimination they all faced from the host [Anglo-Celtic] society' (Patrias 1990).

The new immigrants had different behavioural patterns that often offended the sensibilities of the established residents. Sabbath observance was not part of their cultural background, and weddings and social celebrations on a Saturday could extend into the Sabbath. Teetotalism was an unfamiliar concept and the making and consumption of wine was endemic. Former agricultural practices were maintained around the home, even in urban neighbourhoods. 'Most of their houses were surrounded by vegetable gardens, and most families kept livestock. Almost everyone, regardless of ethnic background, had hens and pigs. The Italians kept goats, while eastern Europeans had ducks and geese ... Many families also had cows' (ibid.). Socially, however, the shared experience and lifestyles of the immigrants crossed ethnic lines. 'Urban Crowland was ... too small to permit the emergence of enclosed cultural islands based on shared regional origin. No matter how strong their sense of group loyalty, the members of each ethnic group came into daily contact with other ethnic groups' (ibid.). As the different ethnic groups lived next to each other on the same street, their contacts were as neighbours, in the stores, at work, through the children who attended ethnically mixed public schools and played in the streets, and through social, recreational, and community activities provided by the Catholic and United churches.

The traditions and character of Europe transferred to the Welland Market. 'Here the Italian women in their black gowns and kerchiefs contrast with the brighter colours of the Slavic women with their gay dresses and headgears. The flaxen hair and blue eyes of the Nordics contrasts sharply with the sloe-eyed French Canadian housewives. And occasionally one can hear the clump and clatter of wooden shoes from a Dutch farmer who has come to sell his wares' ('The Welland Market,' *Ontario Today*, 3 June 1961).

Then came the Great Depression, when 'local employers did not hesitate to exploit the desperation of working-class families' (Patrias 1990). The Empire Cotton Mill preferred to employ children between the ages of fourteen and sixteen because they were paid less than adults, and adult textile workers were obliged to accept a 25 per cent cut in wages yet continue to work fifty-five hours a week. At the Electrometallurgical Company, four six-hour shifts were replaced by three of eight hours, with no increase in wages for the longer hours in extreme heat and noxious fumes. At Page-Hersey foremen were rewarded for maintaining production with a reduced number of workers. Atlas Steels cut back on production. These were the sparks that ignited resentment. In 1935 the Crowland Relief Strike began when thirty-nine out of the forty-seven men who worked on the Beatrice Sewer downed their tools. Joined by virtually all relief recipients on public work projects, the workers sought to obtain 'fair compensation' for their labour. As Patrias (1990) has stated, 'What makes the Crowland Strike truly remarkable, however, is that it required collaboration among immigrant workers from a wide variety of European backgrounds.'

It is probable that the same type of adverse living and working conditions among immigrant workers, and cleavage between immigrants and the host society, prevailed elsewhere along the canal, in North St Catharines, South Thorold, and East Port Colborne. Industrialization meant a life that was drab, but better than former conditions in Europe and with more opportunities for social and economic advance. Integration into the mainstream of Canadian urban life, especially during and after the Second World War, was to follow. In the meantime, much suffering and hardship existed.

Port Colborne

The catalyst of harbour improvement was followed by the government elevator, Maple Leaf Milling then Union Furnace, and most significantly after the First World War, in 1918, the town that centred around its enlarged port and harbour facilities attracted a refinery of the International Nickel Company (Inco) to a 75-acre (30.4-ha) lakefront site east of the canal. This company, which soon became the city's largest employer, dominated the town's economic structure and contributed to the rapid growth of population that included a wave of Hungarian and Polish immigrants before the Depression.

Port Colborne was an attractive location for shipping in pure nickel oxide produced in the company's smelter at Copper Cliff near Sudbury and, by either rail or ship, for the other raw materials required to make anodes for use in the production of electrolytic nickel. The nickel was used for strengthening steel, in great demand during the First World War for armour plate and ordinance. Postwar, when new industrial purposes had to be developed, Inco established a research and development department to examine the peacetime application of strengthened metal products.

Nickel oxide was mixed with petroleum coke and charged in open-hearth furnaces, using oil as fuel. As electric power was used extensively in the electrolytic refinery area, proximity to The Falls played an important role in industrial location. There was also the factor that large volumes of water could be pumped from Lake Erie for cooling purposes, and the company developed its own water-supply system to meet these needs. The production of electrolytic cobalt was added in 1954, and over 2000 male workers were employed by the late 1960s.

As at Welland, industrial growth meant social segregation. Reinforced by the canal as a barrier, the east side of Port Colborne was inhabited largely by plant workers who needed to live in close proximity to the factory; the west side became the preferred area for plant managers and supervisors (Salerno 1989). This division was reinforced as the workers were essentially from central and southern Europe, whereas the senior staff were mainly English. The former were Catholic and the latter Protestant, and the differences in job and financial security between the two groups was substantial. Moving from the east to the west side of the canal was seen as a sign of affluence and upward mobility.

An adverse factor in conjunction with the end of the streetcar and the advancing use of the automobile for recreational journeys was the wane of tourism. Inco took over the attractive beach area that had developed at the southern end of the Welland Railway. The special train from Buffalo along the lakeshore ended in the late 1920s. The Humberstone Club was dissolved in 1933, in part because of the Depression and in part through the deterioration of the beach front by silting caused by the extended system of canal breakwaters. This decline is the opposite of the situation that prevailed at Port Dalhousie, and more akin to the end of St Catharines as a spa resort.

A novel enterprise at Port Colborne was that of smuggling into the United States, where Prohibition was more strongly applied than in Canada. During the 1920s, when liquor legally produced or imported into Canada could be exported into the United States, smuggling arose and Port Colborne, because of Lake Erie and the proximity of the U.S.A., was probably more involved than other locations along the canal. Canadian whiskey destined for Cuba supposedly arrived by the carload. Small boats then left from the tunnelled sections of the First and Second Canals next to the business centre, supposedly destined for an overnight return journey to this tropical island or South America!

Further changes in the steel industry in conjunction with the harbour and its shipping facilities included that Canadian Furnace, which expanded during the First World War, declined during the Depression years. In 1950, this plant was bought by the Algoma Steel Corporation of Sault Ste Marie, and in 1962 was renamed the Canadian Furnace Division of that company. New facilities, new sources of raw material, and the use of natural gas to fire the blast furnaces increased the steel-making and steel-finishing capacities to some 700 tons (634.3 tonnes) and some 175 employees. The supply of water was obtained from the canal and the firm's requirements of iron ore were brought in by boat, as was limestone from Michigan because the stone from quarries close to Port Colborne was not of sufficiently high grade for metallurgical use.

At these quarries, Canada Cement, modernized in the 1920s, supplied part of the demand for cement required to build the hydroelectric power plant at Queenston. Its cement was also used for various road-improvement projects by the provincial government and for construction works on the Fourth Canal. The Depression then hit hard. The old plant closed in 1931, but a new plant began operation next year. A spur line brought the cement through the streets of the town for loading onto vessels at the harbour. The new plant, which produced some 12,000 bags of cement a day, closed in 1963. It had in turn become obsolete and the stone resources were becoming exhausted. Crushed stone continued to be produced at the Port Colborne quarries north of the town, mostly for export to the United States from a dock and storage area next to the canal at Rameys Bend. Trucks hauled the stone to this facility.

A new metal activity was the Port Colborne Iron Works, which located on Fraser Street in 1922 to make custom fabricated steel. This output in 1931 included a rivetless water tank, the first of its kind to be manufactured in Canada. Another item was fairleads, which were moulded into the hull of a ship and through which hawsers were passed to moor a vessel.

After wartime prosperity, when food was in great demand, Maple Leaf Milling Company as part of its extensive operations within Canada concentrated on the domestic business and purchased bakeries and bakery chains. Canadian Bread was purchased in 1918, Campbell Flour Mills of Toronto with its popular Monarch brand of flour in 1919, Canadian Bakeries in 1925 to serve the West, and Eastern Bakeries in 1928 to serve the Maritimes. With each acquisition, new companies were formed and the older, smaller mills were closed. At Port Colborne a rebuilding program in the large mill on the west pier at the harbour entrance increased its capacity to 26,000 100-pound (45.4-kg) bags of flour, which made this mill the largest in the British Empire. A second vessel was purchased in the early 1920s (the first had been lost on war service) to bring in wheat from Fort William (now Thunder Bay) and to carry flour between Port Colborne and Montreal.

During the Depression, output was reduced to sometimes two days a week, but full capacity and peak employment resumed from 1940. The demand for mixed animal and poultry feeds increased, and in 1946 the Port Colborne feed plant was rebuilt with an increased capacity. Further modernization followed in the 1950s, but then disaster hit when an extensive fire destroyed the mill in 1960. The mill was rebuilt and opened in 1962 to produce wheat flour and rye flour, with a substantial storage capacity for grain in the elevator and on ships in the canal during the winter months. The Ontario Bag Company in the same building was destroyed by the same fire, but this facility with its extensive female employment was not reconstructed.

Earlier, in 1919, there had been an explosion at the Government Elevator. Managed by the National Harbours Board, this plant was then rebuilt and an annex added that doubled the storage capacity. In a new milling activity, the Robin Hood Flour Mill, part of International Multi-Foods of Minneapolis, located in 1940 on the west bank of the abandoned Third Canal and next to the Fourth Canal at Rameys Bend. Hard wheat was brought in from the Lakehead by water, and soft and durum wheat from Ontario by truck. Bread, cake, and pasta flour were the main products of the mill, together with cake and bakery mixes. The winter storage capacity of ships berthed in the canal was again an important asset. The domestic and American markets were important for both Maple Leaf and Robin Hood, and their location in Port Colborne also avoided the need for vessels to pass through the canal and pay tolls.

In the business of ship repair, H.E. Heighton & Sons was established in 1937 to undertake marine work, which included the repair of ships and their boilers. The winter layover of ships was an important locational factor, and Port Colborne's position as a trans-shipment point between lake and canal steamers was also relevant. Bell Marine and Supply, carrying on the tradition of ships' chandlers, located in 1941 on the west pier to supply the needs of ships. The firm of F. Woods and Son grew with the canal as steam and its machinery became more important. Purchased in 1950 by E.G. Marsh and named after this individual some eight years later, the firm added new fabricating and heavy machine shops for the repair of vessels and other work that included engineering, installation, maintenance, and repair work for heavy industry, thereby providing a range of activities that helped considerably to offset a seasonal unbalance in employment when the canal closed for the winter.

Also directly associated with the shipping business, the Sarnia Steamship Company was founded in 1928, with Captain Scott Misener and John McKellar as partners. In 1936, as Colonial Steamships, the company received a colonial charter; its name was changed to Scott Misener Steamships in 1958. In 1952 Port Colborne was recognized as a Port of Registry for ships, carrying the name of the town to ports around the Great Lakes.

In a different vein (now referring to the leather industry), the Humberstone Shoe Company in 1924 moved to a multi-storey factory at the corner of Elm and Main Streets. With 200 employees, 3000 shoes a day could be produced. The leather from tanneries at Kitchener, Oshawa, and Toronto was first brought by rail to the company's siding, but this mode was gradually displaced by truck transport. This business became the Sunbeam Shoe Company in 1963, at first making all types of shoes but then specializing into children's shoes, then shoes for men and boys for the domestic market. Three shoe companies existed by the 1960s as a spin-off from the Humberstone Shoe operation. Knoll Shoes carried the name of the original founder of Sunbeam Shoes, and the Erie Shoe Company set up its factory on Elm Street in 1938. Leather came from the same sources in southern Ontario, first by rail and then by truck transport. An important factor was the availability of a female labour force, a counterbalance to the male employment in the area's economic base of heavy industry.

Port Colborne, the urban entity at the southern end of the canal that now incorporated Humberstone, had achieved a broad-ranging industrial base that now outpaced Port Dalhousie. By 1970 primary metals, at 68.8 per cent of the total employment in manufacturing, included metal smelting and refining and the production of pig iron as the most important industry. Second in importance was the food and beverage industry at 10.2 per cent, with a focus on grain milling. The leather-products industry, at 6.4 per cent, with its emphasis on shoe manufacturing was followed by transportation equipment, with its emphasis on ship repair work, at 3.7 per cent. Other industries were small in size and relatively unimportant in the employment situation.

Retailing had transferred to the west side of the canal after East Street was acquired for the Fourth Canal, and it also extended away from the canal along Clarence Street, now a through route across the canal. This progress was assisted by the growing influence of the motor vehicle and, after the absorption of Humberstone into the urban matrix of Port Colborne in 1952, by some business transfer from this centre to the superior facilities in the now larger centre at Port Colborne. Humberstone, for example, lost its village office and residents had to pay their various municipal bills at Port Colborne.

The shift of customers from Humberstone to Port Colborne, the only two crossings of the canal along its southern entrance, boosted Port Colborne's retail trade along the automobile route of Clarence Street, a process that in turn introduced decline along West Street facing the harbour. And as the motor car continued to advance in importance, the pattern of urban form that had emerged, then changed, over the decades of canal development was again amended through the arrival of malls and chain stores on the suburban periphery. West Street with its harbour atmosphere declined and necessarily became a focus for urban renewal to reinvigorate a once thriving and prosperous canal locality.

PART 4

The St Lawrence Seaway Authority and the Welland Canals Corridor of Development, Post-1960

14

The Welland Canal as Part of the St Lawrence Seaway

The Fourth Canal became part of the St Lawrence Seaway when this opened in 1959, and was then steadily improved. It was first intended that the single locks be twinned and work started on this project, but this scheme was superseded by proposals for the construction of bypass channels around St Catharines and Welland. The land was purchased for both and has remained under federal ownership, but only the By-Pass Channel at Welland was constructed; the abandoned length of the Fourth Canal through the centre of that city was then gradually transformed into a Recreational Waterway. Meantime, a rehabilitation program conducted by the St Lawrence Seaway Authority along the canal channel extended the expected life of the Fourth Canal up to an anticipated date of 2030, and the land acquired for an improved Fourth Canal at St Catharines was retained for this eventuality.

The Birth of the St Lawrence Seaway

The argument for an international St Lawrence route as a water highway from the Atlantic to the inland heart of the North American continent had prevailed since the Deep Waterways Commission report of 1897, and even though the enlarged Fourth Welland Canal was an obvious success for inter-lake traffic, the 14-foot (4.3-m) depth of the St Lawrence Canals remained a severe bottleneck for foreign and some domestic shipping. The smaller vessels could carry on a trans-Atlantic trade, but not the larger ones that worked on each of the Great Lakes, and many of the vessels that crossed the Atlantic could not advance west of Montreal. This severe impediment to Canada's international trade became recognized during the First World War, but opposition to the construction of the Seaway was long, severe, and bitter (Gregg and Crecher 1927; Patten 1956; Willoughby 1961).

The Seaway was conceived as a joint operation between Canada and the United States. However, staunch objections arose from influential railways, other private industrial sectors, and from American ports including Buffalo and those on the Atlantic Coast in both nations that, fearing the loss of trade, considered they would be affected adversely. During the Depression, rather than pursuing an international scheme of development, American resolve preferred to focus on opportunities within their own nation such as the Tennessee Valley project with its relief of unemployment through dam construction, the prevention of flooding, and hydroelectric production.

Hydroelectricity also played a pertinent role in the eventual achievement of the St Lawrence project. The important point was that, if the river was dammed to provide deep-water navigation, then an abundance of hydroelectricity would be made available. It is interesting that when the First Canal was achieved, its success could be attributed to the provision of water power to drive mill wheels. The scale and the detail had each changed considerably over the previous century, but the fact in both instances was that the canal as a trading waterway needed an influential partner to achieve economic success.

The Treaty of Washington, signed in 1932, failed to win the necessary support in the United States Senate, but circumstances changed after the Second World War. There was a new suggestion to charge tolls for the use of the Seaway; together with the sale of hydroelectricity, these were expected to make the Seaway self-supporting. Other new elements were the fact that the railways could not meet all requirements in a period of rapid industrialization and increasing foreign trade, the decision to develop the rich iron-ore deposits of Labrador, and the acute need for hydroelectricity in Ontario and New York State.

By 1951 it was clear that Canada would act unilaterally to achieve both navigation and electricity, and the St Lawrence Seaway Act and the International Rapids Power Development Act were passed by the Canadian parliament. Three years later the necessary legislation for joint international action was approved by the United States Congress. After further compromises to avoid the duplication of locks, canals, and powerhouses, agreement was reached between the two nations.

Construction was able to start and the first sod for the St Lawrence section of the Seaway was turned in 1954. After over fifty years of often fractious political debate and prolonged negotiations, the actual engineering works took but five years, and the entire Seaway opened for the 1959 navigation season. Only then was the dream of William Hamilton Merritt for the Welland Canal as a small but nevertheless vital component of a larger through route between the Upper Lakes and the Atlantic fully achieved.

An important later addition, in 1968, was the provision of a new link in the Great Lakes–St Lawrence system on the American side at Sault Ste Marie. The new lock, 1200 feet (365.8 m) long, 110 (33.5 m) wide, and 32 feet (9.8 m) deep could pass through large self-unloading vessels with a gross cargo-carrying capacity that could not be accommodated in either the existing Welland Canal or any other of the lower Seaway locks. Historical tradition had repeated itself, with the Welland Canal not being able to receive the largest vessels then in operation on either the Upper Lakes or on ocean passage.

The Fourth Canal provided the dimensions that were used by the Seaway along its St Lawrence section. The new locks were built to the same size as the lift locks on the Welland waterway, but a new method of filling and emptying was devised that reduced the speed of each lockage down to eight minutes. In addition, the railway and highway bridges that crossed the St Lawrence River were provided with the same clearance as over the canal.

On the Welland Canal, as its earlier locks had been constructed to a greater depth than the channel, no works of improvement were required except that the approaches to the ports and the channels between the locks had each to be blasted or dredged and excavated from 25 feet (7.6 m) to the greater Seaway depth of 27 feet (8.2 m). This work was undertaken mostly in winter when the canal was drained, and 2.3 million cubic yards (1.8 m³) were excavated with no interference to shipping. Two dykes created east of Port Weller received this material from the northern length of the canal.

Blasting next to the retaining walls presented a problem, in that material could not be loosened to closer than 14 feet (4.3 m). As this would have meant that a piece of land jutted out underwater and that ships could not moor next to these walls, the solution was to use an ingenious new system of drilling holes next to the walls, providing a cushioning effect through using air-filled canisters to absorb the blast (Chevrier 1959). This blasting was injurious to homes, as for example at Allanburg: 'Drill boats and underwater explosives were used to break up the rock. Many dredges, scows and other work boats were kept busy removing the excavated material. Shock-pressure waves produced by blasting rock in water did much damage to nearby homes. Plaster cracked and fell from walls; wells dried up. The well at the school had to be dug 16 feet [4.9 m] deeper. Indignant citizens held protest meetings at the Township hall' (Vanderburgh 1967). The canal had resumed its impact on community life, this time with deleterious consequences for the residents of the local community.

Improving the Fourth Canal

The Fourth Canal still connected Lake Erie with Lake Ontario, but it now did so as part of an international waterway and the inland waterway system of North America. As such, its links with the canal communities that had grown up in association with the waterway were diminished rather than strengthened.

Work continued in the late 1950s and into the early 1960s on tree-planting schemes on the banks to provide windbreaks as well as a further amenity for the regional landscape. As the Department of Transport noted in 1953, 'an extensive reforestation has been carried out along the banks of the Ship Canal [then the official terminology]. Vast numbers of trees native to the district have been planted and are now mature.'

New also was the construction in 1964 of a headquarters building for the Western (later Niagara) Region of the Seaway Authority. Previously, the Welland Canal offices had been located in central St Catharines next to the Second Canal. They had outlasted the canal in this location until the creation of the Seaway, when local offices were located at Lock 1 and Lock 8, and the canal's administrative offices were again located next to the canal, this time in a prestigious location at the northern end of the Flight Locks. A service centre that included the engineering and maintenance yards was added in 1971.

A further new operational feature on the canal was the development of a Marine Training School on the abandoned section of the Third Canal at Ramey's Bend, Port Colborne. Sponsored by the federal government and staffed by Niagara College, its purpose was to train ship's personnel in how to cope with on-board fires and other emergencies.

The steadily increasing volumes of traffic along the St Lawrence and through the canal led to many schemes for greater efficiency, to remove delays and to hasten the movement of passing vessels (Luce 1965). By 1964 demand on the system as expressed in the arrival rate of vessels came dangerously close to capacity, with delays resulting in long lines of vessels waiting to enter the canal. To offset such delays a coordinated modernization program introduced improvements in operating methods in 1966. When a new centralized traffic-control centre was established next to the Seaway's administrative headquarters, the canal could be surveyed through closed-circuit TV monitors with remote camera controls. Display units covered the activity at each lock and the location of each vessel in the canal. This control system was later extended to cover the eastern end of Lake Erie and western end of Lake Ontario. These traffic-control measures added close to 20 per cent to canal capacity. Transit time for vessels was reduced from twenty-four hours in 1964 to less than fourteen hours by 1969. A further advance was the development of a computer-assisted traffic-control system. Placed in operation during 1971, this new system located in the authority's office building again hastened the flow of traffic. By 1991 the time for a one-way transit of the canal was estimated to be 12.4 hours.

Vessel design presents a further factor of some importance. Since the Seaway opened, steering control, manoeuvrability, and cargo-carrying efficiency have each been improved. Lighter materials and modern, smaller engines have been used so that, as the size of the engine room or the width between the cargo hold and the outer hull of the vessel are reduced, additional space is made available for cargo. Along with hydraulic improvements at the locks and to the canal system, the Seaway has been able gradually to increase the beam size of vessels from 75 feet (22.9 m) in 1959 to 76 feet (23.2 m) by 1976. Generally for a modern, maximum-size laker, an increase in the beam by one inch (2.5 cm) improves the cargo-carrying capacity by roughly 3.1 tons (30 tonnes).

The development of self-unloading vessels, such as those that carried U.S. coal to Hamilton and miscellaneous loads of salt, potash, and stone, provided another means of reducing shipping costs and speeded up the otherwise costly turnaround of vessels. An attempt to move a vessel more rapidly out of a lock by attaching a special tug known as a 'shunter' to the bow of the vessel proved a costly failure; had it been successful, it was estimated that the number of transits might have been increased by 10 per cent.

Over the period from 1959 the maximum depth of vessels passing through the canal was increased steadily from 25 feet (7.6 m) to 25.75 feet (7.8 m) by 1967. The rule of thumb was that for every inch (2.5 cm) of additional draught, the cargo-carrying capacity of each vessel increased by 110.2 tons (100 tonnes). This permitted draught then edged upwards to 26 feet (7.9 m), and was at 26.3 feet (8 m) by the 1992 shipping season. By 1995 it was at 27 feet (8.2 m) from Port Colborne to Ramey's Bend and 26.25 feet (8 m) for the rest of the canal (Canadian Marine Publications 1995). The costly deepening of the total canal system to the full Seaway draught of 27 feet (8.2 m) was by 1995 perceived as an unlikely proposition as the Seaway was not then working to capacity.

By 1960, 84 per cent of the dry-bulk carrier fleet was capable of passing through the Seaway. By 1992 this figure had reduced to only 7 per cent (Atkinson 1992). To resolve this matter by creating a deeper, wider channel is especially complicated, politically and economically, because never before has canal enlargement taken place within conditions of declining trade and uncertainty about the trading future of the water. Previously, each new canal has been associated with an economic climate of expected expansion.

Another approach towards improving the canal's capacity has been to examine how the navigation season might be extended (St Lawrence Seaway Authority 1980; LBA Consulting Partners 1978). The Seaway, including the Welland Canal, has traditionally been closed from mid-December to early April, giving rise to an operating season of about 270 days, with the canal closed to navigation for about a quarter of the year. The three options considered during the 1970s ranged from extensions to the end of January, or to the end of February, to full year-round navigation. The primary benefits of each option, and especially year-round navigation, would be to reduce the expensive stockpiling of raw materials by industries using the canal and to save on the more expensive overland haulage by rail. Neither ports nor shipping would have to close down during the winter months, which would remove the difficulties posed for international shipping and foreign markets, reduce unemployment, and, through the income multiplier of approximately two used on the Great Lakes, extend the economic benefits to the vicinity of the ports.

An adverse factor centred on the Welland Canal, where at least a month is required for lock maintenance and repairs, but this could be offset if the remaining locks were twinned. A bubbler placed across the bottom of the canal near Lock 8 at the Port Colborne entrance prevents the flow of ice into the canal, and ice that formed in the canal might be diverted into the upper length of the abandoned Third Canal from above the Flight Locks.

Icebreaking and winter navigation along the St Lawrence River presented a greater problem than winter navigation through the canal. Here navigation depended upon the uncertainty of severe weather at the end or start of the season. In wide channels the ice could be broken then slipped under other ice, leaving behind an open channel, whereas in a confined channel the ice closed in behind the ice breaker and resealed into a resistant mass. There was also the possibility of potential environmental damage, including shoreline erosion and the destruction of fish and vegetation through the under-ice wave surge caused by large freighters, and the fact that no winter contingency plans existed to prevent winter spills of petroleum and other chemicals.

A misfortune of the Welland Canal is that its Atlantic outlet lies considerably to the north of the climatically favoured Niagara Peninsula in a much colder zone of winter severity. This poses a severe handicap to the capability and effectiveness of the canal, but the local advantages that ensue from these unalterable natural circumstances include winter employment opportunities through the repair and servicing of vessels laid up in the canal and the possibility of carrying out canal works during the closed season of navigation.

Overall, the studies of extending the shipping season demonstrated that the idea was feasible, but the costs increased considerably as the opening period for the Seaway system was extended through the three options under consideration. By the 1990s, in the novel circumstances of declining trade flows, such extension remains only a distant possibility. 'It simply makes no economic sense for the Seaway system to remain in operation during the winter period of cold weather and heavy ice ... What the Authority is concentrating on is to provide at least 8.5 months of safe, trouble free and efficient navigation, while giving consideration to gradual, incremental extensions to any season based upon the weather, facilities, costs, and the amount of business' (Atkinson 1992).

It will be recalled that of the eight locks along the Welland Canal only Locks 4, 5, and 6, the Flight Locks, are doubled, and that Locks 1, 2, 3, and 7 on the northern reach have single chambers. The twinning of these four locks to double chambers was approved by the federal government in 1962, when 320 acres (129.5 ha) of land were acquired for this purpose. The further implications of this twinning proposal were that, taking into account the anticipated increase in highway traffic flows and probable developments on the land next to the canal, several additional crossings of the canal were deemed to be required. In St Catharines these included an additional bascule bridge at Lakeshore Road, a tunnel at Carlton Street, and a new lift-bridge at Homer, but neither the Garden City Skyway nor the bridge on Glendale Avenue were expected to be affected. At Thorold a tunnel or a high-level structure was proposed, together with removal of the Peter Street Bridge and the Highway 58 crossings of the canal. To the south, the bridge structures were not affected by twinning at the Lift Locks, but it was expected that higher volumes of vehicular traffic in a conflict situation with ship traffic on the canal would require a tunnel or a high-level structure at Allanburg. No changes were envisaged on the lesser-travelled route across the canal at Port Robinson.

A tunnel was required at Welland and also a four-lane lift-bridge at or slightly to the north of the Main Street Bridge that served the town centre. The Lincoln Street Bridge might remain, but without improvement as its traffic would be diverted to either the tunnel or the Ontario Street Bridge, where a new four-lane structure might be required. To the south, the twinning of Lock 8 and higher traffic volumes at Port Colborne required a tunnel midway between the existing bridges on Highway 3 and on Clarence Street at the town centre. These two crossings at Port Colborne would each remain.

Bridges and Tunnels across the Canal

In 1965 the Ontario Department of Highways started on the two tunnels required to relieve traffic congestion. The project at Carlton Street in St Catharines was discontinued when the twinning program was announced. At the second site, the Thorold Tunnel that opened in 1968 replaced the bascule bridges over the canal at Peter Street and on Highway 58 that had previously crossed the canal. The intermediate railway bridge had been removed when the N.S.&T. streetcar system ended its operations. At Peter Street, the bridge over the railway remained, as did Government Road for access along the canal to the Flight Locks, Glendale Avenue, and the General Motors plant. Access to the industries between Thorold and Allanburg on the east bank of the canal was now by tunnel, then Highway 58.

Thorold, severely confined across its eastern flank by the canal, now had but the one crossing of the waterway. The Thorold Tunnel, constructed in the dry over three winters when the canal was de-watered, used the cut-and-cover process of construction: that is, the canal was placed back over the reinforced concrete tubes that carried four highway lanes in two separated tubes. A bridge at this location would have required steeper grades than for a tunnel, longer and more extensive approaches, and more land acquisition and would have exerted a greater impact on urban character (Acres 1968; Christensen 1976).

The cost of a tunnel was less than for a high-level bridge, the type of structure that had previously only been used where the QEW crossed the canal. With a length of 2400 feet (731.5 m) and width of 74 feet (22.6 m), the tunnel's cost was borne two-thirds by the Ontario Department of Highways and one-third by the St Lawrence Seaway Authority. An interesting feature is that the five feet (1.5 m) of cover between the top of the tunnel's roof and the canal bed carries the warning on nautical charts not to drop an anchor at this point!

At Port Colborne, the lift-bridges remained on both routes, but consultants (Dillon 1972) and the city argued for a six-lane, twin-tube tunnel along the axis of Killaly Street to avoid congestion and provide an uninterrupted crossing on Highway 3, the major route inland from the Peace Bridge and the international border for American visitors. The interesting outcome was the construction of a second bridge on this highway at the opposite end of Lock 8. Traffic was now no longer held up by shipping. One of the two bridges was always open, and when a ship was entering or leaving the lock, highway signs directed road traffic to the other bridge.

Other changes to the canal crossings were accidental. In 1968, at the head of Lock 2 (Carlton Street) in St Catharines, a Japanese freighter so damaged the bascule bridge that it had to be dismantled for traffic to resume on the canal. Despite the possibility of a tunnel being built, this bridge was replaced and was in use again within eighteen months. Six years later in 1974, when an American ore carrier, the S.S. *Steelton*, sailed into and knocked down the lift-bridge at Port Robinson, the precedent of replacing a demolished bridge was not followed. The village received immediate fame as a local tourist attraction, but this was short-lived as the debris was soon removed. The bridge remains were towed to Port Weller Dry Docks, dismantled, and sold to Atlas Steels.

More consequential was the division of the village into two separated halves by this accident. Instead of a replacement bridge, the Seaway Authority provided a link between the two sides by a six-seater passenger ferry during the summer months, the only such service on the canal, and a shuttle bus with an hourly service linked the two halves of the village through the winter. One argument against a replacement bridge was the possibility that existed of a highway tunnel between Allanburg and Port Robinson; another was the very high replacement costs for a bridge across a wide section of the canal with no locks. Further, a much tighter economic situation prevailed here than at Carlton Street, and the expected flows of highway traffic were lower. The pleas of the local community to replace the bridge were of no avail against such arguments.

The impact on the small canal-centred community was substantial. Instead of crossing directly over a bridge, the circuitous highway routes between the two sides were 8.7 miles (14 km) via Allanburg and 13.8 miles (22.2 km) via Welland. Amputation left the library, credit union, post office, community centre, churches, transit system to Thorold and St Catharines, fire hall, and most of the volunteer firemen living on the east side. The loss of drive-by highway traffic caused the only store to discontinue evening hours and Sunday openings and one garage closed. Church attendance was lower. Cab fares increased, as did the cost of visits by appliance men. 'The loss of our bridge affects us personally. It has lowered the value of our properties, increased the costs of travelling to and from work, disrupted social patterns and shopping habits formed over a period of years. Our community resents the "ships before people policy" of the St Lawrence Seaway Authority' (Michael 1975; Swart 1976). Of the local industries, one recorded that its freight costs were seriously affected and another that trucking companies, customers, and salesmen were vociferous in their condemnation of the decision not to provide a replacement bridge or tunnel, and a training and employment centre reported increased costs and time of travel.

A village that had lost a historic church and ten houses a few years earlier to canal relocation for the Fourth Canal naturally felt dismayed and chagrined with the Seaway Authority, and the small community had little recourse against the greater weight and powers of the federal government. To them, the canal had created then destroyed the life of the village. Even the notice asking for claims with respect to losses was printed in the *Globe and Mail*, a paper not usually sold in the village, and it was too late to file a claim for the increased costs of travel when these became known (Michael 1975).

Of historic interest and a reminder that Chippawa was once an integral part of the earlier, more extensive canal system, is that the bascule bridge across the Welland River–former canal had not been operated as a movable bridge for twenty years. This structure was converted to a fixed span between 1961 and 1963.

A Canal Bypass at St Catharines–Thorold

The work that had started on the lock-twinning project of Lock 8 at Thorold was cancelled in 1965 when an even more grandiose possibility was announced. Effectively marking the start of the Fifth Canal, over 2000 acres (809.4 ha) of land were expropriated to provide a canal with larger and fewer locks between Thorold and Lake Ontario. The new channel would relieve the pressures that existed at the Flight Locks, and remove the need to duplicate the single locks that encumbered the system. The ascent across the Ontario Plain and the escarpment would now be through three or four Super Locks that would be deeper, wider, and longer and capable of handling much larger vessels. Envisaged was a lift of at least 82 feet (25 m) at each lock, a usable length of 1200 feet (365.8 m) and a minimum depth of 35 feet (10.7 m). To construct a channel with three to four locks, the present entrance channel at Port Weller would be removed eastwards by about half a mile (0.8 km). Existing bridges over the canal would be replaced by road and railway tunnels that would be constructed as part of the canal works in the dry. The canal works would also require an extensive new pondage area at the edge of the escarpment to serve the locks, and all would doubtless be designed to provide new features of tourist appeal.

An even grander vision is that of two parallel elevator locks having a single lift of 327 feet (99.7 m). As stated by its proponent, 'The lock would basically consist of two large troughs or chambers moving up and down between guide towers and would be located at the foot of the Niagara Escarpment ... One trough would be hydraulically balanced against the other, permitting one ship to lock "up" and one to lock "down" at the same time' (Ter Horst 1973). Design would be comparable to the lift locks on the Trent-Severn Waterway at Peterborough.

The main disadvantage would be the requirement for a very deep cut between Lake Ontario and the edge of the Niagara Escarpment, but this would become a major tourist attraction to view the largest lock in the world and its international shipping, and this deep cut would be but two-thirds the depth of the Guillard Cut on the Panama Canal. Pipe dream or not, this impressive idea might be compared with that of William Hamilton Merritt to achieve a canal in the first place, or with the failure at the Deep Cut that led to the damming of the Grand River and the transfer of its water flow across Wainfleet Marsh to Port Robinson.

After this scheme of Super Locks was announced, a new industrial slip was developed at Thorold to the south of the Ontario Paper Company where the Weaver Coal Company was located. Exolon was also interested in this project. St Catharines proposed to develop a harbour between the piers of the old harbour at Port Weller and the new canal. This would be enclosed by a breakwater, creating a basin around which there would be space for industrial growth served by rail access from the Canadian National (former N.S.&T.) system. This scheme also had the advantage of providing docking facilities outside the line of the canal and therefore not prejudicing the flow of shipping in the waterway. A survey undertaken by St Catharines Chamber of Commerce indicated that sixty-five industries in the peninsula might make use of these docking facilities.

An issue of concern to the city and residents of Thorold was that the location of Lakeview Cemetery to the east of the canal with 7000 interments was required by the Seaway for the Super Lock proposal. In exchange, the city was given 55 acres (22.3 ha) about half a mile (0.8 km) east of the present site. As the new extension lay immediately next to the old site and was designed as a mirror image of its predecessor, the bodies in the old graveyard would be disinterred then moved to the new location when the canal works started. Funds were provided by the Seaway for the operation and maintenance of the old cemetery and to improve road access along Davis Road to the new cemetery. Meantime, all burials except for those in pre-sold lots in the old cemetery took place in the extension, and the 4000 vacant lots in the old cemetery could no longer be sold. The agreement, signed in 1972, was for a fifteen-year period.

Further results were the acquisition of a school and uncertainty about the future of the St Catharines Airport. The realigned canal came to within 1000 feet (304.8 m) of the runway, and its flight paths crossed the intended canal. Also some 2000 acres (809.4 ha) of valuable fruitlands would be removed from active production. The QEW would cross the new channel by a tunnel and, with the Garden City Skyway still required to cross the Fourth Canal, this would be, local humour noted, the only high-level approach to a tunnel in the world!

The land for this mega project was appropriated by the federal government, but the Canadian Transportation Commission cautioned delay in approving the project. New automated ship-handling techniques were then being introduced, larger and more efficient bulk carriers were coming into use, and competitive systems such as pipelines, container cars, or unit trains might transfer cargoes from the Seaway.

The bypass project has not been implemented because of the unanticipated decline in the Seaway's trading operations, but with an upsurge of this trade and the national financial situation, it might be revived at some future date. In the meantime, the land has remained under federal ownership pending the now distant possibility of the Super Locks. At Lakeview Cemetery, the agreement to disinter and relocate the bodies ended in 1987, but with deferment of the Super Lock proposal and as the re-burial process has not been undertaken, Thorold requested a review of the earlier agreement (Thorold 1992). The city now had to carry the costs and the tax burdens of maintaining two cemeteries: the old where capital works were required for continual care, but with an embargo on new burials to provide the necessary funding, and the new with active burials and the regular sale of its lots.

Meantime cemetery legislation had changed. In 1992, through a New Cemeteries Act, the Ontario government had passed over the responsibility for the upkeep and maintenance of grave markers and the maintenance of cemetery lands to the owners of cemeteries. The problem at Lakeview was that as the vacant lots could not be sold to pay for these extra costs, a self-sustaining operation was not possible. The expropriation of cemetery lands had now become much more difficult than in the late 1960s when the Seaway had envisaged both acquisition and the re-interment of the bodies.

Further, the deferred Super Lock proposal would now almost certainly have to pass through the new series of lengthy and costly environmental hearings promulgated by the provincial and federal governments. As the possible use of larger vessels 1000 feet (304.8 m) in length on the canal would also require modification to all the harbours and channels used throughout the Seaway system and vessels of wider beam would require an enlarged channel along the modern canal, the environmental implications would probably not be restricted only to Thorold. Neither the Seaway nor any previous canal has ever been involved in such procedures.

If a new canal was built to carry vessels with a beam of 105 feet (32 m), the canal channel would have to be widened to more than 350 feet (106.7 m) to allow for two-way traffic, and the excavations at either end of the waterway up to the locks would probably also raise environmental queries. With North America's economic future and the state of marine technology being impossible to predict with any accuracy, the Seaway has had no alternative but to retain the land it has acquired for potential widening purposes.

At Thorold, the city in 1992 argued for the return of the old cemetery to its control and requested a capital contribution towards the upgrading of roads, water mains, and landscaping. The city recommeded, if a new canal was eventually needed, a return to the approved concept of twinned locks rather than proceeding with the futuristic idea of Super Locks (Thorold 1992; Longo 1992). Meantime, the Seaway has retained its right to use the old cemetery lands for the future expansion or replacement through age of the Fourth Canal.

The intended larger canal also raised important questions about Lock 8 at Port Colborne. As its dimensions were insufficient for use by larger-sized vessels, the Seaway has undertaken studies to either enlarge it or remove it completely. The latter approach, which would extend the upper reach of the canal to Thorold, was considered 'a practical move to make as the banks of the canal in most areas were above the high water level of the lake' (Campbell 1973). Regardless of the changes to be made at Lock 8, the same author noted that 'major modifications to Port Colborne would be required to permit larger vessels to enter the canal.' perhaps through widening the channel above Lock 8 and the removal of Bridges 20 and 21. At that time a tunnel was contemplated at Port Colborne, but the possible enlargement of the canal was an additional factor and the significance of relieving the city centre of its through highway across the canal was not part of the Seaway's calculations. However, the increasing size of vessels was leading towards some consideration of a Fifth Canal, with inevitably important consequences for the canal communities that would be directly affected by the plans of the canal engineers.

The Welland Canal By-Pass

Often referred to as the start of the Fifth Canal, the bypass needed at Welland to increase the capacity and efficiency of the operations through the Welland Canal has a very different history. The problems of canal navigation to be offset included the slow, winding channel of the Fourth Canal that passed next to the city's core and the pronounced delays to road traffic at the highway bridges.

The alternatives examined by the Seaway included updating the existing channel, constructing a bypass, or designing a new canal route south from Port Robinson to Lake Erie some 2 miles (3.2 km) east of Port Colborne. The advantages of least disruption to existing development, lower costs of appropriation, and less excavation into hard bedrock carried the day for the bypass route. This decision also required neither additional harbour protection nor new harbour facilities, nor a replacement guard lock at Port Colborne.

The land was expropriated in 1965. In contrast with the acquisition at St Catharines where fruit trees and vineyards predominated, mixed farming on poorer-quality soils prevailed at Welland, though one farm was in the process of development to a feed lot and another to a golf course. When surveyed, farmers in both localities have indicated that the amount of compensation paid by the Seaway was reasonable to generous rather than unfair, with farmers at the St Catharines end of the canal being happier than those in the Welland area (Shantz 1969). Most dissatisfaction arose when farmers could not replace their farms with the money received; with the time lapse of over a year between the date of appropriation and the time of settlement, farm prices had risen in the meantime. Whether or not the full-time farmers relocated to other farms depended upon factors associated with age. Some retired or withdrew from farming, whereas among part-time farmers those who did not relocate had owned their farms for the longest period of time. Presumably, the compulsory purchase of their land was not disadvantageous, and only a few tried to obtain non-farm employment.

Contracts were awarded for the clearance of land in 1966 and construction of the By-Pass started next year. Six thousand five hundred acres (2630 ha) of land were purchased or expropriated by the Seaway Authority. The new channel required the removal of some 65 million cubic yards (49.7 m³) of earth, clay, rock, and silt. The By-Pass was 8.3 miles (13.4 km) long. It provided a navigable width of 350 feet (106.7 m), sufficient for two vessels to pass, and a depth of 30 feet (9.1 m) (Jackson 1975b; Petrie 1973; Sainsbury 1974; St Lawrence Seaway Authority 1973). Widening of the channel to the north between Allanburg and Port Robinson, the traditional Deep Cut of the canal's history, was completed later in 1983 to extend the length of canal where vessels might pass.

The Seaway had intended to carry out work on the By-Pass by wet dredging, but Welland considered that this method would be detrimental to the environment both during the dredging process and in later years because of the ponding and the drainage that would be necessary. It was feared that mosquitoes would breed and that the land would not be stable for many years. Negotiations led to excavation by the dry method, which left the land more suitable for immediate use and in a fit condition to provide the basis for a natural park.

The materials excavated to achieve the By-Pass were placed in predetermined landforms on both banks. These hills rose 60 feet (18.3 m) above the level of the canal and provided a windbreak for passing vessels. The planting scheme defined a wide range of sizes and shapes of spaces. Envisaged were landscaped levees and the establishment of a forest cover with indigenous plant materials that would create continuous, attractive ridge and plateau planting (Project Planning Associates 1968).

Some two million trees were to be planted and a regional park for public use was to be provided, but as only some limited topsoiling and grass seeding were undertaken, no immediate advantage was taken of this potential amenity even though the new landforms that had been created. 'Unfortunately these dreams gathered dust on forgotten shelves amidst Ottawa's apathy, leaving a ravaged landscape. Vision alone is only part of the formula of fruition' (Barnsley 1993/94).

In 1993, twenty-five years later, QUNO (the former Quebec and Ontario Paper Company) announced that the barren landscape would be revitalized by planting two million trees on 3459.4 acres (1400.0 ha) along a 7-mile (11.2-km) strip. This was a twenty-year project to rehabilitate the clay with nutrient-rich biosolids left over from the company's papermaking operations at Thorold. Land rehabilitation would advance at the rate of about 148.3 acres (60 ha) per year, creating a forest of spruce surrounded by hardwood trees such as maple, oak, ash, and poplar (QUNO 1993). In addition, the planting could be undertaken without the costly application of some 6 inches (15.2 cm) of topsoil over the area.

Unsurfaced private roads were provided by the Seaway on both sides of the By-Pass channel for their use as canal service roads and to provide access to features such as electric installations and emergency tie-up facilities. These roads were Seaway property, and except for access to the dock and wharf at Welland, public access was banned by signs and gates. The Seaway's task was to move vessels from lake to lake, and the former idea of creating a landscaped parkway along the banks of the canal was not considered owing to the costs of grading, surfacing, and repairing the roads.

As had happened previously on the Fourth Canal at the centre of Welland, the By-Pass had to cross the Welland River. This was accomplished by constructing a syphon under the canal at the Port Robinson end of the By-Pass channel. Completed in 1971, it contained four tubes and was 638 feet (115.8 m) long. The river was diverted to this new syphon and a new river channel was constructed south of Port Robinson on the east side of the canal. The original section of the river retained its flow by drawing 50 c.f.s. (1.4 m³) of water from the canal, thus guaranteeing that residents in Port Robinson backing onto the curtailed length of river would have neither dead water nor an empty ditch next to their properties. The Welland River was now carried twice under the canal, at the city centre under the old Fourth Canal and at Port Robinson by a modern version of the same type of structure. Also at Port Robinson, an island as previously at Welland had been created between the old and the new river channels. To the south, Lyons Creek had been crossed by the By-Pass and its through flow of water eliminated. The lower channel received a small volume of compensation water from the canal, and the Seaway contributed to a holding-tank project.

Physically, the By-Pass lay slightly beyond the eastern flank of Welland's urban development. In 1961 the municipal boundaries had been extended by over a mile (1.6 km) in every direction. Further expansion in 1970 carried these boundaries some 2 miles (3.2 km) further east across the By-Pass, then under construction. This meant that in terms of land use, the area to the east of the By-Pass now comprised about one-third of the city's administrative extent. The inner city, which for all of its life had been divided by the canal and had always had communication problems across that canal, was now further divided by the By-Pass. It might here be noted that a section of the Fourth Canal next to St Catharines became part of the municipal boundary for that city, but this policy of using the breadth of the canal to define the urban administrative limits was not again pursued at

Welland. In this instance, the town was provided with space for urban expansion within its own municipal limits, but to the east of the canal and separated from the established areas of urban development.

Two tunnels were at first designed to cross the By-Pass, one for a highway on East Main Street and the second for railway purposes close to the southern boundary. These proposals were changed by civic intervention to create two highway tunnels, one was situated at East Main Street and the other, in conjunction with the railway tunnel, was added to the Townline Road. All other road crossings that had previously extended across the site of the By-Pass became dead ends. Apart from the two tunnels, no other east–west highways crossed the new length of canal.

The four-lane divided highway tunnel on East Main Street was 724 feet (220.7 m) long and 75 feet (22.9 m) wide. It had a 6 per cent grade on its approaches and the overall length including its approaches was 1044 feet (318.2 m). Half the cost was born by the Ontario Ministry of Transportation and half by the Seaway Authority. The second highway tunnel, this time in conjunction with three lines of rail track, carried the two lanes of Highway 58A under the new channel close to the alignment of the old Townline Road. Both highways linked across the city to Highway 58 on the west side of the old Fourth Canal. East of the By-Pass, provincial Highway 140 opened in 1972 to link Welland and Port Colborne along this side of the new channel. It replaced Canal Bank Street on the west side of the canal between Welland and Port Colborne, which was intersected by the By-Pass Channel.

Public utilities operated by Ontario Hydro, Provincial Gas, Bell Telephone, and the city of Welland had also to be relocated. The highest transmission towers built in Ontario carried hydro cables with a clearance of 120 feet (36.6 m) over the By-Pass. Other wires, sewers, and water mains connected under the By-Pass so that the area of Welland that lay to its east within the municipal boundary might be capable of development from the city. A further utility provision was to install a lighting system along both sides of the new channel as a night-time safety measure.

Welland gained in the provision of new public open space. When Memorial Park on the line of the new channel was purchased by the Seaway for its canal works, a part of this money was invested in the St George's Park development in Welland South to provide an Olympic-sized pool and changing rooms. The remnant section of Memorial Park was later purchased back from the Seaway. The other part of this previous park was used as a disposal area for material excavated from the canal.

A new dock, 1000 feet (304.5 m) long, that could accommodate two of the largest vessels using the canal was built on the city's side of the By-Pass. Promoted by the chamber of commerce, it was designed to serve Welland and its industries and replaced two such former facilities on the Fourth Canal. It was anticipated that Union Carbide, the Steel Company of Canada, and Atlas Steels would use this slip to export various products. A container facility was also anticipated, as was the classification of Welland as a port and the use of this dock for the winter refitting of vessels. In reality, the dock has been used only occasionally and the industrial expectations have not been fulfilled.

The By-Pass made much of Welland's older urbanized area into an island. Situated between the old Fourth Canal and the new By-Pass channel, this distinctive urban form within a canal setting was now connected by tunnels to the east and bridges to the west to link the central portion of the city with the inland areas of the Niagara Peninsula.

Changes to the Railway Network

Exerting as profound an impact on Welland as the By-Pass were the changes made to its rail operations. The city with its proud motto of Where Rails And Water Meet was the focal point of the railway network at the centre of the peninsula. Before the By-Pass was constructed, all lines of track converged on two rail bridges that crossed the canal south of the city. Bridge 17 was a lift-bridge close to the former city boundary; Bridge 15 was a swing-bridge that turned from a pier in the centre of the canal channel and presented a severe hazard to passing vessels as they negotiated this obstacle.

The alignment selected for the By-Pass was crossed at five separate locations by active railway tracks owned by the Canadian National, Penn Central, and Toronto, Hamilton and Buffalo Railways. The bankruptcy of the Penn Central Railway in 1970 provided an additional complication that required special arrangements for the supply of rails and ties and for some of the contract work, but these did not delay the schedule to open the By-Pass by 1973.

The rail system was changed drastically (Jackson 1975b; St Lawrence Seaway Authority 1972, 1973). All lines were now rerouted to cross the By-Pass at the Townline Road-Rail Tunnel, (Maitland 1974). Some 100 miles (160 km) of relocated railway track at its single-track equivalent were redirected to this tunnel. The tunnel itself, 1080 feet (329.2 m) long, carried three lines of rail track. As long and heavily loaded trains were anticipated along these lines of track, the incline of approach was determined at 0.75

per cent, that is, a fall of 9 inches (2.9 cm) for each 100 feet (30.5 m), a much more slender grade than for the East Main Street highway tunnel.

The tunnel and its approaches required a deep cut across the landscape south of the urban area that was some 2.5 miles (4 km) long on both sides of the By-Pass. This trench, which divided Welland's island site into two segments, had a central depth of 65 feet (19.8 m) below average ground level, and the excavation tapered in width as this depth reduced away from the tunnel. The banks were broken by a series of berms to help stabilize the banks and to reduce erosion. The trench itself, a second confining excavation in Welland's urban landscape, was crossed by four road and two rail bridges.

A further implication of diverting all rail tracks to the Townline Tunnel was that Welland lost its central passenger terminal and the adjacent yards. A new marshalling yard, railway station, and freight yards were constructed west of the city in Wainfleet Township. Had the railway network been constructed with a second tunnel under the abandoned canal channel, then these yards would have been constructed at an even greater distance from the city and the isolation of the city from its railway network would have been more severe.

Of importance for the changing character of urban form, the city was also left with several lengths of abandoned track that had in the past carved the city into segments, helped in the creation of ethnic neighbourhoods, formed barriers to local road transportation, and contributed to blight through their presence and the mixing of industry and often low-quality housing along the lines of tracks. As with the canal By-Pass and the abandoned length of the Fourth Canal, transportation routes that had once helped to create and shape the city now had to be redesigned for other purposes.

The essential argument in this book is that successive decisions about the canal have extended outwards to create then change the associated canal communities. This impact has been centrifugal from the canal to its neighbouring lands throughout the history of the canal. The By-Pass project at Welland was among the greatest and the largest in the history of both the canal and the city but, although the urban implications were extensive, the project was designed with canal engineering to the fore. As bemoaned by the city's consultants when preparing the Official Plan for Welland in 1971, 'In view of the extensive studies on rail relocation made for the Canal Project ... [t]he planner can only lament the fact that these studies were done without proper guidance for planning the City as a whole' (Proctor, Redfern, Bousfield, and Bacon 1971). As always, the city had to take on a new form that was dictated to it by the need to improve the canal.

The Canal as a Water Supply System

A new and more efficient summit level for the canal now bypassed the city of Welland. The water flow from Lake Erie was carried in part through Lock 8 at Port Colborne during the summer months, but during the closed season for navigation the lock gates were closed and no water then passed through this lock. At all seasons the main supply of water to the canal was controlled by a weir on the former Third Canal under the bridge on Main Street (Highway 3), a canal feature recalled in the name of Weir Road, which parallels the old canal.

The canal has been used increasingly for water-diversion purposes (table 9). When the Fourth Canal was built, and during the 1930s, the total diversion was about 2500 cubic feet per second (70.8 m³). After the Long Lac and Ogoki diversions and the construction of the No. 2 DeCew Falls generating plant, this diversion rate had increased to 7700 cubic feet per second (218 m³) by 1951. Over the next twenty years it fluctuated between 5900 and 8200 c.f.s. (167.1 and 232.2 m³), a variation that can be attributed primarily to the water level of Lake Erie.

Higher diversions were possible during the periods of high lake levels, as during the early 1950s and the early 1970s, and lower levels of diverson occurred during the periods of low lake levels from the late 1950s until 1966. 'Thus over the last 20 years, there has been a gradual and noticeable increase in the Welland Canal diversion. Aside from the fact that Lake Erie water levels have been high, the major reasons for the increase include navigation requirements domestic and sanitary requirements, as well as power requirements' (International Niagara Committee 1982).

Important is the increasing amount of water that has been diverted into the Welland River for dilution and water-quality purposes. This water is then either used in the Queenston Power plants or is returned to the Niagara River through the Chippawa–Grass Island Pool above The Falls. Flow requirements for navigation with larger vessels and an increased tonnage had increased from about 85 c.f.s. (2.4 m³) in earlier canal years to over 1000 c.f.s. (28.3 m³), with up to 1400 c.f.s. (39.6 m³) by the early 1980s on a monthly basis.

To provide power to DeCew Falls, water was diverted from the canal at Allanburg through two channels to Lake Gibson that supplied the two power plants. These diversions between 1952 and 1972 were at a mean rate of about 6000 c.f.s. (169.9 m³), though leases and agreements permitted a maximum of 6430 c.f.s. (182.1 m³) on an annual basis. With completion of

TABLE 9 The major users of canal water, 1981

Components	Source	Return water body	Purposes	Amounts (cfs)
Water entering canal				
1 Lock 8	Lake Erie	Summit reach	Navigation	190
2 Supply weir	Lake Erie	Summit reach	All	9600
			Total	9790
Discharging into Welland River				
3 Regional Niagara	Canal	Welland River	Domestic	18
4 Atlas Steels	Canal	Welland River	Industrial	11
5 Union Carbide	Canal	Welland River	Industrial	0.7
6 Port Robinson	Canal	Welland River	Domestic	1.1
7 Lyons Creek Pumping	Canal	Welland River	Water quality	4.9
8 Welland Waterworks	Canal	Welland River	Water quality	160
9 Syphon Culvert	Canal	Welland River	Water quality	500
10 Port Robinson Culvert	Canal	Welland River	Water quality	25
			Total	720
Discharging into Beaver Dams Creek, Lake Gibson, Power Glen				
11 DeCew Falls Plant	Canal	Lake Ontario	Power	6900
12 St Catharines	Canal	Lake Ontario	Domestic	45
13 Beaver Wood Fibre	Canal	Lake Ontario	Industrial	7.1
14 Hayes-Dana Ltd.	Canal	Lake Ontario	Industrial	1.8
15 Exolon Co.	Canal	Lake Ontario	Industrial	3.9
			Total	6960
Discharging into the Second Canal				
16 Supply Weir	Canal	Canal above Lock 2	Bypass	49
17 Ontario Paper	Canal	Canal above Lock 2	Industrial	49
18 Thorold	Canal	Canal above Lock 2	Domestic	15
19 Domtar	Canal	Canal above Lock 2	Industrial	4.2
20 Kimberly-Clark	Canal	Canal above Lock 2	Industrial	4.2
			Total	120
Discharging into the Fourth Canal below Summit				
21 Flow through Lock 7	Canal	Lake Ontario	Navigation	780
22 Flow through Weir 7	Canal	Lake Ontario	Navigation	520
23 General Motors	Canal	Lake Ontario	Industrial	46
24 Seaway Powerhouse	Canal	Lake Ontario	Navigation	260
			Total	1610

Source: International Niagara Committee, 1982

the Welland By-Pass and its improved capacity, new agreements increased this volume to 6887 c.f.s. (195 m³), with more being diverted during the non-navigation season than during the summer months.

This water diversion is part of Canada's allotment for power purposes under the terms of the 1950 Niagara Treaty. It is returned not to the Niagara River but via Twelve Mile Creek to Port Dalhousie, where a small generating plant located next to Lock 1 of the Third Canal provides power to St Catharines before the return of this water to Lake Ontario. Downstream, to the east of Kingston, further use is made of this water in the St Lawrence power plants.

Institutionally, the diversion of water from Lake Erie for navigation purposes started in 1881. This right of Canada was reaffirmed in the Boundary Waters Treaty of 1905. The Niagara Treaty of 1950 recognized that navigation had preference over power. By 1981 the largest proportion of water entering the canal was used at DeCew Falls for power generation. The second largest user was the St Lawrence Seaway Authority for use in the Seaway's power plant, which withdrew water from the canal above Lock 7 and returned it to the canal below Lock 4. This amount varied but was about 20 per cent of the amount used for navigation purposes.

The third largest use was for water-quality management, mostly at Welland to dilute sewage entering the Niagara River, but also at Port Robinson and Lyons Creek to replace flows that were cut off when the By-Pass at Welland was constructed. Industrial cooling water provides a fourth category, with diversions to Atlas Steels and Union Carbide in Welland, to Beaver Wood Fibre, Hayes-Dana, Exolon, Ontario Paper, Domtar and Kimberly Clark at Thorold, and to General Motors in St Catharines.

A fascinating aspect of the canal scene is that the flow of water in the canal and its use for navigation is no longer the primary use. There has emerged over recent decades a close relationship between the water levels in Lake Erie, the flow of water over The Falls, and the use of canal water for industrial, water-supply, and dilution purposes. The canal has become a multifaceted resource that now has to be managed with more than shipping purposes in mind. One suggestion has been that the canal might be used to drain water from Lake Erie at periods of high water, and thereby limit lakeshore erosion and damage to shoreline property (St Catharines *Standard*, 15 February 1986).

The Seaway Moves towards the Twenty-First Century

Shipping stopped using the abandoned Fourth Canal through Welland at the end of the 1972 season. The 9.1-mile (16.6-km) section no longer required for navigation contained some 1280 acres (518 ha) of land of which about 345 acres (139.6 ha) were covered by water (Carvell 1973b). The many issues for resolution included the supply of water to industry and the city's waterworks, highway bridges that crossed the canal, the need to improve the city's east–west traffic circulation, and the envisaged potential of a scenic recreational waterway with landscaped banks as an integral part of urban redevelopment. The City of Welland had grown with the canal for nearly 150 years, and now had to face a novel and uncertain future with the canal no longer at the centre of its economic activities. This challenge will be discussed in chapter 18.

It is important to remember that the series of decisions to achieve the By-Pass were Seaway / federal government decisions. These decisions included the building of the new channel, the abandonment of the old route, the rerouting of the railway network, the curtailment of water flow in the old channel, and the new bridges and tunnel that were provided. As stated in a consultant's study, the 'changes to these transportation modes ... have for the most part been imposed on the City of Welland by the decisions of external agencies without due consultation or due consideration of their impacts on the functioning, the environment, or the quality of life of the City' (Diamond and Myers 1975). The consultants continued, 'As yet the new canal has created no significant benefits for the city [except perhaps for the new docking facility]; indeed its effects so far have been detrimental in that grants in lieu of taxes have declined.' A bypass had been created for shipping and the city had to respond.

Returning to the Seaway as a whole, structural failures became a cause of acute concern when in 1984 the downstream lift-bridge at Valleyfield, Quebec, jammed in a half-open, half-closed position across the Beauharnois Canal. As neither highway nor shipping traffic could pass the obstruction, the flow of marine traffic was blocked for nineteen days and, when navigation resumed, 61 upbound and 104 downbound vessels were awaiting transit. On the Welland Canal in late 1985, a Liberian-registered grain ship, S.S. *Furia*, was trapped in Lock 7. The lower part of the lock's west wall had collapsed into the chamber, pinning the ship against the other side of the lock. Navigation was halted for twenty-four days. With 53 downbound vessels and 77 upbound vessels trapped, the Seaway was again closed.

To avoid the recurrence of such incidents and to restore the Seaway's tarnished reputation as a safe and reliable transportation route, a seven-year Welland Canals Rehabilitation Programme was introduced in 1986.

The aim was to ensure the maintenance of a safe and efficient service into the twenty-first century. This work was remedial. It included cutting back then reinforcing and restoring the lock walls, stabilizing the canal banks and the approach walls above and below the locks, and carrying out some bridge modernization.

No drastic changes were introduced to the canal landscape, but the labour-intensive projects created some five hundred jobs over each year of the scheduled winter operations. As the canal with its new By-Pass at Welland had virtually been rebuilt to ensure its efficient operation into the twenty-first century, a life span of up to approximately the year 2030 was now confidently anticipated by the Seaway Authority (Stewart 1992). The land purchased east of St Catharines for a new canal with Super Locks has been retained under federal ownership to meet this eventuality.

The latest potential change, in the political climate of the mid-1990s, is that the Seaway may be privatized (*Globe and Mail*, 13 July 1995). National marine strategy is that of retreat from the direct operation of transportation facilities, which could mean that Port Colborne harbour might be managed by private or municipal interests. The two scenarios for the canal are that its administration might be taken over by a few large engineering companies or by a consortium of users including the shipping companies, steel mills, and grain handlers. No details were available as 1996 dawned and the implications for the canal and its riparian communities are unknown, but the end of government ownership would be a return to the circumstances that prevailed when the Welland Canal Company introduced the First Canal across Niagara's pioneer landscape of the 1820s, and a reversal of over 150 years of canal history since the Act of Union.

15

The Trading Scene and the Regional Economy

As each earlier canal has been followed by industrial expansion and as this pattern was the prognostication upon completion of the Seaway, the expectations were high that expansion would occur in the communities along the Fourth Canal. These dreams were not fulfilled, even though an increasing volume of trade passed through the canal in larger-sized vessels. Even so, most vessels stopped neither to load nor unload, but shore facilities for repairs and servicing were still required and the canal remained an important component of the regional economy.

Trade through the canal increased until 1979, but then declined rapidly. Meantime, tolls had been established on the canal when the St Lawrence Seaway opened and, although they were expected to cover costs and were frequently raised, income failed to keep abreast of costs. This combination of diminished income from passing vessels and the decline in trading volumes posed a considerable set of new difficulties for the Seaway Authority.

The Anticipation of Industrial Expansion.

Given the flow of goods expected through the canal and the new trading connections that were about to be established with western Europe, a new wave of industrial prosperity was anticipated (Kates 1965). Locally, 'millions of dollars were spent by the steel industries in Hamilton. Industries such as Port Weller Dry Docks, General Motors, Ontario Paper Company, Union Carbide, Page-Hersey Tubes and other industries along the canal all expanded their activities when the Seaway opened. More than likely industrial expansion also took place in many other areas in the peninsula because of the presence of the Welland Canal. Unfortunately, it is not always possible to record similar moves by smaller enterprises, with the result that the total input of a project is not always quantifiable' (Campbell 1973).

Optimism prevailed. A newspaper heading asked, 'Will Seaway Make Thorold the Manchester of Canada?' (St Catharines *Standard*, 6 November 1954). As anticipated by the Greater Niagara Chamber of Commerce, 'Considered opinion indicates that within a radius of fifty miles of Toronto and Hamilton, favourable growth will occur as a result of the Seaway. This means continued yet accelerated industrial development of the Golden Horseshoe area at the western end of Lake Ontario from and including the Niagara Frontier' (Young 1956).

On the Welland Canal itself, now interpreted as 'a passage rather than a port for the transfer of goods,' the chamber argued that a port authority should be established, and five basins along the canal were considered suitable for industrial expansion: south of Port Colborne on the west side of the canal; between Port Robinson, the Welland River, and the canal; near the Flight Locks at Thorold; near where the canal was crossed by the Queen Elizabeth Way; and to the south of Port Weller.

Another possibility was to reuse the eastern arm of the former canal along the Lower Welland River. This idea required a lift lock some 7 feet (2.1 m) high to cover the difference in levels between the Fourth Canal and the river at Port Robinson, and then the deepening, widening, and straightening of the river to as far as Chippawa in order to achieve commercial and industrial development on some 1500 acres (607.1 ha) of land with about 18 miles (29 km) of prime harbour frontage.

Canadian Carborundum considered that the Welland-Chippawa River Project was 'absolutely necessary for the maintenance and expansion of the Carborundum or abrasive industry on the Niagara Frontier.' This company brought in bauxite from South America, and was experimenting to reduce the costs of transportation with shipments via the 14-foot (4.3-m) St Lawrence Seaway, which it 'found to be very satisfactory' to reduce the costs of transportation. 'It would bring in and could use five or six thousand tons [4536 to 5443 tonnes] per boatload during the navigation season,' and would export to England and overseas (Young 1956).

As Niagara Wire Weaving had a fragile product, it was 'very important to reduce the number of transfer points. They ship to Cuba, Australia, Newfoundland and Britain ... [and the] raw material coming from Scotland and Germany would be brought in to the plant at Niagara Falls through the Welland Canal and the Welland-Chippawa River.' Norton, a company that manufactured abrasives, doubted that the project would pay off the capital investment. Nevertheless, the company 'thought that in general it would be good for the community,' and considered that it 'would be to the advantage of the company for import and export to Montreal and St John bringing in bauxite from Dutch and British Guiana.'

Provincial Engineering, which manufactured large machinery and transmission towers thought that the project 'would return financially more than the investment ... [They] would bring in steel for fabrication and for other productions. Nine thousand tons [8163 tonnes] in and out in one order.' They also appreciated that ships would not stop to tie up in the canal unless full shipments were available, and that dock facilities to tie up out of the line of shipping traffic were most important. The company stated that they would be better served by water than by rail. They would bring in steel from the Sault, store it through the winter and export 30,000 tons (27,216 tonnes) 'with little physical effort and little cost and there would be no trans-shipment. This would be of great advantage to this company.'

Other companies canvassed included Lionite Abrasives, which brought in calcined bauxite from South America: 'We have been pinched by gradually increasing delivered cost of this material and estimate that any suitable local handling facilities would give us considerable relief of this cost item alone.' And Bright's Wines would have an import of between 6000 and 7000 tons (5443 and 6350 tonnes) annually. The total expectation from these and other companies was that some 417,060 tons (378,357 tonnes) might be handled annually, with 60 per cent of this flow being outbound.

In reality, the envisaged port authority along the canal was not created, and only one major new company, Casco in 1981 at Port Colborne (to produce high-fructose corn, syrup, and high-protein livestock feed), located next to the canal after the Seaway opened. A pier was constructed by this company at the southern end of the abandoned Fourth Canal to moor vessels, and the imports were conveyed by an overhead system direct to the plant, which was close to but not directly on the canal.

Other companies have continued their links with the canal and its service facilities, and ship maintenance, repair, and construction and canal administration have remained important ingredients of the canal scene, but the canal as a harbinger of new industries has not been part of the Seaway story at Niagara. The reasons were assessed when arguments for the development of a port outside the eastern breakwater at Port Weller were examined (Prince 1985). A port or ports on the canal were not then deemed to be suitable because the truck and rail modes of transportation had become more efficient and economical, and carried loads other than the large but declining volume of bulk traffic that passed through the canal in maximum-size vessels. The number of small vessels had declined, their smaller loads were now less viable than those carried by rail or highway truck, and the ports they served had switched to transport by these other modes.

On the industrial front, many companies found that they could no longer afford the inventory costs associated with the seasonal nature of water transportation, and the railways had introduced rates that were competitive with shipping on a year-round basis when storage and inventory costs were taken into account. The cost and service levels of an inland port on the canal would have to compete with the effective and now well-established short Atlantic passage then inland distribution by the railway system from the coastal ports.

Of the tonnage available on vessels passing through the canal, only a small proportion was general cargo that could be off-loaded and used for the development of port facilities along the canal. Of the upbound cargo about 90 per cent was destined to either Chicago, Cleveland, Toledo, or Detroit, and the average cargo now exceeded 11,250 tons (11,000 tonnes). As the downbound cargo was much lower in tonnage, the conclusion was that 'the small number of ports receiving general cargo and the large loads per ship suggest that there would be very limited incentive for a ship to stop at a Welland Canal port as a means of reducing the total number of calls it must make. There are unlikely to be sufficient savings in this regard to justify change from the ... overland transshipment of cargo' (Prince 1985).

Other aspects of the situation were the high costs of constructing a new port facility at Port Weller and the expected continuing increase in volume and reduced costs of the railway-highway competition. Though negative to the possibility of new port facilities, the report considered that Port Colborne might still offer competitive facilities if the harbour was deepened to accommodate the larger-sized vessels that sailed on Lake Erie, and if added berthing space was provided for the winter servicing and repair of vessels. This might be accommodated along the canal slips at each end of the abandoned Fourth Canal, where Casco had located, and at Port Robinson, where some ship repair work had been undertaken at the entrance to the abandoned channel.

The wharfing facilities along the canal were of a great value to a few industries, but they were viewed as a declining asset relative to Hamilton, where in 1951 the Harbour Commission had initiated an extensive program of land reclamation and the construction and renovation of wharves and freight terminals in anticipation of the increased cargo tonnage to be generated by the Seaway.

Trade along the New Waterway

The Seaway was an immediate economic and commercial success (St Lawrence Seaway Authority 1959–65; Government of Canada 1979). Along its Welland Canal section, in 1959 the cargo carried by vessels increased from 73 million tons (66.2 million tonnes) in 1959 steadily each year to 1966, and reached a maximum flow of 27.6 million tons (25 million tonnes) in 1979 (table 10). Tonnage then decreased steadily almost year by year to a low of 35.1 million tons (31.8 million tonnes) by 1993. Two distinctive periods of traffic flow therefore existed, the first of rapid and substantial growth and the second of serious decline that can be attributed to the changed conditions of international and North American trade and to the extensive decline of manufacturing activity across the Western world.

The number of vessel transits diminished remarkably as cargo was carried in vessels of increasing size (St Lawrence Seaway Authority 1959b). The 8072 transits of 1959 had reduced to 6532 by 1979 when the peak load was carried, and to 2927 at the nadir of operations in 1993. Vessels with a length of over 712.0 feet (217 m) increased from less than 200 in 1959 and 1960 to 1895 or 29 per cent of the total during the peak year of operation. By 1993 the proportion of larger vessels had increased to slightly over half the number of vessel transits.

These figures relate to the gradual disappearance of the old 4.3-m (14-foot) canallers and to the introduction of an increasing number and proportion of the larger-sized vessels. For the canal communities this change represented a diminished opportunity for the service, manning, and repair of vessels. To them, with the now through passage of most vessels using the canal, the flow of vessels was of greater importance than the cargoes that they carried. The considerable two-way historic relationships between the canal and its communities had been severely reduced.

Using the peak year of 1979 as an indication of the prevailing situation, twice as many of the downbound vessels at 89.3 per cent then carried cargo against 45.5 per cent of the upbound vessels. By class of vessel, 26.9 per cent were ocean-going vessels (salties), against the majority of 73.1 per cent that were restricted to use on the Great Lakes (lakers). The commodity groups carried were mine products at 45.2 per cent of the tonnage, followed by agricultural products at 39.7 per cent then manufactures at 14.4 per cent.

An important statistic covers the international trade to and from foreign ports beyond the coast of Canada (excluding the trans-border trade with the United States). When the Seaway opened, the proportion of overseas

TABLE 10 Traffic through the Welland Canal section of the St Lawrence Seaway, 1959–1995 (in thousands of tonnes*)

	1959	1960	1965	1970	1975	1980	1985	1990	1995
Vessel transits									
Upbound	4,078	3,802	4,183	3,557	3,101	3,287	1,908	1,794	1,654
Downbound	3,994	3,734	4,201	3,565	3,031	3,280	1,918	1,783	1,641
Total	8,072	7,536	8,384	7,122	6,132	6,567	3,826	3,577	3,295
Cargo tonnage									
Upbound	8,766	7,596	18,097	19,208	17,138	11,986	10,888	11,961	12,462
Downbound	16,215	18,939	30,365	37,911	37,157	47,620	30,964	27,436	26,914
Total	24,981	26,535	48,462	57,119	54,295	59,606	41,852	39,397	39,376
Toll revenue	1,221	1,318	0**	3,533	3,720	23,610	23,269	31,752	37,257
Agricultural prod.	7,999	8,810	16,762	19,203	20,782	28,257	17,135	12,801	14,948
U.S. grains									
Wheat	501	709	1,112	1,556	3,576	3,645	1,959	1,871	2,971
Corn	1,005	1,179	3,490	2,624	2,818	5,909	2,071	1,178	2,344
Soybeans	476	832	1,525	2,637	1,520	1,590	747	319	1,624
Barley	545	528	679	1,071	180	737	170	847	173
Other grains	615	450	422	263	152	1,446	379	12	333
Total	3,142	3,698	7,228	8,151	8,246	13,327	5,326	4,227	7,445
Canadian grains									
Wheat	3,052	3,365	6,847	6,338	8,297	10,650	9,720	7,495	5,184
Barley	698	764	708	2,763	2,165	1,529	571	279	33
Other grains	577	537	941	864	1,013	1,916	1,034	656	2,309
Total	4,328	4,666	8,496	9,965	11,474	14,095	11,325	8,430	7,526
Mineral products									
Coal	4,394	3,989	6,515	9,763	7,701	7,293	5,808	6,266	3,998
Iron Ore	6,274	7,127	14,637	14,598	14,940	11,418	6,789	7,483	6,329
Stone	657	826	794	891	1,072	1,153	815	1,782	1,412
Salt	124	144	566	982	1,356	1,418	1,521	1,513	2,310
Other	693	647	1,125	1,449	1,834	2,736	2,192	2,547	2,351
Total	12,141	12,733	23,637	27,683	26,903	24,018	17,125	19,591	16,400
Processed commodities									
Fuel oil	750	667	957	1,372	1,033	1,665	629	939	605
Man. iron & steel	345	545	2,525	3,750	2,013	1,309	3,095	2,764	3,333
Chemicals	128	122	177	269	343	600	608	685	544
Cement	131	111	150	174	192	296	309	407	1,194
Other	2,655	2,713	3,254	3,795	2,556	3,006	2,866	2,143	2,287
Total	4,009	4,158	7,063	9,361	6,137	6,876	7,507	6,938	7,963

Source: St Lawrence Seaway Authority 1959b, with additional information, including data for 1995, from St Lawrence Seaway Authority (Ottawa)

*Historical cargo data published in short tons before 1978. This data has been converted to metric equivalences in this and later tables. To obtain the short-ton equivalences, multiply the metric quantities by 1.1023.

**No tolls collected during the 1965 season

tonnage was 18.9 per cent. By the benchmark year of 1979, it had increased to 26 per cent or 19 million tons (17.2 million tonnes), but then reduced rapidly to less than 20 per cent during the 1990 and the 1991 seasons.

It must be stressed that the economic benefits of trade flow through the canal cannot be limited to the canal's environment in the Niagara Region. 'Measuring the impact of the Welland Canal on the Niagara Region would be equivalent to measuring the impact of the Trans-Canada Highway on one city or county. Its value has to be assessed in much broader terms. Without the Welland Canal, wheat prices could be higher, the cost of electricity could increase and the price of steel products could certainly be higher' (Campbell 1973). These factors of national concern are not covered in this book. (For one appreciation of this larger impact see Government of Canada 1979.)

The Change in Commodity Flows

The traditional movement of grain shipments has been from Western Canada by rail to Thunder Bay, then by ship through the Welland Canal either by bulk carriers to the Lower St Lawrence River or in foreign vessels for direct delivery to customers abroad. Alternative possibilities for export shipping, but with lesser amounts by 1979, were the year-round, west coast ports of Vancouver and Prince Rupert, and (with less again) Churchill, Manitoba, on Hudson Bay, which had an extremely short shipping season. There was also some movement by rail direct to eastern ports during the winter months. In addition, American grain exporters used domestic and foreign ships for their exports through the Welland Canal, but the major U.S. flow was down the Mississippi River by barge to the Gulf ports.

By its peak of operations in the late 1970s, the Seaway carried well over half of Canadian grain exports in a single crop year, and this cargo was from 30 to 40 per cent of the total bulk commodity flow on the canal. By 1982–3 the Great Lakes fleet moved 57.2 per cent of Western grain; some 40.8 per cent went through the west coast ports of Vancouver and Prince Rupert. Five years later, 41 per cent of export grain went east and 56 per cent went west. Projections were that by 1995–6 only 30 per cent of the export grain would pass through the Welland Canal (Atkinson 1992; Carr 1970; Cresswell 1994; Misener 1981; William Hamilton Merritt Lecture Series 1986–90).

The movement of wheat to west coast ports was encouraged as the Canadian government and several Western provincial governments subsidized the construction of new grain cars, elevators, expanded terminal capacity, and railway tunnels to transport Canadian wheat to Pacific Rim nations, including Japan, and other Pacific Rim nations. Traditional markets in North Africa and the Middle East were also increasingly served through the west coast. At the same time, extensive agricultural subsidies in the European Common Market dramatically reduced the amount of grain shipped through the Welland Canal, from a maximum of 18.8 million tons (17.1 million tonnes) in 1983 to less than half this amount by 1992 (Leitch 1993). In the early 1980s the United States also imposed a grain embargo on the sale of grain to the Soviet Union that adversely affected the flow through the canal. When this embargo was lifted, U.S. grain passing through the canal had diminished in volume – another instance of the effect of external events on the life of the canal and its communities.

World market conditions had changed drastically, but the Western Grain Transportation Act of 1983 exacerbated the situation by providing rail rate subsidies and disincentives for the use of other modes of transportation. All costs for moving grain were now pooled and then an average tonne per mile cost was applied. A shipper by rail had the same cost per mile whether the grain moved uphill on the high-cost route west through the mountains to the Pacific or along the lower-cost, downhill gradient to Thunder Bay. Opposition to this subsidy by the marine industry was consistent, but it was not removed until the national financial budget of 1995.

As a consequence of these adverse provisions for the flow of trade through the Welland Canal, Thunder Bay became 'a residual port which only receives export grain when there are capacity problems at west coast ports or it is a matter of customer preference ... A grain transportation system without modal bias is vital for the future of the Great Lakes / Seaway System' (Atkinson 1992). It is truly remarkable how distant economic events and complex political decisions within and beyond Canada have exerted their adverse impact on Niagara's economy. With Thunder Bay now a residual port and the American trade favouring the subsidized Mississippi route, the impact of declining cargoes is transferred downstream along the Seaway system to the Welland Canal. As the sharp drop in grain exports was not offset by the increased movement of other commodities, the services provided for passing vessels also diminished substantially.

In Niagara, the local grain trade has also changed drastically. After 1932, when the Fourth Canal opened, Port Colborne ceased to be a major transfer point for grain as ships could now pass through the new, larger canal, and elevator facilities at Prescott gradually took over this role. Later, when the Seaway opened in 1959, Prescott was bypassed as larger vessels could now traverse the whole Seaway from Thunder Bay to the ocean. As the wheat trade through Thunder Bay declined, facilities built on land downstream in order to accommodate demands stemming from marine transportation became ephemeral, raising severe local issues such as unemployment and a changing urban-industrial structure as the canal communities were increasingly bypassed by the through passage of vessels.

With regard to the movement of minerals through the canal, one compelling reason for the construction of the Seaway was the movement of iron ore from Quebec and Labrador to American steel mills around the Upper Lakes. Iron ore was also sent to the Ontario market at Hamilton, though this flow did not pass through the canal. The alternative source of ores from the Mesabi Range in Minnesota on the American side of Lake Superior detracted from the utility value of these Canadian deposits. Large vessels that could not pass through the canal provided ore movements to steel mills around the Upper Lakes, whereas the smaller Seaway-sized vessels moved the coal, iron ore, and limestone to the steel works at Hamilton.

Attrition to both movements took place with the recessions of the 1980s and 1990s in the North American economy, including the severe contraction in the sales of American automobiles, the restructuring of the American steel industry, which involved the closure of plants at Lackawanna near Buffalo, and the transition in steel production to new southern locations in the United States. Iron ore passing through the canal reduced from 20.4 million tons (19.9 million tonnes) in 1977 to 4.4 million tons (4.3 million tonnes) by 1992.

Combining these grain and mineral movements through the canal, a deteriorating traffic situation was heightened by the fact that ships had traditionally carried grain downbound through the canal to the St Lawrence, then returned with an upbound load for the steel mills around the Great Lakes. This backhaul was greatly reduced by recession in the steel industry and, as many vessels had to return empty, this increased the cost of carrying wheat.

Coal movements have also changed (Government of Canada 1979; Misener 1981). Metallurgical coal was brought by rail to ports south of Lake Erie, then shipped across the lake for use in the steel industry at Hamilton by Dofasco and Stelco. The development of Nanticoke on the north shore of Lake Erie transferred some of this movement from passing through the canal to a trade that was limited to Lake Erie. Thermal coal used by Ontario Hydro was also shipped from U.S. mines to ports on Lake Erie and then to generating stations primarily on Lake Ontario, but an increasing reliance on nuclear generation has reduced this movement. Coal destined to Europe from Pennsylvania and its nearby states might also flow through the canal rather than be carried through U.S. territory because of the shorter haul by rail to Lake Ontario than to the American Atlantic ports. Coal and oil depots along the canal have also diminished as less coal was used by industry and for domestic purposes, and the expanding pipeline system diminished the movement of oil through the canal.

Other commodities passing through the canal included coke from overseas to American steel mills, petroleum products to smaller markets from the oil refineries and pipeline terminals, and sulphur as a by-product from oil and gas production to overseas markets. Potash from Saskatchewan might move to the agricultural areas of southern Ontario. Cement for building and highway construction, and sand, gravel, and stone as low-value, bulk materials might also move through the canal, as did salt from the Goderich and Windsor areas for industrial use and winter use on roads.

Another element of marine change was that containerized sea transportation of general cargo, which included a wide variety of manufactured goods ranging through machinery and equipment to consumer products, was at first important for the Seaway because ocean-going vessels were able to sail to ports round the Great Lakes. As container traffic became an expanding operation across the Atlantic between Europe and North America, it used large-sized, specialized container vessels that were too large to transit the Seaway. The railways, with water-competitive rates plus year-round operations to ice-free Atlantic ports such as Halifax and Saint John, and the ability to offer savings to industry through eliminating the need for winter storage, now had a powerful competitive advantage.

This led by the early 1980s to the end of container traffic through the Welland Canal and to the discontinuance of the general cargo service that Canada Shipping Lines and Manchester Liners once operated. As Ontario ports could not compete with the east coast ports, left for the canal were primarily the large-volume traffic using the maximum-size bulk carriers and the inflow of cargo to a few large industries (Ontario Ministry of Transportation 1983). 'The promotion and marketing of the Seaway, how best to reduce the total cost of carrying a cargo from point of origin to destination (including the limitations on the draft of vessels and the costs of a nine-month season) and how to retain or even to expand the diminished flow of traffic had become the facts for pertinent resolution and a series of marketing studies (Savage 1989).

The circumstances of the sharp decline in the grain and mineral movements were not at first thought to be either serious or of long-term consequence. The expectation was that growth would soon resume and that further improvements would still be required for the canal. As stated in the

Seaway's Annual Report of 1980, 'The long term trend traffic is definitely upward, and by the end of this decade our economists estimate that something like 99.2 million tons (90 million tonnes) of cargo will be offered for transit through the Welland.' Doubts were raised that this volume of cargo could be serviced through the existing facilities, even if enough ships were available. Despite these projections, the reality was that the canal by 1990 carried rather less than 44.1 million tons (40 million tonnes).

A second example of the difficulty of making accurate estimates of future trading flows is that when the Great Lakes / Seaway Task Force developed a traffic forecast from 1979 to 1985 at a time when traffic in the system was increasing, the estimate was that the canal would carry a downbound tonnage in the range of from 67.8 to 73.3 million tons (61.5 to 66.5 million tonnes), that is, an increase of 18.2 to 22.6 million tons (16.5 to 20.5 million tonnes) over 1979 (Misener 1981). The reality was depression and a fall in cargo to 31.8 million tons (30.9 million tonnes). And against the expectation that upbound tonnage would increase, the actual performance involved a substantial decline.

The point is not the inaccurate nature of these forecasts, hindsight wisdom being easy, but the rapidity and the unexpected nature of the changes that occurred across such a wide range of commodities that were basic to the effective operation of the canal and its larger Seaway component. It is important also to recognize that forecasts, even though later found to be incorrect, have to provide the basis for the new engineering works that are planned along the canal. The canal had to be repaired, improved, and maintained even though its trade was in decline, and these canal works were necessary even though the number of vessels to be accommodated were in decline. As stated by a federal subcommittee on the St Lawrence Seaway in 1994, 'Nobody has suggested that we should shut down the Seaway. In fact, just the opposite. Everyone, including all western interests, has said that Canada must maintain the current east-west grain handling and transport system ... Indeed, the Seaway is of vital importance to the iron ore mines of Labrador and Quebec, the ports of the Lower St Lawrence, the ports of the Great Lakes, the manufacturing, agriculture and mining industries in the heartland of North America, and equally important, the western grain handling and transportation industries. In other words, the Seaway is not a parochial central Canadian issue. It is a "Canada" issue and should be seen as such by all of the various interests that benefit from the system' (Comuzzi 1994).

The canal across the Niagara Peninsula is but a small part of this wider picture of many related activities, and the qualities of their associated freight movements, both within the Niagara Region and across the Great Lakes–St Lawrence Seaway Transportation System. The region's urban systems in the past have inevitably had to respond to wider dictates that are far beyond the control of local communities and regional government, and the same must be anticipated in the future (Mayer 1976; Sainsbury 1976).

Tolls and Lockage Fees

User tolls were incorporated in the legislation that established the Seaway Authority at a level based on estimates of expanding Seaway traffic, and with the intent of repaying capital with interest over a fifty-year period and at the same time covering operating and maintenance costs. In theory this toll policy would amortize the construction costs, pay the interest, and cover all the operating expenses. Tolls were also applied to the Welland Canal, toll free since 1903, by the Seaway Authority.

Ever since, this issue has been extremely controversial (Corbett 1992; Comuzzi 1994; Kingsley 1958). Shipowners and groups associated with trade through the canal such as the chambers of commerce and marine organizations, including the Dominion Marine Association and the Great Lakes Waterways Development Association (GLWDA), have argued the principle of maintaining free and open waterways for the purposes of commerce. In a contrasting vein, the federal government through the Seaway Authority has sought increased tolls to offset the shortfall in income against the construction, maintenance, and operating costs of the canal.

Arguments against tolls on the Welland begin with the fact that, after completion of the St Lawrence Seaway, the Welland Canal was defined by the Seaway as a deep waterway. Yet no locks had been added during the period when the Seaway was constructed, the American canal at Sault Ste Marie which also fitted this description remained toll free, and there was the threat to Canada of impeding the free flow of trade along the St Lawrence waterway as the highway leading to the centre of the continent. Further, 'Many industries have established themselves along the Welland Canal ... where they could have the advantage of low-cost waterborne transportation. To now impose tolls is, in a sense a breach of faith. An important facility which was held out to these industries as an inducement for them to locate and expand in their present position has been taken away' (GLWDA 1960).

In 1963 the association argued: 'The St Lawrence Seaway from Montreal to Lake Ontario is the only waterway in North America which is subject to tolls. In our view, this constitutes discrimination against users of a transportation facility of vital importance to the Canadian economy.' Two years later, its protest membership in the meantime having increased, the association again argued that 'tolls are discriminatory, and thus work against the interests of the whole Great Lakes economy, and its vast and increasingly-populous hinterland ... Continued adherence to the principle of recovering through tolls capital costs, plus interest, and the escalating costs of administration, maintenance and operation means that any fall-off in the level of cargo throughput would inevitably necessitate a higher level of tolls.'

The original Seaway agreement between Canada and the United States applied only to the Montreal–Lake Ontario section, where revenue was divided according to construction expenses, 71 per cent to Canada and 29 per cent to the United States. This breakdown lasted until 1967, when it became obvious that the Seaway project would never become self-liquidating. The United States with its tradition of opposition to toll rates on inland waterways refused to raise tolls, but Canada's share of the tolls was then increased to 73 per cent.

Meantime, on the Welland Canal, with its earlier period of construction than for the St Lawrence Canals, the costs had been borne wholly by Canada, which owned and managed the Welland Canal through the agency of the St Lawrence Seaway Authority. Here, from 1959, rates were based on the gross vessel tonnage and the nature of the commodities that were shipped. Tolls were suspended over the years from 1963 to 1966, but a progressive lockage fee was introduced in 1967. This fee, at $20 for the passage of each vessel through each lock, increased in graduated steps every two years from $20 to $100 by 1971. To pass a vessel through the eight locks on the Welland Canal therefore increased from $160 in 1967 to $800 by 1971. These changes were related to neither the general increase in price levels nor to the costs of canal improvements, and in 1975 the Seaway's annual report stated that they did not cover 25 per cent of the operating and maintenance costs.

By 1976, the financial situation of the Seaway had changed adversely, as its outstanding debt had increased. It was then apparent that the users could not pay, as had been anticipated at the outset, all of the operating costs, the repayment of construction, and the interest thereon. Tolls would have to be raised fourfold to satisfy the Seaway's financial obligations, and this increase would lead to the diversion of traffic to other modes of transportation. The changes made instead were the conversion of outstanding loans to equity held by the federal government. Unpaid interest was converted to an interest-free loan that would be forgiven, and tolls were raised. The increased user charges were phased in for the 1978 navigation season,

then increased over the next two seasons. The intention was to cover operating costs, but vessel transits halved during the 1980s, which, combined
with the decline in toll revenues and increased operating expenses, greatly
aggravated the situation.

As stated in the Seaway's report on operations for 1982, with no major
problems of weather or interrupted navigation to beset the navigation season, 'the number of commercial transits dropped ... The resulting drop in
tonnage brought in a substantial reduction in toll revenues forcing the
Authority to re-assess its work project priorities. A number of major maintenance works and capital improvement projects were postponed or curtailed and other restraint measures were undertaken ... [Traffic] was
recorded at the lowest level since 1974 ... The decline in Seaway traffic in
1982 is mainly attributed to the prolonged world-wide economic recession
and, in particular, to the very substantial drop in steel production in the
United States ... The 1983 outlook is not encouraging.'

On the Montreal–Lake Ontario section, tolls remained frozen at their
1983 level for six years through to 1989. On the Welland section, the major
source of the Seaway's debt situation, a toll increase of 15 per cent over the
level in 1985 was 'reluctantly implemented ... In an effort to cover at least a
part of the substantial deficit incurred on that part of the system tolls were
increased by an average of eight per cent in each of 1987 and 1988' (St
Lawrence Seaway Authority Annual Report, 1986–7). The annual report
for 1990–1 recorded that there would be a further annual increase in tolls of
5.75 per cent during each of the years 1991, 1992, and 1993.

Tolls for passing a vessel through the length or a part of the canal introduce a new differential into the industrial location equation. As tolls were
levied only for passing through a lock, the harbour entrances at Port Weller
and Port Colborne were immune from tolls for traffic from their immediate lake, and only one lock was involved in the north up to Lock 2 (Carlton
Street) in St Catharines or in the south between Lake Erie and Lock 7 in
Thorold. In these locations, cargoes might be received, processed, and
exported either with no tolls or by paying the fees for but one lock. Casco
in Port Colborne, had taken advantage of this favourable situation.

Many political and economic issues are involved in the levy of tolls.
They provide the Seaway with almost its entire source of revenue, but the
critical factor is at what level would an increase of rates to meet higher costs
jeopardize the competitive advantages of the canal to certain industries for
transportation? Would there be a transfer of some or all of their traffic to
rail or highway competitors? To what extent are the other transportation
modes subsidized from the public purse, and how might an even playing

field between all modes best be achieved when the lowest costs of marine transportation depend upon using the largest vessels with full loads in both directions?

The root issue is that the canal has to remain competitive on a route that is neither exclusive nor necessarily predestined to expand and survive. A part of the canal trade has transferred to the west coast, which was subsidized through the Canadian Grain Transportation Act. The subsidized Mississippi system does not have tolls, and rail cars moving south or to either of the east and west coasts on the deregulated U.S. rail system are 'proving to be a factor to be reckoned with' (Stewart 1990). As an example of changing circumstances, tolls and lockage fees plus the inventory costs associated with seasonal navigation were factors both in the switch from shipping to rail for the inward and the outward movements of cargo at the Ontario Paper Company (later QUNO) and the closing down of its subsidiary, the Quebec and Ontario Transportation Company.

By the early 1990s, the toll situation on the Welland Canal was composed of three elements (St Lawrence Seaway Authority, *The Seaway Handbook*). Vessels paid a composite toll for passage of the Seaway, which was a charge against the gross registered tonnage of each vessel; a charge per each metric ton of cargo, with this being higher for general cargoes than for bulk cargoes and grain; and a further charge per lock for cargo vessels, with this being higher for loaded vessels than for those in ballast. Pleasure craft were charged per vessel for the transit of each lock, and passengers were charged on a per capita basis for the passage of each lock. These costs have been seen as a curb on passenger travel on the Great Lakes in that they inhibit cruise ships that might dock in the canal and offer coach tours to places of interest. A flight to Toronto, then a two-week cruise visiting ports around Lakes Erie and Ontario is not an impossible dream in the days of expanding world travel using cruise ships as a base for land travel.

The Welland Canal was designed and is managed as a commercial waterway and, although small passenger craft and tourist vessels do make some limited through use of the waterway to move from one lake to another, this is not the canal's prime purpose. Seaway policy is that priority must be given to bulk cargo vessels, and that a mixture of small craft with such vessels would present a severe safety hazard. The Welland Canal can be compared with neither the Trent-Severn nor the Rideau systems of navigation in this respect. Neither marinas nor marine service stations have developed along the line of channel, though both exist in or close to the lake ports and along the abandoned length of the Fourth Canal.

The major Seaway task with respect to its traffic flows has become that of maintaining if not expanding its diminished market share for the movement of bulk commodities. To this end trade promotion has been undertaken to encourage an increasing interest in the Seaway by European shippers, with a delegation in 1984 being the first to combine Canadian and U.S. Seaway officials together with provincial and federal representatives. A New Business Incentive Tolls Program introduced in 1990 provided a 50 per cent reduction on tolls for certain shipments to improve transits and tonnage through the slow summer months from June to August. As this project brought in an estimated 821,214 tons (745,000 tonnes) of new cargo, an enhanced scale of rebates was made available for each of the 1991, 1992, and 1993 navigation seasons.

The requirement of financial self-sufficiency in the operation of the Welland Canal extends into the work of the Great Lakes Pilotage Authority, established as a subsidiary of the Seaway in 1972, which entails an additional cost for the passage of certain vessels through the canal and the Seaway system. Canadian ships and masters with experience of the canal are exempt from pilotage fees, but fees are levied on foreign vessels to ensure safe navigation. These arrangements are criticized because of the high cost involved and because they are perceived as being a tax on imports and exports.

As part of the cost-saving measures within the Seaway, in 1991 the crews that secured vessels to the lock walls were reduced from four to three at each lock, a marine provision but one that added its mite to the regional situation of severe unemployment. As was the situation in many other business and industrial activities, the Seaway also reduced its staff in the regional offices at St Catharines to four hundred full-time employees, almost half the level of a decade earlier.

The Regional Importance of the Seaway

By 1970 when 7122 vessels and 63.1 million cargo tons (57.2 million tonnes) passed through the canal, only a small proportion of this volume either originated or had a destination in the canal. This situation was recognized in a background study for the regional plan, which stated, 'It becomes apparent that the Canal is basically a "through" transportation facility and is relatively unimportant as a transportation facility for the Region itself. The canal is therefore more important as an employment and income generating activity just like any other industry, rather than as a transportation facility from the Region's standpoint' (Read, Vorhees & Associates 1971).

When a study undertaken in 1977 by the Niagara Regional Chamber of Commerce assessed the details of this economic activity generated by the canal, at least 4651 residents attributed their wages from employment to the marine industry. The largest single employer was Port Weller Dry Dock (710), followed by the St Lawrence Seaway Authority (650). Vessels passing through the canal provided employment for a further 2700 with homes in the communities bordering the waterway.

Then there were the suppliers and contractors, their suppliers and sub- contractors, and the taxi services that moved crew members and pilots to and from the ships. An electrical appliance store sold one hundred television sets and a department store outfitted ten ships with carpets. In an industrial estate close to the canal, offices and warehouses supplied marine pumps, electrical and mechanical equipment, food, and other supplies to the vessels. Shipping companies had located in the same area, and others had their offices elsewhere in the region. Even excluding those industries that had located in the canal corridor because of the canal, and excluding foreign vessels, the overall conclusion was that 'the marine industry is the second largest in the Niagara Region.'

Ten years later, the direct impact of the Seaway had diminished to some 3000 jobs generated on vessels through servicing, employed by the Seaway, engaged as ships' provisioners, suppliers, and chandlers, occupied in ship-building, repair, and conversions, providing laundry services, and employed in electronics. Through this decline the Seaway had slipped to fifth place among the region's sectors of industrial employment. 'Employment in our shipbuilding and ship repair industry has been savaged due to the lack of new vessel construction. The other economic restraints most shipping companies have had to introduce means that only the most pressing and urgent repairs are being undertaken' (Niagara Region Seaway Task Force 1988).

The economic undertow was serious. By the 1990s, 'regional councillors were calling for tangible support of Great Lakes shipping, the Niagara shipbuilding industry and the St Lawrence Seaway – especially the Welland Canal – as "integral components" of Canada's economic and industrial structure' (Dolynsky and Harding 1994). It was urged that neither ship-building nor the facilities available through the canal system, both seen as significant employers for many residents in the Niagara area, should be allowed to deteriorate further. In the contrasting vein of new economic opportunity, Dolynsky and Harding also anticipated some alleviation of decline in the marine industry through the expansion of tourism, with 'the need for a tourist focal point associated with the Welland Canal and the

Marine Industry. Tourism opportunities related to the Welland Canal have now been identified as one of the most important development initiatives to promote Niagara's tourism industry.'

With a reduction by one-third in the number of ships using the canal from the study of 1977 to the 1990s, and as a contribution towards improving the situation, the shipping industry was re-structured when five of the largest shipping companies on the Great Lakes amalgamated. In 1991 Great Lakes Bulk Carriers united Canadian Steamship Lines, Misener Holdings, and Richardson, and Seaway Bulk Carriers was formed as the joint marketing and transportation arm of Algoma Central and Upper Lakes Shipping. In 1994 Seaway Bulk Carriers signed an agreement to purchase Great Lakes Bulk Carriers, a business decision seen as essential for retaining a viable marine industry on the Great Lakes.

Despite the many misgivings that existed through the considerable decline in its volume of through trade, the Fourth Canal by 1994 remained an important commercial artery for the movement of vessels. It was also of direct relevance for the economy and the social well-being of its adjacent communities, and it had a new but encouraging role to play through its tourism and heritage ingredients as the twentieth century came to an end.

On the canal and to navigate the lakes, from the time that a ship enters the system to the time that it reaches a destination in Lake Superior, at least twelve government agencies are involved in the process. These involve Canadian and American regulations, pilots, two Seaway authorities, and costs in Canadian and American dollars that have been described as 'a bureaucrat's dream and a user's nightmare' (Comuzzi 1994). Given the tenor of the times, with cost cutting and the need for a more effective and efficient government being the order of the day, as well as the North American Free Trade Agreement seeking parity in trade between Canada and the United States, change will almost certainly be introduced. The Comuzzi Report of 1994 recommended 'a Binational Agency to manage and operate the Great Lakes / St Lawrence Seaway system,' but M. Guimont, official opposition critic for Transport, argued that 'the St Lawrence Seaway is situated in Canada's territory and taxpayers in Canada and Quebec paid for most of its construction. Why should Canadians share its management with Americans? ... We do not have to relinquish part of the administration of the St Lawrence Seaway to foreign interests in order to make it more effective.' This observation appropriately places the future of the Seaway within an international context, and is especially apt given that the Welland Canal was initiated to safeguard Upper Canada against commercial rivalry from the American Erie Canal.

16

An Urban-Industrial Corridor of Development

Urban and industrial development within the canal corridor took place within a changing structure of municipal government as the former small canal communities grew in size, amalgamated, and became cities. Regional government was established in 1970 over the corridor and its surrounding area in the region. Within this context, the major direct manufacturing and service contributions of the canal are recognized, with tourism being increasingly viewed as an important new economic incentive for the canal corridor.

A New Administrative Structure

The physical expansion of the canal communities into and across their adjacent rural areas continued during the 1950s (Krushelnicki 1994). The city of St Catharines took over a few small areas in Grantham Township every year from 1953 to 1955, and Merritton a more substantial area in 1954, when it broadened to the west to meet the suburb of New Glenridge in South St Catharines. Welland expanded south into Crowland Township (Higgins 1956), but neither the town of Port Dalhousie nor the town of Thorold expanded over this decade.

The culmination of urban outgrowth under pressures from the automobile, population growth, inward migration from abroad, and suburban development climaxed in 1961, when the city of St Catharines amalgamated with the towns of Port Dalhousie and Merritton and several isolated parts of Grantham Township to form a new, more extensive administrative unit that retained its earlier name (Proctor and Redfern 1960). One large administrative unit centred on the canal now extended as a continuity across the Ontario Plain between Lake Ontario and the Niagara Escarpment. The era of three former independent canal communities – four if Port Weller is included – had ended. The change in population was from 37,984 within the old boundaries of St Catharines to 84,472 within the newly created municipality. This figure included over 7000 from Merritton and Port Dalhousie and some 15,000 from suburban spread into Grantham Township.

This regrouping of the former canal communities did not incorporate Thorold, despite the close economic, transportation, and social links that existed, and even though the Hardy Report of 1955 had argued that Thorold and parts of Thorold Township should be included as an integral part of an enlarged St Catharines. This rationale was not pursued because of the different vested interests that were involved, and in particular because the town line between Thorold and Merritton was also the county boundary between Lincoln and Welland Counties.

Within Lincoln County, which feared diminished importance and status, the dispute over amalgamation was bitter. It was somewhat more acceptable within Port Dalhousie, where links were strong between the city and the port, with attributes that included Lakeside Park and the Henley Rowing Regatta. As these facilities, although outside the city's municipal boundary, were generally perceived to be part of St Catharines, the citizens of Port Dalhousie at first supported amalgamation. But when a vote on the issue was taken in 1960, support was obtained from only 43 per cent of the residents. The fight for independence was more bitter at Merritton where, after an inflamed and acrimonious debate, only about 1 per cent of the population voted in favour of amalgamation.

The Welland Canal had created not only communities, but staunchly independent communities, each with their own elected officers and officials, police, fire brigades, municipal services, schools, and social organizations that had grown with pride and loyalties that had developed over their 130 years of previous history. Even though amalgamation was expected to be more efficient in the provision of services, localism still played an important role. It was anticipated that the bigger unit would be more distant and remote from the local decision-making process.

To the south, the city of Welland grew to the west in 1950, and then more extensively outwards in all directions to incorporate suburbs that had developed across parts of Crowland, Humberstone, Pelham, and Thorold Townships. The population of 15,382 at 1951 within the old boundaries had increased to 36,079 within the new dimensions by 1961. At the southern entrance to the canal, Port Colborne had consolidated to the east and to the west in 1950. It then absorbed the neighbouring village of Humberstone in 1952, which created one large municipal unit based on the port.

The new urban centre in turn obtained the status of a city in 1966, joining St Catharines and Welland in this category and enjoying both independence and separation from the county system of government.

When Thorold's incorporation as a city followed in 1975, all the administrative units along the canal now had the important municipal status of city. Along the old and the new canal channels the cities of St Catharines, Thorold, Welland, and Port Colborne now juxtaposed to form one urban unit with distinct parts, the whole area between Lake Ontario and Lake Erie being referred to as the Welland Canals Development Corridor as a reflection of both its historic and its modern, urban-industrial character.

Throughout this corridor of industrial and urban development, the administrative units were now larger than had existed previously in the canal's history. Each city also contained more than one of the earlier canal communities within its structure, and each had also absorbed extensive tracts of surrounding farmland and adjacent rural acres. It was now an urban-dominated situation, very different from the earlier scene of a pioneer-rural-canal oriented setting.

The urban population of the four canal cities continued to spread-eagle across their extended areas of municipal space, and urban advance over agricultural land was encouraged by the outward extension of services and utilities from the earlier canal cores of these urban units. Dispersal outwards across the landscape on a grand scale was taking over from the earlier pattern of small initiating canal settlements that were now but a part of the corridor's wider urban agglomeration.

The resultant problems caused widespread concern and led to the formation of a Niagara Region Local Government Review Commission (Krushelnicki 1994; Mayo 1965, 1966). Major issues included the severe fragmentation of authority, pollution, a shortage of recreational space, and the provision of services including transportation, police, the administration of justice, and fire protection. Public health and welfare, as well as elementary and secondary schools, presented further problems, as did extensive ribbon development and sprawl into the Niagara fruitlands.

The Welland Canal was viewed by Chief Commissioner Mayo as a potential location for a scenic drive. 'While the provision of all-day parks is important and becoming more so, it is also important to provide scenic drives, comparable to those of the N.P.C. [Niagara Parks Commission] on which people can drive to other parts of the peninsula. Potential routes would follow ... the canals ... There are a number of areas along some suggested routes – by the side of the canals, for instance – which could be developed into beautiful recreation and picnic sites.' With this being the period when the canal bypass projects at St Catharines and Welland were under active consideration, Mayo continued, 'We consider that the St Lawrence Seaway Authority will build virtually a whole new Welland Canal System in the next few years. And that by 1972 many miles of the present canal system will be sold to one or other level of government for recreational purposes. What a wonderful opportunity this will be for the Region. If the N.P.C. or the Parks Branch of the Provincial Government does not undertake this venture we suggest it would be a suitable undertaking for a regional municipal authority' – a wise recommendation still under examination some thirty years later.

The Review Commission argued for an upper-tier level of regional government over the two counties and four cities (namely St Catharines, Welland, Port Colborne, and Niagara Falls) along the canal. At the lower tier, each centre was awarded additional land for urban expansion. The municipal functions of government were divided between the regional and the local levels of administration, and the counties were abolished. 'The establishment of parks and beaches and recreation facilities ... and the framing of a regional policy for recreational facilities' were viewed as regional responsibilities of government, and 'the preparation of an official plan for the Region should be regarded as a matter of urgency.' The new regional organization came into operation in 1970 and henceforth responsibilities for the canal communities were divided between two masters, the upper-tier regional municipality and four lower-tier city units.

Regional Planning Identifies the Canal as a Recreational-Ecological Resource

With regional government in place, work started on a series of background studies for the region's official plan. These recognized the canal, the Seaway Authority, and its extensive control of land as key factors in making provision for recreational advance. 'Along the existing canal, there are several areas which could be developed for recreational use, and some areas which are presently used for recreational use. The City of St Catharines maintains picnic facilities on Seaway property, and the Department of Lands and Forests leases a water fowl area at Mud Lake, Port Colborne. The Seaway Authority maintains a viewing station and information booth at Lock 3. A marina is on leased property on the eastern breakwall of Port Weller in Lake Ontario' (Philips 1972). That land owned by the Seaway was often leased to other government and private agencies for land-based activities, providing an important aspect of the canal scene in each municipality. Overall, the modern canal was seen to offer 'a series of interrelated recreational areas. It lends itself well to the corridor concept of recreational development, similar conceptually to proposals for the Niagara Escarpment.'

With respect to this environment, the Niagara Escarpment and Development Act of 1973 laid out the process of management to achieve a Niagara Escarpment Plan and established a commission for its achievement. At the turn of the century the Niagara Parks Commission had been established to conserve amenity, but this process of independent control and management with recreation and amenity considerations to the fore was not pursued along the Welland Canal. The Seaway Authority as the prime operating agency was in control. Its essential function was to secure the passage of vessels through the canal system, and it had neither a recreational nor a landscaping division, and no program of public participation.

The canal had introduced new landforms that included lakes, ponds, spoil heaps and extended shorelines. Streams had been blocked to create wetland areas. Referring to these environmentally sensitive areas, Philips (1972) in a background study for the region's official plan stated that 'the preservation and improvement of distinctive natural features should be an integral part of the recreation planning process.' Four fragile and unique biological areas were identified in the canal zone. Malcolmson Park at Port Weller, described as 'a botanical delight' and 'a mature nursery of exotic and rare plant species,' had been planted in the 1920s during the final stages of canal construction to beautify the canal. Allanburg Weir Swamp provided 'nesting sites for waterfowl and other bird species.' The duck pond at Port Robinson was 'inhabited by Wood Ducks' and 'provided a nesting area for hawks and owls.' Mud Lake between Welland and Port Colborne, with '20 species of birds which nest here and are known to nest nowhere else in the Region,' was 'the only known location in the Region for the plant species of Bog Twayblade Orchid and Nodding Ladies' Tresses,' and 'provides nesting sites for Pied-Billed Grebes, Coots, Least Bitterns, Virginia Rails, Soras, Gallinules and various duck species.'

Philips then noted the outstanding factors that contributed to the canal's potential as a major focus of recreational activity: the technology of the locks and the lift-bridges; the history of the canal and the remnants of the first three canals; the extensive areas of land under government ownership; and many of the associated lands being areas of scenic attraction or great natural interest. After the many points of cultural, scenic, historic, and natural interest had been identified, it was proposed that 'the attractiveness of the Canal would be greatly enhanced by the marking and promotion of a

406 St Lawrence Seaway and Canal Development, Post-1960

scenic drive from Port Weller to Port Colborne, similar in purpose to the Escarpment Scenic Route [not achieved] or the Niagara Parkway.'

The Niagara Peninsula Conservation Authority in 1972 confirmed this rich assessment when it stated, 'The large amounts of land associated with the Canal offer opportunities for recreational development without the cost of acquisition ... Many of the Canal associated lands are areas with scenic attraction, or areas which have great natural interest.' More than commerce and trade are involved in the canal scene, but one is reminded again that the canals were constructed and are managed for these prior purposes. The other more modern considerations are of interest primarily for residents and visitors to the canal communities.

When the Regional Niagara Policy Plan was prepared in 1973, its major theme was to establish the limits of urban growth, a dual process that involved attempts to save agricultural land and curb the excesses of urban growth over these acres. This led in 1974 to an Urban Areas Boundaries Map which indicated that urban development would normally only be permitted within these boundaries and in part took advantage of the Fourth Canal to restrict urban expansion. The canal provided the eastern limit for urban growth generated by St Catharines, except where the city had overlapped the canal at Port Weller Dry Docks and General Motors. South of Thorold the canal provided the western limit for growth in the industrial area that had developed along the canal towards Port Robinson. At Welland the old Fourth Canal was used as a divide between the eastern and western sections of that city, and the city overstepped the By-Pass Channel to the east. At Port Colborne the Fourth Canal again functioned as a divide through the city.

The canal in its length from its breakwaters in Lake Ontario to Lake Erie, and including the abandoned Fourth Canal at Welland except for its Central Reach, was identified as an Environmental Area under public ownership and viewed as one of 'the primary areas for the provision of recreational, open space and park areas in the Region. These areas form an interlinking network of corridors, which must be protected both to meet the increasing desire of residents and visitors for recreational and scenic areas and also as a continuing and irreplaceable natural heritage.' Later, the policy plan noted that 'The Welland Canal has potential as a continuous open space corridor ... It is important that [this open space, together with various river valleys,] be recognized and that development in these areas be carefully controlled before they are lost forever' (Regional Municipality of Niagara 1973).

At the lower-tier municipal level, in the official plan for St Catharines the parkway system included linear belts from Port Dalhousie to the Niagara Escarpment along the routes of the First and Second Canals, the escarpment section of the Third Canal, and continuously along the Fourth Canal. Here 'a continuous system of open space is to be created and maintained that will traverse the whole City with an environment of trees, grass and natural wilderness readily accessible to the majority of inhabitants.'

'As far as is possible, the system is to be continuous and lands will be added wherever possible to improve upon the continuity and to gain direct pedestrian access from adjacent lands. Pedestrian walkways and other open space, particularly along watercourses, will be joined to the system wherever possible.' This provision for open space along the canal lands was continued to the south through Thorold, where lands along the present canal and former canal lengths were designated as parks and open space. 'The City recognizes the importance of the Welland Canal as a transportation facility. Council also recognizes the opportunity to develop the Welland Canal Corridor for open space and recreational uses. Council encourages the use of the canal-associated lands for park and open space uses.'

As the same type of public open space–recreational provision existed in the plans for Welland and Port Colborne, the region and the local municipalities were now unanimous in their assessment of the provision that should be made along the modern canal and the remnant sections of the predecessor canals that had survived as bodies of water. The canals had changed from their earlier circumstances as a commercial artery with important industrial and urban overtones to being a linear resource asset that now featured recreation and open space as special attractions.

The Canal-Industrial Link

The direct association between the modern canal and its industrial enterprises was examined in 1967 by Wieler. At its northern entrance, Port Weller Dry Docks had taken over the shipbuilding activities of the area as the trend to larger vessels took place. Its central location on the Seaway system was also ideal for ship-repair work that included foreign vessels that had suffered mishaps. The seasonality of labour was avoided by using the winter months to build large vessels and the summer months for repair work.

McKinnon Industries, a subsidiary of General Motors and the largest single company on the canal, employed some 4500 residents from the surrounding area. Sand and coal for use in their foundry were shipped in by the canal, which also supplied water for cooling purposes. A more flexible, daily truck service was used to export the engines that were produced, otherwise storage facilities would be required next to the canal to stockpile the engines for a large enough shipment by boat, as well as in Oshawa to stockpile the engines that could not be used immediately.

Thorold, though smaller than St Catharines in population, had a larger manufacturing representation that used the canal. Exolon manufactured various carbides and oxides for abrasive and refractory use, importing coke from the United States and bauxite from South America via the canal, and also in part exporting via the canal. Domtar manufactured building paper and roofing felt from old paper. The company used water from the canal in its manufacturing processes, and exported via the canal to Montreal and Quebec.

Ontario Paper used the canal to the greatest extent. As with the two previously mentioned companies but with much higher volumes, it was a major user of water from the canal and received more than twice the volume than did McKinnon. Its fleet of eight ships imported pulpwood from Baie Comeau in Quebec and exported newsprint to Chicago; coal was imported from the United States. With an employment of about 1300 persons, this company was the backbone of the local pulp and paper industry.

Beaver Wood Fibre had previously produced paper and used the canal, but now produced tissue and toilet paper made of scrap paper. The canal was used only as a source of water, as both the raw materials and the finished goods were shipped by rail. Hayes Steel to the south, which manufactured auto parts such as clutches, rear axles, and drive-train components in the former Pilkington Glass factory, relied on the canal, but to a lesser extent, for the supply of water and the import of coal. Trucks were used for the export of its finished products.

At Port Robinson, both James United Steel and National Refractories used the canal as a water source and for the supply of coal, but truck and rail were used to receive goods and supply the surrounding market. At Welland, which Wieler described as a 'canal oriented city with many industries using the canal for various purposes besides water,' Union Carbide manufactured carbon electrodes for a world market and shared docking facilities with the Steel Company of Canada. Quartzite and manganese ore were imported from South Africa and South America. Coal and oil were shipped in by the canal, while carbons destined for various areas of the world were exported by the same route. Stelco, its dockside companion, also imported coal and exported large pipes. Atlas Steels, although not situated on the canal, nevertheless used it to import raw materials such as coal and oil. R. and I. Ramtite imported chemicals from Europe and South

Africa by the canal, and exported cement bricks and adhesives for various metals by the same waterway. Together, these canal-associated companies at Welland employed about half the labour force engaged in that city.

At Port Colborne, the Maple Leaf and the Robin Hood Flour Mills had located next to the canal for transportation purposes, but the increasing size of vessels and their direct movement through to the St Lawrence had diminished the former attractiveness of this marine setting. Of the metal industries, Algoma Steel used the canal for its water supply, to obtain metallurgical coke from the United States, and for the import of various ores, with manganese ore predominating. Its production of pig iron was also shipped out to various destinations via the canal. International Nickel, though again not situated directly on the canal, used a pipeline to its facilities for the import of bunker oil from Venezuela and other foreign sources of supply. Some of its ores were brought in via the dock of Algoma Steel and water was obtained from the canal, but its shipment to the European market were by rail through New York.

With respect to the industrial service enterprises along the canal (*Shipping Reports* 1960–77), bituminous coal was imported from Pennsylvania and elsewhere in the United States, then distributed locally within the regional area through four coal companies. Empire Hanna had located on the east and west piers of the harbour at Port Weller, Weaver Coal serviced the Thorold area, the Rochester and Pittsburg Coal Company operated at Welland, and the Valley Camp Coal Company on the west pier at Port Colborne. They each handled from 100,000 to 150,000 tons (90,720 to 136,000 tonnes) per year, major contracts were received annually from companies such as McKinnon Industries and Ontario Paper, and huge piles of coal occupied their respective wharves.

Diversification, and the need to make maximum use of the docking facilities and machinery, also led to stockpiles by Empire Hanna of zircon ore imported from Australia to Port Weller for use by the National Lead Company in Niagara Falls, New York. After the Seaway allowed in larger vessels, which encouraged a cheaper commodity flow, bauxite was imported by various abrasive companies on both sides of the Niagara River. Salt for winter use on the roads had been considered, as well as iron ore, but demand for the latter was less than for scrap iron. Wieler concluded that industries found the canal 'an ideal means of supplying these materials and they are now using the canal to a greater extent. Much of this has occurred because the opening of the Seaway has allowed larger ships to come directly to this area.'

Oil was another major commodity shipped in economically by water.

St Catharines Fuel Oil, with three major oil dumps next to the canal, imported petroleum from Venezuela to storage tanks next to the Garden City Skyway. In a given year some 45 million gallons (170 million litres) of bunker and fuel oils were imported, plus smaller volumes that were supplied by boat to Thorold and to Atlas Steel and Union Carbide in Welland. Interlake Oil (Shell Oil) and Jerry's Fuel Oil had nearby tanks. They shipped in smaller volumes of Canadian oil from Sarnia at the western end of the Trans-Canada pipeline, and used trucks during the winter months. Another Shell-owned company on the west pier at Port Colborne supplied the local industries in that area.

Several companies serviced vessels, using a truck service and the delay involved in passing a vessel through the canal and its locks to service the ships. A laundry company picked up dirty clothes at one end of the canal and returned the cleaned goods at the other end. A dairy had a refrigerated truck that supplied ships with their products. Two suppliers provided ships with food and hardware supplies. Two other companies provided a dry-cleaning service. These companies each worked out of St Catharines. A Port Colborne company offered food supplies at its end of the canal. As a vessel arrived at Lock 1 or Lock 8 it might well be greeted by vans bringing in these supplies, or by taxis and cars sent to meet a crew member who might go ashore to spend a short time with his or her family.

These service industries operated only during the navigation season, but seasonal winter work prevailed when ships docked at Port Colborne or Port Weller through the winter months. The Port Colborne companies that might benefit included a machine shop, a boiler maker, two ship-repair works, an electrical-repairs outfit, and a firm of carpenters. At Thorold a company repaired instruments, and at Port Weller there were the dry-dock facilities. During the summer months repair work might be undertaken while a ship was in transit through the canal; the worker(s) might even be carried on board for a period.

These employment facilities were in addition to the St Lawrence Seaway Authority, with some 775 persons in its employ – 125 in administration and 650 working on the canal. Seasonally, various projects such as the repair of lock gates, deepening and widening of the canal, and the building or repair of tie-up walls and docks provided winter employment, and extra labour had often to be hired.

The industrial landscape along the canal included various stacks and piles, sometimes under cover but often in the open and enclosed by wire fences. There was the storage of coal and sand at Port Weller, and fuel oil in tanks along the canal in St Catharines. Thorold had its piles of pulpwood and also coal. Welland stored coal, coke, and fuel oil, and Port Colborne had flour in its elevators, fuel oil in tanks, and open piles of pig iron, coal, and crushed stone.

Different types of storage can be discerned. For instance, there were the service industries where demand was related to the size and distribution of the population, in which the inflow of bulk raw materials was off-loaded at central points then distributed to local markets within the peninsula; examples included coal and fuel oil and the privately owned industrial docks in Thorold. A different type of industrial movement was the inflow of raw materials for use in the manufacture of finished or semi-finished products such as the inward movement of grain, iron ore, or pulpwood. Another type of industrial movement was the outward flow of industrial products such as crushed stone from the Onondaga Escarpment at Port Colborne or rolls of paper from Thorold.

Stacks and piles are usually viewed as objectionable and visually unattractive features of the urban landscape. Clay (1973), recognizing their economic importance, offers a different viewpoint: 'Looming quietly as a backdrop in thousands of neighborhoods is a huge, piled-up-mass of something-or-other. It just sits there quietly. Nobody pays it heed. Bulldozers snort around and nibble at its hindquarters. Trucks come and go, raising dust, paying tribute, leaving deposits, scurrying away.'

Changing Population and Industrial Circumstances, 1971 and 1991

The four canal cities had a total population of 190,604 by 1971. St Catharines dominated the scene with a population of 109,722, or 57.6 per cent of the total. Welland, the second in population size, had 44,397, or 23.3 per cent of the total. Of the smaller units, Port Colborne with a population of 21,420 held 11.2 per cent of the total population, and Thorold with 15,065 people had 7.9 per cent of the total.

All four cities were heavily industrialized. St Catharines had the largest employment in manufacturing sector with 30,015 employees. Next came Welland with a quarter of this total, followed by Port Colborne with 3585 employees and Thorold with 2525 employees. The canal corridor and each of the canal cities were among the most intensive areas in Canada for manufacturing industry.

St Catharines was dominated by transportation equipment (7065 employees) and in particular vehicle and parts manufacture (6615). Next in importance were the paper and allied industries (1655), almost all in pulp and paper mills, and metal fabrication (1435), in boiler and plate works,

metal stamping, wire and wire products, and in hardware and tool manufacturing. Reflecting the city's central position in the Niagara Fruit Belt,
the food and beverage industries (980 employees) followed, emphasizing
fruit and vegetable processing (330), dairy products (205), and wine production (200). Next came the metal industries (525), including various
types of machinery and equipment manufacture (425); the electric-prod-
ucts industries (440), including electrical industrial equipment (350); pri-
mary-metal industries (420); printing, publishing, and allied industries
(375); and non-metallic mineral products (280). St Catharines, with this
extensive range of manufacturing industry within its environment had the
most diversified economic structure of all the canal cities.

Thorold also had its pulp and paper mills (1225 employees), as the paper
and allied industries had developed across the edge of the Niagara Escarpment using water from the abandoned Second Canal for industrial purposes. Also of note in this city were the manufacture of transportation
equipment (415) and non-metallic minerals (210).

In Welland at the centre of the peninsula the emphasis was strong on the
primary metal industries (3345 employees), the industrial classification that
included as specialties iron and steel mills (2300), steel-pipe and tube mills
(650), and smelting and refining (270). The metal-fabricating industries
(795) followed, then textile industries (725), mainly cotton yarn and cloth,
and motor vehicle and parts manufacturers (630). Employing fewer than
500 were the manufacture of industrial chemicals (410), and the machinery
industries (365), which included agricultural machinery (260). The rubber-
products industries (335), the food and beverage industries (290), including
bakeries, and the electrical-products industries (275) followed.

Port Colborne's most important contribution was in the field of primary
metals (1850). Then came the food and beverage industries (310) at a significantly lower level, closely followed by the leather industries (305), mostly
in the manufacture of boots and shoes, then transportation equipment
(245).

By 1991, twenty years later the sequence of the canal cities by their population size remained unchanged, but their rates of growth had varied considerably. St Catharines and Thorold had each expanded by about 17 per
cent. Welland had also grown, but at about half the rate of the two northern centres. Port Colborne, with the opposite experience of a declining
population, had diminished by 12.4 per cent to 18,766 by 1991. Overall, the
combined growth for the four cities over the two decades was 22,918 or 12
per cent, which may be compared with the provincial average of 30.9 per
cent or the national average of 21 per cent. The canal centres had grown

substantially, but at a lower rate than either the province or the nation. This pattern was also a reversal of the situation that had prevailed until the mid-1950s when the Niagara Region tended to increase at a greater rate than either the province or Canada.

Drastic alterations had taken place in the urban structure as the manufacturing base that had developed over the decades changed remarkably. At 1971 the per centage of all employment in the manufacturing sector ranged from a high in Welland of 44.1 per cent to a low in St Catharines of 41.5 per cent. By 1991 this proportion had diminished considerably in each canal centre. Welland retained the highest proportion in manufacturing employment, but at the reduced per centage of 26 per cent. This city was followed by Port Colborne at 24.6 per cent, then by Thorold at 23.1 and St Catharines at 20.2 per cent.

Though this sector was much diminished, it is important to emphasize that the canal communities still retained a creditable range of diverse manufacturing industries, a fact well illustrated by noting that the provincial proportion of the labour force employed in manufacturing, by comparison with the preceding data, was lower at 17.3 per cent. Manufacturing employed a workforce of 13,155 in St Catharines, 6066 in Welland, 2105 in Port Colborne, and 2090 in Thorold, thus providing the considerable total of 23,410. Although in relative decline and not so ebullient as in previous years, the canal zone remained of exceptional importance for industrial endeavours as an imperative feature of its economy.

The world recession and the changing circumstances of international trade that had so jeopardized cargo flows through the Welland Canal had hit with equal force at the manufacturing activities that had developed along the canal. Large and small industries, Canadian and American plants and other national subsidiaries, with each type of manufacturing industry, had been affected adversely.

In late 1982, when an unemployment rate of 19.1 per cent (supposedly the second highest in Canada) was recorded in the Niagara Region, a special committee of Regional Council was formed to investigate the designation of the Niagara Region for assistance under the federal Industry and Labour Adjustment Program (Marshall 1982). Evidence obtained from the trade unions indicated a decline from 777 to 606 employees between 1980 and 1982 in industries associated with the United Steelworkers of America that included Atlas Steels, General Signal, and Welmet. Thompson Products had laid off 193 from its total of 754 hourly rated employees. At Hayes-Dana the Thorold plant had laid off 1208 out of 1923 employees and its office in St Catharines had reduced staff from 96 to 45.

Companies associated with Local 523 of the United Electrical, Radio and Machine Workers of America, with a maximum employment of 2571, had at mid-November 1982 only 921 working at Page-Hersey, Welland Tubes, Union Carbide, General Drop Forge, and Welland Forge, against 1479 on recall. The United Textile Workers of America in Welland and St Catharines had reduced from a membership of 544 in November 1980 to 395 two years later. When businesses in Welland, Fonthill, and Port Robinson were contacted, 231 positions had become redundant and 2100 people were on indefinite lay-off.

Recession was severe, hitting hardest at heavy industry that had been encouraged to locate along the canal. The repercussions on family life, social services, and household expenditures were often stringent and persistent. By the census of 1991, in St Catharines, Welland, and Port Colborne the male and female unemployment rates were each substantially higher than the provincial average and the participation rates for both sexes were each lower than the provincial proportions. At Thorold these circumstances were marginally better than the provincial figures, but as Thorold had the smallest employment of the four canal communities, this in no way affected the average situation that prevailed along the canal corridor. Recession, depression, unemployment at higher than either the national or the provincial levels, economic hardship, the loss of jobs, reduced hours of work, wage freezes and cutbacks, downsizing, restructuring, free trade with the United States, enforced early retirement, reduced pension packages, and severe overspending and deficit problems at the federal and the provincial levels of government had each taken a severe toll on manufacturing employment.

As the primary industries of agriculture, fishing, forestry and quarrying had also declined over the two decades from 1971 to 1991, there was relative employment growth in the service industries, including trade, government, education, finance, health, and social services. Retailing, including the steady advance of shopping centres, schools and colleges, hospitals, small business offices, and strips of commercial development are to be found in suburban locations in several places throughout the canal corridor. They have each contributed greatly to the employment situation, and in turn have both undermined the versatility and strengths of the established city centres, which grew with service activities in association with the canal. Outward dispersal has contributed significantly to the growth, size, and outward spread of the canal communities, but not to the vitality of their pre-existing, canal-oriented, business and commercial areas.

Using the specific example of Port Colborne to illustrate the severity of the adverse economic situation, the manufacturing share of the labour force declined from 50.8 per cent in 1961 to 32.4 per cent in 1986. The growth in the non-manufacturing groups from 16.4 to 25 per cent can be attributed to developments in the tourist industry and retirement services (Salerno 1989). In the manufacturing sector, a diverse array of new small industries added about 760 new jobs in the food industry as poultry played an increased role in dietary habits, and three new companies in fabricated metals added a further 180 new employees. By contrast, the industrial losses can be attributed primarily to decline in milling, cement, and primary metals. Employment in Maple Leaf Mills had reduced from some 180 persons at 1960 to 96 by 1980 through a combination of equipment upgrading the loss of the bag-making operation, and decline in the export trade. Robin Hood Multifoods halved its employment from 310 to 172 persons for the same reasons. When the Canada Cement closed its plant in 1965, this resulted in a loss of some 120 jobs.

Primary metals, which in the 1960s accounted for nearly 70 per cent of the total employment, dropped to less than 20 per cent of all manufacturing employment by the 1980s. Algoma Steel expired in 1977, with the loss of 400 employees, as the demand for Canadian pig iron declined, and the Inco refinery reduced from over 2000 employees in the 1950s and 1960s to a skeleton of less than 400 employees by 1988. The repercussions of these declines in manufacturing spread across the economy. Loblaw's closed a supermarket in 1977, and the property market lapsed in the construction of new residential, commercial, and industrial units as the number of employees in basic and service activities declined.

It is obvious that the communities created then built up by the canals will have to change their economic base as both their manufacturing industries and their service support from the canal decline. One obvious approach is to make greater use of existing assets, including the scenic attractiveness of the modern canal, the heritage character of the old canals, and the surviving community buildings of historical or architectural importance.

The Journey to Work and Housing Provision

The impact of economic depression within the canal communities is cushioned to some extent by the proximity of other nearby centres and their interlinkage through the automobile. Home and workplace are not necessary found within the same town. The issue has become regional rather than narrowly focused within each community, as some displaced workers are able to secure employment elsewhere. Based on the census data for 1991, St Catharines was the outstanding centre for employment in the Niagara Region. Its workforce employed 35.1 per cent of the employed labour force by place of residence within the region, and for the four canal cities this proportion rose to 58.2 per cent. The second major focal point of employment attraction was Niagara Falls, with 21.4 per cent of the employed labour force.

The majority of people who lived in the canal cities both lived and worked in that environment. The numbers were St Catharines, with 40,670; Welland, 12,840; Port Colborne, 4120; and Thorold, 2045. The per centage of people who lived in the same place where they worked was highest at 68.5 per cent in St Catharines, slightly lower in Welland at 60.3, and 52.6 in Port Colborne. The anomaly of Thorold, at 24.3 per cent, is that this city might well be described as an industrial and residential suburb because of the strong two-way commuting links with adjacent cities.

The largest external movements from a place of residence in St Catharines were to Niagara Falls, 2505; Thorold, 2445; Niagara-on-the-Lake, 1935; Lincoln, 1685; and Welland, 1030. The inward movements to St Catharines were most pronounced from Niagara Falls, 4730; Thorold, 3910; Welland, 2550; Pelham, 1790; Niagara-on-the-Lake, 1665; Lincoln, 1510, Port Colborne, 565; and Fort Erie, 550. Except for Niagara-on-the-Lake and Lincoln, the inflow to work in St Catharines was greater than the outward flow to external places of residence, thus again emphasizing this city as the employment capital of the canal communities.

At Thorold, 3910 of the city's residents worked outside the city in St Catharines. Next came movements out to Niagara Falls, 655, and Welland, 465. The inward movement to a place of work in Thorold was highest from St Catharines, 2445; Niagara Falls, 1180; and Welland, 680. A point of interest is that the outflow to work in St Catharines and the inward movement to work in Thorold from that city were both appreciably higher than the actual number who lived and worked in Thorold.

Residents of Welland undertook daily commuting out to Niagara Falls, 1145; Thorold, 680; and Pelham, 625. The greater influx of workers from external communities arrived from Pelham, 1725; and Port Colborne, 1500. St Catharines, Niagara Falls, and Wainfleet each contributed about 1000, and Wainfleet, Fort Erie, and Thorold each some 500 workers. Port Colborne presents a different character in that the outward movement was now greater than the inward movement. The strong outflow was to places of work in Welland, 1500, and then St Catharines, 565. The more slender inward movements were from Fort Erie, 325; St Catharines, 240; and Niagara Falls, 180.

This data has indicated that, in addition to their distinct identities, the canal cities were now united by the dual bonds of an inflow from the other canal centres and the fact that some of their working populations lived externally. As ties and links to and from the adjacent communities also existed, the earlier days of independent canal communities had been replaced by one large-scale, interlocking urban agglomeration. Growth of population and industrial advance in conjunction with ease of travel through the spread of paved roads and the ubiquitous use of the motor vehicle had promoted the considerable divorce that now existed between place of residence and place of work.

Further powerful inducements to urban growth have been provided through small family units occupying low-density housing. By tenure, two-thirds of the dwellings in St Catharines and Welland were privately owned. This proportion increased to about three-quarters in Thorold and Port Colborne. In all four cities, the average household size was in the range from 2.5 to 2.7 persons. When the structural type of dwellings is taken into account, then St Catharines and Welland had the lowest proportion of single, detached dwellings at 60.5 and 66.3 per cent respectively. These proportions increased to 73.5 per cent at Thorold and to 76.4 per cent at Port Colborne.

Apartment buildings in both their number and proportionally were most strongly represented at 23 per cent of all dwellings, in St Catharines, where they were about equally divided between those with five or more storeys and those with less than five storeys. At Thorold and Welland, a higher proportion of the apartment units had fewer than five storeys, whereas at Port Colborne this situation was reversed and more units had five or more storeys. Row housing was minimal everywhere, except in St Catharines, where its proportion of total dwellings was 5.5 per cent.

By mother tongue, slightly over 80 per cent of the residents in the cities spoke English, except for Welland where the proportion dropped to 71.3 per cent. Sizeable minority languages in St Catharines included German, French, Italian, and Polish, in that order. Italian as the mother tongue was important in Thorold, and French and Italian in Port Colborne.

Religious affiliations also varied. Adherents to the Protestant faith predominated in St Catharines and Port Colborne, whereas there were more Catholics than Protestants in Thorold and Welland. These two faiths together exceeded 85 per cent of the population in each canal centre. There were churches, often with former close associations to one or other aspect of the canal, in all the inner downtown areas; but with urban advance into the suburbs many new churches had also followed this redistribution of the population.

The Tourist Industry

Tourism, an expanding industry in the Niagara region, is based on the region's physical attributes. Identified when the region's official plan was prepared were the Niagara River, two lakeshores, the varied terrain, and the Welland Canal. There are also the historical and heritage attractions, especially at Niagara-on-the-Lake but including the various sections and features of the canal communities that have been referred to in this book. Societal aspects of tourism include the increasing ease and cheapness of travel, higher real incomes, longer hours of leisure, and a more educated population that would appreciate the characteristics of distinctive environments.

Competition for the tourist dollar is severe at the world level, with North America being but one of many potential global destinations. Rivalries exist within Canada as each province, region, and town compete avidly for the tourist trade, and in the Niagara Region the main foci of attention centre on The Falls, Niagara-on-the-Lake, and the Niagara River Parkway. The encouragement of tourist development along the Welland Canals is viewed as a side event compared with these major attractions, but the attraction of visitors to the canal environment would assist materially in extending the stay of visitors within the region and offer a major secondary attraction to the Niagara River frontage.

As conceived in the 1988 long-term plan of the Niagara Parks Commission, there would be a linking network of tourism and activity based on the Niagara River Parkway and the Niagara River Recreational Trail as recreational routes between Niagara-on-the-Lake and Fort Erie. Key linkages to a parkway along the Welland Canal would be by routes along the two lakeshores, the Welland River, and the Niagara Escarpment. 'So that Niagara will form one continuous attraction for visitors, as much attention will be given to the linkages between the sites as to the sites themselves. The resulting network will encourage tourism, extending benefits into the Region and Province' (Moriyama and Teshima 1988).

Tourism is a large, complex, and often nebulous industry. When Mayo in 1966 examined the administrative structure of the Niagara Region, he noted that seven different classes of authority were involved. They included the St Lawrence Seaway Authority, federal and provincial agencies, and the Niagara Peninsula Conservation Authority, and ranged through the several levels of government to local councils, local boards, and private operators. Mayo argued for a comprehensive regional tourist policy and a regional authority to achieve that policy.

There were three different approaches: the official plan prepared by the Region's Planning and Development Department with its concern for the land base of tourism; the Regional Niagara Tourist Council's efforts to encourage and promote the tourist industry through the preparation of brochures and the establishment of tourist booths; and the Niagara Region Development Corporation's work to encourage development and to promote its economic and employment advantages. Each has to some extent focused on the Welland Canals and the opportunities they offer, both locally and as a prime regional asset. Coordination and consistency of approach were improved in 1995, when the tourist council and the development corporation were united into a single branch of the regional government.

Viewed as an industry, tourism has been described as 'big business in Niagara, creating considerable employment both directly (e.g. jobs in the accommodation and food service industries), and indirectly (e.g. jobs in businesses that provide support services to hotels, motels, and restaurants)' (Regional Planning and Development Department 1983a, 1985). This income generated by tourism is spread across a wide variety of establishments ranging through accommodation, the food and beverage industries, amusement and recreation, transportation, and service stations to retail sales, with almost every establishment enjoying some degree of benefit from both the passing trade and the local market.

As with all industries, each of these tourist activities offers the extended advantage of the income multiplier effect as the money earned is spread through the economy. This varied between the various sectors of tourism, but it has been assessed regionally at an average of 1.64 (Regional Planning and Development 1985). 'This seems to represent between 10% and 20% of the employed labour force in the area. For comparison it appears that about a third of the employed labour force works in manufacturing.'

The factors of importance now become the amount spent by each visitor, the distribution of this expenditure across the various sectors of the economy, and the length of stay by each tourist (Regional Planning and Development Department 1983b). Short-stay visitors spend less than those on a more extended stay, and they are the most numerous because the area lies within easy daily travel distance for the population living in the towns of southern Ontario, western New York State, and to the south of Lake Erie.

The influx of tourists is also seasonal, the period of concentrated flow being even shorter than the navigation season along the canal. This raises the need for year-round attractions such as theatres, museums, concerts, university lectures, and the like of elderhostel programs, conventions, and coach tours. There are also seasonal tourist features such as the Winter Festival of Lights, the greenhouses at Niagara Falls, and attractive off-season rates in the tourist accommodations, but the canal might contribute more through a parkway that would draw visitors to its full length, inspection tours of moored winter vessels, museum and exhibition displays, tours of the canal communities, and regular visits to their commercial and industrial establishments.

Every tourist attraction has to be publicized, promoted, and advertised. An image has to be established over the years so that the place and its features are known and appreciated as an interesting place to visit. Niagara and its association with canals are especially favoured in this respect. The name Niagara is well known because of the Horseshoe and American Falls on the Niagara River, through the historic character of the Niagara Frontier, its forts, and battlefield sites, and its qualities of further appeal that include the Niagara Escarpment and the Niagara Fruit Belt. International canals of world renown also carry a strong image. Whether it be the Welland, Suez, Panama, Kiel, or the Gota Canal, each is an acclaimed tourist magnet.

The Welland Canal system is and has been urban waterfront. In the words of an Ontario provincial study, 'Throughout North America, waterfronts are enjoying a new prominence in the urban fabric as generators of commercial revenue, as places to live and as focal points for tourism development and recreational activities ... Urban waterfronts are a very special community resource which can provide unique and exciting opportunities to serve the diverse needs of many different groups. Just as downtown business areas have been improved in many municipalities across Ontario, urban waterfronts can be better utilized to respond to particular needs and conditions' (Ministry of Municipal Affairs 1987).

This provincial publication indeed highlighted many features along the Welland as examples that might be followed elsewhere. The entry sign on Lakeport Road at Port Dalhousie is presented with the observation that 'controlling signage can assist in creating a more distinct and cohesive image for a waterfront commercial area.' The Henley Regatta is noted as a popular summer event, and the Lincoln Fabrics Mill next to the Third Canal is used to demonstrate that industrial activities should not be neglected: 'many waterfronts are characterized by vacant buildings and abandoned land and equipment. These vacant and under-used lands ... are a community resource that is often completely ignored.'

The restoration at Port Dalhousie of Lock 1 of the Second Canal and the open iron fence that presents views onto the marina in the former lower harbour are used to illustrate how 'outlooks may need to be created when public access must be restricted.' Outside Port Dalhousie, at Lock 3 on the Fourth Canal, the viewing platform is used to provide an example of how 'observation decks allow visitors to watch the activity [of passing a vessel through a lock] in safety.' At Welland, the retention of the Main Street Bridge over the abandoned length of the Fourth Canal 'provides a visual focal point linking the main street and the canal.'

Although the Welland Canal has many relevant examples of wider application, much design, planning, and landscape work still remains to be undertaken along its urban waterfront. Here the consideration of future possibilities might benefit considerably from the lessons of improvement that have been achieved elsewhere, which offer suitable direct comparisons. Design and its achievement are part of a never-ending operation of continuous appraisal and re-evaluation of relevant solutions achieved elsewhere.

An excellent comparative example might be the Lachine Canal to the west of Montreal. Similarities to the Welland are that it bypassed boiling rapids. The Lachine Canal was completed in 1825 around the Lachine Rapids on the St Lawrence River, only four years earlier than the Welland Canal bypassed The Falls on the Niagara River. As along the Welland Canal, an urban-industrial nucleus then arose with industries, homes, churches, stores, schools, and an immigrant population. In 1847, Lachine received the Montreal and Lachine Railroad, built parallel to the canal between these two centres, its comparison being the Welland Railway, constructed along the Welland Canal. The Lachine Canal that had created the village, the town, and then the city of Lachine also contributed materially to the industrial and commercial growth of Montreal. Like its Welland counterpart, the Lachine Canal was also enlarged, improved, and rerouted along a parallel alignment over the period from 1843 to 1848 that reflects the Second Welland Canal. This progress through the canal and its associated availability of water power had a significant input downstream from the locks on the social and economic development of Montreal, which might be compared with the Thorold–Merritton–St Catharines corridor on the Welland.

A second enlargement of the Lachine Canal, the equivalent of the Third Welland Canal, was undertaken during the last quarter of the nineteenth century, and this remained Montreal's premier axis until the 1950s and the

1960s. The opening of the Seaway in 1959 led to the demise of through navigation on the Lachine, and in 1970 the canal was closed to shipping, three years before the abandoned Fourth Canal at Welland was replaced by the By-Pass Channel.

Although the two canals offer engineering and industrial similarities at different ends of the St Lawrence waterway, dissimilar options have prevailed with closure. At Lachine by the 1980s, in a joint effort of the municipal administration, the Seaway, and the federal government, through Parks Canada, landscaped park space serving a variety of recreational activities had been provided next to each of the earlier canals, and bicycle and walking trails followed the canals from end to end.

In addition to a museum complex and an interpretation and visitor reception centre, the fur trade is interpreted in a renovated stone warehouse that, as a National Historic Site, offers permanent exhibitions, theme activities, guided tours, and illustrated publications. Many industrial, commercial, and railway buildings as an expression of the canal's heritage have also been restored, and a detailed historical guide in French and English provides walking and bicycling tours of the sites of interest that have been reclaimed (Services Récréatif et Communautaires 1986; Sicotte and Thériault 1987).

Very few of these developments have obtained along the Welland system. No comprehensive master plan for its waterfront areas exists. Progress towards historical recognition and amenity provision has had to rely upon sometimes diffident municipal enthusiasms and, despite the fact that the Welland Canal was at least as significant as a national transportation route to the continental interior as its Lachine equivalent, Parks Canada has not yet been involved in any of the works of redevelopment and restoration that are required.

Municipal parks have been provided along some lengths of the old canals, but the Welland Canal has no museum where the sole function is to interpret the marine, technological, municipal, and industrial history of the canals. Nor does any unified overall approach to development along the canal's urban waterfront exist. In fact, despite the provision of canal-access roads by the canal authorities, these are private, and no through road for use by the public follows either the old or the new canal from lake to lake.

These discrepancies in action cannot be attributed to any lack of understanding possibilities and urging their achievement. When ideas to advance tourism were put forward by the regional municipality in 1984, several activities along the Welland Canal were indicated (Regional Planning and Development Department 1984). 'Given the increased interest in physical activities and the established tourist attraction of the Welland Canals ... the subsequent development of a related Regional recreational/parks system should be given high priority. The potential of the Feeder Canal and the possible integration of that system into a walking/biking/parks network connected with the Merritt Trail should also be considered.' Suggesting that 'the Old Welland Canal has great potential as a recreational resource,' this report foresaw end-to-end hikes once the Merritt Trail bicycle/walkway had been established ... Wind gliding and waterskiing activities could be held on the abandoned Fourth Canal at Welland and, despite Highway 406, an attractive parks system might be developed along much of Twelve Mile Creek.' The locks of the Third Canal received favourable mention as 'an interesting recreational feature,' as did the idea of a 'half-day boat trip through a couple of locks [which] would provide an interesting opportunity to observe the Canal from a boat.' Ideas have not proved a problem. It is their execution that has not always been achieved.

In the words of a presentation to the Provincial Parks Council, the Welland Canal 'begins and survives as a waterway for the shipping, trade and commerce of pioneer settlers. It expands as a national and international waterway of world importance. It becomes a major agent in urban evolution, as expressed in the historical geography of its riparian settlements across the Niagara Peninsula. It exhibits features of exceptional interest such as the locks, bridges, channels and dimensions which survive from the past. It exhibits a century and more of engineering and advancing technology from 1824 up to the present. There are vegetational, water and landform features of considerable interest. It is endowed with many existing and potential recreational attributes' (Jackson 1975a).

As the author continued, 'The Welland Canal, to state the issue in simple and succinct terms, has become an intrinsic resource asset of great regional, provincial and national importance. Those important features that survive from the past and remain at the present should not be neglected. They have their role and contribution to make as Canada advances towards its emerging future, and are an integral part of that future' (Jackson 1976). Finally, Kevin Lynch, author of *What Time Is This Place?*, has stated: 'Throughout the world, but particularly in the economically advanced countries, fragments of an obsolete physical environment are lovingly preserved, or restored so that they may be preserved, as relics of time gone by ... We prefer a world that can be modified progressively, against a background of valued remains, a world in which one can leave a personal mark alongside the marks of history.'

17

The Canals as Heritage and Amenity

The canals as heritage and as an open space–recreational resource are themes of mixed success since the 1960s in the development process. Advances and delays, defeats and successful accomplishments will each be described in this chapter, to be followed by case studies of the abandoned Fourth Canal at Welland in chapter 18 and progress towards the achievement of a lake-to-lake Welland Canals Parkway in chapter 19.

Amenity and Aesthetic Considerations

Amenity – 'concerned with the essential pleasantness and aesthetic qualities of the urban environment as a satisfactory place in which to live, work, and spend one's leisure time' (Schwilgin 1974) – provides one important justification for planning controls over development in the public interest. Like the term aesthetics, which also incorporates judgment over matters of quality, attractiveness, and order, the concept of amenity is difficult to define with precision and equally as difficult to achieve or control.

Applying the amenity concept to the past and modern canal systems includes, on former canals, the historical appreciation of their surviving heritage features and those of their communities, and, along the modern canal, its engineering calibre, its ships, and shipping activities (Jackson 1981). On the land, public open space, green areas, parkland and woodland, landscaped settings, and the provision of recreational space, together with the buildings and structures of historic and architectural appeal within the canal communities, also offer amenity attributes.

Every municipality along the canal may claim some success for one or more of its waterfront actions along the canal since the 1950s, but more has been left undone or inadequately achieved. When the earlier canals were abandoned, sections were filled at Dunnville, Thorold, and St Catharines. Water flow has been retained along lengths of the Second Canal across the Niagara Escarpment and within the abandoned channel of the Fourth Canal at Welland. The Feeder Canal has been partially filled and no longer carries a through flow.

When the various canal sections were closed, thought was not given to feasible propositions such as retaining a small flow of clean water within the banks or along a narrowed channel, or retaining the canal and its setting as linear green space through the urban environment. The old canals were not respected for any amenity or resource values they might possess. They had become unimportant remnant features of a bygone age that might conveniently be filled, then reused in the circumstances of the day for a miscellaneous series of activities that have included space for parked vehicles, outdoor storage, and the extension of adjacent property ownerships. Locks have been buried, a shopping plaza in south St Catharines involved the demolition of a lock keeper's cottage, and former stone mills have been taken over for storage or been demolished and replaced by modern buildings.

Generally, neither the municipalities nor the development industry have seen any point in retaining or reusing the past as it advanced into the future, and the weak voices of LACACs (Local Architectural Conservation Advisory Committees), local citizens, and heritage groups have carried little to no weight in the decision-making process (Fraser 1979; Taylor 1992b). The emphasis has been on development rather than conservation.

Approaches to National Recognition of the Canal System

In 1960 the Welland Canal Park Committee formed by George Seibel, president of the Kiwanis Club of Stamford, prepared a brief urging that the area then referred to as the Third Canal Lands extending from Locks 11 to 17 of that canal across the Niagara Escarpment be made into a national park and historic site (map 13). A marine museum of canal history with scale models of the first three canals and the types of vessel using each system, showing how each were hauled or navigated through the system, was advocated.

The museum would have working models of each type of lock, maps, and photographic displays. An observation tower for an overview of the Flight Locks and its vessels from the eastern side was later suggested. Picnic, camping, and fishing areas would be provided on the surrounding areas of flat wooded land zoned as parkland in Grantham Township and owned

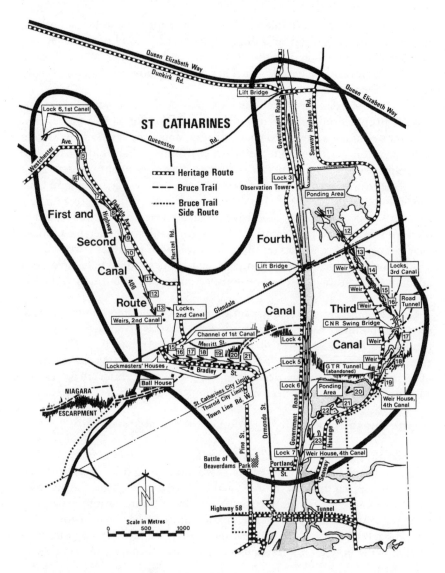

Map 13. *The Escarpment Locks Heritage Area.* A heritage proposal for the escarpment area by Parks Canada. (1980)

by the Department of Transport. The proposed site covered several hundred acres with immediate access from the then new Queen Elizabeth Way, and included the abandoned stone railway tunnel under the Third Canal, a resource that was incorporated into the design as an access route between different sections of the site.

Correspondence indicates that the Seibel brief was supported by the area municipalities, their chambers of commerce, historical societies, the Niagara Peninsula Conservation Authority, two cabinet ministers, and local MPs and MPPs. Architectural plans with estimated costs for the site layout, museum, viewing stand and tower were submitted to the minister of Northern Affairs and Natural Resources, and referred to the Historic Sites and Monuments Board of Canada. Their review stated: 'The practical problems involved in restoration would undoubtedly be more than the Department could reasonably justify for the purposes of commemorating the canal system. It was estimated by the inspecting party that there would be as much maintenance on the complete set of locks in that location as would be involved in the entire Halifax Citadel.'

In 1962 it was resolved that, 'in the opinion of the Board, the Welland Canal System is of national historical importance and the Board recommends that one section of the locks of the third system in the St Catharines area be preserved and restored; that a scale model showing the evolution of the canal systems from the beginning to the St Lawrence Seaway be provided as an interpretative device, and that the Minister take whatever steps he deems appropriate to accomplish this.' Restoration would involve the provision of gates, and the ground in the vicinity of the locks would be 'landscaped around so as to give the public an opportunity to inspect closely, but safely. The scale model would suffice if further interpretation of the historical area were required.'

A companion approach was the submission of a similar brief in 1964 to John P. Robarts, premier of Ontario. This led to the consideration of the escarpment area around the locks of the Third Canal as a regional park; funds were made available through the Treasury Board for a consultant's report and Project Planning Associates were appointed to design the layout. Their master plan included a water transport museum, an observation tower at the Flight Locks, picnic areas, sports fields, and camping areas. After its submission, committees were appointed to study the various aspects of the proposal, but nothing then happened on the ground.

Meantime the federal government determined that the maintenance of all the lock structures would be too costly. One lock with its water system might have sufficed, and the waterfalls created by water rushing through the sequence of gateless stone locks themselves provided a considerable scenic attraction. A scale model with plans, photographs, and interpretive displays might suitably have replaced the need to reconstruct all locks. But after consideration of the situation, the National Historic Sites and Parks Branch made the decision that 'the canals had been sufficiently commemorated by a plaque erected in 1924 at Allanburg' (Parks Canada 1980).

A second unsuccessful example is provided by Martindale Pond. Here was a popular recreational attraction. The Canadian Centennial Regatta had been held on this course in 1967, and with it the first North American Rowing Championships. A dredging project had been completed just previously in 1964 to provide a water depth of 10.3 feet (3 m) along a course that was 2734 yards (2500 m) long and 109.4 yards (100 m) wide, the only Class A course approved in North America by FISA (Fédération International des Sociétés d'Aviron), the governing body for world rowing. After further dredging in the late 1960s, in 1970, for the first time outside Europe, Port Dalhousie hosted the World Rowing Championships, a five-day regatta. Canal antecedents included the physical setting of the pond between high wooded banks, protection from adverse crosswinds for the rowers, a flow of water that could be controlled mechanically from the DeCew power station, and a significant history as a canal pond from 1840 on.

In 1995, against competition from Lucerne, Switzerland, St Catharines was awarded the 1999 World Rowing Championships by FISA. Improvements are required, this time to provide a uniform depth of 3 metres (3.3 yards) throughout the width and length of the course. A part of this successful award can be attributed to the development of training facilities at Brock University and to collaboration between the rowing authorities in that certain regatta events have been held on the Recreational Waterway at Welland, but the important component for this narrative is the grand transition that has taken place on former but now abandoned canal waters.

Access from the Lower Harbour to Martindale Pond and the Muir Dry Docks was closed at the end of the Seaway's 1968 navigation season for all vessels, even small yachts and powered pleasure craft. The idea of a connecting marine railway capable of handling vessels weighing up to 15 tons (13.6 tonnes) and 50 feet (15.2 m) was considered by the Seaway, but rejected as the demand was not deemed to be sufficient and access by power boats with their wakes and waves would have disturbed the fragile sediment at the base of the pond.

Martindale was for rowing and sculling only, and in 1969 the City of St Catharines prepared a brief for the pond and sections of the Lower Harbour to be developed as a National Historic Park and recreational area. A museum was proposed, but the intended restoration of the Third Canal lock and a multipurpose recreational area based on the Henley Regatta course did not receive support from Parks Canada despite the many historical and present-day credentials. The excuse of the Allanburg cairn was again used, and '[the Martindale] project was abandoned due to the Branch's previous decision that the canals had already been sufficiently commemorated' (Parks Canada 1980).

Commercial vessels had now disappeared from Martindale Pond and through cumulative practice over the years the water was used almost exclusively for rowing (City of St Catharines 1963). The pond had become a spillway for the DeCew hydroelectric generating station, which had the advantage of introducing clean water; in addition, the rate of water flow could be reduced by Ontario Hydro, which by arrangement at regatta times meant that races could be rowed under virtually dead-water conditions (McCollum 1980). Sewage-treatment facilities had removed most of the waste from the course, and when a series of detailed environmental studies were undertaken of the bottom sediments (Arcturus 1994), it was demonstrated that the Henley course could be improved to host the 1999 World Rowing Championships. Progress has continued, but a national historical designation would have provided a helpful contribution towards action.

The remarkable official argument, now used at both the Third Canal locks and at Martindale, that a single cairn was sufficient to recognize a major national resource is at marked variance with the approach elsewhere. Nearby Niagara-on-the-Lake has over forty historical plaques and the Niagara River Parkway is resplendent with such records of past events. For the federal government to argue that historic areas in the landscape should be marked by but one plaque is to devalue both the many achievements that have taken place over a century and a half along the Welland Canals and the Canals' many contributions to Canada's development.

Where the Third Canal crossed the escarpment, and where Seibel had proposed a major heritage park, the possibilities were pre-empted in 1966 when the Seaway announced its proposals to construct a bypass around St Catharines. This shelved the park proposals, but to the east of this possible park and next to the QEW, harness racing in the form of the Garden City Raceway had been introduced in 1964 by the Ontario Jockey Club. After the closure of this operation in 1976, a large-scale scheme of private land development was proposed. The possibility of a public park was replaced by a private-enterprise plan that made provision to retain and preserve the locks within a landscaped setting. Extending from the QEW up to the Fourth Canal, several hundred acres were intended for amenity, recreation, and open-space purposes. The plans for this mega project included a health spa, business convention centre, a golf course, and an arts centre. Space was reserved at the Flight Locks and along the intended By-Pass should the need for this canal bypass materialize in the future.

This scheme was terminated because of the economic depression during the eighties, but it clearly demonstrates how a park could have been provided in conjunction with the intended By-Pass through reserving this route, focusing on the abandoned locks, and not permitting any form of permanent development along the intended line of canal with its new Super Locks. The park would have had a triple series of attractions: the abandoned and now gateless locks of the Third Canal, the Flight Locks of the Fourth Canal, and the abandoned CN railway tunnel. It might even incorporate the Flight Locks of the Second Canal (map 13). And after the completion of the possible Fifth Canal, the park would then contain the magnificent spectacle of four canals with their flights of locks representing different technological eras. What an exciting theme – thus far deferred but remaining a feasible proposition as the land has been retained by the Seaway under federal ownership pending the future possibility of a new channel, now envisaged circa 2030.

The Role of Private Individuals and Organizations

Promotion of the merits of the canal heritage, of William Hamilton Merritt for his achievements, and of the values of the modern canal has been pursued by various individuals and organizations, with varying degrees of success, but at least always recognizing the canal as a quality resource in addition to its prior role as a transportation artery.

An early but important advance towards this national recognition was the issue in 1974 of an eight-cent stamp for first-class mail. Promoted by Louis Cahill of the Ontario Editorial Bureau (now OEB International), this stamp recognized the creation of the Welland Canal Company in 1824 and then 150 years of continuous canal history. The stamp depicted a canal lock at the crest of its flight across the Niagara Escarpment at Thorold; Merritt, as the canal's founder, is superimposed over this scene. Next year a federal plaque unveiled at Lock 3 in St Catharines commemorated Merritt as the merchant and industrialist who was primarily responsible for raising the funds to achieve the First Canal and who then continued to promote its successful achievement until his death in 1862.

In his ongoing search for recognition of the canal by all levels of government, Cahill established Welland Canal 150th Anniversary Inc. to commemorate the opening of the First Canal in 1829 and to reflect upon the significance of this event for Canadian history and development. The president of this organisation was the chairman of the regional municipality, and its officers included the mayors and businessmen from all the canal communities. Programs throughout the region were organized by local committees during 1979. The region and the local municipalities were now acting together as a unit to promote the Welland Canals and their many contributions to regional life.

Following this celebration, the Welland Canals Foundation, successor to the 150th Anniversary, has continued to promote the canals and its founder. This private group in conjunction with the Seaway Authority has organized regular marine ceremonies that include the Top Hat welcome for the first ship to pass through the canal each year and an annual Merritt Day to commemorate the first lake-to-lake journey through the canal. Other events have included canal days within each community as a summer festival feature, the organization for visits by the Tall Ships, and the training of canal guides for tour groups, a program later taken over by the Niagara Region Tourist Council. The Foundation has prepared media publicity for canal events and publications on marine and community aspects of the canal, and William Hamilton Merritt has been represented by actor David MacKenzie at public events associated with the canal. The canals and their history have been promoted by lecture series, speakers at various events, the preparation of a teaching manual to encourage visits by school groups, and the conduct of field excursions along the canal and within its communities for visiting groups.

Later private initiatives have included the formation in 1977 of a popular citizen's group, the Welland Canals Preservation Association (WCPA), to promote the conservation of the old canals and the development of their tourist and recreational potential. A report for a bikeway trail network along the old canals in St Catharines was published in 1978, followed in 1983 by a proposal for a lake-to-lake heritage parkway (Mroz et al. 1983). Achieved in 1985, the unsurfaced Merritt Trail provided a 28 mile (45 km) hiking route between Port Dalhousie and Port Colborne. In 1988 a link extension along the Feeder Canal from Welland to the Grand River was blazed by the Bruce Trail Club. This connected with the Merritt Trail at Welland and with the Bruce Trail at Bradley Street in St Catharines.

WCPA arranged to clear accumulated debris and partially restore Lock 1 of the Second Canal at Port Dalhousie. The John Howard Society as a community-service project in conjunction with the Correctional Institute in Thorold undertook this rehabilitation work, and the lock was then placed in a small landscaped park. Another attempt at preservation was the excavation of buried Lock 24 of the First Canal in Merritton on the canal's escarpment length. An archaeological report was prepared on the structure of the lock (Historica Research 1988), but the timbers were then buried as a 'safeguard' against damage. Their preservation would have been a costly but feasible alternative.

In 1988, WCPA worked with the Regional Municipality Niagara and the municipalities to interlink the several canal communities and the northern length of the Fourth Canal by a signed road route. Described as the Scenic Canals Drive, signs were erected to direct motorists on a two-hour tour between Port Dalhousie and Port Colborne. For the first time there was now a marked route to take tourists along the canal, but the itinerary was indirect along main roads, it did not follow the canal, and it carried neither displays nor directions to points of interest. Notable heritage sites such as Welland Vale, where the canal system was initiated, and the lock sequences of the Second and Third Canals across the Niagara Escarpment were not included on the selected route.

In another private initiative, a group in the Wainfleet-Dunnville area in 1979 prepared a feasibility study about how the neglected potential along the disused channel of the Feeder Canal might best be used to serve heritage, amenity, and recreational purposes (Rehabilitate the Old Feeder Canal Association 1979). Though the Feeder was infilled at the Dunnville end, water nevertheless remained along the channel in a series of small discon-nected, and often stagnant, ponds. These sections of the old canal, together with the resource features in the former canal communities of Dunnville, Stromness, Port Maitland, and Wainfleet, offered recreational and heritage opportunities, but little except for some cleaning and weeding of the channel has followed. By 1995 Wainfleet had initiated a historical park and was con-templating the reconstruction and operation of a canalside mill, a barge towed by oxen, and a wooden swing-bridge over the canal.

Mention might also be made of the Canadian Canal Society, founded in 1982 at Port Dalhousie. As stated in the preamble to its revised constitution of 1994, 'This Society is founded and functions on the proposition that Canada's Canals are a priceless heritage that can be enjoyed by all with an interest in Canada's past, present, and future.' The major contribution of this and the other canal groups has been to keep alive the idea of the canal as an important historic feature, a role also pursued in the various museums and libraries along the canal.

Historical Plaques

Generated mostly through the initiative of local groups, provincial plaques have recognized Lock 6 of the First Canal in Centennial Gardens, St Catharines, though the inappropriate surroundings of Indian totem poles bear no relationship to the canal history of that city. Somewhat better in location and purpose are plaques at their respective city halls that mark the founding of Port Colborne and St Catharines. Port Robinson is recognized close to a lock on the Second Canal as an important terminus of the canal where ships were constructed. A plaque to Louis Schickluna, the ship-builder, unveiled in 1979 close to his now buried shipyard in St Catharines, provides a further introduction to the industrial history of the canal.

At Thorold the site of the first cotton mill in the city and the province, initiated through cotton imported by the Second Canal, is memorialized. So too is the founding of the town during the construction period of the First Canal. It is regrettable that the latter plaque, though located next to the canal, is immodestly placed in a park that has been named in unbecoming fashion after the Battle of Beaverdams during the War of 1812, an engagement fought on land that now lies to the east of the present Thorold Tunnel. As at St Catharines, the city's history in relation to the canal has been misrepresented.

Certainly, the few plaques that exist provide some belated recognition for earlier events in association with the canal. They have a useful interpretive value, but they have not been assembled into an explanatory book as a tourist guide to features of interest. Nor do plaques replace the need to retain features that continue to be lost, and many important canal features lack similar appreciation.

Even the site of Merritt's mill is not recognized. This site on Twelve Mile Creek was purchased by Merritt in 1816. It became the Union Mills, using Lock 2 of the Second Canal, and then Welland Vale Manufacturing, Canada's largest axe company. Here, in addition to the vital Merritt associations, might suitably be located a display or exhibition to illustrate the crucial role that the canal as a purveyor of water power has played in the industrial evolution of southern Ontario.

Next to downtown St Catharines, the mill of Canadian Hair Cloth constructed in the late 1880s expanded into and in 1910–11 took over an earlier woollen mill of the early 1840s. Both were fed from the Hydraulic Raceway that had been provided by the Welland Canal Company. Here a plaque plus the excavation of a short length of the raceway to indicate its former dimensions might demonstrate the major role that this raceway has played in the industrial evolution of St Catharines: revealing, mechanically, how the water system operated and, economically, the types of industry that were attracted.

Plaques recognize the Shickluna shipyard and the Welland mills in Thorold, but most of the industrial heritage has yet to be acknowledged. For instance, the growth and modern character of the paper industry across the escarpment provides a theme that might suitably be presented for its educational values. As a leading company with important long-term, contributions to the Canadian economy, the paper industry has also made advanced efforts in pollution control.

The visually important landmark industrial buildings such as Lincoln Fabrics and Canada Hair Cloth require special recognition for their distinctive architectural expressions, which typically include the use of locally quarried stone that provides points of strong community identification within the urban landscape, as do the Morningstar Mill at DeCew Falls, where the canal's railway incline was intended, and Independent Rubber in Merritton. The distinctive housing provided in Welland by Plymouth Cordage for its workers is another notable landmark. Industrial history using the mills and the factories, the machinery and the equipment that have survived from the past provides points of interest of which greater account might be taken as urban development advances.

In terms of engineering, even the world-famous lock series across the escarpment of the Second, Third, and Fourth Canals remain without identification and explanation by federal or provincial plaques, although the Fourth Canal has been recognized by the Niagara Escarpment Commission. No guide or map is available that leads the interested visitor from one to the other of the canal attractions, and even the route of the Scenic Drive that supposedly follows the length of the canal does not pass by the closely spaced sequence of locks from the Second and Third Canal eras where they cross the escarpment.

The few historical plaques along the canal are greatly outnumbered by those in either Niagara-on-the-Lake or along the Niagara River Parkway. The disgrace of the current adverse situation for due recognition of the Welland Canals is clearly indicated when a federal publication about heritage canals is studied (Parks Canada 1982), where it is stated: 'Each year Canada's heritage canals are used by thousands of skaters, skiers, joggers, cyclists, sightseers ... and boaters too.' Those mentioned include, in Quebec, the Chambly and St-Ours canals built to overcome the rapids of the Richelieu River, the Lachine Canal near downtown Montreal, and the Carillon and Ste-Anne Canals on the Ottawa River route. Ontario is represented by the Sault Ste Marie Canal, the Trent-Severn Waterway, and the Rideau Canal, but the Welland Canals are not included.

Further, along these national heritage canals, many of their associated features have been preserved. Publications and displays extol their historic virtues. Administration buildings, museums, and visitor centres make use of former canal buildings, and there are exhibits of maps, photographs, films, and artifacts connected with the canal's history. Elsewhere marine railways and locks have been retained, repaired, and improved. Block-houses, blacksmith shops, and lockmaster's houses have been restored to the period of occupation and reused to house exhibits, often to house arte-facts of the period. The text of Heritage Canals is resplendent with state-ments such as 'an outstanding colonial engineering feat ... one of the world's foremost construction achievements ... popular recreational corri-dor.' Descriptions of canals refer to the fact that they offer 'a scenic route for pleasure craft' and that they are 'a major tourist attraction.'

The Welland Canals, though they are older, had to offset the much higher and more pronounced physical impediment of the Niagara Escarp-ment, and are of equal or greater merit as construction feats, are not repre-sented as a national heritage endowment. As a consequence many of their earlier features continue to be lost or to decay, or they are destroyed to make way for new developments. The reasons for this relative neglect of the Niagara circumstances include the costs of restoration, owing to the large size and number of the locks that are involved, and more especially that the fact that the present Welland canal as an active, functioning water-way is viewed as the star attraction rather than the historic waterways and their communities.

An imperative factor in this relative neglect is that the function of the Seaway Authority is not tourism, but the movement of ships. Tourism and amenity were also not part of the canal's design and construction, but were introduced at a later stage, almost as an afterthought. At the local level, the process of urban change has also been conducted through development-oriented municipalities, with heritage and conservation concerns and the aesthetic quality of development not being high on their list of priorities. In the pursuit of business interests, canal preservation is viewed as a sentimen-tal concern.

The Canal Scene

The modern canal offers the fascination of passing vessels. The public has a particular interest in visiting the locks, and especially the Flight Locks, the major drawing card in the canal corridor. Yet, even in this example of the canal as a prime tourist resource, the main public viewing platform has been located on leased Seaway land at the single Lock 3 operation, rather than at the more magnificent double series of triple Flight Locks. At both locations the Seaway has introduced chain-link fencing to reduce its public liability and visitors are excluded from close proximity to the locks for safety reasons. At the Flight Locks it is not now easy for a spectator to photograph the scene, as a wire-mesh barrier has been erected, and private automobiles are not allowed to use the canal's service roads. Precautions introduced by the Seaway have, in part, detracted from public enjoyment of the scene.

Along the canal, much remains to be achieved. With worries about trespassing and possible litigation should an accident occur, areas next to the canal have been closed to the public, and public access has been restricted by chain-link fencing, gates, and notices that state 'No Admittance' and 'No Entry' for either people or vehicles. By 1995 these restrictions applied with particular severity to the two piers at Port Weller, along the east side of the old Fourth Canal between Welland and Port Colborne, and along the west side of the Fourth Canal next to the By-Pass Channel. Prosecutions under the St Lawrence Seaway Authority Act have followed (Egerter 1995). The Seaway is technically correct in its actions, but the canal banks are viewed by the public as open space that may freely be entered. The legal issue of possible liability is in conflict with the tourist-amenity aspects of the canal.

The anomaly that has worsened over the years is that access along the banks has been provided by the Seaway to service vessels and engineering structures as might be required, but not to the public. Signs stating 'Use at your own risk' and advising cyclists to walk their bikes over bridges have attested to the willingness of the Seaway to ignore its technical right to deny admittance. Meantime, for years there has been advocacy for a lake-to-lake parkway along the canal (chapter 19), with a continuing argument that federal ownership should provide this opportunity. Here is an area where the conflict in attitude between the various levels of government must somehow be reconciled to meet the often expressed need for public access to and along the canal banks.

A few lengths of the canal service roads including Government Road and sections of the Haulage Road in St Catharines were surfaced in the early 1960s and later successively became the property of the municipality during the 1980s and early 1990s, but similiar actions did not extend to the roads along the canal on both sides north from Port Colborne. The abandonment of the Fourth Canal through Welland when the By-Pass opened and the renewed idea of a Welland Canals Parkway will ensure that consideration be given to surfacing these routes as highways for public use.

Municipal and Regional Actions

The municipalities and developers have made several contributions towards the conservation of the canal heritage, but this story is incomplete and often harmful to canal interests. Every canal community can record the loss of buildings and canal structures that might, but not always could, have been saved. Taylor (1992b) recorded this attrition in St Catharines. In addition, many surviving buildings have been so altered as to be unrecognizable in their original form and purpose. For example, both St Catharines and Thorold once had several lock keeper's cottages of 1840 vintage across the escarpment, but only two now survive. The restoration abilities exist, but they have not usually been applied in a developer-oriented economy where the local historical societies are more discussion than action groups.

Nor have public actions proceeded with any alacrity. At Port Dalhousie the two sides of the outer harbour remain unconnected. An intended fisherman's wharf on the east side of the harbour next to the Port Dalhousie Yacht Club has not materialized, and neither road nor boardwalk nor park space encircles the harbour close to water's edge to link all activity areas as proposed (Lanmer 1984). The city has enclosed the beach area with visually unexciting chain-link fencing. Nor has a floating boardwalk across Martindale Pond to link with the second lock of the Third Canal and to replicate the towpath of the Second Canal been constructed. An interesting development, and a reminder of the canal's water-power function, is however the construction of a small hydroelectric power station in Lock 1 of the Third Canal (St Catharines Hydro-Electric Power Commission 1986; Woolton 1980).

At Port Colborne an outdoor marine museum promoted by the Welland Canals Foundation has not advanced. This project has remained dormant despite the provision of a site by the Seaway Authority, a location on the main road (Highway 3) through the city, the availability of artifacts, and the preparation and costing of alternative designs. At Welland, after the By-Pass was constructed, the intended planting scheme was not initiated until twenty-five years had elapsed.

Some heritage projects are ill designed. An example is the lock at the Junction where the Feeder Canal met the main line of the canal at Welland and the former summit level of the canal system. A small park has been set around this lock, but the stonework is not visible, the outline of the lock is unclear, the site cannot readily be seen or approached from the adjacent highways, and the scant information provided on the plaque reflects neither the importance of this location within the canal system nor the crucial role of the Feeder Canal in the canal achievement. Vandalized seats testify to the disregard paid to this site and the need for its redesign.

In St Catharines, where Government Road follows the canal and heralds a parkway, the route has been treated by the municipality more as an arterial highway than as a high-calibre scenic route. Where continuity along the canal should prevail, Government Road has been turned away from the canal at Queenston Street to become a traffic intersection with lights, whereas previously a separated traffic system existed, as across the Niagara River Parkway on the entry into Canada over the Rainbow Bridge at Niagara Falls. Further, low-quality commercial development lines Queenston Street, the major access route to the canal parkway from the QEW, and to the north the continuity in the flow of recreational vehicles has been reduced by traffic signals that impede the flow of traffic along the canal. Despite the Parkway's evocative name, emphasizing scenic, landscaped open space, in both instances engineering rather than amenity considerations have taken precedence.

In the official policy plan for the Niagara Region, the Welland Canal is presented as one of 'the primary areas for the provision of recreational, open space and park areas in the Region. These areas form an interlinking network of corridors, which must be protected both to meet the increasing desire of residents and visitors for recreational and scenic areas and also as a continuing and irreplaceable natural heritage.' Later it is stated that 'the Welland Canal has potential as a continuous open space corridor. Similarly the stream valleys of the Region have great recreation potential as well as ecological value. It is important that they be recognized and that development in these areas be carefully controlled before they are lost forever.'

As these sentiments are also reflected in the municipal plans, a slowly changing attitude towards the Welland Canals may be discerned since the late 1960s. Heritage contributions have in particular been made at Port Dalhousie and Port Colborne where, in both instances, the old waterfronts, their commercial buildings, and former canal features have played an important role in urban renewal initiatives. The same argument applies at Welland, where recreational uses have been encouraged along the abandoned Fourth Canal. Mud Lake to the south, where spoil from the Fourth Canal was dumped to encourage wildlife, was leased to the Ministry of Natural Resources and then to the Niagara Peninsula Conservation

Authority, which, in conjunction with Ducks Unlimited, has developed a respected conservation area with trails around the lake and bird-watching facilities. Another example of conservation is that in 1988 Thorold purchased property in Port Robinson containing a lock of the Second Canal for the lock walls to be retained as a visual attraction within public open space.

Although no major exhibition centres cover either the canal's marine or its industrial history, the museums at St Catharines, Welland, and Port Colborne have instructive displays that depict the history of the canal and that community. Most canal centres have walking tours of their urban environments featuring salient architectural and historical sites. The municipal libraries have extended their local and canal collections, as has also Brock University through its Special Collections and the Map Library.

Marinas as a reflection of the canal's marine heritage have been introduced in the harbour and inside the canal breakwaters at Gravelly Bay in Port Colborne (Fenco 1986) and at Port Dalhousie within the outer harbour and to the east of its protective breakwaters. Viewing platforms of the canal and its locking procedures have been located at Lock 3 in St Catharines and Lock 8 in Port Colborne. Both are important visitor attractions, and the viewing area at St Catharines has helped to draw the city's museum, a Lacrosse Hall of Fame, and the possibility of expanded tourist facilities to that location, which in 1993 drew about 450,000 annual visitors.

In 1976 the neglected area of the former nurseries and greenhouses in St Catharines that in the 1930s had been created to landscape and to act as a windbreak for the Fourth Canal was leased from the Seaway to the city. Renamed Malcolmson Park after Mary Malcolmson who founded the city's first Canadian Girl Guide troop in 1910 (rather than after a canal personality such as A.J. Grant who completed the Fourth Canal), the area was purchased in 1979 for transition to a natural area as a city park. A canal feature had been retained, but in a rather accidental manner.

Many Actors in the Decision-Making Process

Previously, an abandoned length of canal might be transferred to its new owner, usually the municipality, for a nominal sum. However, society had become much more complex when the By-Pass was constructed around Welland and when former canal lands were declared surplus to Seaway needs in the early 1990s. The transition from unified ownership under federal management to new activities now involved many different departments at the various levels of government and their often diverse interests.

As an indication of this administrative complexity, when in 1980 Parks Canada in conjunction with the Ontario Ministry of Culture and Recreation and the Ontario Ministry of Natural Resources prepared a concept plan for the systematic conservation, interpretation, and public enjoyment of the canals' heritage resources, nine federal departments were identified as having major programs that might contribute towards the implementation of the plan. In addition, the Department of the Environment offered four different program possibilities and the Department of Public Works another two. At the provincial level, nine departments, each with their sub-programs, were then noted, together with six municipalities and several private agencies (Parks Canada 1980).

And when the Ontario Ministry of Municipal Affairs in 1987 listed the approvals required for work on waterfronts, thirty-four agencies were enumerated. 'Waterfronts are some of the most regulated of lands, largely because of their transport function and their location at the interface of land and water. The authority over the surface of navigable waters ... rests with the federal government. Authority over the use of the lake or river bed rests with the provincial Ministry of Natural Resources. The control of land uses on the shoreline rests with the municipalities.' Table 11 introduces the range of interests that may be involved when the future of Martindale Pond is under consideration.

Canada through its legislation and through its several levels of government has erected a formidable array of rules, regulations, and policy determinants that have somehow to be coordinated if development is to proceed. The leadership role in any large-scale undertaking is often uncertain, and divisions by functions within the same department of government do not assist the process. Criticism of this bureaucratic jungle has increased greatly from the 1960s into the 1980s, leading to the reform of the provincial planning system that by the 1990s had 'evolved into a difficult, complex and complicated system that held back millions of dollars of viable and environmentally sound development projects' (Ministry of Municipal Affairs 1992).

The problem of how to proceed, or even where to start, is aggravated by periodic elections at each level of government. Their outcome might change any program, the rules and regulations that are involved, the availability of supporting grants, and the officers and elected officials with whom one has been dealing. Consistency and continuity over an extensive period is difficult to achieve as the relevant acts, budgets, policies, and priorities may each be changed. This continuity becomes even more difficult to achieve when different options and possibilities exist and progress has to

TABLE 11 Departments and government agencies with an interest in development around Martindale Pond (1980)

Agency	Responsibilities with examples
Federal	
Public Works	Leases to public and private agencies including Canadian Henley Rowing Corporation; City of St Catharines for Lakeside and Rennie Parks; Ontario Hydro for control dams and the spillway between the pond and harbour; St Catharines Seaplane Services; Dalhousie Yacht Club and Snug Harbour for the operation of marinas; Lincoln Fabrics for their factory building; Bell Canada for underwater cable access; Provincial Gas for underwater gas lines across the pond; and various private citizens.
	Maintaining lands inherited from the Seaway Authority and other sources in a park-like manner on a permanent basis.
	Disposal of surplus lands via priorities to the provincial government, then municipal government, and finally to private companies or individuals.
Department of Fisheries and Oceans	Federally owned harbours including the construction of breakwaters and dredging the harbour; and providing wharves and ramps for recreational boating.
	Fisheries and the collection of usage charges.
Department of Transportation and Communication	Aids to navigation, including lighthouses.
Department of the Environment	The flow of water from Lake Erie to Lake Ontario through Twelve Mile Creek, including some jurisdiction over Ontario Hydro.
Canadian National Railways	Successor to the N.S.&T. streetcar service; certain portions of the abandoned right-of-way have been retained, and sections leased to the Welland Canals Preservation Association to achieve the Merritt Trail.

(continued)

TABLE 11 (concluded)

Agency	Responsibilities with examples
Provincial	
Ministry of Natural Resources	The goal is to provide opportunities for resource development and outdoor recreation and to manage and conserve public lands and waters, including a forestry program and the management of fish and wildlife populations.
Ministry of the Environment	Control over industrial pollutants, including the discharge of waste-water effluent into the Second Canal, private and municipal water supplies, sewage, and water treatment.
Ministry of Culture and Recreation	Heritage resources might be affected.
Ministry of Transportation and Communications	Major highways including the QEW, its interchanges, exits, and signage as the major approach route to the old and modern canal systems.
Niagara Peninsula Conservation Authority	Conserving, restoring, developing, and managing the natural and heritage resources, including a flood control program, storm-water regulation, wetland preservation, and the management and reforestation of private lands.
Ontario Hydro	Control over the flow of water and levels in Twelve Mile Creek and Martindale Pond through its use as a spillway for the Decew Power Station.
Region	
Regional Municipality, City of St Catharines	Representing the interests of private organizations, owners, and groups such as the Port Dalhousie Quorum.

Source: Based on McCollum 1980

Note: When an interdepartmental meeting to discuss surplus lands was held in 1976, federal representation included the Ministry of State for Urban Affairs, Parks Canada, Department of the Environment (Lands Directorate and Small Crafts Harbours), Central Mortgage and Housing, and Department of Public Works.

be steered through different layers of committees and many sectors of government, as when the abandoned length of the Fourth Canal at Welland became available for other types of development.

The change in administrative control upon redevelopment is from the unified operations of the Seaway Authority as a powerful single-purpose federal agency and the largest landowner in the Niagara Peninsula to the municipal realms of competing authorities and the very large number of government agencies that have been alluded to at each of the federal, provincial, regional, municipal, and local levels of jurisdiction. The unitary powers of the Seaway Authority, with its prime focus on marine transportation and an internal decision-making process not subject to public scrutiny and with the powerful backing of federal funding has been replaced by a complex medley of many diverse actors on the scene (table 11). Now, with no one agreed script, but rather a public process where the many different sets of possibilities had somehow to be agreed upon then formulated for action, without necessarily any commitment to the required funding by senior levels of government, the decision-making necessarily became slow and cumbersome. To this was added a series of planning and design consultants, with their burgeoning reports (which will be examined in chapter 18 for Welland and chapter 19 for steps towards the potential achievement of a parkway along the canal).

18

The Reuse of the Abandoned Fourth Canal at Welland

The transfer in ownership and management from an active international waterway to a passive urban recreational–open space resource is no easy task. It involves many distinct and often competing interests within government, and also differences in attitude as the various possibilities are examined and made subject to public scrutiny. This chapter introduces the administrative, technical, and civil aspects of change at Welland upon completion of the By-Pass. The progress or rejection of proposals for development along the abandoned channel were complicated and time consuming. Each had to weave its way through the interlocking maze of departments at the four levels of government, and every government agency with an interest in land use, planning, engineering, conservation, environment, parks, recreation, transportation, and highways made its respective inputs. Citizen participation at each level and over each concern also played a considerable role as each idea or project advanced or failed to advance through the administrative machinery.

The New Opportunity at Welland

The possibility of reorganizing and invigorating the urban environment of an industrial city divided into distinct sections by its river, canal, and railways through the relocation of the canal to the east and the railways to the south has been a topic of concern since the By-Pass was announced in late 1965. The Institute of Land Use at newly founded Brock University offered research assistance to the Seaway Authority to help in resolving the urban, recreational, and public open-space issues that might then be foreseen. However, the emphasis in discussions between the City of Welland and the Seaway centred on the By-Pass, where a second highway tunnel and a dock were obtained. The future use of the abandoned channel might be envisaged, but nothing could be undertaken on the ground until navigation ended, and not until the early 1970s did intensive discussions start with respect to the abandoned channel.

As stated in a consultant's report, 'there is felt to be great potential for improving the city's outward appearance as well as the sense of civic pride, by re-utilizing the abandoned canal and rail lands to create a pleasant and urbane centre. The opportunity for shaping the city's future history has been called "a life-time chance" for Welland ... The canal has been a source of civic pride, but not, in the ordinary sense, a source of recreational pleasure. This abandonment of the canal however offers great potential for improving this feature of the town' (Diamond and Myers 1975).

Twenty years later, the outcome included some successes, but there had also been a considerable degree of frustration and disenchantment. As recorded in the Welland *Tribune*, 'The city's claim to the old Welland Canal lands has drowned amidst more urgent fiscal pleas, pouring years of studies, hopes and dollars down the drain ... For two decades, local municipal and federal politicians have dangled the carrot of the abandoned canal lands before its voters ... We're going for the Guinness Book of World Records with these negotiations' (Putter 1995).

Should the Abandoned Channel Be Retained as a Water Body?

An issue in early 1970 was whether or not to retain the canal as a water body. The Ontario Department of Highways and the regional municipality wished to drain the channel and use its bed for the southern length of Highway 406, the intended link along the Welland Canals Corridor between Port Colborne and the QEW at St Catharines. This route at that time reached only to the north of Welland and its continuation to Lake Erie was via Highway 58. The extension of Highway 406 along the canal bed was supported by the Seaway Authority. As the authority's prime desire was to dispose of the now unwanted channel, its transfer to the provincial government for highway purposes provided a simple solution. Only a single transfer of property ownership and management between the two senior levels of government was required. As the automobile was very much in the ascendancy in the 1970s, this solution would contribute materially towards the solution of Welland's severe problems of traffic circulation. Public opinion, however, averse to this automobile-oriented solution, strongly favoured the retention of water in the abandoned channel and the redevelopment of the canal as a Recreational Waterway.

When policy changed to retaining the former canal as a water body, now in two separated sections divided by the western railway cut to the Townline highway-railway tunnel, the Seaway at first intended to retain standing water in these two reaches and to treat these closed lengths with alum to maintain their quality. The city objected and insisted that a through flow of unpolluted water be retained in each of the reaches, a solution that they argued could be achieved by carrying water over, above, or around the Townline Cut in a pipe or syphon.

The city's idea was also that the canal might be filled partially along its sides and the banks sloped to leave a series of small lagoons where recreational areas might be provided (Welland *Evening Tribune*, 27 and 28 November 1972). This solution was found to be difficult, costly, and complex due to the depth of the bed, the effect of the canal within its banks as a stabilizer of groundwater, and the condition of the banks (Trow 1975). Except at span and causeway crossings, filling was not recommended.

The solution to the water situation also involved consideration of the syphon. Its normal flow was negligible, and in quantity and quality the river was unable to supply the waterworks. However, in severe weather conditions such as Hurricane Hazel, the peak flow created a differential across the syphon from west to east of about 5 feet (1.5 m). The removal of the syphon would reduce the danger of upstream flooding, a gravity water pipe from the canal could supply the waterworks, and a causeway across the old canal would relieve traffic congestion at the town centre. This would represent a return to the circumstances before 1829, when neither a syphon nor an aqueduct existed.

The issue of chemical treatment was resolved when experiments in an abandoned length of the Second Canal south of Thorold concluded that treatment was unnecessary because of the depth and width of the channel (Shannon and Vachon 1973; Albery et al. 1975). In addition, water in the reaches at Welland was pulled to industrial intakes, and churning action from the propellers of vessels and the wake of their passing across the necks of the abandoned canal created a flow of water into or out of both ends of the abandoned channel. The water would not become a stagnant pond as had at first been feared. Rather, the two curtailed lengths of the canal might each be appreciated as long, deep lakes that were capable of renewing and maintaining themselves on a four-day turnaround basis.

The solution adopted for the supply of water to industrial and municipal consumers was ingenious. The Townline Cut blocked the flow in the southern length, but this did not affect the supply to John Deere. To supply the northern and the central lengths, including the critical supply to the waterworks, the new By-Pass was used to carry water *north* to Port Robinson; a reversed flow was introduced with the abandoned channel carrying water back *south* to the centre of Welland. This reversed flow was used to supply Atlas Steels and the waterworks, and holes were drilled through the top of the syphon to supply the Welland River with its required diversion water to offset pollution.

Highways, Bridges, and Traffic Circulation

Severe traffic congestion existed on the approach roads to the business centre. Only three bridges here crossed the Central Reach of the canal between the river syphon and Townline Road. Main Street carried three lanes, two out from the business district and one for the inward movement of traffic, and the approach from the west was also poor. The Lincoln and Ontario-Broadway Bridges were narrow lift-bridges constructed in the early days of the automobile. Traffic flows were aggravated by the outward movement to department stores in St Catharines and then to the Seaway Mall in South Welland, in both instances drawing residents from the east to the west side of the canal. Vehicles from industries east of the canal, the majority with trucking connections to the QEW, had also to pass through the centre of the town, and likewise many internal work trips from the new residential areas in the north to industries east of the canal had to pass through the business centre.

After studies and public reaction had demonstrated that Highway 406 as part of the regional arterial grid system could no longer follow the abandoned channel, discussion focused on two alternative routes for this highway, one to the east of the city next to the new By-Pass and the other across the western suburbs (Acres 1974). The alternatives were evaluated for their respective contributions to the city and, after intense public debate, the western route next to the By-Pass was selected. This decision meant that both the abandoned channel of the Fourth Canal, the adjacent Welland River, and the intermediate Merritt Island had each to be crossed by a new highway bridge or bridges.

Pending the achievement of Highway 406, and as no highway bridge crossed the canal between Allanburg and the centre of Welland (Bridge 12 at Port Robinson had been destroyed by a freighter), the solution was to extend Woodlawn Road west from Highway 58. Like Highway 406, Woodlawn also required a bridge to cross the barriers to movement presented by the abandoned channel, Merritt Island, and the Welland River.

Recalling now that the abandoned canal was at a higher level than the river and set within spoil banks along its western side, the provincial Ministry of Transport at first intended a low bridge 9.8 feet (3 m) above the water for Highway 406. As this height restriction would have limited the recreational use of the channel below by vessels and jeopardized the construction of a parkway next to the abandoned canal, after an intensive Raise-the-406-Bridge-Campaign the ministry in 1990 yielded to public pressure, and a bridge at a height of 18 feet (5.5 m) was designed to cross the old canal, Merritt Island, and the Welland River as one continuous unit. This more costly decision also delayed the completion of Highway 406 until 1995, and Woodlawn, where no height limitations were involved, was the first of these two bridges to be completed.

At the town centre, Bridge 13 on Main Street through public acclaim was retained in a fixed down position. The twin towers and the bridge's lattice work continued to dominate the urban scene and to provide a significant focal point for the central business district. To relieve traffic congestion, an additional highway crossing was added at Division Street. The Main and Division Street Bridges now each operated on a one-way basis in conjunction with the pattern of traffic circulation in the urban core (Damas and Smith 1975).

To the south, the causeway or plug on the approach to the Townline railway-highway tunnel was used to provide an extra highway crossing of the abandoned canal. Division Street Bridge (1981), Lincoln Street Bridge (1982), and the Broadway-Ontario Street Bridge (1995) were each conceived as a causeway and span, the municipality urging four-lane crossings to encourage the recreational waterway and pedestrian flows along the canal banks (Parker 1988).

Traffic circulation within the centre of Welland was eased greatly by this provision of new crossings over the abandoned canal, recreational potential along the canal banks was enhanced, and the Woodlawn–Highway 406 provision relieved the previous difficulties of access to the isolated promontory of land extending north from the centre of Welland between the old and the new channels that was bereft of approach roads except from the city's business centre. The city had gained. It now had a much freer traffic flow in its business centre than before the By-Pass was constructed, and more land was available for development than previously through the improved access.

The Disposal of Former Seaway Properties

In 1974 the abandoned sections of the old Fourth Canal were transferred from the Seaway to the Department of Public Works (DPW) as the department responsible for the management and disposal of excess federal properties. DPW set up a tri-level committee that included the federal and provincial governments and the city's representative on regional government. The assumption was that the latter individual could represent the interests of both the city and the region, but the committee soon became a four-level committee on which both the municipal and the regional levels of government were represented. The terms of reference were 'to examine ways and means by which the long term public interests in the properties, and redevelopment alternatives might be identified' (Technical Advisory Committee 1975). Their task was to take over the administration and control of former Seaway lands through Welland pending the sale of these assets, but this committee was dissolved when DPW decided that surplus canal lands would be disposed of through individual agencies, including municipalities and developers, rather than through the mechanism of a joint intergovernmental agreement (Parks Canada 1980).

The Welland Canal Advisory Group was then established by DPW in 1981 as a citizens' advisory board. Its tasks were to make recommendations on the long-range recreational uses of the former canal, the impact of the proposed uses on water quality in the canal, and the administration for the future development and use of these properties. Their recommendations included that 'no land will be sold for future commercial or private development' and that 'there will be complete public pedestrian and bicycle access along the entire canal' (Wright 1982). The waterway north of the Main Street Bridge was to be reserved for non-motorized recreational facilities, and all future canal crossings were to be open spans rather than causeways. The suggestion that Main Street Bridge be preserved as a historic landmark was supported, as were many other specific recommendations including water skiing in the south and landscaping along the banks and as a buffer for industrial uses.

Land was developed in accordance with the group's land-use plan. Primarily on Merritt Island, public access was improved with roadways, parking facilities, pedestrian walkways, boat ramp facilities, playgrounds, information signs, benches, picnic areas, lighting, and washroom facilities. Merritt Island was landscaped, upgraded, and managed at the level of an urban park.

The advisory group was replaced in 1984 by a group that in 1985 was named the Welland Canals Parkway Development Board (WCPDB). Now set up to advise DPW on matters pertaining to the sale and development of the former canal lands, it advertised some 1004.4 acres (406.5 ha) for sale in

1985. Its activities included the review and update of the conceptual plan completed by the Welland Canals Advisory Group, a focus on the alternative methods of implementation, and a revised mandate to promote redevelopment of the lands and to develop terms of reference for the sale or lease of land.

Meantime, the city in 1985 went 'on record as being interested in acquiring surplus St Lawrence Seaway Authority lands,' and established its own Welland Canals Land Review Committee of senior officers and officials to work with and to review all work undertaken by DPW and WCPDB. The latter was again reorganized in 1986, when the focus changed to one that encouraged the commercial use of property that included housing, institutional, commercial sports, and commercial uses. Landscaping and amenity improvements were then de-emphasized. Conflicts and differences of opinion now emerged between the contrary ethics of development and conservation.

The Seaway retained the authority to maintain an adequate supply and level of water flow in the channel. It also retained ownership of the land at each end of abandoned channel as canal reserve areas so that the Seaway might continue to lease land along these marine slips for the winter berthing and repair of vessels, and for industrial clients who might wish to locate there and use the canal for transportation purposes. By contrast, Welland viewed these areas and the total length of the abandoned channel as potential green space for public open space–recreational pursuits.

These differences in attitude became marked in 1989 when DPW prepared a marketing study that proposed the sale of certain crown lands along the old Fourth Canal for development (Proctor, Redfern, et al. 1989). As these lands were zoned as open space in the city's official plan, redesignation would be required at both the regional and municipal levels of government if some alternative type of development with a higher market value was to be obtained. The marketing study recognized thirteen 'Opportunity Areas ... [where] Site developments have been carefully selected to improve community recreational resources, to provide a broader base of housing stock and to take advantage of opportunities to promote tourism. It is anticipated that the developments proposed will stimulate redevelopment on adjacent private lands. Public access has been maintained throughout to a minimum depth of 10 metres [32.8 feet], and notwithstanding of any development proposed on adjacent lands existing paths and walkways will be maintained as they exist.'

This study and the attitudes it represented were rejected totally by the city. 'The implementation of the government scheme would *not* be in the best interests of the City and corridor communities. In fact, the effect of several of the proposals would be in direct conflict with the City's and Region's long term planning objectives to preserve an invaluable heritage resource, open space amenity and tourist development potential resource. Some of the development proposals would substantially reduce public access to the Old Canal and significantly change the character and ambiance of the Old Welland Canal as a recreational waterway through the City of Welland ...

'Equally disturbing was the attitude in the report ... that the old Welland Canal Lands can be viewed as a local issue and disposed of in isolation, without regard to the negative impact on the future development of a Welland Canal heritage and tourism corridor between Lake Erie and Lake Ontario.

'... It was felt that a more long term approach to the development of these lands should be taken as opposed to the fragmentation of our valuable heritage resource through disposition of various parts of it, without consideration of the negative impact on the future potential of the remaining lands and waterway' (Welland Planning and Development Department 1989).

The City of Welland then passed a resolution 'not to endorse the Public Works Canada [DPW's new title] Report, ... as this fragmentation of selected sites from the overall canal land area would drastically diminish the opportunity for the creation of a long term plan for the retention of a Welland Canal Heritage and Tourism Corridor between Lake Erie and Lake Ontario, and the establishment of such a major heritage resource would have far greater economic and cultural benefits to today's and future residents of Welland, the Region of Niagara and of Canada, than the immediate monetary gains from the sales of these selected sites.' As this attitude was supported by the regional municipality, the issue was clearly that of development versus obtaining the best recreational–public open space use for the land, with Welland at the same time seeking 'to authorize a transfer of ownership of the Old Welland Canal lands from Public Works Canada to Parks Canada.'

When the PWC proposals were subjected to environmental assessment procedures, the report argued that 'the potential negative impacts focus mainly on the physical factors associated with the proposed development. They include such impact as degradation in water quality, settlement of land, visual impact along the canal, reduction of access [to the canal banks] and destruction of historical archaeological resources.' These issues 'are largely, if not wholly, mitigable through architectural and engineering design and adherence to current regulations and standards.' Even so, 'conflict with local and regional planning standards remains an issue. These will have to be addressed as part of the planning and development process for each site, as will the potential loss of existing "valued" land uses and decreased possible access along the Old Canal' (Environmental Services 1990). Positive results such as new recreational opportunities for the canal, the potential for improved aesthetics along portions of the canal, and possible long-term increase to the economic potential of the area were also identified.

The focus of WCPDB again changed in the early 1990s to one that would facilitate the transfer of former canal lands to the City of Welland, and in 1992 the Treasury Board announced a $4 million compensation package to the city. This offer was rejected by the city and WCPDB was disbanded (City of Welland 1992). As argued by the city based on financial-impact studies (ARA 1991; Sandwell 1991), 'the costs of required repairs, rehabilitation and maintenance of this major land area, which consists of 686 acres [277.6 ha] in a 4.5 mile [7.5 kilometre] long strip approaches $23 million if these works and maintenance are performed in accordance with City parks' standards, or $12 million if in accordance with Public Works Canada's standards, which we believe is inadequate.' The costs assumed that the Seaway would maintain ownership of the waters and two sets of financial-impact figures were examined: the PWC and the Proctor and Redfern / City Planning perspective in the Land Use/ Marketing Review of 1989. Expected costs included physical liability for possible collapse or replacement of the aqueduct, the repair and maintenance of the canal banks and their retaining structures, pollution of the canal water, and responsibility for contaminants in the canal. Works and maintenance of the now landscaped Merritt Island were also included, together with a road on the western side of the canal between Welland and Port Colborne.

The city then requested a lump sum of $25 million, or $3 million a year for 10 years to cover these developments and for a partnership to achieve recreational facilities. 'The best way for the City to ensure the property of the old canal lands would be for the City, its sister municipalities along the Old Canal, the Region and the federal Ministries of Environment (Parks Canada), Public Works, Employment and Immigration, Small Business and Tourism, Fitness and Amateur Sports to form a partnership to finance, promote, and operate the old canals for outdoor recreation and tourism pursuits' (Neathery 1993).

The justification for this approach was based on the region's severe economic depression, its high rate of unemployment, and the increasing welfare rolls. 'Many of our unemployed workers will require retraining and education upgrading in order to acquire marketable skills needed to find employment ... We believe that one way to provide new employment now, is to repair and rehabilitate sections of the old Welland Canals and either develop or provide the opportunity to develop outdoor recreation, historic and tourism facilities along the former waterways.'

Based on a series of studies (Coopers and Lybrand 1993), development proposals on canal lands in Welland included a world-class rowing facility that would attract 30,000 people a year and 50,000 to major events such as world championships or university games. A KOA family-oriented campground with 210 sites could achieve a 65 per cent occupancy over a 150-day camping year (Bodo 1990), and demand existed for a 240-slip marina facility (Ogilvie 1989). A golf course would capture some 31,000 rounds of golf a year (Doral 1990), and potential existed for a 200-room hotel. These possibilities were based on the Recreational Waterway as a major basis for Welland's changing urban future, but the request for this funding was rejected by the federal government in early 1995.

Although the By-Pass Channel had opened in 1973 only eight years after the acquisition of land, nearly three times that length of time had elapsed and the urban situation still had many uncertainties as a legacy from its canal past. History would seem to indicate that it is easier and quicker for engineers backed by powerful funding and working within the confines of the federal government to construct a bypass than it is to achieve community interests in urban development through the multiform layers, regulations, ambiguities, and conflicting interests of modern government. Change from an active commercial waterway has been slow, painfully slow. The By-Pass was completed in a few short years. By early 1995, over twenty years after this construction feat, many of the urban repercussions were still under scrutiny.

Development along the Canal Banks

Administratively, the Northern Reach of the abandoned canal extended into Thorold, the Southern Reach into Port Colborne, and only the Central Reach lay wholly in Welland. This channel had no locks, but its structures included the syphon to carry the Welland River under the canal. The decommissioned length also included a drainage channel from the Feeder Canal to the Welland River, and the channel itself fell readily into a Northern Reach between the syphon and Port Robinson, a Central Reach between the syphon and the Townline Cut, and a Southern Reach from this cut to Ramey's Bend.

These reaches were different in character and each offered different sets of opportunities. The Northern Reach included spoil heaps excavated from various canal channels on the west bank, and the elongated Merritt Island between the abandoned channel and the Welland River. Atlas Steels drew water from this length, as did the Welland Waterworks to supply the city with its domestic water supplies and dilution water for the Welland River. The canal was also crossed at its northern end by a new CNR crossing as part of the rerouted railway system, and later by the Woodlawn and Highway 406 bridge crossings, which effectively eliminated navigation between the new canal and Welland's city centre. An open flow of canal water had of necessity to be retained as the Welland River had neither a sufficient flow nor the purity that was required to supply Atlas Steels and the city's waterworks. In the city's official plan, 'The Canal banks and associated lands are designated as open space and will be developed for park, active recreational, passive open space and wilderness purposes' (Proctor and Redfern 1972; see also Pearson 1965).

The Central Reach was where the Feeder Canal had crossed the Welland River to introduce Welland's urban history. The town then grew industrially and commercially out from this focal point that became the city centre, and then south along the canal banks as new industrial enterprises were attracted to the growing town. Economic activities included two industrial docks and the supply of industrial water to the plants of Union Carbide and Page-Hersey Tubes, which lay next to the canal.

The city's planning strategy was 'to concentrate first on the downtown area in order to attract new commercial development, and to try to provide a stimulus for further interest. It is also intended to secure an early improvement in the [highway] crossing of the canal in the vicinity' (Proctor and Redfern 1972). As another requisite, the city's plan wished to maintain an open-space link along the canal, with residential and associated commercial and recreational uses to the south.

Along the Southern Reach, the west bank carried the Mud Lake Conservation Area within an area of former spoil dumps; the east bank, now an isolated spur of land between the new By-Pass and the former channel but previously on a through road to Port Colborne, had become a somewhat lengthy residential cul-de-sac. Bridges 17 and 18 for Forks Road and a railway crossed the former canal. The retention of a marine slip by the Seaway Authority was viewed as an encouragement to industrial development on the west bank at the Port Colborne end of the abandoned channel. John

Deere was also supplied with industrial water from this reach, which the draft official plan viewed as 'an area of mixed residential and industrial uses ... The Canal will be maintained ... as a passive open space feature with landscaped banks, picnic areas and trails and will be integrated into the general open space system of the city.'

All three reaches were designated by the city as a continuous Recreational Waterway in accord with the city's official plan, which stated: 'The whole length of the old canal including all three reaches is viewed basically as a linear open space feature.' Steady progress towards creating such a waterway after the abandonment of the canal channel for commercial purposes included its use for rowing, canoeing, and water skiing events during the summer months. Boat ramps were provided, together with new opportunities for swimming, cycling, hiking, and picnicing. Designed areas of public open space with walkways were also created on both banks, including on Merritt Island, which became a public park.

The South Niagara Rowing Club leased facilities at Dain City from the Seaway, and that section of the old Welland Canal became designated as the site every second year for the Ontario Rowing Championships. The possibility of extending this course south for 328.1 feet (100 metres) also exists. On the Central Reach to the north of the Townline Cut, the Welland Water Ski Club took advantage of the sheltered, straight stretches of water for the rowers and used the banks as vantage points for spectators.

A considerable degree of urban advance and improvement has also taken place along the west side of the Central Reach. Condominium, high-rise residential projects and a multi-storey hotel have been introduced between Prince Charles Drive and the landscaped extent of public open space created along the Recreational Waterway. These new developments often have favoured views towards the waterway and its open-space setting, as well as pedestrian access to the scenic amenities provided along the canal bank. Much advantage has been taken of the reinvigorated waterway to provide a positive new milieu for modern urban developments in the vicinity of the former canal.

On the east side of the canal next to the city centre, a public promenade has been created next to the canal north from Main Street Bridge and a green setting provided for the preserved stone aqueduct of the Second Canal. North of the syphon, this green space connects with the public open space of Merritt Island, where a paved recreational trail was constructed for the full length of this island.

Within the city centre, colourful Giant Murals on the walls of buildings have depicted the history of the canal in relation to the city (Gibbs 1991).

Their size is grandiose, some being 80 feet (24.4 m) long and others three storeys high. The first mural was commissioned in 1986 by the owners of the Seaway Mall on Niagara Street south of the city centre, but by 1991 their number had increased successively to twenty-eight, four in the vicinity of the mall and twenty-four in the downtown area. The idea came from Chemainus on the east coast of Vancouver Island, where a declining lumber community redeemed its failing economic base and became a tourist attraction through a similar initiative.

At Welland the murals depict many scenes associated with the canal and the city. A link with Chemainus through a painting by Dan Sawatzky depicts the arrival of immigrants for canal construction circa 1910, then the new factories offering permanent jobs and a reason to settle in the town. Plymouth Cordage and its female labour force is portrayed as Welland's first major industry, and Morwood's store as a major retail outlet active on Main Street for over a hundred years. Towpaths, lift-bridges, the Townline Railway Tunnel, tugboats that worked on the canal in the 1920s, the building that housed the canal engineers working on the Fourth Canal, and the steam engines that helped to build the canal are all recalled, as are community scenes such as educational developments and the N.S.&T. streetcar system.

These murals, distinctive as art and as a public outdoor expression of the community's canal-urban history, make a significant contribution to Welland's urban-renewal process and encourage the city's nascent tourist industry. Coupled with events such as the annual Rose Festival in June and the associated cultural events that include sporting activities, rose exhibitions, and the Ethnic Day sponsored by the Welland Heritage Council, Welland is starting to make a gradual transition from an industrial past to a more tourist- and amenity-oriented future, a process of urban transformation that is centred on the change from an active canal next to the city centre to a Recreational Waterway in the same location.

Whither the Urban Commitment?

During the period of these events, other forces of change have exerted a substanial adverse impact on the city centre. The automobile and the municipal-regional decision-making process have encouraged suburban malls to flourish, while the central part of Welland as a business centre has declined seriously. This situation of change is not, of course, unique. It has prevailed strongly in most communities across North America.

An amenity recreation–public open space future is severely hampered by the fact that the former canal service road along the west side of the canal between Welland and Port Colborne has remained in an unimproved state. It enjoys a high-quality location next to the Recreational Waterway, but by 1994 it was not capable of any sustained public use because of its deplorable unsurfaced condition, deep potholes, and restrictions on public access by the Seaway Authority. As stated by DPW, 'The existence of such a significant amount of land held in public ownership affords a unique opportunity to exploit its potential, fully allowing social, economic and environmental considerations to be taken into account' (Technical Advisory Committee 1975). The assessment was correct, but the statement has not led to a surfaced route along the canal as the costs of this provision could not be borne by the two municipalities.

Visions for the urban future by the Welland Canals Advisory Group and by promotional groups such as the Greater Welland Chamber of Commerce have included stocking the old canal with fish for angling enthusiasts, converting an obsolete freighter to a marine museum providing hotel and restaurant facilities, cruising the canal with passenger vessels to the Flight Locks and a round journey to the harbour at Port Colborne, and recreating an authentic nineteenth-century canal village. Such dreams are possible, but a radical change in outlook is needed to take advantage of the potential offered by the Recreational Waterway during a period when recreation, tourism, and leisure pursuits are more to the fore than during the city's earlier history.

Much has changed since the By-Pass opened. Industry now operates in a less polluted environment. In 1995 the city hosted over four days the 24th annual Canadian Street Rod Association Annual at the Welland Sports Complex and Niagara Regional Fairgrounds. A two-day Food Festival was held in 1994 at the Market Square. In relation to the canal's continuing importance, an observation area with walkways and picnic areas was, in 1996, in the process of being added next to the By-Pass at the eastern end of the Main Street Tunnel. In 1995 the Canadian Canoe Association hosted the Canadian championships on the rowing course of the abandoned Fourth Canal. Much remains to be achieved, but the start of the new process in urban evolution was the By-Pass channel around rather than through the business centre of Welland.

19

Towards a Welland Canals Parkway

When in 1970 regional government was established across Niagara, planning, tourism, and economic development became functioning departments in the new municipality, and a series of reports studied the old and the modern canals for their potential. Each noted the virtues of a parkway system along the old and the new canals, but only limited action has followed. A new possibility arose when the Seaway Authority announced the sale of land surplus to their navigational requirements, which reawakened interest in the parkway possibility, leading by 1995 to a regional-plan policy amendment to reserve space along the canal for a parkway.

The Seaway Places Non-Operational Land on the Market

When surplus lands along the length of the St Lawrence Seaway were announced for sale in the early 1990s, the justification was the Authority's need to be self-sufficient and to offset the declining tolls and cargo tonnages. As stated in its annual report for that year, the Seaway 'took action to sell properties not required for operational purposes or future expansion.' This change was not limited to the Welland canal, but was applied at Cornwall and elsewhere along the Seaway.

Changes to Seaway policies in 1991 permitted mortgages, and approved the extension of leases from ten to forty-nine years in order to encourage long-term leases affording an attractive revenue and to promote the possible sale of surplus lands. Leases reflected current market value and, as land development did not previously form a part of the Seaway's mandate, the Authority assembled its own team of real-estate professionals to manage, lease, and sell its real-estate holdings. These changes and the reduction of grants in lieu of taxes to achieve financial gains to the Authority were not warmly received by the municipalities concerned.

The Seaway, like every other business organization, had to reduce its costs in order to remain in operation. The difficulties for urban development in the Niagara region were that, unlike the arrangements that have been described at Lachine, no prior arrangements for land-use planning or design had been made before the Seaway declared its surplus land for sale. Nor did either the regional municipality or the individual municipalities proceed immediately with the preparation of an overall plan as to how these former canal lands might best become part of the urban domain and be used for regional and municipal advantage. The canal communities now of necessity had to face a future that was not envisaged when their official plans were prepared, and with much more limited funding, legal, and land-development powers than had been available previously to the federal government.

To encourage the new development opportunities in previously unforeseen locations, municipal zoning and the official plans might have to be changed, as might the urban area boundaries that limited the extent of urban expansion. As has been indicated with the abandoned length of the Fourth Canal at Welland, the municipalities tended to favour open space–recreational opportunities, whereas the Seaway Authority favoured development that reaped the highest financial return. Moreover, with this warning sign of potential problems ahead, the achievement of a continuous parkway along the canal might be jeopardized by a single major development on the canal bank. Until a right of way was protected, a severe concern existed that parcels of land might be acquired and used in a way that did not complement the development of a parkway. At Welland this possibility, in late 1992, led to an Interim Control By-Law over lands along the By-Pass and the Townline Tunnel (By-Law 9989), which effectively froze any change in use for two years, but reduced the value of the affected properties. The land had to remain under existing agricultural and rural type uses until the expiry of this by-law.

With the sale of surplus lands, the canal service roads now offered a revised opportunity to recognize their scenic and heritage potentials through the establishment of a continuous lake-to-lake parkway. As along the Niagara River Parkway, development outside the bounds of the parkway might take advantage of the opportunity to overview open space along the canal, and itself contribute to enjoyment of the parkway. Bridge crossings owned by the Seaway might also be modified and upgraded, as had occurred along the abandoned Fourth Canal at Welland.

Some form of new provision might also have to be made to accommodate water-supply agreements with industry, the hydro authorities, and the municipalities, including the discharge of water into other watercourses to retain their flow. Harbours and breakwaters might present new opportunities for marina development. Leases might have to be rearranged, roads widened, and rights-of-way across property maintained. The sudden transition from a commercial canal to the municipal and community use of the same areas for recreation presented both a new set of opportunities and an array of difficult policy decisions for the urban and regional structures of government.

An example of the administrative dilemma that existed is provided by the issue of the Peter Street Bridge in Thorold during 1993 and 1994. The Seaway wished to dispose of this structure as no longer relevant to its marine operations. Outwardly, only the simple transfer from one public body to another was involved, but the bridge also needed repairs and the municipality was unable to take over these costs without funding from the senior levels of government. The bridge had also to be widened or realigned to provide improved access to the Flight Locks. As it existed, with its narrow lanes and turns, it could not be used by coach traffic, and access to the Flight Locks was from the north only. A world-famous scene required a better approach for tourist viewing.

In the two years of contentious debate that followed, heightened by critical observations of the process in the local newspaper, one Seaway official even intimated that Government Road next to the canal might have to be closed and become a private road. This was correct because the bridge and road were Seaway property, but to the public the reference was to a highway that provided access to the world-famous Flight Locks, a tourist route of considerable importance to view the escarpment section of the canal, part of the potential lake-to-lake scenic highway, and a route in heavy use for the daily ebb and flow of traffic by residents, including the journey to work between Thorold and General Motors.

The public was angry over the closure and uncertain future of this bridge, and in 1994 the responsibility for its future was transferred to the municipality against federal and provincial grants towards the construction of an improved bridge. This arrangement for city ownership was accepted reluctantly by the municipality. Some two-thirds of the cost came from the federal and provincial governments and the remaining third from the city's reserve fund. The title transfer was then delayed as new questions arose about Thorold's responsibilities for road improvement along Government

Road past the Flight Locks, but the construction contract for the bridge was awarded in 1995.

The municipality and the region had eventually obtained an improved bridge, but the length and rancour of the proceedings certainly did not demonstrate the expected liaison and cooperation between the various levels of government that might reasonably have been expected as land was relinquished from Seaway control to the neighbouring communities. The problem was not one of need, but who bore the cost and in what proportions when a route considered to be of national importance was relinquished to municipal control, a very similar issue to that which had prevailed at Welland over the decades.

The root of the matter lay not so much in the obvious need to retain and realign the bridge for use by the public and as an essential link in the envisaged Welland Canals Parkway, but rather in the divisions of authority within government, the complications of the administrative world, and the critical financial question as to which level or levels of government, and in what proportion, carried the costs. The structure of government had divided the essential unity of the canal from its associated highway and land-development possibilities that have been the primary consideration of this book.

Reports Towards the Achievement of a Parkway System

A series of government reports from the 1970s on have studied the old canals for their heritage relevance, but only limited actions have followed. In three reports prepared by the Ontario Ministry of Culture and Recreation in 1975, the Welland Canals were 'judged ... to be of national and provincial historical significance because of their role and significance in the Great Lakes–St Lawrence Shipping Route, and in the economic development of Southern Ontario' (Greenwald 1976). The surviving canal structures along the earlier canal systems were specified and described 'because of the obvious historical importance of the Welland Canals and the general lack of knowledge about the surviving resources.' The resource analysis identified ten important canal areas (* indicates the priority localities), at Port Dalhousie*, Welland Vale–St Catharines, Second Canal–Centennial Park–Town Line*, Thorold, Third Canal at the Escarpment*, Summit Level, Deep Cut, Welland, Port Colborne, and the Feeder Canal. Preservation alternatives were examined in each area, and potential programs and funding agencies listed.

These provincial reports led to an intergovernmental committee involving Parks Canada, the Ontario Ministry of Culture and Recreation, and the Ontario Ministry of Natural Resources in the preparation of a joint concept plan for the old Welland Canals (Parks Canada 1980). Heritage areas were now recognized at Port Dalhousie, Burgoyne Bridge, the escarpment locks, Port Robinson, and Port Colborne. As stated in the joint report, 'In order to create a concept plan which would function as a single system it was seen as imperative that these five thematically different heritage areas be linked by some form of road network' (ibid.).

Four automobile routes were then suggested to provide the necessary linkages between the heritage areas: a route along the First and Second Canals; another along the Third and the old Fourth Canal; a route along the present Fourth Canal and its By-Pass at Welland; and an interim Welland Canals Heritage Route. In addition, a series of bicycle routes and lanes were proposed to provide an alternative means of reaching and viewing canal features, and at each heritage location, the development opportunities included projects associated with conservation, interpreting the story of the Welland Canals, recreational possibilities, and visitor services.

Towards implementation, the plan's preface stated: 'It was anticipated that the concept plan would be realized by the coordinated involvement of many levels of government and the private sector through a formal agreement. It became evident however, that due to existing financial and manpower restrictions the various departments would be unable to meet this goal.' The plan was therefore introduced to serve as a guide for the work of the various agencies concerned with the conservation of the old canals.

Faced with this failure of the federal and provincial governments to pursue procedural actions, an Ad Hoc Welland Canals Group was formed. Known as the 'Group of 19,' it met to pursue the possibility of implementing the concept plan and included representatives from the various levels of government. Meetings during 1981 and 1982 identified the functions, the staff structure, and the financial structure required to coordinate resource development and management through the Welland Canals Corridor. An impasse was reached when no clear direction was available about whether the province was interested in and/or capable of involvement. This led to an Interministerial Task Force on the Welland Canals that involved the provincial ministers of Citizenship and Culture, Tourism and Recreation, and Natural Resources (Greenwald 1984).

This task force also failed to achieve action, but in 1985 the Regional Planning and Development Department produced a report that estimated the benefits of developing Niagara's historic waterways centred on a continuous circular drive linking the canal corridor with the Niagara River

Parkway. It concluded that 'the annual economic potential of $4.0 million seems realistic and possible' (Regional Planning and Development Department 1985). Existing and alternative routes were identified, together with those lengths that required upgrading. The concept included link roads with the Niagara River Parkway along Lakeshore Road next to Lake Ontario and Highway 3 next to Lake Erie and a continuous route along the Welland Canals. This possibility, a repeat of the program envisaged by the Niagara Parks Commission in the 1930s, was reformulated when this commission in the late 1980s prepared its long-term plan (Moriyama and Teshima 1988).

The Welland Canals Corridor Development Guide

In 1988 a team of consultants, with the assistance of a Canada-Ontario Tourist Agreement and advice from a technical committee representing the region and its local municipalities, prepared an extensive report, the Welland Canals Corridor Development Guide (WCCDG), for the regional municipality. This report, the culmination of much previous work, depicted the Welland Canals Corridor as 'a unique form of urban waterfront – a waterfront created by man to by-pass nature's more turbulent waterfront, the Niagara River ... The time has come to re-focus the community's attention towards the Canal for yet another economic transformation – from transportation spine to tourist attraction. And yet, it is the Canal's past historical, economic and human transformations which will provide the basis for its tourist-oriented future ... There is a need to connect the physical manifestations of the past in an integrated and interesting way in order to tell the compelling story of the Welland Canals' (Marshall Macklin Monaghan 1988).

The approach focussed 'on the development of a select number of "attractive destinations" where both public and private resources can create the critical mass needed to establish the Welland Canals Corridor as a tourist destination.' Anchor destinations considered to be the strategic points of entry and exit to the canal corridor existed at Port Dalhousie, Port Robinson, and Port Colborne. Here would be located interpretative centres with different themes unique to that community, and the expectation was that there would be a strong commitment to the preservation and restoration of important heritage resources in these locations.

Urban cultural parks as 'magnet attraction' destinations were proposed across the escarpment, at St Catharines and at Welland. The concept was that five major goals would be pursued in these locations: the preservation of historic buildings and their settings; economic development through public and private investment in adaptive reuse and interpretive attractions; recreational use of the settings for active and passive pursuits; education of the public about the history of heritage settings; and public participation through programming and special events intended to create places for residents and tourists alike to be together (map 13).

These anchor and magnet destinations were to be linked together by a scenic drive. As argued in the WCCDG, 'it is critical that there be a physical and thematic linkage through the Corridor. It is therefore proposed that a scenic drive be developed through the Corridor which links the magnet and anchor destinations as well as provides access to secondary resources. This route could potentially add another link to existing auto routes such as the Niagara Parkway and Talbot Trail, providing a continuous, circuitous route through the Region. The auto routes would be complemented by hiking and cycling routes which in turn, would link with existing facilities such as the Bruce Trail.' Two scenic routes were then proposed, the Schooner Route, which followed the course of the old canal, and the Bridge Route, which followed the existing Fourth Canal.

A recreational waterway that would use short sections of the existing canal at its southern end and rebuild parts of the old canal system was also proposed. It would follow Twelve Mile Creek from Port Dalhousie to DeCew Falls, cross the escarpment by a marine railway to Lake Gibson, then follow the present canal to Port Robinson and its abandoned length through Welland to Port Colborne. Links would be established via the Welland River to the Niagara River, and via the Feeder Canal to Lake Erie at Port Maitland. Described as 'imaginative and unique,' and with issues such as overhead clearances, fluctuating water levels, and conflicts between users (e.g., through the Henley Regatta course) to be resolved, this waterway would reflect the earlier circumstances of the First Canal and 'introduce recreational use of the water within the historical development of the Welland Canals Corridor.'

The character of the corridor would be expressed by developing four broad themes. 'The Niagara Frontier,' in the pre-canal era up to the 1820s, would introduce the physical setting, including the two lakes and the Niagara River, and their defensive role. 'Scaling the Escarpment,' from the 1820s to the 1870s, would portray the early period of canal building, the people who achieved the canal, and the technologies that were used. 'The Powerhouse of Canada' from the 1870s to the 1920s, would cover hydroelectric power, the development of industry, and the associated growth of the canal communities. 'Continental Divide,' from the 1930s to the present, would demonstrate how historical processes have shaped the region and its modern character.

Implementation of this plan involved four different possible organizational alternatives: the Welland Canals Society, an independent private organization formed in 1986 as an offshoot of the Welland Canals Preservation Association; a regional government agency emerging from the restructuring of the Niagara Region Tourist Council and the Niagara Region Development Corporation in conjunction with the regional planning and transportation departments to form a Regional Tourism and Development Corporation; a Welland Canals Conservation and Development Commission that would be a new single-purpose body akin to the Niagara Escarpment Commission; and a Niagara Peninsula Mega-Commission that would unite all federal, provincial, regional, and other agencies into one super-agency responsible for all aspects of economic development and tourism in the area.

WCCDG favoured the formation of a Conservation and Development Commission, and recommended a Welland Canals Interim Management Team to coordinate the implementation of the guide's proposals and to bridge the period between its publication and the establishment of an organization to implement them. Instead, the Regional Municipality established a five-member Co-ordinating Committee to review the implications stemming from the guide. Their report stated that the guide 'was very successful in raising the profile of the [Welland Canals] Corridor,' but serious concern was expressed about the viability of many of the guide's specific development proposals. 'The proposals perhaps served to illustrate the Corridor's potential, but they cannot be used as "The Plan," or as "The Basis" for selling private or public investment.' The challenge was seen to be 'guiding what already is moving, as opposed to implementing the Guide' (Co-ordinating Committee 1989).

Concern about the viability of the guide was also expressed in many submissions. For example, the Welland Canals Foundation stated that 'certain aspects of the Development Guide seem impractical in terms of either their excessive cost and/or the technical problems presented in their achievement ... Projects capable of achievement and those making greater use of existing resources should be given priority ... We are ... suspicious of various aspects of the Mountain Locks Park ... The creation of a marine harbor on the former Second Canal at St Catharines seems improbable, as does the Recreational Waterway ... Absolute priority must be given to the existing Henley Regatta Course ... Its long-term operation since 1903 should in no way be prejudiced by introducing alien attractions. Does any need exist to make

Twelve Mile Creek navigable to DeCew Falls? ... We have studied the Welland Canals Guide and hope that this report will not become just another unused canal document. The Guide contains many valid points which should be implemented. In particular we stress the admirable concept of a Scenic Parkway' (Welland Canals Foundation 1988).

When the Co-ordinating Committee focused on implementation, it considered that the functions of market identification and promotion for the corridor's development, and the need to foster private and senior government intervention, were already undertaken within regional government by its Tourist Council and the Regional Development Corporation. When representatives of the tourist-related departments at the senior levels of government were interviewed, neither saw their role as the builder or the operator of tourist attractions. 'Both governments will look to Regional Government for co-ordination of proposals as widespread as the Corridor's.' In this vein, the Co-ordinating Committee recommended that 'Regional Council establish a Welland Canals Corridor Development Committee,' and that 'a new senior officer position' be created to support this committee.

The Welland Canals Society

The recommendations of neither WCCDG nor the Co-ordinating Committee were accepted by Regional Council, and the critical task of implementing the guide was awarded to the Welland Canals Society (WCS). The Memorandum of Understanding between the region and the WCS included the requirement for a semi-annual report to Regional Council outlining progress in identifying issues and promoting canal development. Funded in part by the region and more extensively by the federal government, the Society's mission statement, adopted in 1989, was 'To provide leadership and assistance to the public and the private sectors in achieving heritage-sensitive and tourism/recreation related economic development in the Welland Canals Corridor.'

WCS revitalized a former industrial building next to the Second Canal in St Catharines as its administrative offices and works area, but then moved to more central and prestigious office space. The society adopted WCCDG as its working guide and became involved in several small projects. WCS also regarded itself as the sole negotiator with the Seaway Authority and the various levels of government, an approach that was perceived within the organizations active in some canal-related schemes as intrusive. When financial support from the region was withdrawn in 1991, WCS folded.

As stated by the Port Colborne–Wainfleet Chamber of Commerce, which severely questioned the credibility of WCS, 'In the past year, there have been significant staff changes, a disproportionate number of resignations from the board, lack of support from certain key players and disinclination of the W.C.S. Board of Directors to exert responsible leadership ... [We] can not, in good faith, recommend that the Welland Canals Society be the implementor of the Welland Canals Development Guide' (Salmon 1989). The depth and vigour of the WCCDG proposals were beyond the understanding of WCS, and few consequences of its work were visible on the ground (Pugliese 1990).

The reasons for the failure of WCS were administrative, and not related to the tourist potential of the Welland Canals Corridor. The fault lay with the region not accepting the advice provided by the WCCDG. It was unreasonable to expect that a private organization independent from any agency of government would be able to initiate a series of developments, then coordinate the various components and steer each through the various procedural operations of government at each of its four levels. The lesson of the WCS failure is that progress towards a parkway can be achieved only as an integrated exercise within government, headed by competent professional staff and backed by a strong citizens' advisory group.

A New Set of Parkway Proposals

The idea of a parkway system within the Welland Canals Corridor was revived in 1992 when the sale of surplus land by the St Lawrence Seaway made possible its achievement. Also, Niagara's industrial economy had then regressed severely through economic depression, high unemployment, and the decline or closure of many industries. The argument was now made that a parkway would make fuller use of existing resources and the heritage potential that existed for tourism, and also help materially towards the necessary restructuring of the regional economy. In addition, it would contribute materially towards the provision of a Niagara Greenways Network then under discussion, with bicycle trails, scenic routes, and pedestrian trails.

A private report submitted to the region and its municipalities introduced the Welland Canals as an important but underappreciated resource, and urged immediate positive action through the use of federal and provincial infrastructure grants that were available to improve and refocus the economy. Two distinctive signed routes were advocated. 'The first, the Welland Canals Parkway, would provide a continuous automobile route between Port Weller and Port Colborne. This project would follow closely the banks of the present canal. It would be linked carefully with the canal's shipping, trading, and engineering abilities, and introduce numerous opportunities for government and private agencies to portray the strengths of the Welland Canals and their associated communities. The second route, The Welland Canals Heritage Route, would in part follow the [existing and inadequate] Scenic Drive, but improve on this, and add signage, displays and other attractions to more effectively associate it with the earlier canals and their heritage communities' (Jackson 1993b).

These two routes might initially be combined into a single lake-to-lake project, then incorporate other elements at a later stage to provide a comprehensive parkway system, with trails, an automobile route, open space, viewing locations, signs, exhibition areas, and displays. It was stressed that 'all the highway and development proposals are job-creating, of permanent economic advantage to the region and specifically related to the area's existing resource endowments. They are firmly based on the strengths that already exist' and on the proposals that have been mooted for their development over the past seventy years.

Meantime, the Niagara Region Tourist Council proposed to Regional Council 'that the spirit of the Welland Canal Development Guide be resurrected and that a committee be struck to deal with parkway related issues' (Hardy 1993). This brief noted: 'The one thing that remains lacking is a connected drive along the canal that would link together all the attractions and historic sites. The idea of developing a parkway along the Canal is not new, but the time has come to place it back in the public domain. A connected parkway would also have access to all canal attractions and could link up with other scenic drives to form a circuit that could in itself be a tourist attraction.'

The outcome was that in 1993 the Regional Municipality established a Welland Canals Parkway Development Committee, which by early 1994 had prepared a Proposed Welland Canals Parkway Development Plan. The parkway that was proposed started in Port Weller, followed the modern Fourth Canal past the Flight Locks to Allanburg and Port Robinson, and was then routed along the abandoned Fourth Canal through Welland to Port Colborne. Supplementary and secondary routes were envisaged to link and associate the parkway with all the canal and with other attractions in the area. A Welland Canals Parkway Implementation Committee as a partnership between the region and the municipalities along the canal was presented as the administrative structure to proceed towards achievement (Department of Planning and Development 1994a). When this report was circulated to the municipalities, other government agencies, and some public groups for their observations, some '200 responses were received ... indicating a very high level of interest in this topic. The comments were overwhelmingly supportive of the concept with strong encouragement for immediate action. It was hoped that this report would not be another study to be put on a shelf but rather a blueprint for further development of a Parkway' (Department of Planning and Development 1994b).

The region in late 1994 then proposed an amendment to the regional plan that envisaged 'two corridors, along the east and the west sides of the Canal, within which one parkway would be established,' and indicated 'the need for priority work to be undertaken to identify the route between the Peter Street Bridge in Thorold and Highway 20 and to design the route between Forks Road in Welland and Port Colborne' (Department of Planning and Development 1994c). There was also 'the need for a program of activities which would involve a financial commitment to be included in the Region's Capital Works Program to promote the creation of the Parkway.'

The parkway concept was again proceeding, this time with municipal and public support. At two hearings, public support was again expressed. The inclusion of equestrian trails was requested in addition to the provision of bicycle tracks and pedestrian trails within the canal's curtilage. Action was needed to promote the reuse of the closed west pier at Port Weller for institutional or housing development and/or with camping-recreational facilities under provincial, conservation-authority, or private management. Other contributors urged the need to renegotiate for a possible federal park where the Third and Fourth Canals crossed the escarpment, and presented the idea that the Seaway might operate within the setting of a lake-to-lake national park. Many presentations argued that the canal service roads should be opened to the public and suitably surfaced to provide linear access to the modern canal and the abandoned channel at Welland.

The regional initiative to reserve land next to the canal for a parkway by means of a Regional Policy Plan Amendment provides a necessary first stage to these other achievements. The final report and recommendations for a Welland Canals Parkway were approved by the Welland Canals Parkway Development Committee and then by the Planning Services Committee of Regional Council in March 1995, and became By-Law 7994-95, which was submitted to the provincial Ministry of Municipal Affairs.

At the same time, the Seaway Authority announced a new organizational structure of area-based service teams. 'The Canal has been divided into three distinct operating areas, each with a manager. These Area Managers with their teams will be totally responsible for all aspects of the operation in their area' (Trépanier 1995). With the terms of reference including cooperation with the regional municipalities, for the first time in the history of the modern canal a direct face-to-face mechanism had been created for informal communication between the federal agency responsible for the canal and the municipalities. Hopefully, this new administrative structure will be able to resolve such contentious issues as Peter Street Bridge and the years of delay that have occurred along the Recreational Waterway in Welland. As Trépanier remarked, after commending the region for moving the Welland Canal Parkway development project forward, 'We understand the importance of this parkway for the development of tourism in the Niagara Peninsula. We will do our best to assist in the *Welland Canal Parkway* project without compromising our basic good business and stewardship principles.'

In early 1995 an extensive area of land next to the viewing area of Lock 3 was rezoned from neighbourhood residential and light industrial uses to tourist-commercial and major open space, 'which primarily allows parks, conservation uses, and operations of the Welland Canal that generally retain the open space nature of the land' (City of St Catharines 1995). A month later ground was broken for Seaway Village, a 9-hectare (22-acre) tourism development recreating a seaport village, expected to be completed by 1996 and in turn fostering other tourist-related developments both in its vicinity and along the parkway (Smith 1995). With Regional By-Law 7984–95 and municipal and Seaway support, the ball was now rolling slowly but favourably forward towards the long-awaited but now expected achievement of a Welland Canals Parkway, and recognition on the ground for the over 170 years of canal-community achievement that has been depicted in this book.

The latest development in September 1995 was the appointment by Regional Council of a consulting team, IMC Consulting Group of Cambridge, Ontario, with terms of reference to identify a preferred route for a two-lane driving parkway between Lakes Ontario and Erie, with an accompanying trails system to accommodate hikers, cyclists, and equestrians. Its mandate included the preparation of design drawings to illustrate the concept plans and types of facilities being proposed, estimated costs of construction, and advice on future directions regarding the administration, maintenance, financing, and ownership of the parkway system.

Again, after a further series of public hearings (there have been a dozen or more during the course of preparation and after the publication of the several parkway studies), by May 1996 the IMC consulting team had identified a preferred route for a Welland Canals Parkway and its associated trail systems. The driving route followed the west side of the modern Fourth Canal from Port Weller to the Townline Tunnel in Welland, where it crossed to the east side of the canal By-Pass. Continuing along the east side of the By-Pass to Port Colborne, it then crossed into the centre of that city and extended south to the marina developments on Lake Erie. The intention was to include a bicycle lane and pedestrian walks along the driving route, and other trails close to the canal in other locations to accommodate walkers, cyclists, equestrians, roller-bladers, and snowmobilers.

Epilogue:

The Changing Canal Scene

All aspects of the Welland Canals have changed dramatically over the years since their introduction to Niagara's landscape in 1829. This truism applies to the canal as an engineering feat, to its trading flows and patterns, to its role in community development, and to the character of these communities. Edmund Spenser referred to 'the ever-whirling wheel of Change,' and Benjamin Disraeli wrote: 'Change is inevitable. In a progressive country change is constant.' Against the background of William Hamilton Merritt and the canal's border location, the more notable examples of change are summarized in this concluding section.

The Role of William Hamilton Merritt

Verses by Howard Engel (1978) introduce the basic contributions of William Hamilton Merritt as the instigator and promoter of the canal endeavour.

> The first time I heard
> Merritt talk about
> his much vaunted canal
> was in the back room at Dr Chase's
> old store, where the *Post*'s
> printing office now is.
> We listened to him for over an hour
> and to hear him talk
> it was only a piddling feeder ditch
> two and a half miles long
> running from the Chippawa Creek

to the west branch of the Twelve.
The way he described it
he and I could have dug it
in an afternoon without unbuttoning our vests.
And the results of this puny effort
would keep my run of stones spinning
winter and summer, no freshets
overflowing my milldam in spring
or summer drought closing me down.
We all put in a little seed money
and I went with him to measure the ground.
According to the figures
we needed a cut of about thirty feet deep
over the height of land.
Now to tell the truth
I didn't put much stock
in Merritt's talk about shipping
but I was canny enough to see there'd be
hydraulic power to spare
so I was hooked like a trout
in my own millpond.

. . .

A string of forty little locks
he wore round his neck
a rosary of his accomplishments
a testament to his will.

He kept the whole thing going
kept the workers digging with their picks
kept the money men entrammelled
until it was finished
not as he would have wanted it
but well enough
like a house of cards
it might fall if you breathe
it stood there by the force
of Merritt's personality
the water level tenuous

the logs above water
already beginning to rot
but they did manage a celebration
as two sloops climbed up the stairs
to the distorted sounds of Handel
above the vanquished waters.

The Impact of the American Border on the Canal and Its Communities

When the canal was planned in 1825, the intention was to introduce it along a route determined by the physical geography and drainage pattern of the Niagara Peninsula. The Grand River was to be linked to the Welland River, and a relatively short cut across the Niagara Escarpment and its vicinity would then connect with the north-flowing Twelve Mile Creek and bring the canal down to the level of Lake Ontario. Only the name Welland Canal has survived from these worthy goals.

When the canal opened in 1829, its route instead was from Lake Ontario via Twelve Mile, Dick's Creek, an unanticipated channel through Merritton and Thorold across the Niagara Escarpment, and the Lower Welland River into the Niagara River at Chippawa. In 1833 this canal was extended through a man-made cut to Port Colborne on Lake Erie. This lake-to-lake route formed the basis for the Second, Third, and Fourth Canals. As each new canal has retained considerable sections of the previous version, this process of change yet continuity may be suggested from the Deep Cut between Allanburg and Port Robinson and the harbour at Port Colborne, both with unbroken histories of canal association that date from the mid-1820s and the early 1830s, respectively.

The canal's location has always been close to but inland from the American border along the Niagara River. This border initially meant uncertainty, including doubts about whether a canal should even be constructed, given the hostile presence of the United States on the opposite bank and then the inability to use the Lower Niagara River as the northern outlet for a canal crossing the Niagara Peninsula. Had there been no American Revolution and no boundary, the logical route for a 'Niagara Canal' would have been across the now American side around The Falls between the Upper and Lower Niagara River – an option that remained alive until the Robert Moses Power Station eliminated this possibility. As it was, the Erie Canal provided the spur for constructing a competing Canadian canal to the British port of Montreal via the St Lawrence River rather than to the American port of New York via the Hudson River.

The canal was born in the period of tension that followed the War of 1812 and, as the nearby boundary with the United States changed from a frontier 'of conflict to one of contact (Watson 1943), railway bridges crossed this boundary and the physical divide of the Niagara River. American and Canadian lines converged on the peninsula and diverged into the United States and across Canada. Here was a major competitor to the canal system, but the response was the development of the canal-oriented Welland Railway, which carried canal traffic and helped bring about the eclipse of the contending Erie Canal on the opposite side of the Niagara River.

Later, when tariff barriers were erected around the new Canadian nation after Confederation in 1867, the proximity of the Niagara Peninsula to the international boundary and the ease of movement to and from the United States again played a significant role in enhancing the canal zone as a locale for industrial development. Now with rail access to the major American centres of industrial activity, the 'Iron Horse' helped to introduce American capital, entrepreneurial activity, and external control to the region. American-owned manufacturing plants developed along the Welland Canal to serve either the Canadian or the imperial market with goods processed or produced in Canada. Protective tariffs encouraged the movement of American industry to nearby locations across the border, the Niagara Peninsula and its canal zone became a fringe locality tributary to American interests. Hansen (1970) and Marshall (1936) have portrayed the canal zone, with its large number of American companies and branch plants, as an American industrial outpost in Canada, and Jackson and White (1971) have described the area as an 'extension of the adjacent geography of the midwest United States.' The setting was also on the fringe of the expanding urban milieu of southern Ontario, pivoting on the Toronto metropolitan area (Spelt 1973) and influenced strongly by this powerful urban concentration.

The Canal as a Flow of Water

The canal itself as a body of flowing water was supplied from the Grand River through the Feeder Canal from Dunnville. This source changed to Lake Erie in 1881 and has remained from this lake to this day. In-course sources of water supply have included Twelve Mile Creek to the First and Second Canals in St Catharines and, on the present system, reservoirs and the beheaded upper streams of Beaverdams Creek to the locks and reaches at the northern end of the Fourth Canal. A disadvantage of providing a direct water connection between Lake Ontario and Lake Erie has been that the lamphrey eel and later the zebra mussel were able to penetrate the Upper Lakes. Channels have been gouged into the lakes at each end of the canal, and the breakwaters have led to sedimentation along each lakeshore.

The water in the canal, with an expanding volume through time as the channel was widened and deepened, has been used for a variety of purposes. These started with the ubiquitous mill fed by mill races from the canal. By the turn of the century, the canal was used as a domestic source of supply for its riparian settlements and extensively to supply water for cooling and industrial purposes in many of the large-scale manufacturing plants that developed along its course. The canal then encouraged the generation of electricity at municipal and industrial plants, and later its waters were diverted to a municipal reservoir at DeCew and to hydroelectric plants at Welland and DeCew. These hydro developments serviced from the canal were taken over by electricity production at The Falls, and then from Queenston Heights, but the initial advantages that had accrued to the canal communities were important for their industrial progress because of the low cost and the abundance of local power supplies. With advances in settlement and the pollution of local sources, the modern canal now supplies drinking water to the municipalities along its length and to an area north of the escarpment from Niagara-on-the-Lake to Grimsby.

The Marine Trading Scene

The channel followed by the canal as a body of flowing water has changed in length, width, depth, and the profile of its banks, and from in part following natural streams and rivers to being wholly a man-made cut incised into the landscape with an occasional embankment upon completion of the Fourth Canal. Major viewing areas have been provided at St Catharines and Port Colborne and one is under construction at Welland, but numerous casual viewing areas also exist along the canal banks. Perversely, access to the banks along the unpaved service roads is now restricted by the St Lawrence Seaway Authority because of the dangers to shipping through trespass and the deposit of unwanted materials in the canal. An observation tower and a national historic park were promoted at the Flight Locks during the 1960s but not fulfilled – an irony as this scene is of international calibre through its scenic character, the unique value and heritage of the engineering works, and the flow of passing vessels.

Locks and weirs that control the flow of water through the canal create a specialized landscape, with upstream ponds, basins, or reservoirs, and a series of level expanses of water between the lakes. This new environment was created in particular across the Ontario Plain, then the Niagara Escarpment, where the locks and hence the ponds were closely spaced, at Martindale Pond in Port Dalhousie, and between Allanburg and DeCew, where the ponds served as supply reservoirs for the municipal then regional waterworks and hydro generation.

As a man-made feature, the canal also changed permanently the character of the physical environment. It permitted land in the Wainfleet Marsh and between Welland and Port Colborne to be drained and become fertile agricultural land. Excavated material from the channel and from where the locks were constructed has been used to raise the height of land along the banks of the channel and to extend the land into the lake at Port Colborne and Port Weller. At Port Dalhousie the natural sandbar has been lengthened and widened. Elsewhere spoil heaps have created new hills in the landscape and at Mud Lake the infilled land has become a wildlife conservation area.

Advances in marine technology have steadily extended the canal's trading possibilities. The First Canal typically carried a vessel that was 100 feet (30.5 m) long with a cargo capacity of 165 tons (149.7 tonnes). The lock measurements had increased to a length of 140 feet (42.7 m) and a cargo capacity of 750 tons (680.4 tonnes) by the date of the Second Canal (Cowan 1935). The Third Canal extended these dimensions to a length of 255 feet (77.7 m) and a cargo capacity of 2700 tons (2449.4 tonnes), whereas the more grandiose Fourth Canal introduced vessels with a length of 859 feet (261.8 m) and a cargo capacity of 25,000 tons (22,680 tonnes) at a draught of 24 feet (7.3 m). As ship design has changed, the largest cargo to pass through the Welland Canal took place aboard the M/V *Patterson* in 1990, when 32,599 tons (29,574 tonnes) of iron ore were carried from Pointe Noire, Quebec, to Cleveland, Ohio.

The total of all freight passing through the canal was almost one million tons (907,200 tonnes) by Confederation. It first reached two million tons (1.8 million tonnes) in 1909, and three million tons (2.7 million tonnes) in 1913 during the era of the Third Canal. By 1932, when the Fourth Canal opened, the tonnage was at 8.6 million tons (7.8 million tonnes), a figure that had doubled by the early 1950s. It reached an all-time high of 73 million tons (66.2 million tonnes) in 1979, declining to 42.3 million tons (38.4 million tonnes) during the 1994 season.

The Fourth Canal as part of the St Lawrence Seaway opened in 1959 with 8072 vessel transits through the canal in that year. This volume, in part through the use of larger vessels and in part through declining trade, had diminished to 3378 by 1994. The shipping industry had changed radically in the form, type, and size of vessels using the canal. Even so, it continued to provide employment to many people with homes along the canal.

The total of toll revenue had increased from $1.2 million in 1959 to over $20 million twenty years later, and over $30 million in 1988, in part through higher lockage fees. This contentious issue involves national transportation policy and competition from railways and highways, their levels of subsidy, and also the extent to which the public or bulk carriers should carry the costs and the relative situation vis-à-vis competitive U.S. systems of movement. As world trading movements have also changed remarkably over recent decades, traffic flows and revenue serve as a reminder that the canal is subject to the vagaries of national and international pressures. Still, the impact of the canal is local on the communities along the canal and the regional corridor of development that has been created along the canal.

By the modern era, port facilities at Port Colborne, Thorold, St Catharines, and Welland had been reduced to being stopping points for loading or unloading. Their total tonnage was well below 10 per cent of that passing through the canal, and goods unloaded were more important than goods loaded. Unloaded products included bituminous coal at Thorold; iron ore and coal at Port Colborne; gravel, sand, coal, and iron ore at St Catharines; and manganese ore and coal at Welland. Then came agricultural products, mostly wheat at Port Colborne, where tonnage reduced drastically with the Fourth Canal as vessels could now pass through the port and the canal downstream without stopping. Some manufactured goods were unloaded at St Catharines, and fuel oil to tanks at each of the canal towns for distribution within their local catchment areas.

Cargoes were loaded, mainly at Port Colborne from the Onondaga limestone quarries destined for the United States. There was also some wheat to Montreal, though most was not reloaded into canal vessels but, distributed as flour by rail or truck to the Ontario markets. Newsprint came from Thorold, soybeans from Welland, and classified as an export from St Catharines were vessels repaired or constructed at the Port Weller Dry Docks.

Water transport remained economical for bulk, long-distance loads, but this traditional flow was confined to ports around the Great Lakes and across the Atlantic, destinations of little importance to the many companies that had markets in southern Ontario or exports destined for the United States. Road transport, with its convenience, speed, flexibility, and ability to handle small door-to-door loads, had become the dominant mode of transport for these companies. The railways, still important by the 1950s, were also important for the receipt of imports by many companies, and one quarter of all companies used this mode for their exports. The Fourth or Seaway Canal, now less important in providing transportation facilities for local industries, remained of high national importance for its trading flows. As vessels had each to be constructed, serviced, supplied, repaired, and manned along the canal, these marine industries in the canal corridor remained a significant but diminishing ingredient in the regional economy as more vessels passed through the canal, often provisioned and serviced at their main home and destination ports.

Manufacturing Industries

The industrialized canal communities have certainly been innovative (Watson 1943). Being in the forefront of economic activity has included Ontario's first cotton, sulphite, and silk mills, and the first electric-light factory. There have been many advances in the electro-metallurgical, electro-chemical, and pulp and paper industries, the canning of vegetables and fruit, and the bottling of wine. Shipbuilding, ship repair and design, the electric streetcar, and considerable progress during the early days of automobile industry must also be emphasized.

Mills and mill sites fed from the canal started this process of achievement. Every canal lock, where the canal crossed a stream at a higher level and where the flow in the canal might be diverted into a mill race provided an opportunity for this use of water power. The grinding and sifting of grain, especially flour, preceded the canal and was the first local staple of industrial development. But flour milling then moved to the canal with its greater flow, and hence the potential of larger mills, a greater number of stones, and an increased output through more stones. The canal also encouraged the change from wheat as a local source of supply brought to the mill door by horse and wagon to a supply brought in by grain vessels, and then its export in barrels to destinations downstream and abroad. At a later period, when ladened grain vessels were lightened to pass through the canal, this transferred milling advantages to St Catharines and Port Dalhousie, and vessels used on the canal and lake might also be provisioned, repaired, and serviced at these ports. When, during the early period of railway development, terminal elevators were constructed at Port Colborne, this port took over from St Catharines as the milling centre in the peninsula. Also, as wheat output from the Prairies and its flow through the canal increased, elevators were constructed elsewhere and the flour-milling industry expanded in western centres such as Goderich and Midland.

When the Third Canal with its loaded grain vessels bypassed St Catharines, severe depression hit that city and numerous mills were abandoned. This change, however, also meant that buildings and vacated mill sites were available for use by other industries that wanted nursery space for some new activity, or a building in an area next to the canal, or water rights. The diminished wheat trade provided new industrial opportunities, and the city recovered.

Sawmills, once abundant and also drawn to the canal through its greater water-supply availability than at sites on the perennial streams of the peninsula, also passed through different eras of technological change. Through the canal and its supply route along the Feeder Canal, the catchment area for timber supplies was extended upstream to the Upper Welland and the Grand River valleys. Downstream transport favoured the development of sawmills and plants processing their associated products, as well as the production of machinery and shipbuilding along the canal. In particular, the water powers available across the Niagara Escarpment were extensive and the manufacture of wood products ranging from sailing vessels and barges to furniture, domestic products, and building materials arose close to and often in conjunction with the sawmills.

As this industry changed its character from sawing timber and dressing and cutting logs for pulp and paper, larger sites with abundant supplies of water were required. This again favoured the length of canal between Thorold and St Catharines, where the greatest volume and head of water anywhere on the canal system existed. As the pulp and paper industry also required an enormous input of power, the proximity of the canal to the hydroelectric resource of the Niagara River proved a further distinct advantage for industrial change and expansion.

Textile mills and factories first obtained their wool from sheep on local farms, and used the area's streams as required for power and for washing purposes. Canal sites provided advantages for both activities, and also for the ease of imports from the United States, England, and Australia, depending upon the quality and the type of wool that was required. As the woollen mills made the transition to large-scale plants that produced hosiery, clothing, and carpets, the textile industry branched out into cotton blended with horse or goat hair, with cotton imported from the southern areas of the United States. Water transport, including via the canal, permitted and encouraged these new industrial possibilities.

Contributing to these advances were hydroelectric power from the canal and later from The Falls. Rail access to and from the United States and within Canada were also important ingredients, as were a skilled local work force, the entrepreneurial abilities of management, and skilled technicians to erect, use, and repair the machinery that was required in each plant. Industrial advance, which could no longer be viewed in isolation, relied extensively on the many associated business activities that existed nearby within the canal corridor of development, and upon the interchange of activities and services as these were needed by each plant. It was promoted considerably by local boosterism and aided and abetted by protective tariffs that worked to promote the advance of American industry in locations, such as along the canal, that were close to the international border.

With respect to the iron and steel industries that developed in the canal centres, the early use of bog iron with local charcoal as a fuel was displaced by imported American materials for use in the foundries and blacksmiths' shops and for a whole range of domestic iron products, mill machinery, agricultural machinery, and bent goods for the manufacture of wagons, coaches, and sleighs. From here it was but a short step to the production of automobiles and the many parts that they required. The canal at several locations was used as a source of power for industrial water and as a suitable channel for the disposal of effluent. Coal and iron, absent from the local economy, were brought in from Pennsylvania or Ohio, either by rail from Buffalo or by boat across Lake Erie or through the canal to Welland and St Catharines.

Shipbuilding and ship repair even preceded the opening of the first canal, for vessels constructed in St Catharines could pass down Twelve Mile Creek and into Lake Ontario through the lock that had been constructed at Port Dalhousie. Shipbuilding arose at several points along the canal, at Port Dalhousie and St Catharines as indicated, at Welland and Port Robinson where the demand included vessels that plied the Feeder Canal and the Upper Welland River, and at Thorold and Port Colborne to a lesser extent. Production has included sailing and propeller vessels, tugs and barges, and almost every type of vessel used on the canal and the adjacent lakes and rivers. Shipbuilding reduced with the end of the sailing vessel, but survived at Port Dalhousie and Port Colborne, and then at Port Weller after the Fourth Canal opened, with these dry docks being the sole surviving yard on the Canadian side of the Great Lakes by the 1990s.

Industrial Transformation

The Welland Canals Corridor has attracted much industrial development over the decades, and all the major components of industrial development and achievement have changed remarkably. As Watson (1945a) remarked,

'Whether we look at individual industries, or at industrial towns and cities, we see throughout the Niagara Peninsula a story of constant change, as men at different periods of their history have discovered the new techniques, or taken advantage of the different opportunities in the geography of the region.'

At first a direct relationship existed between the location of mills and water power from the canal, but coal, hydroelectricity, and then oil each gradually eroded the previous substantial influence of water on the processes of industrial location and growth. The canal link remained as coal then oil were brought in by the canal, and hydroelectricity was produced from the canal at certain locks and by the diversion of water to DeCew. Over time these changes were nullified by technological advances. By 1960 coal was much less used for industrial power, and its stack piles disappeared as industries modernized and turned to new and cleaner sources of power. The generation of electricity from the canal was displaced by the abundant power resource at The Falls, and the delivery of oil by tanker barge was replaced by a national system of pipelines that also reduced the importance of the Erie gas field in the south of the peninsula.

Even so, many of the canal sites and industrial buildings that had relied on water for power and transport retained a manufacturing activity, and many of the new industrial arrivals used the canal for one or more of their transportation, water-supply, or locational advantages. Raw materials continued to be imported via the canal and exports were dispatched through wharves located on the canal banks.

Water had a further attribute, as large volumes were required for industrial processing and cooling purposes when large-scale manufacturing industry was attracted to the canal banks. This element of the canal advantage for industrial location applied especially to the industries processing primary metals at Port Colborne and Welland, and to the pulp and paper industry that developed from Thorold to St Catharines along the northern length of the canal.

Gradually introduced into the earlier situation of reliance on transportation by water were railways, including the N.S.&T., its spur lines, and their freight-handling capacity, then modern highways and truck transport. The canal worked closely with the lake-to-lake Welland 'portage' railway, and could deliver grain to New York either via Lake Ontario and the Oswego Canal, at a lower cost than via the Erie Canal, or to Montreal via the Grand Trunk railway. At this time the Erie Canal was losing its effectiveness to the railways radiating out from Buffalo, but the Welland Railway saved the canal from the same fate. It added elevators at Port Dalhousie and Port

Colborne, provided the steamer lake service between Toronto and Port Dalhousie, and directly aided industry at St Catharines through the construction of mills and the expansion of shipbuilding activities on the canal.

A significant part of this expanding industrial process was the inter-association of several economic activities, both within the canal centres and between them, as industrial plants used the supplies, products, and skills available in other plants. As milling, shipbuilding, paper making, engineering, and later the automobile and auto-parts industries each used items from other manufacturers, there emerged not only a large number of specialized industrial establishments, but a complex and integrated system of industrial inter-linkage that together created an energetic industrial complex along the Welland Canals Corridor.

The Canal Communities

St Catharines on the main line of the canal as an agricultural centre, Humberstone as a straggling hamlet, and Chippawa as a port on the Niagara River each preceded the canal. St Catharines lay on the major east–west road of the peninsula, Humberstone on the lakeshore road that followed the higher, drier land of the Onondaga Escarpment inland from Lake Erie, and Chippawa on the Niagara Portage as the traditional route between Lake Erie and Lake Ontario.

The other canal centres started as labour camps to house in communal barracks and family shacks the labourers who worked on the canal. These accommodations were located at Port Dalhousie and where the canal climbed the Niagara Escarpment. The villages of Allanburg and Port Robinson arose at each end of the Deep Cut. Welland was born where the Feeder Canal crossed over the Welland River by an aqueduct, and Port Colborne where the First Canal dropped through a lock into the level of Lake Erie. Along the Feeder Canal, the dam across the Grand River gave rise to Dunnville and Byng, and Wainfleet emerged at its centre.

Progress from shanty towns to active local service centres was achieved rapidly. The norm was for growth to start at a mill site, with local services such as a church, school, store, tavern, blacksmith, and post office providing the economic base as the community grew. This growth was in response to the products of the agricultural hinterland and the trade and servicing requirements resulting from the canal: supplying stores for the vessels and their crew; providing services such as shipbuilding and ship repair; furnishing stables, animals, and tow-boys to pull vessels through the canal and, later, cordwood to feed the boilers of passing vessels; providing wharves, docks, and harbours where vessels might load and unload, or berth through the winter months.

The most advantageous locations for these community developments were at the two lake ports, with their regular arrival and departure of vessels, turning basins, and winter facilities where repairs and servicing might be undertaken. St Catharines had the advantage of providing the most comprehensive array of services, including the administrative offices of the Welland Canal Company. Port Robinson was at the centre of the canal and the point of junction with the Lower Welland River, Welland (with Helmsport at the junction with the Feeder Canal) was where produce from the valleys of the Grand and Upper Welland Rivers flowed towards the canal system, and Dunnville was the entrepôt for the Grand River valley.

With the dual economic base of mills and services, a movable bridge to cross the canal was an imperative ingredient for local success. The canal in between these bridges exerted a barrier effect on land transportation, and routes and traffic were obliged to focus on the bridges that were constructed. The canal villages with their mills, services, and bridges became nodal points in the landscape, with this nodality in turn enhancing their service capability. The village entrepreneurs, the mill owners, and the storekeepers must not be omitted from this schematized pattern of urban growth. Every centre is associated with prominent names who have left a distinguished mark on urban progress.

Urban form was determined by the location of the bridge crossing and the main street approaches to this bridge, with the commercial centre generally emerging on one side of the canal only, a statement applicable to Port Dalhousie, Merritton, Thorold, Allanburg, Port Robinson, Welland, and Dunnville. Port Colborne by contrast grew around its harbour, and Humberstone and St Catharines along their respective main streets. At St Catharines this development also resulted in a one-sided canal community, as St Paul Street ran parallel to the deep canal valley that lay south of the business centre. Urban growth from these slender beginnings was in parallel streets and along the canal, where the new industrial developments took place. Physically, the road grid in response to the canal was not based on the north–south, east–west patterns of the township survey systems, but on roads that ran parallel to the canal that merged with the township system of survey roads by a noticeable change of direction.

Apart from the canal, the growth of each community was in response of its mill and then its industrial developments. By 1960, the Port Dalhousie–St Catharines–Merritton–Thorold agglomeration was the most concentrated urban industrial unit along the canal and within the Niagara Region.

By the volume of employment in each industrial category, the emphasis lay on transportation equipment, especially vehicle parts; the paper and allied industries, with several pulp and paper mills; primary metals and metal fabricating, which extended over a wide range of plants; and the food and beverage industries, including wineries, canneries, and the preservation of fruit.

Industrial sites reflected the canal in Port Dalhousie, at Welland Vale, next to the centre of St Catharines, and along the canal valley through Merritton to Thorold and Allanburg. Railways rather than the canal were reflected in sites around the main-line railway station and to the north-east of the city, where the QEW also exerted a strong influence on industrial location. Large manufacturing plants had moved to or were established in fringe locations on the outskirts of the town, and the city had then subsequently grown so that these industrial areas were set within residential environments and not grouped in a particular sector of the city.

Welland, including Port Robinson, at the centre of the peninsula, was the second largest industrial environment. Here the industrial structure was highly specialized in primary metals with iron and steel, steel-tube, and pipe mills. Non-metallic industries followed, with ferro alloys, carbon electrodes, and artificial graphite being produced by one company. A few large companies and American-owned industry prevailed, as did a linear north–south pattern along the canal, which here included the full sequence of four canals in close proximity to each other.

At Port Colborne, the smallest of the canal communities and fourth after Niagara Falls in the regional sequence, primary metals were the most important industry, mostly in metal smelting and refining followed by the production of pig iron. The availability of water transport to bring in raw materials was of particular importance, but direct association with the canal had diminished because of the through trans-shipment of grain and the increasing proportion of through vessels, which restricted the ship-repair industry. Even so, Port Colborne functioned as a port longer than any other location on the canal and, unlike Port Dalhousie, retained this significant marine status after the Fourth Canal opened.

The growth of these distinct but integrated industrial complexes has included both local demands and pressures from outside Niagara. The years of the two world wars proved to be periods of pronounced growth, but also of advantage have been Canada's links to the British Commonwealth and the Canadian and American spread of settlement to the west and the north, with the canal communities and southern Ontario being an important source of supply for these new and expanding patterns of settlement and receiving their resources. The manufacturing activities have grown, but important segments have also declined or died as new processes, materials, technology, products, or markets came to the fore. Even so, the total industrial complex along the canal grew steadily in terms of employment and output as the nineteenth century advanced into the modern era.

A further growth factor has been the role of industrial incentives and boosterism as towns competed for industry in order to obtain the most favourable reward. This was an acceptable practice until the privilege of being able to grant tax and site incentives was withdrawn from the municipalities in 1958 by provincial legislation. Even so, the canal towns have continued to compete against each other for industrial advantage, despite the introduction of regional government in 1970. They started as individual units in separated locations and have remained independent units, each with their local pride, achievements and enthusiasms, honoured names, and distinctions of urban form and character.

Mills created nodes of development close to the locks and the raceways of the canal's hydraulic system, but as this space became exhausted industry gradually extended linearly along the canal banks, from St Catharines along the valley of the Second Canal, then south of Thorold, and from Welland south towards Port Colborne. As a consequence, the canal communities have inched gradually towards each other and linked slowly but steadily together into one extensive urban unit along the canal banks.

At the same time, the transition was from nodes to the peripheral growth of each settlement as houses and other service developments were added steadily to the towns and villages that grew along the canal. The pattern of settlement changed slowly but steadily under these forces from a series of independent communities to more of an intertwined unity and an urban-industrial system that was related closely to the canal membrane. The original settlements and the industrial axis were both reinforced as industries and transportation developments took place and changed the canal landscape.

The spacing of communities along the canal became an important factor in their growth. Port Dalhousie and Merritton became subordinate to St Catharines and Humberstone to Port Colborne. And when the prized accolade of county achievement was awarded, this superior administrative status went first to Welland and then to St Catharines, which were encouraged to expand as regional service centres. Thorold, between the two and off-centre in Welland County, declined before the more substantial advance of Welland. As communication between the canal centres was improved by railways, the streetcar, and then by modern highways and the growing use of the motor car, the tendency was to visit the nearest larger centre with its greater choice of service facilities. From Humberstone it was easy to travel to Port Colborne, and from Port Dalhousie and Merritton the local journey was to St Catharines. Meantime, neither Allanburg nor Port Robinson could sustain a high level of service provision and both declined as centres.

The canal communities have gained and lost from the canal endeavour. It has not been a consistent story of collective advance. Chippawa then Dunnville and the canal communities along the Feeder Canal were the first to be divorced from their canal foundations, followed by Merritton, Thorold, and the business district of St Catharines as the Third Canal reached fruition. Port Dalhousie lost its canal basis of importance when the Fourth Canal opened, and Allanburg and Port Robinson were in essence bypassed and declined from the era of the Third Canal onwards.

Physical changes to the canal environment have also reverberated back into the communities served by the canal. The reconstruction of the Fourth Canal under the aegis of the St Lawrence Seaway Authority led to a canal By-Pass around Welland and to the abandonment of the already established route through the centre of that city. At Port Colborne the transfer of the entry lock for the Second Canal from the lakeshore to the head of the harbour had jeopardized early commercial development, and the construction of the Fourth Canal later removed the east-side commercial street that faced the canal.

The canals, when abandoned, have taken on many new forms. They have been infilled and graded to accord with the level of the adjacent land, then used for urban purposes including vehicle parking, outdoor storage, and some public buildings, such as a fire station at Thorold and a linear area of light industry across north-east St Catharines. The Feeder canal has become a series of isolated, sometimes stagnant ponds, and the abandoned Second Canal from Thorold to St Catharines an objectionable sewer. Elsewhere, the flow of water has been maintained, providing a home for the Henley Rowing Regatta at Port Dalhousie and contributing to a recreational waterway at Welland.

At first the size of the canal village and its service activities depended on the population living in the village and its nearby farming communities. The docks, wharfing, service, and repair facilities provided along the canal encouraged local businesses, which, together with the mills, were the mainstay of the canal communities. A contrary factor was that, as the proportion of through vessels increased and as steam then diesel vessels came into operation, the connection between the passage of vessels and the canal communities was reduced. The exception obtained at the terminal ports, where there was also the advantage of winter tie-ups when services and repairs might be undertaken.

The most recent attribute of the canal is a growing awareness of the tourist-recreational potential that exists along the banks of the old and the modern waterways. This outlook began in the late 1920s when the Fourth Canal was nearing completion. The idea of a lake-to-lake boulevard was then advanced by the canal authorities and extensive tree planting was undertaken towards this achievement. The Niagara Parks Commission then envisaged a continuous parkway along the Niagara River linked to another parkway along the canal. After many reports and much failure to act, progress towards a Welland Canals Parkway was initiated by the regional municipality in 1995. The canal scene, old and new, is urban waterfront.

The growth of the canal communities has relied on an influx of population from the rural surrounds and elsewhere in Canada (mostly Ontario), through most of the nineteenth century on mainly British (English and Scottish) immigration from abroad, and later on immigration from southern and eastern Europe. The canal played its role in diversifying this pattern, at first through using an Irish labour force and then through the job opportunities provided for the duration of its later construction projects. The canal communities were changed by these means, from predominantly British and Protestant to a wider ethnic and cultural mosaic. The following description of the arrival and settling of an immigrant in Welland is likely typical: 'My father came as a young man, ... through Ellis Island ... in 1915 ... And then he travelled through the States and worked his way up into Canada and then [in 1928] he worked on the [Fourth] Welland Canal. He had a rough time. Then he sent for my mother ... She had a very difficult time because she had no relatives – nobody here. So it was very difficult for her. But she managed also. They had their own Italian groups and they stuck together. And they all helped each other. They had their halls and dances and meals together. And it really worked well for them' (Burns 1994).

The communities were distinctive – both in their types of immigrants and in their manufacturing base and the work opportunities they offered. They were also distinctive in their setting, with different agricultural and mineral resources in their vicinity. For example, limestone obtained from the Onondaga Escarpment was either used in or shipped from Port Colborne, whereas in the north the Niagara Escarpment was quarried for building stone and to construct the locks of the Second and Third Canals. Clay for bricks was to be found in several locations, and brick making became important in St Catharines and Welland.

Minerals also provided St Catharines with an early advantage as salt was a pioneer product in great demand to flavour and to preserve, leading later to bottled mineral waters, spa baths, and resort hotels. In the south natural gas and the sandy beaches of Lake Erie provided industrial and recreational advantages that were lacking in the north, until sand was introduced from the Fonthill area at Port Dalhousie to extend its sandbar and promote its summer recreational abilities. The roles of country town and local marketing centre favoured the north over the south of the peninsula, for the soil, in depth and quality, was richer along the Ontario Plain than the Erie Plain; the land was also more closely settled, and the population higher.

Regionally, St Catharines during its early canal years was larger than Hamilton. It had more factories, but as the twentieth century advanced, Hamilton enjoyed the advantages of a wider catchment area, better distribution facilities, and greater promotional and marketing strengths as Canadian and American settlements expanded west of the Great Lakes (Weaver 1982; Dear et al. 1987). The same might be said about the attendant growth of Buffalo, Niagara Falls, and Tonawanda in the United States relative to the American West (Brown and Watson 1981; Jackson and Stein 1995; Whittemore 1977). The American industrial strip along the eastern side of the Niagara River in association with the Erie Canal was greater by far than the more subdued development achieved along either the river or the Welland Canal on the Canadian side.

Industrial expansion along the Welland Canal also meant that the growing strengths of its canal communities were increasingly off-centre within the Canadian context. They were sandwiched between Buffalo and Hamilton when viewed from the North American perspective of urban-industrial advance and were on the fringe of Canadian experience as settlement advanced to the west and north of the nation. The canal from its inception to this day has always had to operate as part of a much greater series of external influences shaping its character, and the same is true of the communities it created. The canal is both a significant agent of land development and a waterway of international importance. The two forces intertwined have created its modern character.

Bibliography

Note: The Main Library at Brock University, St Catharines, in particular its Special Collections Division, holds most of the canal and other rare local reference materials noted below.

Abbott (An Emigrant Farmer). 1844. *The Emigrant to North America*. London, Eng.: Blackwood

Acres Counsulting Services. 1974. *Highway 406 (Welland) Feasibility Study*. Niagara Falls: The Company

Acres, H.G., & Company. 1968. *Thorold Tunnel*. Niagara Falls: The Company

Adams, J.R., ed. 1958. *Welland Centennial, 1858–1958*. Welland: Corporation of Welland

Addis, F. 1979. 'Shipyards of the Early Welland Canals (1828–1933).' In J. Burtniak and W.B. Turner, eds, *The Welland Canals*. Proceedings of the First Annual Niagara Peninsula History Conference, Brock University, St Catharines

Aitken, H.G.J. 1952. 'The Family Compact and the Welland Canal Company.' *Canadian Journal of Economics and Political Science* 18

– 1954. *The Welland Canal Company: A Study in Canadian Enterprise*. Cambridge: Harvard University Press

Albery, Pullerits, Dickson and Associates. 1975. *Water Quality and Remedial Action for Abandoned Welland Canal at Welland*. Don Mills, Ont.: The Company

Aloian, C. 1978. *A History Outline of Port Dalhousie, 1650–1960*. Port Dalhousie: Port Dalhousie Works

ARA Consulting Group et al. 1991. *A Financial Impact Analysis of the Proposed Transfer of the Old Welland Canal Lands in the City of Welland*. Toronto: The Company

Archaeological Services. 1990. *A Survey of Historic Structures – The Welland Canal Industrial Corridor*. St Catharines: Welland Canals Society

Arcturus Environmental Limited. 1994. *Henley – Rowing Course / Martindale Pond Environmental Site Assessment for Canadian Henley Rowing Corporation*. Niagara Falls: The Company

Ashdown, D. 1988. *Railway Steamships of Ontario*. Erin, Ont.: Boston Mills Press

Atkinson, H. 1979. 'Aspects of Engineering of The Welland Canals.' In J. Burtniak and W.B. Turner, eds, *The Welland Canals*. Proceedings of the First Annual Niagara Peninsula History Conference, Brock University, St Catharines

Atkinson, K. (Chairman, Sub-Committee on the St Lawrence Seaway). 1992. *The Future of the Great Lakes / St Lawrence Seaway System: Report of the Standing Committee on Transport*. Ottawa: House of Commons

Atkinson, M.B. 1928. 'Bridges over the Welland Ship Canal.' *Engineering Journal* 12.2

Atlas Steels Limited. 1948. *Steelmaking at Atlas Steels Limited*. Welland: The Company

– 1952. *Annual Report 1952*. Welland: The Company

Barber, T.L. 1960. *Pilkington Brothers and the Glass Industry*. London: George Allen & Unwin

Barnsley, R. 1987. *Thomas B. McQuesten*. Markham, Ont.: Fitzhenry & Whiteside

– 1993/94. 'Turning Barren Land into Lush Parkway.' In *Planting a Forest*, QUNO, 1.2

Barnsley, R., and J.H. Pierce. 1989. *The Public Garden and Parks of Niagara*. St Catharines: Vanwell

Bassett, J.M., and A.R. Petrie. 1974. *William Hamilton Merritt: Canada's Father of Transportation*. Don Mills: Fitzhenry & Whiteside

Beckley, M. 1994. *Ontario's Tourism Industry: Opportunity – Progress – Innovation: The Report of the Advisory Committee on a Tourism Strategy for the Province of Ontario*. Toronto: Ministry of Culture, Tourism and Recreation, Queen's Printer for Ontario

Bell, W.J., and A.E. Berry. 1935. 'Sanitary Situation in Sections of the Old Welland Canal.' *Canadian Engineer* 68

Bender, F. 1950. 'The Story of Chippawa.' In *Official Program, Chippawa Centennial, 1850–1950*. Niagara Falls: Printed by F.H. Leslie

Bill to Improve and Amend the Communications Between the Lakes Erie and Ontario, By Land and Water. 1799. Niagara: Printed by S. And G. Tiffany; reproduced by the Library, Brock University, St Catharines, 1968

Bird, J.H. 1971. *Seaports and Seaterminals*. London: Hutchinson

Bird, M., and D. Bird, and C. Corbe. 1974. 'The Foster Glass Works.' Typescript. St Catharines: The Authors

Bleasdale, R.E. 1975. 'Irish Labourers on the Cornwall, Welland and Williamsburg Canals in the 1840s.' MA thesis, University of Western Ontario, London
– 1978. *Class Conflict on the Canals of Upper Canada in the 1840s.* Ottawa: Canadian Historical Association
Bodo, N.L. 1990. *Camp Ground Development – Old Welland Canal.* Welland: The Author
Bloomfield, E.P., and G.T. Bloomfield. 1983. *Urban Growth and Local Services: The Development of Ontario Municipalities to 1981.* Guelph: Department of Geography, University of Guelph
Bond, R.C. 1964. *Peninsula Village: The Story of Chippawa.* Niagara Falls: Lindsay Press
Bonnycastle, Sir R.S. 1846. *Canada and the Canadians in 1846.* London: Colburn
Bouchard, L.J. 1946. 'Welland.' *Société Historique de Nouvel Ontario* (Sudbury) 10
Bouchette, R.S.M. 1867. *Welland Canal – Statement Showing the Number of Trips Made by Steamer, Sailing and Other Vessels, also Their Freight ...* Ottawa: Customs Department
Bourton, E.W. 1930. 'Port Robinson, Then and Now.' Typescript. Port Robinson
Bradstreet's Book of Commercial Ratings. 1911. New York: The Company
Brochu, P. 1989. 'Impact of the French Community on the City of Welland.' Unpublished Applied Urban Geography research project. Department of Geography, Brock University, St Catharines
Brock, R.W. 1930. 'The Welland Ship Canal and Its Place in the Progress of the Niagara Penninsula.' *Industrial Canada*
Brock University Library (St Catharines), Special Collections. 1841–80. Commissioners of Public Works, Register B of Canal Leases, Deeds, etc. St Lawrence Seaway Authority
– c. 1850. Welland Canal Book 2, Lock 5 to Port Robinson
Brooks, S., P. McAleese, and D. Wilson. 1988. *Sculls & Shells: A St Catharines Tradition.* St Catharines: Canadian Henley Rowing Regatta
Brosius, H. 1875. *St Catharines, 1875.* Chicago: Lithographed by C. Shober
Brown, G.W. 1926. 'The Opening of the St Lawrence to American Shipping.' *Canadian Historical Review*
Brown, J.G. 1850. *Essay on the Advantages of the Canals to the Farmers of Canada.* Toronto
Brown, R.S., and B. Watson. 1981. *Buffalo: Lake City in Niagara Land.* Buffalo: Windsor Publications in cooperation with the Buffalo and Erie County Historical Society
Burghardt, A.F. 1969. 'The Origin and Development of the Road Network of the Niagara Peninsula, 1770–1851.' *Annals of the Association of American Geographers* 59.3

Burns, J. (Guest Curator) 1994. *The Italian Experience in Welland*. Welland: Welland Historical Museum
- 1995. *The Historical Community in Welland*. Welland: Welland Historical Museum
Burtniak, J., and W.B. Turner, eds. *The Welland Canals*. Proceedings of the First Annual Niagara Peninsula History Conference, Brock University, St Catharines
Burton, A. 1989. *The Great Days of the Canals*. Newton Abbott, Eng.: David and Charles
Cameron, E.G. 1929. 'Construction Methods on the Welland Ship Canal.' *American Society of Civil Engineers, Papers*
Campbell, M.S. 1973. 'Speech at the Symposium Held at Welland on April 25, 1973.' Typescript. St Lawrence Seaway Authority, St Catharines
Campbell, W. 1972. *Canada Post Offices, 1755–1895*. Lawrence, Mass.: Boston Quarterman Publications
Canada Statistical Abstract and Record for the Year 1888. 1889. Ottawa: Department of Agriculture
The Canada Year Book. 1905–. Ottawa: Census and Statistics Office
Canadian Marine Publications. Annual. *Great Lakes Navigation*. Montreal: CMP
Canal Commission. 1871. 'Letter to the Honorable the Secretary of State from the Canal Commissioners Respecting the Improvement of the Inland Navigation to the Dominion of Canada. In *Sessional Papers*, no. 54, vol. 4.6 (Ottawa)
Carr, D.W., & Associates. 1970. *The Seaway in Canada's Transportation: An Economic Analysis*. 2 vols. For the St Lawrence Seaway Authority. Ottawa: The Consultants
Carr, J. 1992. 'Hydro-Electric Power Developments in the Niagara Peninsula.' In J.N. Jackson and J. Burtniak, eds, *Industry in the Niagara Peninsula*. Proceedings of the Eleventh Annual Niagara Peninsula History Conference, Brock University, St Catharines, 1989
Carter, D. 1923. 'The First Welland Canal.' Port Colborne *Citizen*, 29 March to 8 November
- 1928. 'Outlines Position of New Welland Canal.' *Toronto Mail & Express*, 27 March
Carter, M. 1985. 'Willson, Thomas Leopold. In *The Canadian Encyclopedia*. Edmonton: Hurtig
Carvell, J.T. (Chairman) 1973a. *Report: Future Disposition of the Section of the Fourth Canal Which Will Be Replaced by the By-Pass*. Ottawa: St Lawrence Seaway Authority
- 1973b. *Task Force Report on Disposal of Canal Lands*. Ottawa: St Lawrence Seaway Authority

Chevrier, L. 1959. *The St Lawrence Seaway*. New York: St Martin's Press

Chief Engineer's Branch. 1929. *Report on the Physical Conditions Existent at Various Industries, Operating Under Lease or Attornment from the Department, along the Old Welland Canal and Hydraulic Raceway Between Locks No. 1 and No. 25*. Ottawa: Department of Railways and Canals

Christensen, C.J. 1976. *History of Engineering in Niagara*. St Catharines: Niagara Peninsula Branch, Engineering Institute of Canada

Clay, G. 1973. *Close-up: How to Read the American City*. New York: Praeger

Clinton, G. 1918. 'Evolution of the New York Canal System.' *Buffalo Historical Society Publications*

Commissioners, State of Illinois. 1871. *Sessional Papers no. 54*, vols. 4 & 6. Ottawa

Committee on Transportation. 1875. *Paper on the Subject of Enlargement of the Welland Canal ...* St Catharines Board of Trade

Comuzzi, J. (Chair, Sub-Committee on the St Lawrence Seaway, House of Commons). 1994. Report to Honourable Douglas Young, Minister of Transport, 25 October (referred to as Comuzzi Report)

Cooley, C.H. 1894. *The Theory of Transportation*. Washington: American Economic Association

Coombs, A.E. 1930. *History of the Niagara Peninsula and the New Welland Canal*. Toronto: Historical Publishers Association; revised 1950

Coopers and Lybrand Consulting Group. 1993. *Critical Review of Your Recommended Federal Endowment Payment Request for Old Welland Canal Lands*. North York, Ont.: The Company

Co-ordinating Committee, Welland Canals Corridor Interior Management Team. 1989. *Report to Niagara Regional Council*. St Catharines: The Committee

Corbett, R.A. 1992. *The Future of the Great Lakes / St Lawrence Seaway System*. Report of the Standing Committee on Transport. Ottawa

Cowan, P.J. 1935. *The Welland Ship Canal between Lake Ontario and Lake Erie, 1913–1932*. London, Eng.: Offices of Engineering

Craick, W.A. 1917. 'The Industrial Evolution of Welland.' *Industrial Canada*, November

Craig, G.M. 1963. *Upper Canada: The Formative Years 1784–1841*. Toronto: McClelland and Stewart

Creighton, D.G. 1956. *The Empire of the St Lawrence*. Toronto: Macmillan; published 1937 as *The Commercial Future of the St Lawrence, 1760–1850*

Creighton, O. 1830. *General View of the Welland Canal*. London: Miller

Cresswell, P.R. 1994. 'Suffering Industry Is on Course for Recovery.' Address to the St Catharines Rotary Club, St Catharines

Cruikshank, E.A. 1887. *The History of the County of Welland*. Welland: Welland Tribune Printing House; reprinted Belleville, 1972

- 1925. 'The Inception of the Welland Canal.' *Ontario Historical Society Papers and Records* 22
- 1932. 'The Conception, Birth and First Steps of the Welland Canal.' In L.B. Duff, ed., *Welland Ship Canal Inauguration*. St Catharines: Commercial Press
Crysler, J.M. 1943. *A Short History of the Township of Niagara*. Niagara: Niagara Advance
Cumberland, B. 1913. *A Century of Sail and Steam on the Niagara River*. Toronto: Musson
Cuthbertson, G.A. 1931. *Freshwater*. Toronto: Macmillan
Damas and Smith. 1975. *City of Welland Redevelopment Proposals*. Willowdale, Ont.: The Company, for Diamond and Myers
Dear, M.J., J.J. Drake, and L.G. Reeds, eds. 1987. *Steel City: Hamilton and Region*. Toronto: University of Toronto Press
Denis, L.G., and A.V. White. 1911. *Water-Powers of Canada*. Ottawa: Commission of Conservation Canada
Department of Planning and Development (Thorold). 1983a. *Tourism in the Niagara Region – Report No. 1: Economic Impact ... Tourist Characteristics*. Thorold: Regional Municipality of Niagara
- 1983b. *Tourism in the Niagara Region – Report No. 2: Marketing, Trends and Tourism*. Thorold: Regional Municipality of Niagara
- 1984. *Tourism in the Niagara Region – Report No. 3: Some Tourist Attractions – Present and Possible*. Thorold: Regional Municipality of Niagara
- 1985. *Tour Niagara's Waterways*. Thorold: Regional Municipality of Niagara
- 1989. *Developing the Welland Canals Corridor, Existing Organizations, Partnership, Focus*. Thorold: Regional Municipality of Niagara
- 1994a. *Proposed Welland Canals Parkway Development Plan*. Thorold: Regional Municipality of Niagara
- 1994b. *Summary of Responses to Circulation of the Proposed Welland Canals Parkway Development Plan*. Thorold: Regional Municipality of Niagara
- 1994c. *Proposed Regional Policy Plan Amendment 94: Welland Canals Parkway*. Thorold: Regional Municipality of Niagara
Department of Railways and Canals. 1909. *Welland Canal: Port Colborne Harbour and Elevator*. Ottawa: The Department
- 1920. *Welland Ship Canal (Under Construction)*. Ottawa: King's Printer
- 1932. *The Opening of the Welland Ship Canal ...* Ottawa: King's Printer
- 1935. *Welland Ship Canal 1934*. Ottawa: King's Printer
Department of Transport. 1937. *The Canals of Canada*. Ottawa: King's Printer
- 1947. *Rules and Regulations for the Guidance and Observance of Those Using and Operating the Canals of Canada under the Jurisdiction of the Department of Transport*. Ottawa: King's Printer

– 1953. *The Canals of Canada*. Ottawa: Queen's Printer

D.G.B. 1979. 'Port Colborne Nickel Refiners.' Typescript. Port Colborne: Inco Metals Company

Diamond and Myers. 1975. *Welland Federal Lands: Planning Study and Conceptual Design*. Toronto: The Consultants, for Public Works Canada

Dillon, M.M. 1972. *City of Port Colborne – Township of Wainfleet Decision Highways: Summary of Feasibility Study Report*. Toronto: The Company

Dolynsky, D., and M. Harding. 1994. 'Down But Not Out.' *Niagara Business Report* (St Catharines), Spring

Dominion Bureau of Statistics. 1950. *Eighth Census of Canada 1941*. Ottawa: Ministry of Trade and Commerce

Doral Holdings. 1990. *Golf Course Development – Old Welland Canal*. Welland: The Company

Douglas, R.C. 1884. *Report on the Necessity of Deepening the Welland Canal*. Ottawa: MacLean, Rogers & Co.

Duff, L.B. 1932. 'Paper Making in Thorold.' In *Welland Ship Canal: Inauguration*. St Catharines: Commercial Press

Duff, L.B., ed. 1930. 'The Original Welland Canal.' In *Welland Ship Canal*. St Catharines: Commercial Press

Dun, Wiman & Co. 1878, 1884. *The Mercantile Agency Reference Book for the Dominion of Canada*. Montreal: The Company

Duquemin, C.K. 1979. *The Historic Welland Canal*. 2 vols. (Outdoor Studies Pamphlets nos. 106 and 109) Welland: St Johns Outdoor Studies Centre, Niagara South Board of Education

– 1980. *A Guide to the Grand River Canal*. St Catharines: St Catharines Historical Museum, Publication no. 1

Easterbrook, W.T., and H.G.J. Aitken. 1958. *Canadian Economic History*. Toronto: Gage

Easterbrook, W.T., and M.H. Watkins, eds. 1967. *Approaches to Canadian Economic History*. Toronto: McClelland and Stewart, Carleton Library no. 31

Egerter, M.E. 1995. 'Seaway Cuts off Walker's Pleasure.' St Catharines *Standard*, 24 April

Engel, H. 1978. 'That Meritorious Work: The Welland Canal.' *Queen's Quarterly* 85.3

Engineer-in-Charge. 1931. *Welland Ship Canal 1931*. St Catharines: Engineer's Office

Environmental Services. 1990. *Environmental Assessment and Review Process: Initial Assessment – Old Welland Canal Lands, Welland, Ontario*. Willowdale, Ont.: Public Works Canada

Exolon Company. 1984. *The First Fifty Years, 1914–1964.* Tonawanda, NY: The Company

Fenco Engineering. 1986. *Port Dalhousie Marina: Report to Landcorp Ontario.* St Catharines: The Company

Forsey, E. 1981. *Trade Unions in Canada, 1812–1902.* Toronto: University of Toronto Press

Fraser, M. 1979. 'Conserving Ontario's Main Street.' Proceedings of the conference at Trent University, Peterborough, on 24, 25, 26 August 1978. Toronto: Ontario Heritage Foundation

Gardiner, J.C. 1913. 'Re Power Development at Jordan.' Letter to Minister of Railways and Canals, 22 January. St Lawrence Seaway Authority (St Catharines), File 33-2-2-2

Gayler, H.J., ed. 1994. *Niagara's Changing Landscapes.* Nepean, Ont.: Carleton University Press

Gentilcore, R.L. 1984. *Ontario's History in Maps.* Toronto: University of Toronto Press

Gibb, J. 1991. *A Festival of Canadian Art: Welland's Giant Outdoor Murals.* St Catharines: Lincoln Graphics

Gillham, G. 1992. *The Ships of Port Weller.* St Catharines: Riverbank Traders

Glazebrook, G.P. de T. 1964. *A History of Transportation in Canada.* Toronto: McClelland and Stewart, Carleton Library no. 11

– 1968. *Life in Ontario: A Social History.* Toronto: University of Toronto Press

Goodman, G.E. 1966. 'A Case Study of Urban Growth [Welland].' Unpublished paper, Department of Geography, University of Western Ontario (London)

Gore and Storrie. 1939. *Report on Disposal of Sewage and Industrial Wastes in Second Welland Canal Area.* For Department of Transport (Ottawa)

Gorman, L.H. 1966. 'Uses of the Welland River: Past, Present and Future.' Unpublished paper, Department of Geography, University of Toronto

Gourlay, R. 1822. *General Introduction to Statistical Account of Upper Canada ...* London: Simpson and Marshall

Government of Canada. 1979. *The St Lawrence Seaway and Its Regional Impact.* Ottawa: Ministry of Supply and Services Canada

Grand, A.J. 1928. 'Chippawa Creek Syphon Culvert of the Welland Ship Canal.' *Engineering Journal* 12.2

Grant, A.J. (Engineer in Charge, Welland Ship Canal). 1927. *Welland Ship Canal.* Welland County Historical Society, vol. 3

– 1931. *Annual Report Fiscal Year 1929–1930.* Ottawa: Department of Railways and Canals

– 1932. 'Welland Ship Canal.' In L.B. Duff, *Welland Ship Canal: Inauguration.* St Catharines: Commercial Press

Great Lakes Waterways Development Association. 1960. *Submission ... to the Minister of Transport and Government of Canada Concerning Tolls on Canadian Waterways and Particularly on the Welland Canal.* Toronto: The Association

– 1963. *Submission ... to the St Lawrence Seaway Authority Concerning Tolls on Canadian Waterways.* Toronto: The Association

– 1965. *Submission ... to the St Lawrence Seaway Authority.* Toronto: The Association

– 1966. *Submission with Respect to St Lawrence Seaway Tolls and Welland Canal Lockage Fees ... to the St Lawrence Seaway Authority.* Toronto: The Association

Greater Port Colborne Chamber of Commerce. 1980. 'The Impact of the Welland Canal / St Lawrence Seaway on the City of Port Colborne.' A brief to the Great Lakes / Seaway Task Force for Ontario, 8 July

Green, E. 1925. 'John De Cou, Pioneer,' *Ontario Historical Society Papers and Records* 23

– 1926. 'The Niagara Portage Road.' *Ontario Historical Society Papers and Records* 23

– 1930. 'Before the Canal.' In L.B. Duff, ed., *Welland Ship Canal: Inauguration.* St Catharines: Commercial Press

Green's Great Lakes and Seaway Directory. 1909–. North Olmsted, Ohio: M.E. Green (later Mitchell and Company's *Marine Directory of the Great Lakes*)

Greenwald, M. 1975. 'Ships Climbing the Mountain: A Brief History of the Welland Canal.' Unpublished manuscript, Ministry of Culture and Recreation (Toronto)

– 1976. *The Welland Canals Study: A Summary.* Toronto: Ministry of Culture and Recreation

Greenwald, M., et al. 1976. *The Welland Canals: Historical Resource Analysis and Preservation.* Toronto: Ministry of Culture and Recreation

– 1984. 'Interministerial Task Force on the Welland Canal.' Typescript. Toronto: Ministry of Citizenship and Culture

Gregg, E.S., and A.L. Crecher. 1927. *Great Lakes-to-Ocean Waterways: Some Economic Aspects of the Great Lakes – St Lawrence, Lakes-to-Hudson, and All-American Waterway Projects.* Washington: Department of Commerce

Groh, I. [1969]. 'The Negroes of the Niagara Peninsula.' Typescript. St Catharines: The Author

Guillet, E.C. 1933. *Early Life in Upper Canada.* Toronto: Ontario Publishing Company

Guy, St D. 1993. 'An Erie Canal for Western Upper Canada: A Forgotten Episode in Ontario's Transportation Evolution.' *Ontario History* 85.3

Hadfield, C. 1968. *The Canal Age.* Newton Abbott, Eng.: David and Charles
- 1986. *World Canals: Inland Navigation Past and Present.* Newton Abbott, Eng.:
David and Charles
Hall, F. 1830. *Report of the Engineer Appointed to Examine the Works upon the
Welland Canal.* St Catharines: Welland Canal Office
Hansen, M.L. 1970. *The Mingling of the Canadian and American Peoples.* New
York: Arno Press
Hardy, G. 1993. *Welland Canal Parkway Development.* Thorold: Region of Nia-
gara Tourist Council
Harris, A.C. (Superintendent). 1917. *Operating Rules: Welland Ship Canal Con-
struction Railway.* Ottawa: Government Printing Bureau
Harris, W.R. 1895. *The Catholic Church in the Niagara Peninsula, 1626–1895.*
Toronto: Briggs
Harvie, F.W. 1950. *Town of Thorold: Centennial 1850–1950.* Thorold: The Corpo-
ration
Hayes-Dana Limited. 1972. *50th Anniversary, 1922–1972.* Thorold: The Company
Heisler, J.P. 1973. *The Canals of Canada.* Ottawa: Department of Indian Affairs
and Northern Development, Occasional Papers in Archaeology and History,
no. 18
Hill, B.E. 1971. 'The Grand River Navigation Company and the Six Nations Indi-
ans.' *Ontario History* 63
Hill, H.W. 1918. 'Historical Sketch of Niagara Ship Canal Projects.' *Buffalo His-
torical Society Publications* 22
Higgins, F.P. 1956. *Proposed Amalgamation with Part of the Township of Crow-
land.* Toronto: The Company, for the Corporation of the City of Welland
Historical and Descriptive Sketch of the County of Welland. 1888. Welland: Sawle
and Snartt
Historica Research (C. Andreae). 1988. *Archaeological Excavation of Lock 24, First
Welland Canal.* London, Ont.: The Company
The History of the County of Welland ... 1887. Welland: Welland Tribune Printing
House
Howison, J. 1821. *Sketches of Upper Canada ...* Edinburgh: Oliver and Boyd
Imlach, W.I. 1900. *An Old Man's Memories: Reminiscences of Port Maitland's
Early History and Life ...* Reprinted from the Dunnville *Chronicle*
International Niagara Committee. 1982. *Report by On-Site Representatives to the
International Niagara Committee on Welland Canal Diversions*
Jackson, J.N. 1969. 'The Welland Canal: Its Initial Impact on Urban Development
in Relation to Alternative Possibilities of Canal Location.' Unpublished manu-
script, Brock University, St Catharines

- 1974. *A Planning Appraisal of the Welland Urban Community: Trends – Transition – Potential.* St Catharines: The Author, for Public Works Canada
- 1975a. 'The Recreational Potential of the Welland Canal System in the Niagara Peninsula of Southern Ontario.' Submission to the Provincial Parks Council, 20 October
- 1975b. *Welland and the Welland Canal: The Welland Canal By-Pass.* Belleville: Mika Publishing
- 1976. *St Catharines, Ontario: Its Early Years.* Belleville: Mika Publishing
- 1981. 'Canalscape Features of the Welland Canals.' Paper to the Annual Meeting of the Canadian Association of Geographers, St John's, Newfoundland
- 1986. 'Names along the Welland Canals.' *Canoma* 12.2
- 1989. *Names Across Niagara.* St Catharines: Vanwell
- 1993a. *Ideas for the Niagara Region: A Welland Canal Parkway.* St Catharines: The Author
- 1993b. *Recommendations for a Proposed Welland Canals Parkway and Heritage System of Development.* St Catharines: The Author
- 1993c. 'The Welland Canals as a Catalyst for Urban-Industrial Achievements.' In *International Waterways, Papers presented at an international meeting held at Buffalo, USA on 20th June 1992.* Bridgewater, Eng.: IWA International Committee

Jackson, J.N., and F.A. Addis. 1982. *The Welland Canal: A Comprehensive Guide.* St Catharines: Welland Canals Foundation

Jackson, J.N., and J. Burtniak. 1978. *Railways in the Niagara Peninsula.* Belleville: Mika

Jackson, J.N., and G.P. Stein. 1995 (pending). *One River, Two Frontiers: Settlement across the International Boundary at Niagara.* Buffalo: Western New York Heritage Institute, Canisius College

Jackson, J.N., and C. White. 1971. *The Industrial Structure of the Niagara Peninsula.* St Catharines: Brock University, Department of Geography.

Jackson, J.N., and M. Wilson. 1992. *St Catharines: Canada's Canal City.* St Catharines: St Catharines *Standard*

Jewett, F.C. (Division Engineer). 1927. 'Memo Re Reforestation and General Maintenance Sections Nos. 1 & 2 Welland Ship Canal, March 31, 1927.' In National Archives (Ottawa), Department of Railways and Canals, RG 43.2159, file 548-1927

Johnson, E.P. 1926. 'Canal Engineering Yesterday and Today.' *Ontario Historical Societies Papers and Reports* 23

Johnson, J.K., ed. 1975. *Historical Essays on Upper Canada.* Toronto: McClelland and Stewart. Carleton Library no. 82

'Junius' (pseud. Oliver Seymour Phelps). 1856. *St Catharines, A to Z*. Reproduction of newspaper articles from the St Catharines *Journal*

Kates, J., and Associates. 1965. *St Lawrence Seaway Traffic Studies*. Toronto: The Consultants, reprinted for the St Lawrence Seaway Authority

Keefer, Rev. R. 1935. *Memories of the Keefer Family*. Norwood, Ont.

Keefer, S. 1852. 'Letter to the Secretary, Public Works, 22 December 1852.' *Journal of Legislative Assembly*, appendix YYY

Keefer, T.C. 1850. *The Canals of Canada: Their Prospects and Influence*. Toronto: Andrew H. Armour & Co.

– 1864. 'Travel and Transportation.' In H.Y. Hind et al., *Eighty Year Progress of Baptist North America* (1864). Toronto: Nichols

– 1920. *The Old Welland Canal and the Man Who Made It*. St Catharines: The Print Shop

Kerr, D., and D.W. Holdsworth. 1987, 1990, 1994. *Historical Atlas of Canada*. 3 vols. Toronto: University of Toronto Press

Kingsford, W. 1865. *The Canadian Canals: Their History and Cost ...* Toronto: Rollo and Adam

Kingsley, A.E. 1958. *Seaway Tolls*. Reprint of articles from St Catharines *Standard*

Koabel, L.A. 1970. 'Port Colborne: The Changing Core.' BA thesis, Department of Geography, Brock University, St Catharines

Koene, W. 1984. *Loyal She Remains: A Fictional History of Ontario*. Toronto: United Empire Loyalist Association of Canada

Krushelnicki, B. 1994. 'The Progress of Local Democracy in Niagara: The Evolution of Regional Government.' In H.J. Gayler, ed., *Niagara's Changing Landscapes*. Ottawa: Carleton University Press

Kurchak, D., and J. Lafferty, ed. 1973. *Canada's Canal Zone: A Brief History of Thorold & A Guide to the Welland Canal*. Thorold: Kiwanis Club of Thorold

Lake Carrier's Association. 1885–. *Annual Report of the Lake Carrier's Association*. Cleveland: The Association Annual

Landcorp Ontario. 1986. *Port Dalhousie Marina*. St Catharines: The Company

Langevin, H.L. (Minister of Public Works). 1871. 'Memorandum to the Canal Commission.' In *Sessional Papers*, no. 54, vol. 4.6 (Ottawa)

Lanmer Consultants. 1984. *Port Dalhousie Harbour Study: Harbour Plan – Report to City of St Catharines*. St Catharines: The Consultants

LBA Consulting Partners. 1978. *The Seaway in Winter: A Benefit–Cost Study*. Ottawa: The Consultants, prepared for the St Lawrence Seaway Authority

Leeson, M.W. (Chairman, Centennial Committee). 1974. *Merritton Centennial, 1874–1974*. Merritton: The Committee

Legal and Commercial Exchange of Canada. 1894. *The Legal and Commercial Exchange of Canada (Mercantile Agency)*. Toronto: The Company

Legget, R.F. 1976. *Canals of Canada*. Vancouver: Douglas, David and Charles
- 1979. *The Seaway*. Toronto: Clark Irwin
Leitch, J.D. 1988. 'The Welland Canal – Who Needs It?' William Hamilton Merritt Lecture Series, Brock University, St Catharines
- 1993. *Meeting the Welland / Seaway Shipping Challenges*. St Catharines: Welland Canals Foundation
Leung, F.L. 1986. 'Direct Drive Waterpower in Canada: 1607–1910.' Presented to the Historic Sites and Monuments Boards of Canada (Ottawa)
Local Architectural Conservation Advisory Committee to the City of Welland. 1992. *Historical and Architectural Reflections of the Founding Peoples of Welland*. Welland: The Committee
Longo, W.C. (Mayor of Thorold). 1992. *Minutes of Proceedings and Evidence of the Sub-Committee on the St Lawrence Seaway on Transport*. House of Commons (Ottawa), issue 6
Lovell, J. 1851. *The Canada Directory for 1851*. Montreal: The Company
- 1857. *The Canada Directory for 1857–58*. Montreal: The Company
Lovell's Business and Professional Directory of the Province of Ontario for 1882 ... 1882. Montreal: The Company
Lovell's Province of Ontario Directory for 1871 ... 1871. Montreal: The Company
Luce, A.M., and P. Sandor. 1965. 'A Systems Approach to the Problem of Increasing the Effective Capacity of the Welland Canal.' Transportation Research Forum, Annual Meeting (New York)
Lynch, K. 1972. *What Time Is This Place?* Cambridge, Mass.
McCollum, J. 1980. *Martindale Pond Development Study: Background Report*. St Catharines: Welland Canals Preservation Association and Canadian Henley Rowing Corporation
McCormick, B.J. 1909, 1913, 1921. 'Turn Wellandward,' Welland *Telegraph*
MacDonald, C., ed. 1992. *Grand Heritage: A History of Dunnville and the Townships of Canborough, Dunn, Moulton, Sherbrooke and South Cayuga*. Dunnville: Dunnville District Heritage Association
Macdougall, A. 1886. *Report of the Proposed System of Sewerage for the Northern Portion of the City of St Catharines*. St Catharines: C. Sherwood
McDougall, J.L. 1923. 'The Welland Canal to 1841.' Unpublished MA thesis, University of Toronto
McGeorge, C.E. 1947. *Report on the Welland and Feeder Canals*. To C.G. Edwards, Deputy Minister, Department of Transport, Ottawa. Chatham: The Author
McGlone and Associates. 1983. *Soils Investigation Report for Embayments, Urban Fishing Program, Welland, Ontario*. St Catharines: The Company
- 1990. *Geotechnical Review, Old Welland Canal Lands, Welland, Ontario*. Thorold: The Company, for Public Works Canada

Macht, W. 1981. *The First 50 Years: A History of Upper Lakes Shipping, Ltd.* Toronto: Upper Lakes Shipping, Ltd.

Mackay, R.S. 1851. *The Canada Directory* ... Montreal: Lovell

McNabb, F. 1969. *Sports History of St Catharines.* St Catharines: Advance Printing

Maitland, S.T. 1974. 'Townline Road/Rail Tunnel Design.' *Engineering Journal* 57

Mansfield, J.B. 1980. *History of the Great Lakes.* Vol. 1. Chicago: J.H. Beers, 1899. Reprinted as *The Saga of the Great Lakes.* Toronto: Coles

Maple Leaf Milling Co. 1960. *1910–1960, Our Golden Anniversary Year.* Port Colborne: The Company

Marshall, H., F.A. Southard, and K.W. Taylor. 1970. *Canadian-American Industry: A Study in International Investment.* New York: Russell and Russell

Marshall, W. 1982. *Unemployment and the ILAP Programme.* A Submission to the Chairman and Members of the Niagara Regional Council. St Catharines: The Author

Marshall Macklin Monaghan Limited et al. 1988. *The Management of the Heritage Economic Resource Welland Canals Corridor Development Guide: A Background Report and Welland Canals Corridor Development Guide.* Thorold: The Company, for the Regional Municipality of Niagara

Martine, G. 1961. *The Role of the Welland Canal in Industrial Locations.* Department of Geography, University of Toronto

Mayer, H.M. 1976. 'Urban Impact of the Great Lakes – St Lawrence Seaway Transportation System.' *The East Lakes Geographer* 11

Mayo, H.B. (Chief Commissioner). 1965. *Data Book of Basic Information.* Thorold: Niagara Region Local Government Review for the Government of Ontario

– 1966. *Report of the Commission.* Thorold: Niagara Region Local Government Review for the Government of Ontario

Meaney, C.F. 1980. 'The Welland Canal and Canadian Development.' MA thesis, McMaster University (Hamilton)

Merritt, J.P. 1875. *Biography of the Hon. W.H. Merrit, M.P.* St Catharines: Leavenworth

Merritt, W.H. 1852. *Brief Review of the Origin, Progress, Present State, and Future Prospect of the Welland Canal.* St Catharines: Leavenworth

– 1857. *A Lecture Delivered before the Mechanic's Institute of St Catharines on the 21st Day of January, 1857.* St Catharines: Leavenworth

Merritt Papers. *See* Papers of William Hamilton Merritt.

Merritton Centennial Committee. 1974. *Merritton Centennial, 1874–1974.* St Catharines: St Catharines *Standard*

Merritton Residents' Neighbourhood Improvement Committee. 1979. *Merritton Redevelopment Plan.* Merritton: The Committee

Merritton, Town of. 1949. *Town of Merritton: Seventy Fifth Anniversary*. Merritton: Town of Merritton

Methodist and Presbyterian Churches, Department of Social Services and Evangelism. 1916. *Report on a Limited Survey of Religious, Moral, Industrial and Housing Conditions*. St Catharines: St Catharines Survey Committee

Michael, B. 1967. *Township of Thorold, 1793–1967: Centennial Project of the Township of Thorold*. Toronto: Armath Associates

- 1975. *Brief Presented to the Honorable Mr. Lang, Minister of Transport*. Port Robinson: The Author

Michelin. 1982. *Tourist Guide of Canada*. Quebec: Michelin Tires (Canada) Ltd.

Michener, D.M. 1973. *The Canals at Welland*. Welland: Rotary Club of Welland

Mika, N. and M. 1985. *The Shaping of Ontario from Exploration to Confederation*. Belleville: Mika Publishing

Mills, J.M. 1967. *History of the Niagara, St Catharines and Toronto Railway*. Toronto: Upper Canada Railway Society and Electric Railway Association

Milne, K. 1958. 'Our City Owes Its Start to the Canal.' In J.R. Adams, ed., *Welland Centennial: 1858–1958*. Welland: Corporation of Welland

Ministry of Culture and Recreation. 1976. *The Welland Canals: A Summary*. Toronto: Ministry of Culture and Recreation

Ministry of Municipal Affairs. 1987. *Urban Waterfronts: Planning and Development*. Toronto: Community Planning Wing, Community Improvement Series, vol. 4.

- 1992. *Ontario's New Planning System*. Toronto: Government of Ontario

Ministry of Natural Resources. 1977. *Short Hills Provincial Park: Master Plan*. Toronto: Ministry of Natural Resources

Ministry of State for Tourism. 1985. *Tourism Tomorrow: Towards a Canadian Tourism Strategy*. Ottawa: Supply and Services

Ministry of Transportation and Communications. 1975. *The Queen Elizabeth Way: Ontario's First Superhighway*. Toronto: Government of Ontario

Misener, P. 1987. 'The People and Economics of the Welland Canals, Yesterday and Today.' William Hamilton Merritt Lecture Series, Brock University, St Catharines

Misener, R.S. (Chairman). 1981. *The Great Lakes / Seaway: Setting a Course for the '80s*. Report of the Provincial Great Lakes / Seaway Task Force, Ontario. Toronto: Provincial Government

Monkhouse, F.J. 1965. *A Dictionary of Geography*. London, Eng.: Arnold

Monro, T. 1872. *Report ... on the Enlargement of the Welland Canal*. Ottawa: I.B. Taylor

Monture, P. (Director, Six Nations Land Claims Research Office). 1994. Personal Letters. Six Nations Council (Ohsweken, Ont.), 20 October and 12 December

Morison, S.E. 1950. *The Ropemakers of Plymouth: A History of the Plymouth Cordage Company 1824–1949*. Boston: Houghton Mifflin

Moriyama and Teshima Planners Ltd. 1988. *Ontario's Niagara Parks: Planning the Second Century*. Toronto: The Company, prepared for the Niagara Parks Commission

Moses, R. 1958. *Memorandum to the Trustees on Report from T.F. Farnells: Paralleling the Welland Canal and Effects on the Niagara Power Project*. Lewiston, NY: Power Authority of the State of New York

Mroz, D., R. Niesink, and A. Wilmot. 1983. *A Proposal for the Welland Canals Heritage Parkway (Phase II)*. St Catharines: Welland Canals Preservation Association

Muller, E. 1993. *Welland Canal Math Trail*. St Catharines: Brock University

Muntz, M. 1997 (pending). *John Laing Weller, C.E., M.E.I.C.: 'The Man Who Does Things.'* Parry Sound: The Author

Neathery, S. 1993. *The Welland Canals: A Resource Worth Investing In*. Welland: City of Welland

Nelson, J.G., and P.C. O'Neill. 1989. *The Grand as a Canadian Heritage River*. Waterloo, Ont.: Heritage Resources Centre, University of Waterloo, Occasional Paper 9

Niagara Parks Commission. 1938. *Fifty-First Annual Report of the Niagara Parks Commission 1936 and 1937*. Toronto: Printed by T.E. Bowman

Niagara Peninsula Conservation Authority. 1972. *Niagara Peninsula Conservation Report*. Vol. 1. Toronto: Department of the Environment

Niagara Region Seaway Task Force. 1988. *Niagara Jobs in Jeopardy: A Report on the Economic Importance of the St. Lawrence Seaway and in Particular, The Welland Canal, to the Residents of the Niagara Region*. Thorold: Niagara Region Development Corporation

Niagara Regional Chamber of Commerce and St Catharines and District Chamber of Commerce. 1977. *Local Impact of the Welland Canal and St. Lawrence Seaway Operations*. St Catharines

Nichols, C.M., comp. 1907a. *Special Souvenir Number Descriptive of and Illustrating St. Catharines ..., Thorold, Merritton, Port Dalhousie and Their Industries*. St Catharines: Standard Printing Company

– 1907b. *Special Souvenir Number of the Telegraph Descriptive of and Illustrating Welland, the Hub of the Peninsula, Port Colborne and Their Environment*. Welland *Telegraph*

Ogilvie, Ogilvie and Company. 1989. *Welland Marina*. Welland: The Company

Ontario Ministry of Transportation and Communications, Ontario International Corporation, et al. 1983. *An Investigation of and Proposal for Commercial Promotion of the Great Lakes / Seaway System*. Toronto: The Ministry

Orr, P.M. (Project Manager). 1978. *The New City of Thorold: A Study of Its Past and Its Buildings.* Geography Department, Brock University, St Catharines
- 1979. *A Tour of Historic Thorold* and *A Selection of Some of Thorold's Historic Buildings.* Thorold: Young Canada Works Project 3040 TK6
Ott, E.F. 1967. *A Condensed History of the Township of Humberstone in the County of Welland to Commemorate Canada's Centennial 1967.* Port Colborne: Humberstone Township
Page, H.R., et al. 1876. *Illustrated Historical Atlas of the Counties of Lincoln and Welland, Ontario.* Toronto: Craig Steam Litho
- 1877, 1879. *Illustrated Historical Atlas of the Counties of Haldimand and Norfolk.* Toronto: Craig Steam Litho
Page, J. (Chief Engineer, Public Works). 1872a. *Report on the Enlargement of the Welland Canal.* Ottawa: Taylor.
- 1872b. *Report of the Chief Engineer of Public Works on the Enlargement of the Welland Canal.* Ottawa: Public Works
Papers of William Hamilton Merritt, 1793–1862. Public (later National) Archives of Canada (Ottawa), Manuscript Division, Merritt family, MG 24, E1
Parker Consultants. 1988. *Environmental Study Report: Broadway Street Bridge on Niagara Roads over Old Welland Canal in City of Welland.* Hamilton: The Company, for Regional Municipality of Niagara
Parks Canada. 1982. *Heritage Canals.* Ottawa: Ministry of the Environment
Parks Canada, Ontario Ministry of Culture and Recreation, and Ontario Ministry of Natural Resources. 1980. *Welland Canals Concept Plan: A Canada-Ontario Study.* Ottawa: Parks Canada
Passfield, R.W. 1989. *Technology in Transition: The Soo Ship Canal, 1889–1985.* Ottawa: National Historic Parks and Sites Branch, Parks Canada
Patrias, C. 1990. *Relief Strike: Immigrant Workers and the Great Depression in Crowland, Ontario, 1930–1935.* Toronto: New Hogtown Press, Woodsworth College, University of Toronto
- 1994. *Patriots and Prolitarians: Politicizing Hungarian Immigrants in Interwar Canada.* Montreal and Kingston: McGill-Queen's University Press
Patten, D. 1956. 'The Traffic Pattern on American Inland Waterways.' *Economic Geography* 32
Paul, L.L. 1983. *The Welland Feeder Canal.* Wainfleet, Ont.: Rehabilitate the Old Feeder Canal Association
Pearson, N. 1965. *Welland Area Plan 1967–1987.* Welland: Welland Area Planning Board
- 1966. *A Planning Concept for the Welland Area.* Welland: Welland Area Planning Board

Petrie, F.J. 1973. *The Welland By-Pass*. St Catharines: For St Lawrence Seaway
    Authority
Philips Planning and Engineering. 1972. *Potential Recreation Areas and Fragile
    Biological Sites: Inventory and Recommendations – Regional Municipality of
    Niagara, Official Plan Studies, Report Number 11*. Burlington, Ont.: The
    Company
Phillpotts, G. 1842. *Report on the Canal Navigation of the Canals*. London, Eng.
Picken, A. 1832. *The Canals, as They at Present Command Themselves*. London:
    Wilson
Plosz, A.J. 1988. *Q & O: Our Story*. St Catharines: Quebec and Ontario Paper
    Company
Plymouth Cordage Company. 1924. *Plymouth Cordage Company: The Hundred
    Years of Service*. Plymouth, Mass.: The Company
Plymouth Senior Public School. 1966–7. 'A Brief History of Welland.' Typescript.
    Welland
Pope, A.J., K.A. Keeleyside, and S.D. Speller. 1993. *An Environmental Evaluation
    of the Lower Welland River*. Toronto: Ministry of Environment and Energy
Port Colborne Fire Department. 1970. *Port Colborne Centennial, 1870–1970*. Port
    Colborne: Moss Press
Port Colborne Greater Chamber of Commerce. 1968. *History of Port Colborne*.
    Port Colborne: The Chamber
Port Colborne – History. 1960. Port Colborne: Port Colborne Public Library
Port Colborne – Waterfront Community Complex. 1987. 2 vols. Port Colborne:
    Port Colborne–Wainfleet Community Futures Committee
Port Dalhousie Quorum. 1989. *A Walking Guide to Historic Port Dalhousie*. Port
    Colborne: The Quorum
Prince, R.D. (Chairman). 1985. *Report of the Welland Canals Development Sub-
    Committee*. Thorold: Niagara Region Development Corporation
Proctor and Redfern. 1960. *St. Catharines–Grantham Amalgamation Study*. St
    Catharines: The Company
– 1971. *Port Weller Sewage Treatment Facilities for the Regional Municipality of
    Niagara*. Toronto: The Company
Proctor, Redfern, Bousfield and Bacon. 1971. *The City of Welland – Official
    Plan Study – First Interim Report – Survey and Analysis*. Toronto: The Con-
    sultants
– 1972. *The City of Welland – Official Plan Study – Second Interim Report – Draft
    Official Plan*. Toronto: The Consultants
Proctor, Redfern and Pannel Kerr Forster. 1989. *Old Welland Canal: A Land Use /
    Marketing Review – Final Report*. Toronto: The Company, for Public Works
    Canada

Project Planning Associates. 1968. *Welland Channel Relocation: Land Form and Planting*. Toronto: The Company, for St Lawrence Seaway Authority
- 1975. *Welland Canal Northern Research Study*. Toronto: The Company, for Public Works Canada
Prospectus of the Welland Railway. 1857a (June 1856). St Catharines *Journal*
- 1857b (September 1856). St Catharines *Journal*
Province of Ontario. 1869. *Gazetteer and Directory*. Toronto: Robertson and Cook
- 1910. *Province of Ontario Gazetteer and Directory 1910–11*. Ingersoll, Ont.: Union Publishing
Public Works Department, Thorold. 1980. *Annual Report No. 10, 1979*. Thorold: Regional Municipality of Niagara
Pugliese, C. 1990. *Future Directions*. St Catharines: Welland Canals Society
Putter, K. 1995. 'Feds axe canal land deal to cut costs.' *The Tribune* (Welland), 22 February
QUNO (former Quebec and Ontario Paper Company). 1993. *Planting a Forest*. Vol. 1.1
Randal, R. 1831. *First General Report from Robert Randal, Esquire*. York: Baxter
Ray, D.M. 1971. 'The Location of United States Manufacturing Subsidiaries in Canada.' *Economic Geography* 47
Raymond, M.J. [1979.] *A History of Wainfleet Township*. [St Catharines]: The Author
Read, Vorhees and Associates. 1971. *Transportation Research and Analysis*. St Catharines: Regional Municipality of Niagara, Official Plan Studies, Report number 6
Real Property Research. 1993. *Property Transfer Assessment – The Old Welland Canal and Adjacent Lands*. Toronto: Public Works and Government Services Canada
Regional Municipality of Niagara. 1973. *Regional Niagara Policy Plan*. St Catharines: Regional Municipality
Regional Planning and Development Department. 1983a. *Tourism in the Niagara Region – Report no. 1: Economic Impact ... Tourist Characteristics*. Thorold: Regional Municipality of Niagara
- 1983b. *Tourism in the Niagara Region – Report no. 2: Marketing, Trends, and Tourism*. Thorold: Regional Niagara
- 1984. *Tourism in the Niagara Region – Report no. 3: Some Tourist Attractions, Present and Possible*. Thorold: Regional Niagara
- 1985. *Tour Niagara's Waterways*. Thorold: Regional Niagara
- 1995. *Final Report and Recommendations, Welland Canals Parkway, Proposed Policy Plan Amendment 94*. Thorold: Niagara

Rehabilitate the Old Feeder Canal Association, Inc. 1979. *A Feasibility Study on the Welland Feeder Canal*. Wainfleet: The Association

Report of Committee on Roads and Bridges. 1867. *Proceedings of the Municipal Council of the County of Welland*. Welland: Telegraph Office

Richardson, K.J. 1907. 'Rates as Per Present Conditions of Welland Canal Compared with Enlarged and Improved Canal with Twenty-One Foot Depth.' In St Lawrence Seaway Authority (St Catharines), File 33-2-2-2, vol. 2.

Robson L., with P. Hutchinson. 1994. *Memories of Morningstar Mills: Mountain Mills at DeCew Falls*. St Catharines: Slabtown Press

Runnalls, J.L. 1973. *The Irish on the Welland Canal*. St Catharines: St Catharines Public Library

St Catharines Board of Trade. 1901. *Annual Report of the St. Catharines Board of Trade for the Year 1900*. St Catharines: The Board

– 1913. *St. Catharines Year Book*. St Catharines: The Board

St Catharines, City of. 1969. *Martindale – Henley Development Survey*. St Catharines: The Council

– 1995. Item no. 276, Report from the Planning Department Dated April 13, 1995, Re: Official Plan and Zoning By-Law Amendments Lock III – Emmett Road Study. St Catharines: The City

St Catharines, City of, and Henley Aquatic Association, et al. 1963. *Royal Canadian Henley Regatta Course Remedial Plan*. St Catharines: The City

St Catharines Historical Museum. 1984. *A Canadian Enterprise. The Welland Canals. The 'Merritt Day' Lectures, 1978–82*. St Catharines: The Museum, Publication no. 4

St Catharines Hydro-Electric Commission. 1986. *Port Dalhousie Generating Station Environmental Study Report*. Niagara Falls: Acres International

St Catharines *Standard*. 1960. *Amalgamation: A Reprint of Articles from the St. Catharines Standard*. St Catharines: The Newspaper

St Lawrence College in conjunction with Lost Villages Historical Society. 1977. *Historical Sketches and Photographs of the Villages after which the College's Buildings Are Named*. Cornwall: The College

St Lawrence Seaway Authority. 1959– (a). *Annual Report*. Ottawa: The Authority

– 1959– (b). *Traffic Report*. Annual. Ottawa: The Authority

– 1965. *Welland Canal: Channel Relocation Study–Port Robinson to Port Colborne*. St Catharines: The Authority

– 1966a. *Proceedings of Public Hearings Held at Ottawa, Ontario*. 2 vols. Ottawa: The Authority

– 1966b. *Summary of Future Traffic Estimates and Toll Requirements*. Ottawa: The Authority

– 1967. *Welland Canal Railway Crossing Study*. Niagara Falls: H.G. Acres

- 1968. *Summary Report on Drainage Control*. Toronto: Swan Wooster
- 1972. *Welland Channel Relocation: 1967–1973, Road – Rail Networks*. Ottawa: The Authority
- 1973. *The Welland By-Pass*. Ottawa: The Authority
- 1980. *Position Paper on Navigation Season Extension*. Ottawa: The Authority
St Lawrence Seaway Authority and St Lawrence Seaway Development Corporation. 1959–. *The Seaway Handbook*. Ottawa: The Authority. Updated annually.
Sainsbury, G.V. 1974. 'Re-routing the Historic Welland Canal.' *Canadian Geographical Journal* 89
- 1976. 'The Seaway Economy: Expectations, Reality and the Future.' *The East Lakes Geographer* 11
Sainsbury, G.V., and R.M. Campbell. 1967. *Tolls and the Seaway*. Tenth Annual Conference on Great Lakes Research and First Meeting of International Association for Great Lakes Research, University of Toronto
Salerno, R. 1989. *The Transformation of a Non-Metropolitan Urban Centre: A Case Study of Port Colborne, Ontario*. Downsview, Ont.: Department of Geography, York University
Salmon, C. (First V-P, Port Colborne–Wainfleet Chamber of Commerce). 1989. 'Re: Welland Canals Implementation Group.' Letter to C.T. Cambray, Manager Policy and Planning, Regional Municipality of Niagara, 9 August
Sanders, R.W. 1919. *What's Doing in Thorold, Ontario and Thorold Township*. Thorold: The Author
Sandwell Swan Wooster Division. 1991. *Welland Canal Impact Study: Capital and Annual Costs Data*. Toronto, North York: The Company
'Sanitary Situations in Sections of Old Welland Canal.' 1935. *Canadian Engineer* 68
Sauer, H. 1973. *The Story of the Erie Glass Company of Canada, Ltd. at Port Colborne, Ontario, 1892–1894*. The author for Glasfax, an Ontario organization for members interested in old glass
Savage, R.W., Consulting. 1989. *Marketing the St. Lawrence and Great Lakes Seaway System*. St Catharines: The Consultant, for Michigan–Ontario Maritime Advisory Committee
Sawle, G.R.T. 1955. *Hometown*. Reprint of articles published in Welland *Tribune*. Fort Erie: The Review Company
Sayles, F.A. 1963. *Welland Workers Make History*. Welland: Winnifed Sayles
Schmjver, G. 1993. *Niagara Manufacturers Survey*. Sharon: Ont.: WCM Consulting
Schwilgin, F.A. 1974. *Town Planning Guide Lines*. Ottawa: Department of Public Works
Scottish Fertilizers. 1931. *High Grade Complete Fertilizers*. Welland: Welland *Telegraph*

Sears and Sawle. 1902. *Souvenir of the Town of Welland*. Welland: Welland *Telegraph*

Seibel, G.A., and O.M. 1990. *The Niagara Portage Road: A History of the Portage on the West Bank of the Niagara River*. Niagara Falls, Ont.: City of Niagara Falls

Services Récréatifs et Communautaires in co-operation with Ministry of Cultural Affairs, Québec. 1986. *Lachine Historical Guide: Discovering Lachine*. Lachine: The City

Shannon, E.E., and D.T. Vachon. 1973, 1974. *The Welland Canal: Water Quality Control Experimentation (Phase 1)*. Burlingon, Ont.: Wastewater Technology Centre, Environmental Protection Service, Canada Centre for Inland Waters

Shantz, R.H. 1969. 'The Expropriation of Farm Properties in the Niagara Peninsula by the St. Lawrence Seaway Authority.' Unpublished MA thesis, Department of Geography, University of Waterloo

Shipley, R. 1987. *St. Catharines: Garden on the Canal – An Illustrated History*. Burlington, Ont.: Windsor Publications

Shipping Reports. 1960–77. Part II, International Seaborne Shipping, and Part III, Coastwise Shipping. Ottawa: Dominion Bureau of Statistics

Sicotte, A., and R. Thériault. 1987. *De la vapeur au vélo*. Translated by P. Adams *From Steam to Cycles: The Lachine Canal Historical Guide*. Montreal: Association Les Mil Lieues

Smith, B.J. 1995. *Seaway Village*. St Catharines: Seaway Village

Smith, G.H. 1935. 'Port Colborne's History Is One of Steady Progress.' In *Programme: The Niagara District Firemen's Association of Toronto*. Port Colborne: Auspices of Alert Fire Company

Smith, W.H. 1846. *Smith's Canadian Gazetteer*. Toronto: H. & W. Rowsell
– 1851. *Canada: Past, Present and Future ...* vol. 1. Toronto: Thomas Maclear

Smith, W.R. 1982. *Aspects of Growth in a Regional Urban System: Southern Ontario 1851–1951*. Downsview: Department of Geography, Atkinson College, York University

Smy, W.R. 1978. *Looking Back*. Reprint of articles from Welland *Tribune*. Port Colborne: Port Colborne Historical Society

Snyder, F.L. 1992. *Zebra Mussels in the Great Lakes: The Invasion and Its Implications*. Columbus, Ohio: Ohio Sea Grant College Program, Ohio State University

Soper, A.P. 1983. 'Escarpment Heritage: The Welland Canal.' *Cuesta* (Niagara Escarpment Commission, Georgetown), Spring

Sorge, L. 1977–89. *Remember When: A Collection of Pictures and Stories Remembering Days Gone By in and around Dunnville*. 12 vols. Dunnville: Dunnville *Chronicle*

Spelt, J. 1972. *Urban Development in South-Central Ontario*. Toronto: McClelland and Stewart, Carleton Library no. 57

- 1973. *Toronto*. Don Mills, Ont.: Collier-Macmillan of Canada

Squires, R.W. 1979. *Canals Reviewed: The Story of the Waterway Restoration Movement*. Bradford-on-Avon, Eng.: Moonraker Press

Stamp, R.M. 1987. *QEW: Canada's First Superhighway*. Erin, Ont.: Boston Mills Press

Statistics Canada. Annual. *Shipping Report*. Ottawa: Queen's Printer

St Denis, G. 1993. 'An Erie Canal for Western Upper Canada: A Forgotten Episode in Ontario's Transportation Evolution.' *Ontario History* 58.3

Steele, P.O., et al. 1979. *Old (2nd) Welland Canal Pollution and Abatement*. St Catharines: Department of Biological Sciences, Brock University

Steele, R.C. 1970. 'The Industrial Evolution of the Town of Merritton, Ontario. Unpublished BA thesis, Department of Geography, Brock University, St Catharines

Stelter, G.A. 1982. 'The City-Building Process in Canada.' In G.A. Stelter and A.F.J. Artibise, *Shaping the Urban Landscape: Aspects of the Canadian City-Building Process*. Ottawa: Carleton University Press

Stelter, G.A., and A.F.J. Artibise, ed. 1977. *The Canadian City: Essays in Urban History*. Toronto: McClelland and Stewart, Carleton Library no. 109

Sterns, F.E. 1928. 'The Lock Gates of the Welland Ship Canal.' *Engineering Journal* 12.2

- 1929. 'Engineering Features of the Welland Ship Canal.' *American Society of Civil Engineers, Papers*

Stewart, G.R. (President, St Lawrence Seaway Authority). 1990. 'The St. Lawrence Seaway Authority in the 1990's.' William Hamilton Merritt Lecture Series, Brock University, St Catharines

- 1992. *Minutes of Proceedings and Evidence of the Sub-Committee on the St. Lawrence Seaway of the Standing Committee on Transport*. Issue 8. Ottawa: House of Commons

Strachan, J. 1820. *A Visit to the Province of Upper Canada in 1819*. Aberdeen: Chalmers

Stretton, R.W. 1950. *Dunnville, Ontario: Centennial Year 1950, 100 Years of Progress*. Dunnville: The Committee

Styran, R.M., and R.R. Taylor. 1980. 'The Welland Canal: Creator of a Landscape.' *Ontario History* 72.4

- 1988. *The Welland Canals: The Growth of Mr. Merritt's Ditch*. Erin, Ont.: Boston Mills Press

- 1992. *Mr. Merritt's Ditch: A Welland Canal Album*. Erin, Ont.: Boston Mills Press

Styran, R.M., and R.R. Taylor, with J.N. Jackson. 1988. *The Welland Canal: The Growth of Mr. Merritt's Ditch*. Erin, Ont.: Boston Mills Press

Swart, M. 1976. 'Replacement of Port Robinson Bridge across the Welland Ship Canal.' Letter to Otto Lang, Minister of Transport, Toronto, The Author (Thorold), 15 July

Sykes, A., and S. Gillham. *Pulp and Paper Fleet: A History of the Quebec and Ontario Transportation Company*. St Catharines: Stonehouse Publications

Talman, J.J. 1976. 'Merritt, William Hamilton.' In *Dictionary of Canadian Biography*, vol. 9. Toronto: University of Toronto Press

– 1979. 'The Impact of the Welland Canals on the Community.' In *The Welland Canals*. Proceedings of the First Annual Niagara Peninsula History Conference. St Catharines: Brock University

Taylor, G. 1943. *Urban Geography*. London: Methuen

Taylor, R.R. 1989. 'Nineteenth Century Industrial Sites: St. Catharines – Merritton – Thorold.' In *Industry in the Niagara Peninsula: Field Trip Guide*. St Catharines: Brock University, for Eleventh Annual Niagara Peninsula Conference

– 1990. 'Merritton, Ontario: The Rise and Decline of an Industrial Corridor.' *Scientia Canadensis* 14.1–2

– 1992a. 'The Growth of Merritton's Industrial Corridor, ca. 1845–1914.' In J.N. Jackson, ed., *Industry in the Niagara Peninsula*. Proceedings of the Eleventh Annual Niagara Peninsula History Conference. St Catharines: Brock University

– 1992b. *Touring St. Catharines in a REO circa 1910–1920*. St Catharines: St Catharines Museum, Publication no. 7

Taylor, R.R., and St Catharines LACAC. 1992. *Discovering St. Catharines' Heritage: The Old Town*. St Catharines: Local Architectural Conservation Advisory Committee

Taylor, R.S. 1950. 'The Historical Development of the Four Welland Canals, 1824–1933. Unpublished MA thesis, University of Western Ontario (London)

Technical Advisory Committee to the Four-Level Committee for the Evaluation of the Abandoned Welland Canal and Adjacent Lands. 1975. *A Redevelopment Study of the Abandoned Welland Canal Lands, Welland, Ontario*. Ottawa: Department of Public Works

Ter Horst, J.G. 1973. *Study of Twin Elevator Locks for the New Welland Canal*. Report to M.S. Campbell, St Lawrence Seaway Authority (St Catharines), 4 May

Thorold and Beaverdams Historical Society. 1897–8. *Jubilee History of Thorold, Township and Town ...* Thorold: Published by J.H. Thompson for the Society: Thorold *Post*. Accompanied by Supplement, 1897–1932.

Thorold Board of Trade. 1953. *60th Anniversary, 1893–1953*. Thorold: The Board

Thorold Centennial Supplement. 1950. St Catharines *Standard*, 26 June

Thorold, City of. 1992. *Lakeview Cemetery Expropriation*. Brief presented to the

Sub-Committee on the St Lawrence Seaway on 5 November. Thorold: The City

Thorold, Ontario, Being the Trade Edition of the Thorold Post for May 16, 1902. Thorold *Post* (Thorold), 1902

Thorold, Town of. 1950. *Centennial 1850–1950.* Thorold: The Town

Thorold Township Board of Trade. 1920. *The Township of Thorold Is the Centre of Electric Energy and Water and Rail Transportation.* Port Robinson: The Board of Trade

Tipton, T.L.M. 1893. 'At the Mouth of the Grand.' *Canadian Magazine,* July

Trépanier, C.G. (Vice-President, Niagara Region). 1995. *Comments at the 1995 Opening of the Welland Canal.* St Catharines: St Lawrence Seaway Authority

Trow, A. 1975. *Proposed Welland Redevelopment Adjacent to Old Welland Canal Townline Road North to the Welland River Crossing Welland, Ontario.* Rexdale, Ont.: The Company, for Diamond and Myers

Tufford, N. 1987. *Tommy Trent's Seaway Manual.* St Catharines: Stonehouse Publications

Turcotte, D. 1986. *Port Dalhousie: Shoes & Ships & Sealing Wax.* Erin, Ont.: Boston Mills Press

Unyi, A. 1995a. *The Empire Cotton Mills and Wabasso Textiles.* Welland: Welland Historical Museum

– 1995b. *Plymouth Cordage: Welland's First Major Employer.* Welland: Welland Historical Museum

Upper Canada Assembly. 1836–7. *Select Committee Appointed to Examine and Enquire into the Management of the Welland Canal.* Third report

Urquhart, M.C., ed. 1965. 'Canals, Tonnage through the Welland Canal, 1867 to 1960.' In *Historical Statistics of Canada,* Series S189–199. Toronto: Macmillan Company of Canada

Van Allen, W.H. 1960. 'Canal Systems of Canada.' *Canadian Geographical Journal* 61

Vanderburgh, O. 1967. *A History of Allanburg and Area.* Allanburg: Allanburg Women's Institute

Waddell, W.H. 1930. 'Welland Canal Forestry Work.' St Catharines *Standard,* 10 December

Warwick, P.D.A. 1978. 'Pioneer Shipbuilder of the Great Lakes.' *Canadian Geographical Journal* 94.3

Watson, J.W. 1943. 'Urban Development in the Niagara Peninsula.' *Canadian Journal of Economics and Political Science* 9

– 1945a. 'The Changing Industrial Pattern of the Niagara Peninsula. *Ontario Historical Society Papers and Records* 37

– 1945b. 'The Geography of the Niagara Peninsula.' Ph.D. thesis, University of Toronto

Way, R.L. 1960. *Ontario's Niagara Parks: A History*. Niagara Falls: Niagara Parks Commission

Weaver, J.C. 1982. *Hamilton: An Illustrated History*. Toronto: James Lorimer and National Museum of Man

Welland Canal Park Committee. 1961. *Welland Canal National Park: A Brief to the Honourable Walter G. Dinsdale, Minister of Northern Affairs and National Resources with Respect to the Proposed National Park Embracing Lands in the Counties of Lincoln and Welland in the Province of Ontario Abutting the Route of the Welland Ship Canal*. St Catharines: The Committee

– 1964. *Draft Brief Welland Canal Park to the Honorable John P. Robarts, Q.C.* St Catharines: The Committee

Welland Canals Foundation. 1988. *Comments on the Welland Canals Corridor Development Guide*. St Catharines: The Foundation

Welland Canals Parkway Committee. 1994. *Proposed Welland Canals Parkway Development Plan*. Thorold: The Committee

Welland Canals Preservation Association. 1978. *A Proposed Bikeway Trail Network for the Old Welland Canals Area in the City of St. Catharines*. St Catharines: The Association

– 1980. *A Guide to Historic Port Dalhousie*. Port Dalhousie: The Association

– 1983. *A Proposal for the Welland Canals Heritage Parkway*. St Catharines: The Association

Welland Canals Society. 1989a. *Presentation to Regional Niagara Co-ordination Committee, Welland Canals Corridor*. St Catharines: The Society

– 1989b. *A Heritage Assessment of Welland Iron and Brass in Welland, Ontario*. St Catharines: The Society

– 1990a. *Preservation through Development*. St Catharines: The Society

– 1990b. Archaeology Project, *A Study of Historic Structures: The Welland Canal Industrial Corridor*. St Catharines: The Society

Welland, City of, to Hon. Gilles Loiselle, President of the Treasury Board. 1992. *Old Welland Canal Land Transfer Endowment Request*. Welland: The Corporation

Welland, Corporation of. 1958. *Welland Centennial: 1858–1958*. Welland: Welland Printing Co.

– 1983. *Welland, Ontario 1858–1983, Celebrating 125 Years of Canadian Heritage*. Welland: The Corporation

Welland, County of. 1867. *Proceedings of the Municipal Council*. Welland: Telegraph Office

Welland County Council. 1886. *A Historical and Descriptive Sketch of the County of Welland ...* Welland: Sawle and Snartt

Welland Development Commission. 1969. *Classified List of Welland Industries.* Welland: The Commission

Welland Historical Museum. 1995. *The Hungarian Community in Welland.* Welland: Soleil

'The Welland Market.' *Ontario Today*, 3 June 1961

Welland Planning and Development Department. 1989. *An Evaluation of the Public Works Canada Study, Old Welland Canal – A Land Use (Marketing Review).* Welland: City of Welland

'The Welland of Today, (as written in 1887).' 1983. In M. Trufal (Chairman, Welland 125th Anniversary Committee), *Welland Ontario 1858–1983: Celebrating 125 Years of Canadian Heritage.* St Catharines: Lincoln Graphics

Welland *Telegraph*. 1902. *Souvenir of the Town of Welland.* Welland: Sears and Sawle

Weller, J.L. (Superintending Engineer). 1910. Letter to W.A. Bowden, Chief Engineer, Department of Railways and Canals, Ottawa, 6 October. In St Lawrence Seaway Authority (St Catharines), File 33-2-2-2 (vol. 3).

– 1912. *Report on Proposed Welland Ship Canal.* To W.B. Bowden, Chief Engineer, Department of Railways and Canals, Ottawa, 2 March. In St Lawrence Seaway Authority (St Catharines), File 32-2-2-2

– 1915. *Report on the Welland Ship Canal.* Ottawa: Government Printing Bureau

Wells, G. 1938. 'Welland When Young.' *Welland County Historical Society Papers and Records* 5

Whazlo, T.W. 1992. *A Walking Tour of Old Port Dalhousie: Where the Canals Began.* Port Dalhousie: Port Dalhousie Quorum

Whitford, N.E. 1906. *History of the Canal System of the State of New York ..., Supplement to the Annual Report of the State Engineer and Surveyor of the State of New York for ... 1905.* 2 vols. Albany: State Legislative Printer

Whittemore, K.T. 1977. 'Buffalo.' In John H. Thompson, ed., *Geography of New York State.* Syracuse: Syracuse University Press

Wiegman, C. 1953. *Trees to News.* Toronto: McClelland and Stewart

Wieler, R.R. 1967. 'The Influence of the Welland Canal upon Its Regional Economy.' Unpublished essay, Department of Geography, Middlesex College, University of Western Ontario (London)

William Hamilton Merritt Lecture Series. 1986–90. Brock University, St Catharines. Papers by W.A. O'Neil, President, St Lawrence Seaway Authority; R. Peter Misener, President, Misener Holdings; J.D. Leitch, Chairman, Upper Lakes Shipping International; J. Norman Hall, President, Canadian Shipowners Association; and G.R. Stewart, President, St Lawrence Seaway Authority

Williams, J. 1985. *Merrit: A Canadian Before His Time.* St Catharines: Stonehouse

Willoughby, W.R. 1961. *The St. Lawrence Seaway Authority: A Study in Politics and Diplomacy.* Madison: University of Wisconsin Press

Wood, J.D., ed. 1975. *Perspectives on Landscape and Settlement in Nineteenth Century Ontario.* Toronto: McClelland and Stewart, Carleton Library no. 91

Woodruff, S.D. (Superintendent). 1867a. *Statement Showing the Annual Rents of Water-Power Leases, and the Rents of Other Properties.* St Catharines: Welland Canal Office

– 1867b. To Secretary of Public Works. *Water Power and Other Property Leases on the Canal.* Annual Report, Schedule 4, 19 December

Woolton, K.D. 1980. *A Study to Determine the Feasibility of Placing a Small Scale Hydro Generator on Twelve Mile Creek at Port Dalhousie.* Department of Urban and Environmental Studies, Brock University (St Catharines)

Wright, W. (Chairman). 1982. *Land Use Concept Plan.* Welland: Welland Advisory Group

Yeates, M., and B. Garner. 1980. *The North American City.* New York: Harper and Row

Young, F.H. 1950. *Historical Directory, Village of Chippawa, 1850–1950.* Fort Erie: The Review Company

Young, N.R.P. 1975. 'The Economic and Social Development of Welland, 1905–1939.' MA thesis, University of Guelph

The Young Port Colborne Club. 1973. *A History of Port Colborne.* Port Colborne: The Club

Young, R.C. (Chairman). 1956. *Preliminary Brief on the Welland Canal Area: Port Development.* St Catharines: Greater Niagara Chamber of Commerce

Credits

The poem on pages 472–4 is quoted with the permission of Howard Engel.

Photographs

Author's collection: Welland Mills (Thorold), lock-keepers' dwelling, St Paul's Church (Port Robinson), sign at Lock 3, Third Canal Lock (Port Dalhousie), penstocks at Decew, Welland River syphon, Garden City Skyway, Townline Tunnel, Algoma Steel plant, General Motors plant

John Burtniak Collection: Dam at Dunnville, Port Colborne harbour

Fred J. Campbell Collection: Fourth Canal construction (Thorold) and Flight Locks

Canadian Illustrated News, 1871: Canal cross-section at St Catharines

Ontario Archives: Neptune's Staircase (ST 406)

Port Colborne Museum: Hopkins swing-bridge (981.4.1)

St Catharines Historical Museum: Statue of Merritt (N 4094), Muir shipyard (N 1051), *Garden City* side-wheeler (N 1544)

Welland Historical Museum: Aqueducts at Welland (984.033.083), lift-bridge towers at Welland (980.250.001)

Maps

Brock University Library, Special Collections: Commissioners of Public Works, 1841–80 (map 2); Welland Canal Book 2, Lock 5 to Port Robinson (map 10)

H. Brosius 1875 (map 12)

Department of Geography, Brock University (map 1)

H.R. Page 1876 (maps 3, 4, 5, 6, 7, 9, 11); 1877–9 (map 8)

Parks Canada 1980 (map 13)

Index

banks, 14, 107, 128, 169–70, 234, 261.
See also community services
barrier effect of the canals, 16–17, 99,
100–1, 110, 235–6, 281, 484
Bate, Thomas, 176
Battle, John, 231
Beadle, Dr Chauncey, 128
Beaverdams, 77, 87, 88
biological areas. See environmental
areas
Board of Works, Province of Canada,
ix, 50
boosterism, 169, 191–3, 201, 219–20,
259–60
break-in-bulk points, 18. See also
docks and wharves, harbours,
ports
bridges as urban-forming elements,
17, 73–6, 484; over the abandoned
Fourth Canal, 447–8, 449, 460–1;
over the abandoned Second Canal,
264–5, 315; over the First and Second Canals 48, 73–4, 92, 99, 100–1,
103–4, 107, 116, 119, 121, 123, 139;
over the Fourth Canal, 196, 278,
282, 284–5, 286–9, 311, 315–16,
365–8, 376; over the Grand River –
see Dam at Dunnville; over the
Third Canal, 145–6, 155, 195–6,
209–10
Brown's Bridge, 87, 100–1
Buffalo: bridge links with the United
States, 139–40, 151, 189, 198; hinterland area in the Niagara Peninsula,
111, 128, 163–4, 189–90, 192, 224;
terminal port for the Erie Canal, the
Niagara River, and Lake Erie, 11, 34,
96, 149, 193, 197, 270
Byng, 109 (map), 108–10, 112, 483
By-Pass channel of the Fourth Canal,

proposed at St Catharines, 368–72,
429–30. See also Welland Canal By-
Pass

Cahill, Louis, 430–1
Canadian Canal Society, 432
Canadian Canoe Association, 457
canal administrations, xi. See also
Board of Works, Province of Can-
ada; federal government, Depart-
ment of Railways and Canals; St
Lawrence Seaway Authority;
Welland Canal Company
canal communities, 3, 4 (map), 20–2,
60–130, 165–265, 301–56, 377–489.
See also Allanburg, Byng, Chippawa,
Dunnville, Helmsport (The Junc-
tion), Humberstone, Merritton, Port
Colborne, Port Dalhousie, Port
Maitland, Port Robinson, Port
Weller, St Catharines, Stromness,
Thorold, Wainfleet, Welland,
Welland Vale
canal engineers and surveyors, 10;
N.H. Baird, 50; James Brindley, 9;
James Geddes, 37; Alexander J.
Grant, 276–7, 293; H.H. Killaly, 50;
John Page, 150; Thomas Telford, 8;
Hiram Tibbett, 33; J.G. Warren, 272;
James L. Weller, 253, 272–3, 276
canal lengths (reaches) across the Erie
Plain, 7, 32–3, 35, 43, 51, 145–7, 274,
282–6, 322, 474, 477; across the Nia-
gara Escarpment, 7, 32–5, 43, 85,
144, 146, 213, 238, 274, 275, 278–80,
320–1, 337, 340, 421, 425, 426 (map),
435–6; across the Ontario Plain, 7,
43, 50, 52, 142, 144–7, 148, 175, 274,
276–9, 421, 474, 477; between
Stromness and Port Maitland, 10;

128, 247–9, 260; amenity, 433–4,
437–9, 461–3; barrier impact, 16, 406;
boosterism, 169–70, 259–60; bridges
and tunnels, 73–5, 146–7, 264–5, 331,
365, 367; bypassed by Third Canal,
148; canal administration, 126, 128,
362; coloured corps, 68–9; dock, 146;
Hydraulic Raceway, 125, 320, 322–3;
municipal status, 128, 165–8, 260,
401–4; name, 87, 126; pollution, 320–
6; population characteristics, 171–4,
316–18, 335–6, 411–14; railways,
246–7; roads, 122–3, 320, 334–5;
tourism, 138, 249, 488; urban form
and growth, 20, 86, 96, 128, 181–2,
262–3, 334–6. *See* community services, manufacturing, mills, Twelve
Mile Creek, William Hamilton
Merritt
St Catharines airport, 370
St John, Alpheus L., 176, 255
St Lawrence Seaway:
– St Lawrence section, 359–61, 477–9
– Welland Canal section: as water-
supply conduit, 378–80; bypass constructed at Welland, 372–7, and
abandonded length of Fourth Canal,
444–57; bypass proposed at St
Catharines, 368–72; canal service
roads, 15, 294, 374, 469; Lakeview
Cemetery, Thorold, 235, 369–71;
modernizing the Fourth Canal, 361–
66; new bridges and tunnels, 366–8;
Peter Street Bridge, Thorold, 460–1;
possible privatization, 382; regional
impact of canal, 398–400; Rehabilitation Programme, 380–2; surplus
land, 206, 448–53, 458–6, 467; tolls
and lockage fees, 394–8; trading
flows, 387–94, 398–400; traf-

fic-control centre, 362–3. *See also*
canal routes to Atlantic seaboard,
trading flows
sand: at Deep Cut, 36–7; Port Colborne, 47–8, 189–90, 192, 280; Port
Dalhousie, 90, 92, 186, 328–9; Port
Weller, 410; St Catharines, 125–6,
478
Sault Ste Marie. *See* canal routes to
Atlantic seaboard
scenic drives, 403–4, 406, 432, 465. *See
also* Welland Canals Parkway
Scott, Frank, 185–6
seasonal operations. *See* winter closure
Seaway Village, 470
Second Canal, xi, 4 (map); channel
improved, 134; inadequacies of, 143;
modernizing First Canal, 49–54;
operated in conjunction with
Welland Railway, 140–1. *See also*
abandoned channels, Feeder Canal
Second World War, 329, 332, 333, 341,
347
Seibel, George, 425–7
shanty towns, 14, 65–8, 90, 94, 97, 110,
113, 116, 118, 483
Sherwood, Samuel, 102
Shickluna, Louis, 127, 176, 197, 251,
254
Simpson, Melancthon, 251, 254
smuggling at Port Colborne, 352
social divisions. *See* ethnic and cultural
diversity, housing segregation
South Niagara Rowing Club, 455
spoil disposal, 15, 36–7, 208–9, 283,
286, 291, 361, 373, 454, 477
stacks and piles, 410–11. *See also* coal,
sand
streetcars, 155, 159–60, 184, 197, 198,
231, 236, 247, 287–8, 309, 319, 482